Advances in Experimental Medicine and Biology

Volume 927

More information about this series at http://www.springer.com/series/5584

Erwei Song
Editor

The Long and Short Non-coding RNAs in Cancer Biology

Springer

Editor
Erwei Song
Sun Yat-Sen Memorial Hospital
Sun Yat-sen University
Guangzhou, China

ISSN 0065-2598 ISSN 2214-8019 (electronic)
Advances in Experimental Medicine and Biology
ISBN 978-981-10-1496-3 ISBN 978-981-10-1498-7 (eBook)
DOI 10.1007/978-981-10-1498-7

Library of Congress Control Number: 2016944002

Printed on acid-free paper

This Springer imprint is published by Springer Nature
The registered company is Springer Science+Business Media Singapore Pte Ltd.

Preface

The studies on long and short noncoding RNAs (ncRNAs) brought mammalian gene regulation into a new era, adding a complete new layer of cell signaling to the holistic gene regulatory network. Prior to their identification in mammalian cells, biologists had focused on regulatory events mediated merely by protein signaling. With the advance of whole genome research, we now know that protein-coding genes represent only less than 2 % of all human genome, while more than 90 % of it transcribes noncoding RNAs that lack protein-coding functions. Although once considered as evolutionary junk or transcriptional noise, ncRNAs emerge as crucial regulators in numerous physiological and pathological processes. The new findings in ncRNAs have changed the classic viewpoint of molecular biology that genetic information flows from DNA to RNA to protein. In the last decade, ncRNAs, especially microRNAs of around 18–22 nucleotides and long noncoding RNAs of larger than 200 nucleotides, have been studied extensively in a variety of diseases.

Cancer is among the first disease models, in which ncRNAs were studied in depth. Recently, the knowledge of ncRNA deregulation in expression or function that leads to cancer formation and development is increasing rapidly. A burst of studies have shown that ncRNAs play an integral role in nearly every aspect of cancer biology and determine carcinogenesis and metastasis. NcRNAs in tissues or in blood are promising cancer biomarkers for risk assessment, diagnosis, prognosis, and monitoring treatment response. More importantly, ncRNAs may serve as therapeutic targets for cancer treatment, while effective strategies for short ncRNA delivery into cancer cells in vivo are being extensively explored. Therefore, ncRNAs are now considered as one of the most important molecules in cancer research.

Early in 1993, the first microRNA (miRNA) lin-4 was identified in *Caenorhabditis elegans*, which targets the complementary heterochronic gene lin-14. Almost a decade later, a study from the Calin's group demonstrated a frequent deletion of miR-15 and miR-16 loci in chronic lymphocytic leukemia, which was the first evidence indicating a significant role of miRNAs in cancer pathogenesis. Currently, miRNAs have been widely implicated in the processes of cancer development by regulating proliferation, apoptosis, invasion, and metastasis of cancers, either as oncogenes (oncomiRs) or as tumor-suppressing genes, and are deeply involved in all of the ten hallmark features of human malignancies. Owing to the unique expression profile and biological function of miRNAs in various cancers, a breathtaking explosion of research was performed in the past decade to investigate their roles in cancer diagnosis as biological markers and in anticancer strategy as novel therapeutic targets.

Different from miRNA, the discovery of lncRNA could be traced back to 1990 when H19 was identified as a functional RNA molecule for imprinting, which was proved to be an epigenetic modulator later on. Evidence for lncRNA involvement in cancer development was provided in 2003, when MALAT-1 was found to be able to predict both the metastasis and survival for patients with early-stage non-small cell lung cancer. In recent years, we have witnessed an exponential growth of data delineating the regulatory roles of lncRNAs in multiple biological features of malignant tumors. Among them, HOTAIR is the most extensively studied lncRNA that exerts protumor effects in a variety of cancers as it promotes the invasion and metastasis of cancer cells. The versatile biological function of lncRNAs in cancer biology makes them attractive diagnostic markers and therapeutic targets for a spectrum of cancers, although it is still early to bring lncRNA-based antitumor therapeutics into clinical practice.

Apart from the linear RNAs with 5′-caps and 3′-tails as termini, ncRNAs can also be circular (circRNAs) as they form covalently closed loop structures without 5′–3′ polarity. The earliest discovered circular RNA dated back to 1976 in plant viroids, but this ncRNA type was not fully appreciated until 2013 when circRNA CDR1as was identified. Two years later, scientists reported for the first time that the abundance of circRNA was negatively correlated with proliferation of colorectal and ovarian cancer cells. Thereafter, more and more studies have revealed the implication of circRNAs in cancer biology and cancer progression. Meanwhile, increasing evidence is emerging that circRNAs in either tissue or serum are of great value as novel biomarkers for cancer diagnosis, while it remains an untapped gold mine for further discovery of circRNA-based therapeutic targets in the treatment of human malignancies.

In view of the above, ncRNAs have emerged as key players in tumor biology and are of great value in cancer diagnosis and treatment. In the last 10 years, an explosive amount of research has shed light on our understanding of ncRNA functions in regulating cancer biology and the potential roles of ncRNAs in antitumor clinical practice, which calls for a timely and systemic review of the knowledge in this field.

It is, however, impossible to cover all aspects of ncRNA research in the field of cancer within a single book. What we present here includes the molecular working modules of ncRNAs in cancer biology, the biological function of ncRNAs in modulating cancer hallmarks, and the potential of ncRNA application in cancer diagnosis and treatment. We hope that this book will be timely and topical in contributing a state-of-the-art understanding of the complex molecular mechanisms of ncRNAs in cancer signaling and the miscellaneous biological functions of ncRNAs in cancer biology.

Guangzhou, China Erwei Song

Contents

Chapter 1
Noncoding RNAs: New Players in Cancers

Xueman Chen, Siting Fan, and Erwei Song

Abstract The world of noncoding RNAs (ncRNAs) has gained widespread attention in recent years due to their novel and crucial potency of biological regulation. Noncoding RNAs play essential regulatory roles in a broad range of developmental processes and diseases, notably human cancers. Regulatory ncRNAs represent multiple levels of structurally and functionally distinct RNAs, including the best-known microRNAs (miRNAs), the complicated long ncRNAs (lncRNAs), and the newly identified circular RNAs (circRNAs). However, the mechanisms by which they act remain elusive. In this chapter, we will review the current knowledge of the ncRNA field, discussing the genomic context, biological functions, and mechanisms of action of miRNAs, lncRNAs, and circRNAs. We also highlight the implications of the biogenesis and gene expression dysregulation of different ncRNA subtypes in the initiation and development of human malignancies.

Keywords Noncoding RNA • MicroRNA • Long noncoding RNA • Circular RNA • Gene expression regulation • Cancer

1.1 Introduction

Cancer is a heterogeneous disease involving disorders in cell proliferation, differentiation, and death and remains a great challenge in modern medicine. Tumorigenesis in human is a multistep process driven by progressive genetic abnormalities. Either loss of function in tumor suppressor genes or upregulation of oncogenes can contribute to the malignant transformation. Traditionally, these genes were thought to exert their functions through DNA transcription and subsequent protein translation based on the Central Dogma of molecular biology. RNA was previously viewed as a mere transmitter of genetic information and was not fully

X. Chen • S. Fan • E. Song (✉)
Sun Yat-sen Memorial Hospital, Sun Yat-sen University,
107 Yanjiang West Road, Guangzhou, China
e-mail: songew@mail.sysu.edu.cn

E. Song (ed.), *The Long and Short Non-coding RNAs in Cancer Biology*,
Advances in Experimental Medicine and Biology 927,
DOI 10.1007/978-981-10-1498-7_1

appreciated until the 1980s, when its significant cellular roles as regulators and effectors were unveiled [1].

According to the Encyclopedia of DNA Elements (ENCODE) project, which is the most comprehensive effort yet for surveying transcription in human cells, protein-coding genes account for only 2–3 % of the total human genome, of which at least 75 % is actively transcribed without encoding protein [2–4]. The discovery of ncRNAs in the 1950s began to undermine the scientific level of the Central Dogma, but this information still failed to eliminate such notoriety as spurious body "garbage" or scrambled transcriptional "noise" for decades. Early in 1969, however, a model was postulated for the participation of ncRNAs in gene expression regulation in eukaryotes, whereby ncRNAs serve as signaling intermediates between sensory genetic elements and receptor elements, thus affecting coding gene production [5]. With the establishment and development of genome-wide approaches, high-throughput genome sequencing technologies, and functional in vivo models, increasing biological roles of ncRNAs in human cancers have emerged as promising regulators of epigenetic, transcriptional, and posttranscriptional gene expression. Generally, a threshold of 200 nucleotides (nt) divides regulatory ncRNAs into short (or small) and long (or large) ncRNAs [6, 7]. Linear lncRNAs can be distinguished from circular isoforms by appearance. Small ncRNAs (sncRNAs) comprise miRNAs, short interfering RNAs (siRNAs) and PIWI-interacting RNAs (piRNAs).

To date, miRNAs are the most extensively studied small subclass of ncRNAs in cancer. MicroRNAs are small, endogenous, single-stranded RNAs ~22 nt long, and they are encoded by eukaryotic nuclear DNA. Although the discovery of miRNAs can be traced back to 1993, when lin-4 was identified in *Caenorhabditis elegans* [8, 9], the term "microRNA" was not introduced until 2001 [10–12]. Thereafter, the Calin laboratory reported frequent deletion of miR-15 and miR-16 loci in B-cell chronic lymphocytic leukemia (B-CLL), indicating the involvement of miRNAs in cancer development for the first time [13]. Subsequent studies began to uncover the function of miRNAs as "oncomiRs," acting as transcriptional and posttranscriptional negative gene regulators by pairing to their target messenger RNAs (mRNAs). Currently, there is a plethora of publications describing the correlations between miRNAs and cancer pathogenesis, diagnosis and treatment.

Long ncRNAs are a heterogeneous group of RNAs ranging from 200 to 100,000 nt in length [9]. Similar to the generation of mRNAs, lncRNAs are products of RNA polymerase II/III (Pol II/III) transcription at any region in the genome, although they feature a lack of extended open reading frames (ORFs), polyadenylated or not [11]. Since the 1970s, a variety of lncRNAs have been identified across diverse species, including plants, invertebrates and mammals [14, 15]. The earliest-discovered gene-specific regulatory role of lncRNAs dates back to the early 1990s, represented by the epigenetic modulators H19 [16] and Xist [17, 18]. Even so, it was not until the late winter and early spring of 2007–2008 when three landmark studies highlighted the tremendous potential of lncRNAs [19–21]. Recent genome-wide association studies (GWAS) have rapidly unraveled lncRNAs as pivotal molecules regulating gene expression at the epigenetic, transcriptional, and posttranscriptional levels, supported by their tissue-specific expression patterns, subcellular distribution, and developmental regulation [1].

Unlike linear RNAs, circRNAs feature covalently closed loop structures without 5′ caps or 3′ tails. The first circRNA was identified in plant viroids in 1976 [22], followed by more extensive identifications in yeast mitochondrial RNAs [23] and hepatitis δ virus [24] within the decade. Although in the 1990s circRNAs were found to locate in exons [25–27], introns [28, 29], as well as intermediates escaping from intron lariat debranching [28–30], they were misinterpreted as by-products of spliceosome-mediated splicing errors [31, 32], with no place in biological processes [33]. In 2013, circRNAs finally attracted researchers' attention due to two simultaneously published studies on CDR1as in *NATURE* [34, 35], and since then, evidence of altered circRNA expression in diverse types of tumors has emerged.

As the role of regulatory ncRNAs in cancers has been a major focus of recent research, we will provide a summary overview of the biogenesis pathways of three major ncRNAs – miRNA, lncRNA and circRNA – and, more importantly, of their significant gene-regulating roles in cancer initiation and development as well as specific functional mechanisms in this chapter. We will also highlight the implications of these ncRNAs in cancer-related clinical applications, such as cancer diagnosis, treatment, and prognosis. Lastly, we will briefly introduce other cancer-associated ncRNAs, including siRNAs, piRNAs, and "classic" housekeeping ncRNAs, such as ribosomal (r)RNAs, transfer (t)RNAs, small nuclear (sn)RNAs, and small nucleolar (sno)RNAs.

1.2 The Landscape of MicroRNAs

MicroRNAs are well characterized as a large class of gene regulators found in plants and animals [36]. Many of them are conserved throughout evolution, from worms to human, with either oncogenic or tumor-suppressive functions [37]. MicroRNAs are grouped into families based on sequence homology, which is found primarily at the 5′ end of mature miRNAs [37]. Therefore, the targets of a miRNA family are likely to overlap among members; thus, a single mRNA can be targeted by multiple miRNAs, and an individual miRNA can have more than one mRNA targets.

1.2.1 MicroRNA Biogenesis and Functional Mechanism

Propelled by the discovery of let-7 miRNA in *C. elegans* [38], scientists soon elucidated the biogenesis pathways of miRNAs. The majority of miRNAs are processed from dedicated miRNA gene loci, while approximately 30 % from introns of coding genes, both of which can be transcribed by RNA Pol II into primary miRNAs (pri-miRNAs) [39]. An individual pri-miRNA can produce either a single miRNA or clusters of two or more miRNAs [39]. These long transcripts are subjected to microprocessing by the double-stranded RNase III enzyme DROSHA and its essential cofactor, the double-stranded RNA (dsRNA)-binding protein DGCR8, leaving ~60–70-nt precursor miRNAs (pre-miRNAs) with imperfect stem-loop hairpins flanked by single-stranded RNA (ssRNA) in the nucleus [40–43]. These intermediates are then exported into the cytoplasm by exportin-5(XPO5)-RanGTP, a nuclear transport receptor complex [44–46]. Subsequently, another RNase III enzyme, DICER1, measuring from the 5′ and 3′ ends of pre-miRNAs [47], triggers the asymmetrical cleavage of the dsRNA hairpin, generating ~22-nt miRNA-miRNA* duplexes (miRNA is the antisense or guide/mature strand, while miRNA* is the sense or passenger strand) with 2-nt overhangs at the 3′ end [48–51]. DICER1, transactivation-responsive RNA-binding protein (TRBP; also known as TARBP2), and an Argonaute protein (generally AGO1) then interact and assemble, triggering the formation of the miRNA-associated multiprotein RNA-induced-silencing complex (miRISC) [52]. During AGO loading, the passenger strand is cleaved, while the guide strand is incorporated into the functional miRISC. It then undergoes a causative procedure for gene silencing, specifically, targeting complementary mRNAs [53].

Mechanically, miRNAs regulate gene expression in a sequence-specific fashion, which either results in mRNA cleavage and degradation or simply accounts for translational repression, depending on the degree of complementarity between the 5′ seed region and the 3′ untranslated region (3′ UTR) of the mRNA [37, 54–56]. In plants, perfect-to-near-perfect base pairing often leads to the cleavage of target mRNAs and subsequent gene silencing in the so-called RNA interference (RNAi) pathway [57–60]. Alternatively, for only partially complementary miRNA binding in the mRNA 3′ UTR, target protein levels decrease, while mRNA levels may or may not be affected [61]. This phenomenon has been reported to occur in mammals [9, 48, 62, 63]. In human, however, miRNAs mainly cause translational inhibition, with infrequent mRNA cleavage and degradation [36].

1.2.2 Deregulation of miRNA Biogenesis in Cancers

As predicted, miRNAs regulate ~90% of human gene expression [64], and their roles have been shown to include cellular regulation in various biological processes, such as the cell cycle [37], proliferation [65], apoptosis [66], invasion [67, 68] and metastasis [69, 70]. Hence, any dysregulation in the pathway of miRNA biogenesis is likely to be associated with cancer development. The current finding that most miRNAs are decreased in tumor tissues compared with normal adjacent tissues also indicates that the miRNA biogenesis pathway might be impaired in cancer at the transcriptional or posttranscriptional level [71, 72].

First, dysregulated miRNA expression can arise from genetic variations that alter pri-miRNA transcription and eventually up-/downregulate target mRNAs that might play pivotal roles in tumors. This mechanism is exemplified by overexpression of oncogenic miRNAs (such as the miR-17–92 cluster) resulting from genomic amplification and the downregulation of miRNAs acting as tumor suppressors (such as the miR-15a/16-1 cluster) from deleted genomic regions [73]. Similar mechanisms involve transcriptional regulation by tumor-suppressive or oncogenic factors, together with the epigenetic alteration of histone proteins and DNA. Such cases include inhibition of the entire miR-200 family (miR-200a, −200b, −200c, −141, −429) by the transcription factors (TFs) zinc finger E-box-binding homeobox 1 (ZEB1) and 2 [74], along with the transactivation of the miR-34 family (miR-34a, miR-34b and miR-34c) by P53 [75–77] and the miR-17–19 cluster by MYC, both of which promote apoptosis in cancer [78, 79]. Moreover, DNA methylation and histone modifications induce the epigenetic silencing of certain miRNAs in neoplasms [80]. In addition, a wide variety of defects, such as somatic mutations in components of the miRNA processing machinery containing DROSHA, DGCR8, XPO5, DICER1, TRBP and other miRNA regulators (such as LIN28 for let-7 blockade and the Hippo signaling pathway), have been revealed to contribute to tumorigenesis [81–87].

1.2.3 Biological Roles of miRNAs in Cancers

Given that the functional mechanism of miRNAs is to target specific mRNAs, their function as oncogenes or tumor suppressor genes depends in large part on the roles of their mRNA targets and the cellular context [88–90]. Generally, miRNAs with tumor-suppressive potency contribute to oncogene overexpression through their poor abundance in tumors; in contrast, miRNAs upregulated in tumors are identified as oncomiRs that assist tumorigenesis by downregulating tumor suppressors. Currently, more specific roles of miRNAs have been associated with the hallmark capacities of human cancers, as proposed by Hanahan and Weinberg [91, 92], which include: (1) sustaining proliferative signaling; (2) evading growth suppressors; (3) resisting cell death; (4) enabling replicative immortality; (5) inducing angiogenesis; (6) activating invasion and metastasis; (7) genome instability and mutation; (8) tumor-promoting inflammation; (9) reprogramming energy metabolism; and (10) evading immune destruction.

1.2.3.1 Sustaining Proliferative Signaling

Unlike normal cells that are strictly regulated by the production and release of growth signals, which control the cell growth-and-division cycle, cancer cells block the resultant homeostasis of cell number, tissue architecture and function by dysregulating these cell cycle regulators, resulting in limitless growth and proliferation [92]. A plethora of miRNAs has been reported to interact directly with growth-promoting or anti-proliferation factors, thereby exerting their tumor-suppressive or oncogenic functions. The critical mitogenic signals in the proliferation pathways in which miRNAs participate involve RAS, Myc, PI3K (phosphoinositide 3-kinase)/

PTEN (phosphatase and tensin homolog), or ABL, as well as members of the RB (retinoblastoma) pathway, cyclin-CDK (cyclin-dependent kinase) complexes or cell cycle inhibitors of the INK4 or Cip/Kip families [93]. The miR-17–92 cluster, comprising miR-17-5p, miR-17-3p, miR-18a, miR-19a, miR-19b-1, miR-20 and miR-92-1, has been found to reside in a wide range of cancer types with high abundance, including lymphoma, lung, breast, stomach, colon, and pancreatic cancer [78, 88, 94]. Among this family, miR-17-5p and miR-20a can be activated by c-Myc and negatively regulate the tumorigenic levels of transcription factor E2F1, shifting the E2F transcriptional balance toward the proliferative E2F3 transcriptional network [78, 95]. Moreover, miR-221 and miR-222 are aberrantly overexpressed in glioblastoma, affecting the cell cycle and apoptotic cell death through the inhibition of $p27^{Kip1}$, a second member of the Cip/Kip family [96, 97]. In contrast, the let-7 family possesses tumor-suppressive potency, including antiproliferative activity. It comprises 12 members (let-7a-1, let-7a-2, let-7a-3; let-7b; let-7c; let-7d; let-7e; let-7f-1, let-7f-2; let-7 g; let-7i; miR-98) located at eight unlinked chromosomal loci [98]. Consistent downregulation of let-7 was found in lung cancer compared with normal adjacent tissue, while overexpression of let-7 could block lung cancer cell growth and proliferation [65, 99]. Mechanically, the let-7 family directly regulates multiple oncogenes, such as MYC, the RAS family (HRAS, KRAS and NRAS) and high-mobility group AT-hook 2 (HMGA2) [100–103], and promoters of cell-cycle progression, including CDC25A, CDK6, and cyclin D2 [65]. Studies from our laboratory further showed that low levels of let-7 in breast cancer stem cells (BCSCs) led to the upregulation of HRAS and HMGA2, indirectly promoting the self-renewal and differentiation of BCSCs [104]. Moreover, the overexpression of let-7 decreased the proportion of undifferentiated cells in vitro and repressed mammosphere formation, neoplasia, and metastasis in NOD/SCID mice. To date, the LIN28/let-7 loop has been demonstrated to regulate almost all of the cancer hallmarks [105].

1.2.3.2 Evading Growth Suppressors

As a complement to sustaining proliferative signals, cancers always manage to evade growth suppression from TP53, PTEN, pRB or the transforming growth factor β (TGFβ) tumor suppressor pathway. Among miR-34, a tumor suppressor miRNA family downregulated in cancers and transcriptionally activated by p53, miR-34a can directly target both TP53 itself and MDM4, a strong p53 transactivation inhibitor [106]. Another transcriptional repressor, Rb1, is posttranscriptionally inhibited by the overexpression of miR-106a in colon carcinoma, failing to cause a proliferative arrest [107]. Overexpression of the miR-106b-25 and miR-17-92 clusters is under the control of TGFβ signaling in gastrointestinal and other tumors, interfering with cell cycle arrest and apoptosis [108]. Aberrant expression of miR-21 was reported in lung, breast, stomach, prostate, colon and pancreatic cancer [109]. A high level of miR-21 is responsible for the repression of PTEN in hepatocellular carcinomas (HCCs), thus facilitating tumor cell growth, proliferation, migration, and invasion [110]. It also accumulates to induce pre-B-cell lymphoma in mice [111], accelerating KRAS-dependent lung carcinogenesis via the Ras/MEK/ERK pathway [112].

1.2.3.3 Resisting Cell Death

To achieve immortality, cancers must resist cell death induced by apoptosis, autophagy and necrosis [92]. The miRNAs miR-15 and miR-16 constitute a small cluster located in 13q14, a minimal cancer-associated chromosomal fragile site in the genome [69]. They were first reported to be commonly deleted in approximately 60% B-CLL patients and could perform a pro-apoptotic role through the mitochondrial pathway via targeting the anti-apoptotic protein Bcl-2 [13, 113]. The addition of exogenous miR-15a/16 resulted in the reduced expression of cyclin D1, a miR-15a/16 target and cell cycle regulator, thereby restoring cell control and increasing apoptosis [114]. Similar observations regarding the deletion of this cluster and the contrasting effect of its reintroduction were also made in multiple myeloma (MM) [115] and prostate cancer (PCa) [116].

MiR-21 has been shown to act as an anti-apoptotic factor in glioblastoma cells [117]. The downregulation of miR-21 retards both breast cancer (BrCa) growth in vivo and cell growth in vitro, which accounts for increased apoptosis and reduced expression of Bcl-2 [118]. A recent study showed that miR-15a/16 could induce autophagy through Rictor, a component of the mTORC2 complex, partly contributing to the inhibition of cell proliferation and enhanced chemosensitivity of camptothecin [119]. Many other miRNAs, including miR-101, miR-30a, miR-34a, miR-204, and miR-375, are also involved in the regulation of autophagy through targeting autophagy-related proteins (ATGs) in various cancers [120].

1.2.3.4 Enabling Replicative Immortality

In contrast to normal cells featuring finite replication, tumor cells are formed on the basis of immortalization. In normal cells, replicative potential is limited by cell division cycles due to telomere shortening, resulting in either senescence or crisis and ultimate cell death [92]. However, tumor cells acquire immortalization through governing telomere maintenance either via the upregulation of telomerase or, alternatively, the lengthening of telomeres [121]. The miR-290 cluster can influence telomere integrity and telomere-length homeostasis by targeting P130/RBL2 (retinoblastoma-like 2 protein), another member of the pRB family, and regulating RBL2-dependent Dnmt expression [122]. Upregulation of miR-155 drives telomere fragility in BrCa cells by decreasing TRF1 (shelterin component TERF1) levels and TRF1 abundance at telomeres, increasing the genomic instability linked to poor disease outcome [123]. Additionally, miR-34a induces senescence-like growth arrest by regulating the E2F pathway in human colon cancer cells [124]. Furthermore, miR-24 promotes cell proliferation by targeting $p27^{Kip1}$ and $p16^{INK4a}$, both of which are CDK inhibitors, during cell division [125].

1.2.3.5 Inducing Angiogenesis

Tumor angiogenesis is a key step for supplying rapidly growing malignant tissues with essential nutrients and oxygen, permitting further tumor growth and progression. microRNAs also play a part in the "angiogenic switch" regulated by pro- and anti-angiogenic factors, such as VEGF-A (vascular endothelial growth factor-A) and TSP-1 (thrombospondin-1), respectively. The endothelial-specific miR-126 directly targets Spred-1, an intracellular inhibitor of angiogenic signaling, to enhance the pro-angiogenic actions of VEGF and FGF and to promote blood vessel formation, thus governing vascular integrity and mediating developmental angiogenesis [126]. In addition, miR-107 inhibits HIF-1β expression and controls the p53 regulation of hypoxic signaling as well as tumor angiogenesis in colorectal cancer (CRC) [127]. Furthermore, the up-/downregulation of the miR-17-92 cluster, miR-378, miR-296, miR-15b/16, and miR-200b also alters the expression levels of VEGF, TSP-1, and growth factor receptors during tumor angiogenesis [74, 79, 121, 128].

1.2.3.6 Activating Invasion and Metastasis

Cancer metastasis is a complex process involving a succession of cellular events, whereby epithelial cells invade through the basement membrane, intravasate into the blood stream, disseminate through the circulation, extravasate to distal tissues/organs, and adapt to the new environment for survival and proliferation. Accumulating studies have unveiled the role of miRNAs as metastasis activators or suppressors [68]. The miR-200 family plays a crucial role in tumorigenesis and also affects each step of the metastatic cascade [74]. Research has indicated that miR-200a and miR-200b retard ovarian cancer angiogenesis by targeting interleukin 8 (IL-8) and chemokine ligand 1 (CXCL1) [129]. Additionally, the miR-200 cluster is a strong inhibitor of epithelial-to-mesenchymal transition (EMT). By targeting ZEB1 and ZEB2, the miR-200 family can further upregulate cellular E-cadherin levels and maintain a cell in a more epithelial-like state [130, 131]. In breast cancer, miR-9 is aberrantly activated by MYC/MYCN and directly targets CDH1, the E-cadherin-encoding mRNA, thus regulating E-cadherin levels and cancer metastasis [132]. By repressing programmed cell death 4 (PDCD4), miR-21 can enhance the metastasis of CRC [133]. Based on animal models, miR-10b, miR-103/miR-107, miR-373 and miR-520c have been found to promote tumor metastasis, whereas miR-31, miR-34a, miR-126, miR-206 and miR-335 exert the opposite effect [134].

1.2.3.7 Genome Instability and Mutation

Multistep tumor progression is largely initiated by the successive genomic alteration of oncogenes, tumor suppressor genes, and miRNA genes. Genetic defects in DNA repair machinery, cell cycle checkpoints and telomeric DNA contribute to genomic instability and cancer predisposition [92]. As previously described, miRNA genes frequently map to genomic regions that are deleted, amplified or translocated in cancer, such that variations of miRNA expression and miRNA-mediated gene regulation are implicated as key processes of oncogenesis and progression [69, 135]. Apart from the abovementioned miR-15a/16-1 (chromosomal fragile site) and miR-17–92 clusters (amplified gene locus) [73], miR-155-induced telomere fragility can augment genomic instability in BrCa [123].

1.2.3.8 Tumor-Promoting Inflammation

Tumor-associated inflammatory responses can make great contributions to the development of hallmark-facilitating programs through multiple functionally diverse but important bioactive molecules in the tumor microenvironment, such as growth factors sustaining proliferative signaling, proangiogenic factors and extracellular matrix-modifying enzymes promoting angiogenesis, invasion, and metastasis, among others [92]. Pro-inflammatory cytokine (such as IL-6/STAT3 pathway)-induced miR-21 overexpression is thought to account for inflammation-induced tumorigenesis in human colon cancer by targeting IL12-p35 [136]. miR-155 is another oncogenic microRNA that can be stimulated by inflammatory mediators, including TNFα and IFNβ, under pro-inflammatory conditions [137]. Increased miR-155 is responsible for the hyperproliferation of B cells, a common hallmark of leukemia and lymphoma [138], and also causes the repression of p53-induced nuclear protein 1 (TP53INP1) in pancreatic cancers [139]. Our work also showed that downregulation of miR-98 and miR-27b by tumor-associated macrophages (TAMs)-derived CCL18 can facilitate BrCa metastasis [140].

1.2.3.9 Reprogramming Energy Metabolism

The increased reliance on glycolysis for ATP generation is another hallmark of cancer cells, which has been termed the "Warburg effect." To fuel extensive cell growth and division, most tumor cells preferentially operate glucose-dependent energy production, bypassing the tricarboxylic acid cycle to convert it primarily to lactate; the glycolytic switch and altered energy metabolism are widespread in cancer cells despite available oxygen [92]. Currently, the role of miRNAs in the regulation of cancer metabolic pathways is emerging. For example, in the case of miR-23a/b targeting glutaminase, its repression by c-Myc results in elevated mitochondrial glutaminase expression and thus glutamine metabolism (ATP and lactate generation) in human P-493 B lymphoma cells and PC3 prostate cancer cells [141].

1.2.3.10 Evading Immune Destruction

Tumors are equipped with multiple mechanisms to evade early immunological surveillance or limit the extent of immunological killing, thereby regulating their susceptibility to lysis [92]. MicroRNAs can modulate the expression of genes that are critically involved in both innate and adaptive immune responses (e.g., miR-181a, miR-223, miR-155, miR-132 and miR-146) [121] and can also exert regulatory functions in the immune response in cancer (e.g., miR-222 and miR-339, miR-155 and miR-17-92 cluster) [142]. miR-222 and miR-339 negatively regulate the expression of intercellular cell adhesion molecule (ICAM)-1, preventing the recognition and cytolysis of tumor cells by cytotoxic T lymphocytes (CTLs) [143]. The miR-17-92 cluster could confer resistance to tumor-derived immunosuppressive factors and promote the type-1 skewing of T cells [142]. The overexpression of miR-155 in activated CD4+ T cells reverses the cancer-induced skewing toward Th2 polarization and enhances effective Th1-type antitumor immune responses by inhibiting IFN-γ signaling [144].

The roles of miRNAs in the hallmark capabilities of cancer cells quite often seem to overlap. A complete list of the miRNAs described herein and additional examples can be found in Table 1.1 below.

Table 1.1 MicroRNAs and hallmarks of cancer

Cancer hallmark	microRNA	Expression pattern	Function	Reference
Sustaining proliferative signaling	miR-17–92 cluster	Up in lymphoma, lung, breast, stomach, colon, and pancreatic cancers	OG	[78, 88, 94]
	miR-21	Up in hepatocellular carcinomas	OG	[110]
	miR-221 and miR-222	Up in glioblastoma	OG	[96, 97]
	let-7 family	Down in lung and breast cancers	TS	[65, 99, 104]
	miR-146a	Down in prostate cancer	TS	[145]
Evading growth suppressors	miR-34	Down in colon cancer	TS	[106]
	miR-106a	Up in colon carcinoma	OG	[107]
	miR-106b-25, miR-17-9 clusters	Up in gastrointestinal cancer	OG	[108]
	miR-21	Up in lung, breast, stomach, prostate, colon, pancreas, and liver cancers, pre-B-cell lymphoma	OG	[109–111]
Resisting cell death	miR-15 and miR-16	Down in B-cell chronic lymphocytic leukemias, multiple myeloma, prostate cancer	TS	[13, 113, 115, 116, 119]
	miR-21	Up in glioblastoma, breast, colon, lung, pancreas, prostate, and stomach cancers	OG	[107, 117]
	miR-101, miR-30a, miR-34a, miR-204, miR-375	Down in cancers	TS	[120]

(continued)

Table 1.1 (continued)

Cancer hallmark	microRNA	Expression pattern	Function	Reference
	miR-29	Down in KMCH cholangiocarcinoma cell lines	TS	[146]
Enabling replicative immortality	miR-290	Down in DICER1-null cells	TS	[122]
	miR-155	Up in breast cancer	OG	[123]
	miR-34a	Down in colon cancer	TS	[124]
	miR-24	Up in cervical carcinoma	OG	[125]
Inducing angiogenesis	miR-126	Up in endothelial cells	OG	[126]
	miR-107	Down in colorectal cancer	TS	[127]
	miR-17-92 cluster	Up in colon cancer	OG	[79]
	miR-378	Up in papillary thyroid carcinoma	OG	[128]
	miR-200, miR-200b	Down in ovarian cancer	TS	[129]
Activating invasion and metastasis	miR-200 family	Down in breast, stomach, lung, skin, liver, ovary, cervix, esophagus, colon, kidney, prostate, and bladder cancers	TS and OG	[74, 130, 131]
	miR-9	Up in breast cancer	OG	[132]
	miR-21	Up in colorectal cancer	OG	[133]
	miR-10b, miR-103/miR-107, miR-373, and miR-520c	Up in breast cancer	OG	[134]
	miR-31, miR-34a, miR-126, miR-206, and miR-335, let-7 family	Down in breast cancer	TS	[134]
	miR98 and miR27b	Down in breast cancer	TS	[140]
Genome instability and mutation	miR-15a/16-1 cluster	Down in B-cell chronic lymphocytic leukemias	TS	[69]
	miR-17–92 cluster	Up in B-cell lymphomas and lung cancer	OG	[147, 148]
	miR-155	Up in breast cancer		[123]
Tumor-promoting inflammation	miR-21	Up in colon cancer	OG	[136]
	miR-155	Up in leukemia and lymphoma, pancreatic cancers	OG	[138, 139]
Reprogramming energy metabolism	miR-23a/b	Down in B lymphoma, prostate cancer	TS	[141]
Evading immune destruction	miR-222 and miR-339	Up in glioma	OG	[143]
	miR-17-92 cluster	Up in activated T cells	TS	[142]
	miR-155	Up in activated CD4+ T cells	TS	[144]

Up upregulation, *down* downregulation, *TS* tumor suppressor, *OG* oncogene

1.2.4 *Clinical Application of miRNAs in Cancers*

1.2.4.1 Diagnostics

Since the tissue and cell types responsible for dysregulated miRNA expression in cancers are increasingly understood, miRNA expression profiles possess great potency as biomarkers for cancer diagnosis, classification, prognosis, and response to treatment.

For many types of cancer, early detection methods or screening tests are not available, are expensive, or are harmful for patients. However, miRNA-based indicators have shown emerging promise in cancer diagnosis, which includes tumor stage classification and reflects the risk of cancer recurrence, disease progression, and patient death. The significant reduction (80 %) of let-7 expression in lung cancer is coupled with frequent occurrence (43.8 %) relative to normal lung tissues [99]. On this basis, patients in the study could be classified into two major groups. Simultaneously, reduced let-7 expression is associated with poor prognosis or shorter survival after resection [99, 149]. Another study showed that low let-7 and high HMGA2 expression are significantly related to poor clinical outcome in advanced ovarian cancer [150]. Moreover, a five-miRNA signature (let-7a, miR-221, miR-137, miR-372 and miR-182*) was found to predict survival and cancer relapse in NSCLC patients after surgery [151]. Other cases include the overexpression of miR-21 in colon adenocarcinoma (poor prognosis and therapeutic outcome) [152], low levels of miR-335 and miR-126 in primary breast tumors (recurrence and poor distal-metastasis-free survival) [153], and miR-122 downregulation in liver cancer (gain of metastatic properties and cancer progression) [154].

Despite the tissue-specific miRNA expression signatures in cancer, miRNAs have also been found to circulate in body fluids (e.g., blood, urine, sputum, or stool), acting as exosome-mediated intercellular messengers in a "hormone-like" manner [153, 155], as exemplified by let-7 [156].

Although these findings are appealing, more large-scale prospective studies, rather than retrospective cohorts, are required to further bring miRNA-based diagnostics into clinical application. In addition, the intratumoral and intercellular heterogeneity of tumors should be fully considered when evaluating the practical application values of miRNAs [157].

1.2.4.2 Therapeutics

The potential of miRNAs as oncogenes and tumor suppressors in tumor formation and progression has generated great therapeutic promise for drug targets. The mode of action of miRNA-based drugs depends on either suppressing their gain of function (for oncomiRs) or restoring their loss of function (for tumor-suppressive miRNAs) [157]. On the one hand, synthetic anti-miRNA oligonucleotides (AMOs) with 20-O-methyl modification have been shown to effectively mediate the inhibition of endogenous oncomiRs in cell culture and xenograft mouse models [158]. Such antagomiRs could be applied to onco-miR-21, with potent suppression of glioblastoma and BrCa cell growth [117, 118]. Additionally, AMO administration to miR-122 in mice and primates could alter the lipid metabolism and hepatitis C viral load, thereby reducing liver damage [159, 160]. On the other hand, as miRNAs are globally suppressed in tumor cells compared with normal tissues [71, 161], the reintroduction of specific miRNAs using mimetics or viral vector-encoded miRNAs could have a therapeutic benefit in reversing neoplasia [158]. For example, the systemic delivery of a miR-26a mimic was shown to reduce tumor size in a murine liver cancer model [162]. MRX34 (Mirna Therapeutics, TX, USA), the first-tested miRNA drug, is a miR-34 mimic designed to restore its tumor-suppressive function in cancer cells, which is currently being evaluated in clinical trials [157].

Novel developments in pharmacological approaches involve the combination of miRNAs or miRNA/siRNA, which exhibit different degrees of antitumor therapeutic effects. In exploring more effective antitumor drugs, the potential of miRNAs (e.g., miR-200b and miR-200c) in chemoresistance also needs consideration and further investigation [74].

1.3 The Landscape of Long NcRNAs

More complex than miRNAs, lncRNAs are far from exhaustively studied and remain intriguing regarding their biology and function. Well-characterized lncRNAs cannot hitherto be excluded by Xist (X inactive specific transcript) and its antisense Tsix (TSIX transcript, XIST antisense RNA) in X-chromosome inactivation (XCI) [163, 164], H19 and Air (antisense to IGF2R RNA) for imprinting [16, 165], HOTAIR (HOX antisense intergenic RNA) in trans regulating genes [166], as well as NRON (noncoding repressor of NFAT) in nuclear import [167]. Beyond these highly evolutionarily conserved examples, however, most lncRNAs are poorly conserved relative to sncRNAs and exert gene regulatory functions through diverse mechanisms that are as yet unknown. Therefore, nomenclature of lncRNAs lacks specification and comprises multiple synonyms (i.e., long RNA, large RNA, macroRNA, intergenic RNA, and NonCoding transcripts), urgently awaiting a single search term that refers to the same biological phenomenon and keeps abreast of the accumulating lncRNA literature [168].

1.3.1 Biogenesis and Classification of LncRNAs

By virtue of the Pol II transcriptional machinery, a majority of known chromatin-associated lncRNAs are transcribed and processed like mRNAs, with additional 5′ capping, histone modifications associated with transcriptional elongation, and polyadenylation [169]. There are also nonpolyadenylated lncRNAs that derive from Pol III promoters [170, 171] and snoRNA-related lncRNAs (sno-lncRNAs) expressed from introns via the snoRNP machinery (with the supplementary production of two snoRNAs) [172]. On the whole, the lncRNAs described thus far map to a wide range of gene regions, including intergenic regions [169], promoters [173], enhancers [174] and introns [175, 176].

In light of the genomic proximity between neighboring annotated genes, lncRNAs can be classified into five broad categories:

(1) Intergenic lncRNAs, also termed large intervening ncRNAs, lincRNAs, or stand-alone lncRNAs, whose transcriptional units are located independently between two coding genes, at least 5 kb from both sides, exemplified by H19 [16], Xist [17, 18] and lincRNA-p21 [177];
(2) Intronic lncRNAs that arise from inside an intron of a coding gene without overlapping exons at either end, as is the case for COLDAIR, located in the first intron of the flowering repressor locus FLC [178];
(3) Bidirectional lncRNAs, which initiate in divergent directions from promoter or enhancer regions, generally within a few hundred base pairs, thus generating enhancer-associated RNAs (eRNAs) [179, 180] and promoter-associated long RNAs (PALRs) [181], respectively;
(4) Sense lncRNAs, which are transcribed in the same direction as coding genes with at least one exon overlapped, such as Gas5 (growth arrest-specific 5) and MALAT1 (metastasis-associated lung adenocarcinoma transcript 1) [182, 183];
(5) Antisense lncRNAs, also called natural antisense transcripts (NATs), which are transcribed from the opposite strand of coding genes and overlap with sense mRNAs at the 5′ (divergent NAT or head to head) or 3′-ends (convergent NAT or tail to tail), such as the transcription of HOTAIR in an antisense manner from the mammalian homeobox transcription factors C (HOXC) locus on chromosome 12q13.13 [184–186].

Obviously, this classification is too simple to cover the whole lncRNAome; cases such as pseudogenes and telomerase RNA (TERC) still lie outside the list. As such, additional categories are required to keep pace with the rapid appearance of new lncRNAs. In terms of size, the characterized lincRNAs often range from hundreds of nucleotides to several kilobases [169]. However, there are exceptionally long lncRNAs (macroRNAs) and very long intergenic noncoding RNAs (vlincRNAs), stretching tens of kb and 1 Mb, respectively [186]. Another problematic case is ANRIL (antisense noncoding RNA in the INK4 locus), which is a lincRNA [187], a NAT [187], and also a circRNA [188]. Although other grouping criteria are emerging based on correlation with coding genes, coding RNA resemblance, interaction with RNA-binding proteins (RBPs), sequence and structure conservation, stability, and cellular functions, among other factors [189], there is still need for an integrative framework of annotation and classification for the whole lncRNAome, particularly with clearly assigned functionality.

1.3.2 Molecular Mechanisms of LncRNAs in Cancers

Representing the larger subset of ncRNAs in number and size, lncRNAs possess evolutionary complexity and regulatory specificity, possibly functioning via the assembly of diverse proteins and interactions with DNA or RNA [168] or simply alone [190]. Here, we review several major but mutually nonexclusive actions of lncRNA executors as signals, decoys, guides, and scaffolds [191]. The corresponding functional mechanisms of certain lncRNAs in cancers will be described in detail in the following chapters.

First, the cell-type-specific expression of lncRNAs in response to a variety of environmental stimuli indicates that their expression is under considerable transcriptional control. In this case, lncRNAs can serve as molecular signals marking space, time, developmental stage, and expression for gene regulation. Thus, RNAs, rather than proteins, exert rapid regulatory effects. On the other hand, lncRNAs that are simply by-products of transcription also act as markers of functionally significant developmental processes. Known examples include lincRNA-p21 and PANDA (P21 associated ncRNA DNA damage activated), which are induced by DNA damage in a p53-dependent manner [173, 177], as well as HOTAIR and HOTTIP, whose expression and signaling depend on anatomic positions [166].

Second, lncRNAs transcribed from the gene regions of promoters and enhancers are likely to perform as decoys that negatively regulate transcription. They can merely sequester RBP targets (such as TFs and chromatin modifiers) away from chromatin, thus preventing the effectors from functional execution. The abovementioned PANDA can also bind and sequester the NF-YA transcription factor, thus repressing NF-YA-induced apoptotic gene expression [173]. Similar cases involve Gas5 for glucocorticoid receptor (GR) [182] and MALAT1 for serine/arginine (SR) splicing factors [183].

Additionally, the role of lncRNAs as guides refers to the binding and recruitment of ribonucleoprotein (RNP) complexes onto specific target genes, followed by certain chromatin changes, either in cis, near the lncRNA genes, or in trans, to distant target genes. The interacting partners with catalytic action, brought on by lncRNAs, include activating complexes such as the trithorax group proteins (TxG), repressive complexes such as the polycomb group proteins (PcG), and combinations of TFs. In this case, lncRNAs, such as Air and Xist, can recruit chromatin modifiers for the cis-acting silencing of adjacent sites [192, 193]. Conversely, lincRNA-p21, in association with heterogeneous nuclear ribonucleoprotein K (hnRNP-K), regulates gene expression through a trans-acting effect [177].

Lastly, the most complex functional pattern is scaffold molecules, which were traditionally restricted to be proteins. Long ncRNAs with different domains can serve as central platforms, assembling a molecular cargo of specific combinations of distinct effector partners in both time and space, leading to the formation of RNP complexes and the flow of information. The lncRNA-RNP may remodel chromatin through histone methylation with the assistance of PcG proteins. ANRIL is an example that directly interacts with both PRC1 and PRC2, which are two members of the polycomb repressive complex (PRC), to cause gene silencing [194, 195].

Since classes of lncRNAs do not have exclusive functions, there are various combinations that can perform multiple biological functions. For example, the roles of HOTAIR vary from signal (transcribed with anatomic specificity) to scaffold (assembly of both PRC2 and LSD1) and guide (targeting the bound PRC2 to its destined DNA sequence in trans) under diverse developmental cues [191]. As such, multifunctional lncRNAs can either deliver vital genomic messages or interact with other effector molecules, further altering gene expression patterns.

With the four lncRNA archetypes providing useful explanations and predictions of disease outcomes, our recent research reported for the first time that there is a novel class of lncRNAs that directly regulate signaling pathways without the assistance of other molecules. NF-κB interacting long noncoding RNA (NKILA), which maps to chromosome 20q13, can bind directly to NF-κB/IκB and competitively mask the IκB phosphorylation sites for IKK, thus stabilizing NF-κB/IκB to form a ternary complex and prevent NF-κB activation [190]. This ability demonstrates that lncRNA NKILA functions in cancer-associated inflammation through direct interplay with the NF-κB signaling pathway [190].

All in all, it is worth noting that the molecular mechanisms by which lncRNAs exert their function are far more versatile than was previously appreciated.

1.3.3 Functional Mechanisms of LncRNAs in Cancers

After a brief introduction to the molecular roles of lncRNAs, we move on to their functional mechanisms in biological processes and neoplastic transformation [196]. In affecting the onset and progression of tumors, lncRNAs fulfill regulatory roles at almost every step of gene expression, from targeting epigenetic modifications in the nucleus to affecting mRNA stability and translation in the cytoplasm [197].

1.3.3.1 Epigenetic Regulation: in Cis and in Trans

The epigenetic control of gene expression is predominantly achieved by DNA methylation, a covalent modification of cytosine, and posttranslational modifications of histone tails, such as acetylation, methylation, and phosphorylation [198], which in essence contribute to heritable changes in gene activity without altering the genomic sequence [199]. Discovered in human tumors in 1983, epigenetic alterations have been widely recognized as an essential mechanism in oncogenesis [200, 201].

Long ncRNAs have been best characterized as epigenetic modulators, among which HOTAIR is the most extensively studied. Despite the diversity of functional pathways in tumors, HOTAIR is the first lncRNA found to regulate chromatin dynamics in trans, that is, in association with chromatin proteins (PRC2) and their catalytic methyltransferase subunit (EZH2, enhancer of zeste homolog 2), to affect gene expression in a trans-acting manner [202]. As mentioned, the major chromatin-modifying complexes PcG were first identified in *Drosophila* and serve as a bridge between lncRNAs and chromatin to modify local histones and modulate genomic programs [203]. However, the details of the mechanisms by which the complexes target specific genomic loci in mammals remain unclear. The HOTAIR target sites encompass the HOXD cluster in chromosome 2 and hundreds of additional loci spread across the genome [204]. With PcG occupied, these genes further undergo H3K27-trimethylation (H3K27-me3), a hallmark of gene silencing [197]. In addition to the binding of PRC2 with a structural domain at the 5′ end, HOTAIR also interacts at its 3′ end with a multiprotein complex formed by LSD1 (lysine-specific demethylase 1A, also called KDM1), REST (RE1-silencing transcription factor), and CoREST and then demethylates histone H3K4 to prevent gene activation [205]. Thus, HOTAIR acts as a modular scaffold onto which both the PRC2 and LSD1 complexes can assemble, allowing the cooperation of H3K27-me3 and H3K4-demethylation, which account for chromatin remodification and ultimately transcriptional inhibition [205–207]. Similarly, ANRIL is analogously overexpressed in cancers to bind both the PRC1 component Cbx7 and the PRC2 component Suz12 (suppressor of zeste 12 homolog) to trimethylate H3K27, ultimately repressing the INK4a/p16 and INK4b/p15 tumor suppressor loci [194, 195, 208].

On the other hand, many lncRNAs act in cis on local chromatin, thereby targeting chromatin modifiers to neighboring genes of the parental allele where they are coded, as in the case of Igf2r (type II receptor of insulin-like growth factor 2) imprinting by lncRNA Airn. This lncRNA recruits the G9a repressive epigenetic complexes to methylate H3K9 residues over the adjacent genomic region, followed by cis-acting silencing of the Igf2r/Slc22a2/Slc22a3 gene cluster on the paternal allele [192].

In addition to histone modifications, lncRNA-mediated DNA methylation is also an important layer of epigenetic regulation in cancer. DNA methylation at CpG dinucleotides is commonly linked to the repression of gene expression involved in genomic imprinting and dosage compensation [209]. IGF2 (insulin-like growth factor 2) is transcribed opposite to, and situated in the vicinity of, lncRNA H19, whose regions exhibit a parental origin-specific expression pattern [210]. Aberrant DNA methylation upstream from H19 promoters is in part responsible for the loss of imprinting of IGF2 in cancers [185, 211–214]. Another imprinted lncRNA, Xist, maintains dosage compensation for ~1000 genes on the X chromosome, including an entire female XCI [215]. Most importantly, the aberrant expression of X-linked oncogenes on the X chromosome potentially contributes to cancer phenotypes [215]. Additionally, antisense Tsix can mediate long-term Xist silencing through the recruitment of DNMT3A for Xist promoter hypermethylation [216, 217]. DNA methylation is catalyzed by the DNA methyltransferases (DNMTs), among which DNMT3A and DNMT3B methylate unmethylated cytosines, whereas DNMT1 recognizes hemimethylated DNA [199]. Promoter-associated ncRNAs interacting with the rDNA promoter mediate de novo CpG methylation to silence rRNA genes by recruiting DNMT3B [218].

1.3.3.2 Transcriptional Regulation

Beyond epigenetic modulation, lncRNAs also serve as master regulators in transcriptional pathways. Essentially, lncRNA transcription may open the chromatin structure and expose binding domains for transcriptional machinery, thereby enhancing the expression of proximal coding genes. On the other hand, the occupancy of transcriptional machinery on a lncRNA gene locus can physically prevent binding to coding neighbors and repress their transcription [219]. In addition to the basal transcriptional machinery, lncRNAs may likewise directly target gene promoters due to their sequence complementarity and communicate with TFs, bypassing chromatin modifiers [220].

LincRNA-p21, a lincRNA activated by p53 upon DNA damage, binds and guides hnRNP-K to its genomic targets, leading to the transcriptional repression of anti-apoptotic genes in a p53-dependent pathway [177]. However, another p53-induced lncRNA that resides upstream of p21, PANDA, exerts the opposite effect by decoying NF-YA from its pro-apoptotic target genes, such as FAS and BIK, thereby contributing to tumor cell survival and chemoresistance [173].

Another tumor suppressor for TF binding, Gas5, binds the GR's DNA-binding domain by the formation of a secondary structure sharing sequence similarity with the steroid-responsive gene promoter [182]. In this case, Gas5 competitively prevents the glucocorticoid response elements from binding to DNA and inducing transcriptional activation to modulate cell survival and induce glucocorticoid resistance [182].

More examples involve cytoplasmic NRON, which modulates the nuclear trafficking of NFAT (nuclear factor of activated T cells) [167] and the relocation of SRs by nuclear MALAT1 to transcription sites [183].

1.3.3.3 Posttranscriptional Regulation

In addition to its involvement in transcriptional regulation, MALAT1 also posttranscriptionally affects the expression of cytoskeletal and extracellular matrix genes by controlling alternative spicing based on interaction with SRs [221].

Apart from mRNA processing, many other lncRNAs act through transport, the translational regulation of mRNA, and the stability control of proteins. Examples include iNOS (inducible nitric oxide synthase), which is stabilized by the interaction of its antisense lncRNA interacting with stabilizing factor HuR [222], and lincRNA-p21, which acts as a posttranscriptional inhibitor of polysome-bound β-catenin and JunB translation [223].

1.3.3.4 Other Roles of LncRNAs in Cancers

In addition, lncRNAs can function as mRNA or miRNA sponges to inactivate growth-promoting, protumorigenic signaling pathways [219]. Long ncRNAs of this kind are branded as competing endogenous RNAs (ceRNAs) possessing one or more sequences similar to other RNAs, whether coding or not. Such examples can be represented by the pseudogene PTENP1 (phosphatase and tensin homolog pseudogene 1) in relation to its PTEN tumor suppressor gene. PTENP1 harbors miRNA binding sites and acts as a competitive “sponge” or decoy for PTEN mRNA-targeting miRNAs, leading to the upregulation of PTEN expression [224].

There are also lncRNAs that serve as precursors for shorter functional RNAs, as exemplified by the primary transcripts for mi/piRNAs [189]. H19 is the host gene for the primary miR-675 precursor [225, 226], and Gas5 hosts ten highly conserved snoRNAs [227].

1.3.4 Biological Roles of LncRNAs in Cancers

Currently, high-throughput genomic technologies, such as microarrays and next-generation sequencing (NGS), together with functional studies help to identify many lncRNAs involved in a spectrum of tumors. The overexpressed and protumoral lncRNAs are referred to as onco-lncRNAs and include HOTAIR, H19, Xist, ANRIL, and MALAT1. In contrast, tumor suppressor lncRNAs with downregulated expression patterns in cancer include lincRNA-p21, Gas5, PTENP1, and lncRNA-LET (lncRNA low expression in tumor). For example, the high expression of HOTAIR with a pro-tumor effect has been implicated in various tumor types, including breast cancer [228], liver cancer [229], colorectal carcinoma [185], gastrointestinal stromal tumor [230], pancreatic cancer [231], non-small cell lung carcinoma (NSCLC) [232], nasopharyngeal carcinoma [233], laryngeal squamous cell carcinoma (LSCC) [234], esophageal squamous cell carcinoma (ESCC) [235], urothelial carcinoma [236], ovarian cancer [237], renal carcinoma [238], gall bladder cancer [239], nonfunctional pituitary adenoma [240], PCa [241], melanoma [242], and endometrial tumors [243].

Accumulating evidence has unveiled a correlation between lncRNAs and the hallmark features of cancer [196], as shown in Fig. 1.1. Here, we briefly summarize the well-known lncRNAs in the essential stages of tumorigenesis and progression.

Using proliferation and metastasis as examples, the overexpression of ANRIL, PANDA, PCAT-1 (prostate cancer-associated transcript 1) and PVT1 (plasmacytoma variant translocation 1) in PCa, GC, CRC and HCC promotes tumor cell survival, growth and proliferation through the mechanisms mentioned in the previous sections [173, 185, 194, 195, 208, 244–246], whereas lncRNA-p21, Gas5, and PTENP1, which are generally downregulated in cancers, contribute to repressed growth and proliferation together with enhanced apoptosis in colorectal, prostate, breast, gastric and renal cancer [182, 185, 224, 247–251]. In regard to tumor metastasis, the role of lncRNAs varies from promoter to suppressor or both: (1) HOTAIR, MALAT1 and PVT1 facilitate tumor invasion and metastasis in a diverse array of human malignancies [185, 221, 232, 234, 235, 241, 243, 252–258] [185, 244, 245, 259, 260]; (2) NKILA is a potential breast cancer metastatic suppressor [190]; and (3) the H19/miR-675 axis promotes glioma cancer cell invasion [261] but suppresses PCa metastasis [262]. Further examples and more details are presented in Table 1.2.

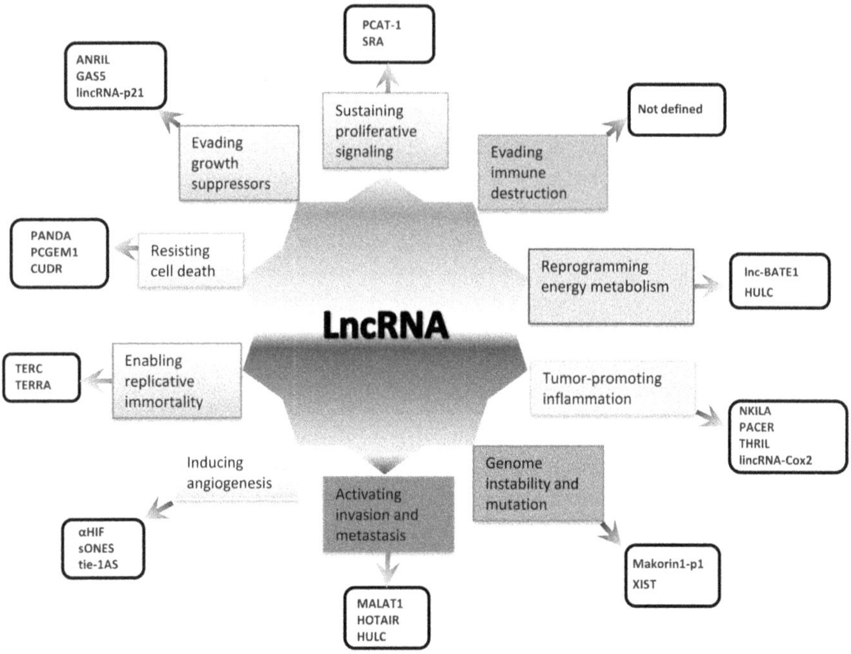

Fig. 1.1 LncRNAs and the cancer hallmarks

1.3.5 Diagnostic and Therapeutic Application of LncRNAs in Cancers

The continued identification of lncRNAs in various cancer types demonstrates that lncRNAs open new avenues for novel diagnostic and prognostic tools or therapeutic targets in the treatment of cancer and resistance induced by antineoplastic drugs.

Currently, new cancer-linked diagnostics and prognostics based upon lncRNA biology are undergoing rapid development. For example, lncRNAs isolated from tumor cells or the circulating bloodstream may provide readily available, inexpensive and stable blood-borne diagnostic markers for the improved detection of cancers and cancer subtypes [219]. The CpG hypomethylation of Xist has been commonly found in serum from male PCa patients compared with normal XY serum, suggesting further clinical applications, such as early diagnosis for cancer [278, 279]. Moreover, the expression level of HOTAIR is higher in advanced tumors than in low-grade tumors, supporting its potential as a novel biomarker in malignant grades [280]. Reduced NKILA expression is correlated with breast cancer metastasis and poor patient outcome [190]. Increased PANDAR expression is associated with poorer overall survival and shorter recurrence, highlighting its potential clinical utility as a promising prognostic biomarker and therapeutic target [263].

Table 1.2 Cancer-associated lncRNAs in tumor development

Type	LncRNA	Classification	Expression pattern	Molecular or functional mechanism	Function	Cancer type	Reference
OG	HOTAIR	Intergenic, antisense	Up	Signal, guide, scaffold	Promote invasion and metastasis	Various cancers	[232, 234, 235, 241, 243, 252, 253]
	ANRIL	Antisense	Up	Scaffold	Promote growth	PCa, GC	[194, 195, 208]
	PANDA	Antisense	Up	Signal, decoy	Promote survival and chemoresistance	BrCa, HCC	[173] [263]
	MALAT1	Sense	Up	Decoy	Promote proliferation, invasion, and metastasis	Lung cancer, HCC, CRC, bladder cancer, kidney cancer, PCa	[185, 221, 254–258]
	TERC	Not defined	Up	Telomerase	Genome stability and replication	Cervical cancer, ovarian cancer, lung cancer, leukemia	[264, 265]
	Air	Antisense	Up	Guide	Gene imprinting	HCC	[192, 266]
	PVT1	Sense	Up	Undefined	Promote proliferation, invasion and metastasis, multidrug resistance	CRC, HCC and GC, NSCLL	[185, 244, 245, 259, 260]
	PCAT-1	Intergenic	Up	Transcriptional repressor	Promote proliferation	PCa	[246]
	HULC	Intergenic	Up	miRNA sponge	Promote growth, modulate lipid metabolism	HCC	[267, 268]
	αHIF	Antisense	Up	mRNA decay	Induce angiogenesis	Renal cancer, BrCa	[269, 270] [271]

Table 1.2 (continued)

Type	LncRNA	Classification	Expression pattern	Molecular or functional mechanism	Function	Cancer type	Reference
OG and TS	Xist	Intergenic	Up	Epigenetic modulator	Genetic instability	OG: male cancer	[272, 273]
			Down			TS: female cancer	[274, 275]
	H19	Intergenic	Up	Epigenetic modulator, miR-675 precursor	OG: promote invasion and metastasis	Glioma cancer, bladder cancer, GC	[261, 276, 277]
					TS: repress metastasis	PCa	[262]
TS	lincRNA-p21	Intergenic	Down	Signal, guide	Promote apoptosis	CRC	[247]
	Gas5	Sense	Down	Decoy, snoRNA precursor	Promote apoptosis	PCa, BrCa, CRC, GC, kidney cancer	[182, 185, 248–251]
	NKILA	Antisense	Down	Inflammatory signaling	Cancer-associated inflammation, metastasis	BrCa	[190]
	PTENP1	Pseudogene	Down	miRNA sponge	Inhibit growth	PCa	[224]

OG oncogene, *TS* tumor suppressor, *up* upregulation, *down* downregulation, *TERC* telomerase RNA component, *HULC* hepatocellular carcinoma upregulated long noncoding RNA

The overexpression of PVT1 correlates with histological grade and lymph node metastasis, indicating its use as an independent predictor for tumor (such as HCC, NSCLC) progression, recurrence and patient survival, and potentially as a therapeutic target [260, 281]. PCAT-1 upregulation reflects the invasion of cancer tissues, metastasis of lymph nodes, advanced tumor stage, and poor prognosis in ESCC, indicating its role as a potential diagnostic and prognostic target in ESCC patients [282]. LncRNA αHIF (but not HIF-1α) transcript expression is associated with poor clinical prognosis in human breast cancer [271].

Despite the application of cancer-associated lncRNAs for diagnosis and prognosis, lncRNA-based antitumor therapeutics remain in the early stage of development. Even so, there has been progress. First, RNAi-mediated gene silencing offers a straightforward approach to selectively silencing oncogenic lncRNAs. The structural motifs of lncRNAs for the recruitment of chromatin modifiers or the formation of triple helixes with DNA could be potential targets of small molecules, which might restore the disturbance of lncRNAs in gene expression [220]. In the case of interplay between HOTAIR and the PRC2 and LSD1 complexes, either targeting endogenous HOTAIR or using small molecular inhibitors of PRC2 could prevent their interaction and thus reduce cancer metastasis [283]. Additionally, synthetically engineered lncRNAs possessing tumor-suppressive effects may be employed through replacement therapy to retard cancer development [219]. Delivery vehicles, e.g., liposomal membranes and nanoparticles, have been designed to deliver these RNA inhibitors. For example, DTA-H19, a plasmid comprising the H19 gene regulatory sequences and diphtheria toxin A, might serve as a potential targeted therapy for both pancreatic and bladder cancers [284, 285]. Antisense oligonucleotides are also emerging as a therapeutic tool to increase tumor suppressor activity by neutralizing inhibitory lncRNAs in the absence of any delivery vehicle [219].

Although they are charming and promising, much work needs to be done before such applications become clinically practical.

1.4 The Landscape of Circular RNAs

In comparison to miRNAs and lncRNAs, circRNAs are less well studied. Recently, however, circRNAs have emerged as a potential novel star among ncRNAs [33] due to limited evidence of translation with ribosomes [286, 287]. In contrast to linear RNAs featuring 5′ caps and 3′ tails as termini, circRNAs show covalently closed loop structures without 5′ to 3′ polarity or polyadenylation at the 3′ ends [288, 289]. Thus, a majority of circRNAs have escaped the general transcriptomic polyadenylated RNA profiling [290, 291]. With the advent and maturation of bioinformatics and RNA deep sequencing technology [292], many circRNAs have been identified in various cell lines and across distinct taxa [34, 175, 287, 293–295].

1.4.1 Formation and Classification

With both polyadenylated RNAs and ribosomal RNAs depleted, circRNA signals were first detected by a genome-wide approach to be accumulated in excised exons or introns [176]. Further works classify circRNAs into three subclasses: exonic circRNAs (ecircRNAs) [296], circular intronic RNAs (ciRNAs) [30] and exon-intronic circRNAs (EIciRNAs) [297].

Unlike the canonical splicing of linear RNA, circRNAs are formed via back-splicing [289]. The biogenesis pathways among distinct subclasses also differ from each other due to a special process termed "alternative circularization" from a single gene locus. Mechanically, back-splicing is characterized by covalent binding of the upstream (5′) splicing acceptor site to the downstream (3′) splicing donor site, producing a covalently closed circRNA and an alternatively spliced linear RNA with skipped middle exons, whose internal introns can be further removed by another canonical splicing event [33]. It has been proposed that back-spliced exon circularization is controlled by RNA pol II and the spliceosome, possibly competing with pre-mRNA splicing for limited splicing factors in a tissue-specific fashion [298, 299]. Two models have been proposed by Jeck et al. and exemplified in recapitulated assays [175, 286, 288, 300, 301]. Model 1 is referred to as "exon skipping" or "lariat intermediate," while model 2 is termed "direct back-splicing." Alternative circularization can occur with different numbers of exons included [175, 188, 286, 289] or with an internal intron included or excluded [289, 297, 300]. Accordingly, ecircRNAs are formed by the head-to-tail splicing of exons. Thereafter, ciRNAs were found to be derived from introns with a 2′,5′-phosphodiester bond, depending on a consensus motif with a seven-nt GU-rich element at the 5′ splice site and an eleven-nt C-rich element upstream of the branch point site [30]. Later, researchers discovered that EIciRNAs were circularized with introns "retained" between the exons [297], but the mechanism remains to be clarified.

Supporting the two models mentioned above, studies have shown that exon circularization correlates with exon skipping [296] and that canonical splicing by cis-elements [286, 289, 294, 302] or trans-factors [294, 298] is involved as well. First, it assists through the RNA pairing of reversely complementary sequences across the flanking introns. RNA pairing can be formed either by repetitive elements, such as inverted repeated Alu pairs (IRAlus) [303], or by non-repetitive but complementary sequences [289]. Moreover, short sequences as small as 30–40 nt have been reported to sufficiently promote circRNA formation [302]. It is worth noting, however, that a selection of RNA pairing within a single intron can facilitate canonical splicing rather than back-splicing to form a linear RNA transcript [289, 304]. Second, protein factors binding to pre-mRNAs can bridge flanking introns together, thereby drawing the splicing donor and acceptor close to enhance exon circularization. Such RBPs include the splicing factors muscleblind (MBL) [298] and quaking (QKI) [305], whereas adenosine deaminase 1 acting on RNA (ADAR1), a double-stranded RNA-editing enzyme that mediates adenosine-to-inosine (A-to-I) editing on IRAlus [294, 306], might directly bind to the dsRNA and antagonize circRNA production [307]. In summary, it remains unknown whether alternative circularization in circRNAs occurs co-transcriptionally or posttranscriptionally, nor is it understood how cis- or trans-acting factors affect exon circularization.

In addition to the sorting of circRNAs by alternative circularization, they can also be categorized based on their locations in mammalian cells and peripheral blood. The first type is cytoplasmic ecircRNAs [34, 286], which accounts for over 90 % of total circRNAs [288]; the second consists of ciRNAs and EIciRNAs, which are located primarily in the nucleus [30, 297]. These two groups are simply divided depending on whether an intron is excluded or included. Another newly identified type is termed "exosomal circRNAs (exo-circRNAs)" [308]. Li et al. have found that this novel class of circRNAs is abundant and highly stable in exosomes of various cancer cell lines as well as human serum, suggesting that serum exo-circRNAs have potential as promising circulating biomarkers for cancer diagnosis. In regard to cancer, ecircRNAs will be introduced as the main player in the subsequent sections.

1.4.2 Properties of CircRNAs

Recent studies have revealed that circRNAs possess different characteristics and distinguished functions from linear RNAs. Based on back-splicing events, circRNAs are:

(1) Generated from covalently closed loop structures without a free 5′ or 3′ end [288, 289].
(2) Resistant to degradation by RNA exonuclease or RNase R, rendering them much more stable than their linear counterparts, offering new diagnostic methods [309, 310].
(3) Widespread and up to 200 times more abundant than linear isoforms in certain cell lines [286, 301], although most of them are not highly expressed.
(4) Expressed in a cell-type-, tissue-, or developmental stage-specific manner, as in the case of hsa_circRNA_21, which was detected in CD19+ but not CD34+ leukocytes, neutrophils or HEK293 cells [34].
(5) Evolutionarily conserved among different species [286, 300, 311], although there are also exceptions in the case of intergenic or intronic circRNAs [30].
(6) Mostly endogenous ncRNAs that are not translatable, except that exogenous circRNAs engineered with internal ribosome entry site (IRES) elements can be translated into peptides in vitro [312] or in vivo [313], although no evidence yet supports the performance of spliceosome-generated circRNAs as mRNAs.

Therefore, circRNAs have a tendency to play important and diverse regulatory roles in gene expression at the transcriptional or posttranscriptional level.

1.4.3 Functions and Mechanisms of CircRNAs in Cancers

Accumulating evidence has shown that circRNAs participate in regulating diverse biological developments as well as disease initiation and progression [292]. In addition to atherosclerotic vascular disease risk, neurological disorders, and regulation of prion diseases [188, 293, 314], circRNAs also perform functions in cancer [315]. Bachmayr-Heyda et al. reported for the first time that circRNAs were globally reduced in cancer cell lines and cancer tissues compared to normal mucosa for colorectal and ovarian cancers and were negatively correlated with tumor cell proliferation [316]. Recent studies also revealed that circRNAs possessed value in cancer diagnosis. Li et al. first found a significantly negative correlation between hsa_circ_002059 and GC metastasis [317], identifying hsa_circ_002059 as a stable biomarker for GC diagnosis. Recently, there was a report on the detection of the presence and enrichment of exo-circRNAs in various cancer cell lines, including colon, lung, stomach, breast, and cervical cancers [308]. The finding of tumor-derived exo-circRNAs in human serum indicates their potential as novel diagnostic tools [308].

Mechanically, detailed studies have demonstrated that circRNAs can serve as ceRNAs (RNA sponges) to bind miRNAs and modulate miRNA-targeted gene expression [35, 298, 318, 319]. EcircRNAs, which are mostly cytoplasmic, contain miRNA response elements (MREs) [286, 300] that are prerequisites for their activity as ceRNAs. Hansen et al. revealed that the cerebellar degeneration related protein 1 (CDR1) gene could translate a natural circular antisense transcript termed CDR1as/ciRS-7 (CDR1 antisense or circular RNA sponge for miR-7) [320]. It was experimentally validated that CDR1as has 74 selectively conserved miR-7 seed sites and can bind densely to miR-7 with AGOs highly occupied [34, 35]. Without being degraded, CDR1as and miR-7 are highly co-expressed and interact specifically in the developing midbrain [34]. Thus, CDR1as acts as a potent miR-7 sponge in the neuronal tissue. Corresponding gain/loss-of-function tests further proved its potency in up-/downregulating the expression of miR-7 target genes [34, 35]. Another similar case is the single-exon circRNA Sry, which derives from the testis-specific murine Sex-determining region Y (Sry) gene [27]. It contains 16 binding sites for miR-138 and serves as a miR-138 sponge [35]. However, a single circRNA does not perform as an individual miRNA sponge. It has been reported that cir-ITCH contains exonic sequences of E3 ubiquitin (Ub) protein ligase (ITCH) and harbors miRNA target sites for miR-7, miR-17 and miR-214 that bind to the 3′ UTR of ITCH [315].

Given that circRNAs can sequester miRNAs by the "sponge" effect to increase gene expression, it is no wonder that circRNAs appear to be related to cancer and other diseases correlated with miRNAs [292]. For instance, miR-7, miR-17 and miR-214 were reported to be highly expressed in a variety of tumor types [321–323], including ESCC [324]. Li et al. detected a reduced expression of cir-ITCH in ESCC in comparison to paired adjacent tissue [315]. Through competitive binding, cir-ITCH performs its antitumor role by increasing miR-7/miR-17/miR-214-targeted ITCH expression, triggering ubiquitin-mediated Dvl2 degradation and further inhibiting canonical Wnt signaling and its target oncogene c-Myc [315]. As miR-7 exhibits both oncogenic and antitumor properties [325], whether the CDR1as/miR-7 axis is involved in cancer development awaits experimental confirmation. According to the abovementioned correlation between circRNAs and proliferation in CRC, circRNAs may also exert functions in many other tumors [316].

To determine whether circular miRNA sponges are a general phenomenon in vivo, subsequent research provided evidence that thousands of other ecircRNAs had little enrichment of target sites for an individual miRNA [39]. For that matter, no more circRNAs displayed their presumed function as miRNA sponges [15]. However, additional regulatory roles for circRNAs had already been reported. As previously mentioned, circRNA biogenesis might compete with pre-mRNA in alternative splicing, and they could exert a transcriptional regulatory role in mRNA production [298]. Chao et al. found that a circRNA produced from the mouse formin (Fmn) gene contains the translation start site and functions as an "mRNA trap," leaving the rest of the linear transcript untranslatable and reducing Fmn protein expression [326]. Many single-exon circRNAs found in human fibroblasts also appear to sequester the translation start site to cause gene silencing [288]. As ciRNAs also have fewer putative miRNA binding sites, they may perform distinct functions in the nucleus, specifically, regulating their parental gene expression via interaction with the Pol II machinery in cis [30, 297, 327]. As a nuclear enriched ciRNA, ci-ankrd52 is found to associate with the elongating Pol II complex at their parent gene loci, resulting in enhanced transcription activity [30]. Analogously, EIciRNAs, such as circEIF3J and circPAIP2, may hold U1 small nuclear ribonucleoproteins (snRNPs) to form EIciRNA-U1 snRNP complexes, which then interact with Pol II at the promoters (cis-acting effect) to promote the expression of the host genes [297]. In a mode referred to as RBP sponges, certain ecircRNAs can function to store, sort, or localize RBPs, mediating the function of RBPs as miRNA sponges do [299, 328]. As a ceRNA, single-stranded CDR1as can directly bind the 3′ UTRs of target mRNAs in a trans-acting manner to regulate their expression [34]. Whether all the complex functionality is linked to cancer remains a current topic of debate, though poorly understood. Current knowledge of circRNAs is merely the tip of an iceberg, with underlying treasures for basic research, biomarker discovery and therapeutic applications.

1.5 Other NcRNAs Implicated in Cancers

In addition to miRNAs, lncRNAs, and circRNAs, there are many other types of ncRNAs involved in cancer initiation and progression. Short interfering RNAs are small double-stranded, non-protein-coding RNAs of 21–31 nt. Similar to miRNAs in biogenesis, siRNAs are generated from dsRNAs by the same ribonuclease and Dicer but in association with a different Argonaute protein (Ago2) [1]. In terms of function, siRNAs are well-recognized players in posttranscriptional gene silencing, especially the RNAi pathway [57, 329]. The perfect complementarity of the siRNAs' antisense strand to mRNA targets mediates RNAi through siRNA-guided RNA cleavage [57]. However, although few intrinsic siRNAs have been identified in human tumors, synthetic siRNAs have been used extensively as experimental tools to manipulate gene expression in basic research as well as in clinical therapeutic approaches to a diverse array of pathologies, including cancer [330, 331].

As the least characterized class of sncRNAs, piRNAs are single stranded, 24–31 nt in length, and are processed from longer precursor transcripts by Piwi proteins [1, 332]. In contrast to miRNAs, piRNAs are less conserved and are mostly detected with Piwi in the germ-line cells of higher eukaryotes [333, 334]. They have been appreciated as essential genetic and epigenetic regulatory factors for germ cell maintenance, genome integrity, mRNA stability, DNA methylation and retrotransposon control. In particular, their involvement in gene silencing has suggested additional roles of the piRNA pathway in cancer [335–338]. In 2011, Li et al. reported the first relationship between piRNAs and oncogenesis: piR-651 was overexpressed in a variety of cancer cell lines, including gastric, lung, mesothelium, breast, liver, and cervical cancer [339]. Subsequent studies have found that a number of piRNAs play a pivotal role in the development of many types of cancers: piR-823 is downregulated in gastric cancers as an inhibitor of cancer cell growth [340] but played an oncogenic role in multiple myeloma, possibly by DNA methylation via DNMTs [341]; overexpression of piRABC represses bladder cancer cell proliferation, colony formation, and enhanced cell apoptosis [342]; piR-Hep1 is upregulated in hepatic cell carcinoma and has been linked to cancer cell viability, motility, and invasiveness [343]; the PIWI/piRNA pathway was detected in breast cancer [344] and epigenetically involved in testicular cancer [345]; and the PIWIL2/piR-932 interaction positively regulates the process of breast cancer stem cells and metastasis through enhancing CpG island methylation of the latexin promoter [346]. The identification of piRNAs targeting key cancer cell pathways also suggests that they exert transcriptional and posttranscriptional gene regulatory actions in cancer [347]. Overall, the roles of piRNAs in human neoplasms need further study, and piRNA-targeted therapeutic strategies for malignancies require additional attention.

The housekeeping ncRNAs have a broad range of roles, such as adaptors (tRNAs) and ribosomal components (rRNAs) during translation, the spliceosome in splicing (snRNAs), and RNA maturation including editing (snoRNAs). Current investigations have found their involvement in cancer. For example, tRNA synthesis is positively regulated by oncogenes (ERK, TORC1, Myc) and negatively modulated by tumor suppressors (Maf1, p53, Rb) through selective effects on Pol III activity, which then influences mRNA translation and tumor growth [348]. Another well-established type of ncRNA is snoRNA, which is mostly known to engage in the posttranscriptional modification of rRNAs [349, 350] and is directly related to mRNA alternative splicing [351]. Small nucleolar RNAs can originate shorter functional sdRNAs (snoRNA-derived RNAs) or miRNAs (sno-miR) [352]. Growing evidence has linked the deregulation of snoRNA expression to human disorders, particularly cancer, revealing the role of snoRNAs as oncogenes and tumor suppressors as well as putative biomarkers for cancers [353, 354].

Therefore, the range of "classic" ncRNA functions in cancer is significantly wider than previously assumed.

1.6 Concluding Remarks and Future Perspectives

In summary, the past several decades have witnessed a steep rise of interest in the study of ncRNAs in human cancers. The pervasively transcribed, evolutionarily conserved, structurally diverse, functionally significant and mechanically complicated noncoding RNA transcripts add a novel and informative layer to our understanding of the complexity of oncogenesis and development. The pathways where distinct subtypes of sncRNAs and lncRNAs are probably intertwined also establish a complex network of interactions and actions required for rapid and fine-tuned gene expression regulation at multiple (epigenetic, transcriptional and posttranscriptional) levels. In contrast to the extensive study of miRNAs, it is still in the early days of assigning clear biological functions to the continually found lncRNAs and circRNAs, let alone understanding their intricate molecular mechanisms. The current challenging problems can be recapitulated in terms of three aspects:

1. To explore lncRNA sequences and secondary and tertiary structures, establishing structure-function relations that define the mechanisms of lncRNA function. In the RNA world, spatial structures are stable and essential for specific interactions with proteins or other nucleic acids to regulate the activity and function of lncRNAs [355]. For example, stem-loops are necessary for the efficient recruitment of PRC2 to Xist RNA [356]. Additionally, Mistral may facilitate the formation of long-range chromosomal loops to recruit the gene-activating complex MLL (mixed-lineage leukemia) [357, 358].
2. To explore the complex regulatory web of RNA-RNA interplay or/and RNA-protein interaction, where different environmental signals are involved. Studies to better understand the molecular mechanisms and signaling pathways in which ncRNAs participate, whether associated with proteins or not, would offer promise for the discovery of new clinical diagnostic biomarkers or the development of precisely targeted therapeutics.
3. To seek more effective strategies to modulate ncRNA expression in a specific manner without perturbing relevant coding genes or to reduce the "off-target" effect of siRNA-mediated gene silencing. Such strategies would also allow the development of ncRNAs as potential therapeutic targets in clinical practice. As previously mentioned, MRX34 is the first ncRNA-based therapeutic mimic of the miR-34 tumor suppressor undergoing evaluation in clinical trials for HCC [157]. Nevertheless, whether the efforts made thus far will be rewarded remains largely unknown.

Although studies concerning ncRNAs are growing fast, few can really address the vast complexity and functionality of ncRNAs in cancer development as a whole.

Long ncRNAs represent a fairly prominent component of the human transcriptome, carrying great biological significance, and are undoubtedly an untapped goldmine for further discovery. It should also be noted that circRNAs need comprehensive exploration, from biogenesis pathways to regulatory functions and molecular mechanisms. To summarize, more in-depth studies should be undertaken to move the field forward at both the basic and clinical levels.

References

1. Gomes AQ, Nolasco S, Soares H. Non-coding RNAs: multi-tasking molecules in the cell. Int J Mol Sci. 2013;14(8):16010–39.
2. Djebali S, Davis CA, Merkel A, Dobin A, Lassmann T, Mortazavi A, Tanzer A, Lagarde J, Lin W, Schlesinger F. Landscape of transcription in human cells. Nature. 2012;489(7414):101–8.
3. Consortium EP. An integrated encyclopedia of DNA elements in the human genome. Nature. 2012;489(7414):57–74.
4. Ium T. ENCODE project writes eulogy for junk DNA. Science. 2012;337:1159–61.
5. Britten RJ, Davidson EH. Gene regulation for higher cells: a theory. Science. 1969;165(891):349–57.
6. Brosnan CA, Voinnet O. The long and the short of noncoding RNAs. Curr Opin Cell Biol. 2009;21(3):416–25.
7. Gibb EA, Brown CJ, Lam WL. The functional role of long non-coding RNA in human carcinomas. Mol Cancer. 2011;10(1):38–55.
8. Lee RC, Feinbaum RL, Ambros V. The C. elegans heterochronic gene lin-4 encodes small RNAs with antisense complementarity to lin-14. Cell. 1993;75(5):843–54.
9. Wightman B, Ha I, Ruvkun G. Posttranscriptional regulation of the heterochronic gene lin-14 by lin-4 mediates temporal pattern formation in C. elegans. Cell. 1993;75(5):855–62.
10. Lau NC, Lim LP, Weinstein EG, Bartel DP. An abundant class of tiny RNAs with probable regulatory roles in Caenorhabditis elegans. Science. 2001;294(5543):858–62.
11. Lagos-Quintana M, Rauhut R, Lendeckel W, Tuschl T. Identification of novel genes coding for small expressed RNAs. Science. 2001;294(5543):853–8.
12. Lee RC, Ambros V. An extensive class of small RNAs in Caenorhabditis elegans. Science. 2001;294(5543):862–4.
13. Calin GA, Dumitru CD, Shimizu M, Bichi R, Zupo S, Noch E, Aldler H, Rattan S, Keating M, Rai K. Frequent deletions and down-regulation of micro-RNA genes miR15 and miR16 at 13q14 in chronic lymphocytic leukemia. Proc Natl Acad Sci. 2002;99(24):15524–9.
14. Gelbart ME, Kuroda MI. Drosophila dosage compensation: a complex voyage to the X chromosome. Development. 2009;136(9):1399–410.
15. Au PCK, Zhu Q-H, Dennis ES, Wang M-B. Long non-coding RNA-mediated mechanisms independent of the RNAi pathway in animals and plants. RNA Biol. 2011;8(3):404–14.
16. Brannan CI, Dees EC, Ingram RS, Tilghman SM. The product of the H19 gene may function as an RNA. Mol Cell Biol. 1990;10(1):28–36.
17. Brockdorff N, Ashworth A, Kay GF, McCabe VM, Norris DP, Cooper PJ, Swift S, Rastan S. The product of the mouse Xist gene is a 15 kb inactive X-specific transcript containing no conserved ORF and located in the nucleus. Cell. 1992;71(3):515–26.
18. Brown CJ, Hendrich BD, Rupert JL, Lafrenière RG, Xing Y, Lawrence J, Willard HF. The human XIST gene: analysis of a 17 kb inactive X-specific RNA that contains conserved repeats and is highly localized within the nucleus. Cell. 1992;71(3):527–42.

19. Wendt KS, Yoshida K, Itoh T, Bando M, Koch B, Schirghuber E, Tsutsumi S, Nagae G, Ishihara K, Mishiro T, Yahata K, Imamoto F, Aburatani H, Nakao M, Imamoto N, Maeshima K, Shirahige K, Peters JM. Cohesin mediates transcriptional insulation by CCCTC-binding factor. Nature. 2008;451(7180):796–801. doi:10.1038/nature06634.
20. Kagami M, Sekita Y, Nishimura G, Irie M, Kato F, Okada M, Yamamori S, Kishimoto H, Nakayama M, Tanaka Y, Matsuoka K, Takahashi T, Noguchi M, Tanaka Y, Masumoto K, Utsunomiya T, Kouzan H, Komatsu Y, Ohashi H, Kurosawa K, Kosaki K, Ferguson-Smith AC, Ishino F, Ogata T. Deletions and epimutations affecting the human 14q32.2 imprinted region in individuals with paternal and maternal upd(14)-like phenotypes. Nat Genet. 2008;40(2):237–42. doi:10.1038/ng.2007.56.
21. Augui S, Filion GJ, Huart S, Nora E, Guggiari M, Maresca M, Stewart AF, Heard E. Sensing X chromosome pairs before X inactivation via a novel X-pairing region of the Xic. Science. 2007;318(5856):1632–6. doi:10.1126/science.1149420.
22. Sanger HL, Klotz G, Riesner D, Gross HJ, Kleinschmidt AK. Viroids are single-stranded covalently closed circular RNA molecules existing as highly base-paired rod-like structures. Proc Natl Acad Sci. 1976;73(11):3852–6.
23. Arnberg A, Van Ommen G-J, Grivell L, Van Bruggen E, Borst P. Some yeast mitochondrial RNAs are circular. Cell. 1980;19(2):313–9.
24. Kos A, Dijkema R, Arnberg A, Van der Meide P, Schellekens H. The hepatitis delta (δ) virus possesses a circular RNA. Nature. 1986;323:558–60.
25. Nigro JM, Cho KR, Fearon ER, Kern SE, Ruppert JM, Oliner JD, Kinzler KW, Vogelstein B. Scrambled exons. Cell. 1991;64(3):607–13.
26. Cocquerelle C, Daubersies P, Majerus M-A, Kerckaert J-P, Bailleul B. Splicing with inverted order of exons occurs proximal to large introns. EMBO J. 1992;11(3):1095.
27. Capel B, Swain A, Nicolis S, Hacker A, Walter M, Koopman P, Goodfellow P, Lovell-Badge R. Circular transcripts of the testis-determining gene Sry in adult mouse testis. Cell. 1993;73(5):1019–30.
28. Kopczynski CC, Muskavitch M. Introns excised from the Delta primary transcript are localized near sites of Delta transcription. J Cell Biol. 1992;119(3):503–12.
29. Qian L, Vu MN, Carter M, Wilkinson MF. A spliced intron accumulates as a lariat in the nucleus of T cells. Nucleic Acids Res. 1992;20(20):5345–50.
30. Zhang Y, Zhang X-O, Chen T, Xiang J-F, Yin Q-F, Xing Y-H, Zhu S, Yang L, Chen L-L. Circular intronic long noncoding RNAs. Mol Cell. 2013;51(6):792–806.
31. Cocquerelle C, Mascrez B, Hetuin D, Bailleul B. Mis-splicing yields circular RNA molecules. FASEB J. 1993;7(1):155–60.
32. Pasman Z, Been M, Garcia-Blanco M. Exon circularization in mammalian nuclear extracts. RNA. 1996;2(6):603.
33. Chen L-L, Yang L. Regulation of circRNA biogenesis. RNA Biol. 2015;12(4):381–8.
34. Memczak S, Jens M, Elefsinioti A, Torti F, Krueger J, Rybak A, Maier L, Mackowiak SD, Gregersen LH, Munschauer M. Circular RNAs are a large class of animal RNAs with regulatory potency. Nature. 2013;495(7441):333–8.
35. Hansen TB, Jensen TI, Clausen BH, Bramsen JB, Finsen B, Damgaard CK, Kjems J. Natural RNA circles function as efficient microRNA sponges. Nature. 2013;495(7441):384–8.
36. Bartel DP. MicroRNAs: genomics, biogenesis, mechanism, and function. Cell. 2004;116(2):281–97.
37. Esquela-Kerscher A, Slack FJ. Oncomirs—microRNAs with a role in cancer. Nat Rev Cancer. 2006;6(4):259–69.
38. Reinhart BJ, Slack FJ, Basson M, Pasquinelli AE, Bettinger JC, Rougvie AE, Horvitz HR, Ruvkun G. The 21-nucleotide let-7 RNA regulates developmental timing in Caenorhabditis elegans. Nature. 2000;403(6772):901–6.
39. Lin S, Gregory RI. MicroRNA biogenesis pathways in cancer. Nat Rev Cancer. 2015;15(6):321–33.

40. Denli AM, Tops BB, Plasterk RH, Ketting RF, Hannon GJ. Processing of primary microRNAs by the Microprocessor complex. Nature. 2004;432(7014):231–5.
41. Gregory RI, K-p Y, Amuthan G, Chendrimada T, Doratotaj B, Cooch N, Shiekhattar R. The Microprocessor complex mediates the genesis of microRNAs. Nature. 2004;432(7014):235–40.
42. Lee Y, Ahn C, Han J, Choi H, Kim J, Yim J, Lee J, Provost P, Rådmark O, Kim S. The nuclear RNase III Drosha initiates microRNA processing. Nature. 2003;425(6956):415–9.
43. Han J, Lee Y, Yeom K-H, Kim Y-K, Jin H, Kim VN. The Drosha-DGCR8 complex in primary microRNA processing. Genes Dev. 2004;18(24):3016–27.
44. Lund E, Güttinger S, Calado A, Dahlberg JE, Kutay U. Nuclear export of microRNA precursors. Science. 2004;303(5654):95–8.
45. Yi R, Qin Y, Macara IG, Cullen BR. Exportin-5 mediates the nuclear export of pre-microRNAs and short hairpin RNAs. Genes Dev. 2003;17(24):3011–6.
46. Bohnsack MT, Czaplinski K, GÖRLICH D. Exportin 5 is a RanGTP-dependent dsRNA-binding protein that mediates nuclear export of pre-miRNAs. RNA. 2004;10(2):185–91.
47. Park J-E, Heo I, Tian Y, Simanshu DK, Chang H, Jee D, Patel DJ, Kim VN. Dicer recognizes the 5 [prime] end of RNA for efficient and accurate processing. Nature. 2011;475(7355):201–5.
48. Bagga S, Bracht J, Hunter S, Massirer K, Holtz J, Eachus R, Pasquinelli AE. Regulation by let-7 and lin-4 miRNAs results in target mRNA degradation. Cell. 2005;122(4):553–63.
49. Hutvágner G, McLachlan J, Pasquinelli AE, Bálint É, Tuschl T, Zamore PD. A cellular function for the RNA-interference enzyme Dicer in the maturation of the let-7 small temporal RNA. Science. 2001;293(5531):834–8.
50. Ketting RF, Fischer SE, Bernstein E, Sijen T, Hannon GJ, Plasterk RH. Dicer functions in RNA interference and in synthesis of small RNA involved in developmental timing in C. elegans. Genes Dev. 2001;15(20):2654–9.
51. Bernstein E, Caudy AA, Hammond SM, Hannon GJ. Role for a bidentate ribonuclease in the initiation step of RNA interference. Nature. 2001;409(6818):363–6.
52. Chendrimada TP, Gregory RI, Kumaraswamy E, Norman J, Cooch N, Nishikura K, Shiekhattar R. TRBP recruits the Dicer complex to Ago2 for microRNA processing and gene silencing. Nature. 2005;436(7051):740–4.
53. Gregory RI, Chendrimada TP, Cooch N, Shiekhattar R. Human RISC couples microRNA biogenesis and posttranscriptional gene silencing. Cell. 2005;123(4):631–40.
54. Chen K, Rajewsky N. The evolution of gene regulation by transcription factors and microRNAs. Nat Rev Genet. 2007;8(2):93–103.
55. Friedman RC, Farh KK-H, Burge CB, Bartel DP. Most mammalian mRNAs are conserved targets of microRNAs. Genome Res. 2009;19(1):92–105.
56. Croce CM, Calin GA. miRNAs, cancer, and stem cell division. Cell. 2005;122(1):6–7.
57. Hannon GJ. RNA interference. Nature. 2002;418(6894):244–51.
58. Llave C, Xie Z, Kasschau KD, Carrington JC. Cleavage of Scarecrow-like mRNA targets directed by a class of Arabidopsis miRNA. Science. 2002;297(5589):2053–6.
59. Palatnik JF, Allen E, Wu X, Schommer C, Schwab R, Carrington JC, Weigel D. Control of leaf morphogenesis by microRNAs. Nature. 2003;425(6955):257–63.
60. Tang G, Reinhart BJ, Bartel DP, Zamore PD. A biochemical framework for RNA silencing in plants. Genes Dev. 2003;17(1):49–63.
61. Osaki M, Takeshita F, Ochiya T. MicroRNAs as biomarkers and therapeutic drugs in human cancer. Biomarkers. 2008;13(7–8):658–70.
62. Olsen PH, Ambros V. The lin-4 regulatory RNA controls developmental timing in Caenorhabditis elegans by blocking LIN-14 protein synthesis after the initiation of translation. Dev Biol. 1999;216(2):671–80.
63. Pillai RS, Bhattacharyya SN, Artus CG, Zoller T, Cougot N, Basyuk E, Bertrand E, Filipowicz W. Inhibition of translational initiation by Let-7 MicroRNA in human cells. Science. 2005;309(5740):1573–6.

64. Miranda KC, Huynh T, Tay Y, Ang Y-S, Tam W-L, Thomson AM, Lim B, Rigoutsos I. A pattern-based method for the identification of MicroRNA binding sites and their corresponding heteroduplexes. Cell. 2006;126(6):1203–17.
65. Johnson CD, Esquela-Kerscher A, Stefani G, Byrom M, Kelnar K, Ovcharenko D, Wilson M, Wang X, Shelton J, Shingara J. The let-7 microRNA represses cell proliferation pathways in human cells. Cancer Res. 2007;67(16):7713–22.
66. Hermeking H. The miR-34 family in cancer and apoptosis. Cell Death Differ. 2010;17(2):193–9.
67. Olson P, Lu J, Zhang H, Shai A, Chun MG, Wang Y, Libutti SK, Nakakura EK, Golub TR, Hanahan D. MicroRNA dynamics in the stages of tumorigenesis correlate with hallmark capabilities of cancer. Genes Dev. 2009;23(18):2152–65.
68. Nicoloso MS, Spizzo R, Shimizu M, Rossi S, Calin GA. MicroRNAs—the micro steering wheel of tumour metastases. Nat Rev Cancer. 2009;9(4):293–302.
69. Calin GA, Sevignani C, Dumitru CD, Hyslop T, Noch E, Yendamuri S, Shimizu M, Rattan S, Bullrich F, Negrini M. Human microRNA genes are frequently located at fragile sites and genomic regions involved in cancers. Proc Natl Acad Sci U S A. 2004;101(9):2999–3004.
70. Zhang L, Huang J, Yang N, Greshock J, Megraw MS, Giannakakis A, Liang S, Naylor TL, Barchetti A, Ward MR. microRNAs exhibit high frequency genomic alterations in human cancer. Proc Natl Acad Sci. 2006;103(24):9136–41.
71. Lu J, Getz G, Miska EA, Alvarez-Saavedra E, Lamb J, Peck D, Sweet-Cordero A, Ebert BL, Mak RH, Ferrando AA. MicroRNA expression profiles classify human cancers. Nature. 2005;435(7043):834–8.
72. Thomson JM, Newman M, Parker JS, Morin-Kensicki EM, Wright T, Hammond SM. Extensive post-transcriptional regulation of microRNAs and its implications for cancer. Genes Dev. 2006;20(16):2202–7.
73. Calin GA, Croce CM. MicroRNA signatures in human cancers. Nat Rev Cancer. 2006;6(11):857–66.
74. Humphries B, Yang C. The microRNA-200 family: small molecules with novel roles in cancer development, progression and therapy. Oncotarget. 2015;6(9):6472–98.
75. He L, He X, Lim LP, De Stanchina E, Xuan Z, Liang Y, Xue W, Zender L, Magnus J, Ridzon D. A microRNA component of the p53 tumour suppressor network. Nature. 2007;447(7148):1130–4.
76. Chang T-C, Wentzel EA, Kent OA, Ramachandran K, Mullendore M, Lee KH, Feldmann G, Yamakuchi M, Ferlito M, Lowenstein CJ. Transactivation of miR-34a by p53 broadly influences gene expression and promotes apoptosis. Mol Cell. 2007;26(5):745–52.
77. Bommer GT, Gerin I, Feng Y, Kaczorowski AJ, Kuick R, Love RE, Zhai Y, Giordano TJ, Qin ZS, Moore BB. p53-mediated activation of miRNA34 candidate tumor-suppressor genes. Curr Biol. 2007;17(15):1298–307.
78. O'Donnell KA, Wentzel EA, Zeller KI, Dang CV, Mendell JT. c-Myc-regulated microRNAs modulate E2F1 expression. Nature. 2005;435(7043):839–43.
79. Dews M, Homayouni A, Yu D, Murphy D, Sevignani C, Wentzel E, Furth EE, Lee WM, Enders GH, Mendell JT. Augmentation of tumor angiogenesis by a Myc-activated microRNA cluster. Nat Genet. 2006;38(9):1060–5.
80. Guil S, Esteller M. DNA methylomes, histone codes and miRNAs: tying it all together. Int J Biochem Cell Biol. 2009;41(1):87–95.
81. Han J, Pedersen JS, Kwon SC, Belair CD, Kim Y-K, Yeom K-H, Yang W-Y, Haussler D, Blelloch R, Kim VN. Posttranscriptional crossregulation between Drosha and DGCR8. Cell. 2009;136(1):75–84.
82. Muralidhar B, Winder D, Murray M, Palmer R, Barbosa-Morais N, Saini H, Roberts I, Pett M, Coleman N. Functional evidence that Drosha overexpression in cervical squamous cell carcinoma affects cell phenotype and microRNA profiles. J Pathol. 2011;224(4):496–507.
83. Melo SA, Moutinho C, Ropero S, Calin GA, Rossi S, Spizzo R, Fernandez AF, Davalos V, Villanueva A, Montoya G. A genetic defect in exportin-5 traps precursor microRNAs in the nucleus of cancer cells. Cancer Cell. 2010;18(4):303–15.

84. Wagh PK, Gardner MA, Ma X, Callahan M, Shannon JM, Wert SE, Messinger YH, Dehner LP, Hill DA, Wikenheiser-Brokamp KA. Cell-and developmental stage-specific Dicer1 ablation in the lung epithelium models cystic pleuropulmonary blastoma. J Pathol. 2015;236(1):41–52.
85. Garre P, Pérez-Segura P, Díaz-Rubio E, Caldés T, de la Hoya M. Reassessing the TARBP2 mutation rate in hereditary nonpolyposis colorectal cancer. Nat Genet. 2010;42(10):817–8.
86. Viswanathan SR, Daley GQ, Gregory RI. Selective blockade of microRNA processing by Lin28. Science. 2008;320(5872):97–100.
87. Mori M, Triboulet R, Mohseni M, Schlegelmilch K, Shrestha K, Camargo FD, Gregory RI. Hippo signaling regulates microprocessor and links cell-density-dependent miRNA biogenesis to cancer. Cell. 2014;156(5):893–906.
88. Volinia S, Galasso M, Costinean S, Tagliavini L, Gamberoni G, Drusco A, Marchesini J, Mascellani N, Sana ME, Jarour RA. Reprogramming of miRNA networks in cancer and leukemia. Genome Res. 2010;20(5):589–99.
89. Voorhoeve PM. MicroRNAs Oncogenes, tumor suppressors or master regulators of cancer heterogeneity? Biochim Biophys Acta. 2010;1805(1):72–86.
90. Farazi TA, Spitzer JI, Morozov P, Tuschl T. miRNAs in human cancer. J Pathol. 2011;223(2):102–15.
91. Hanahan D, Weinberg RA. The hallmarks of cancer. Cell. 2000;100(1):57–70.
92. Hanahan D, Weinberg RA. Hallmarks of cancer: the next generation. Cell. 2011;144(5):646–74.
93. Bueno MJ, de Castro IP, Malumbres M. Control of cell proliferation pathways by microRNAs. Cell Cycle. 2008;7(20):3143–8.
94. He L, Thomson JM, Hemann MT, Hernando-Monge E, Mu D, Goodson S, Powers S, Cordon-Cardo C, Lowe SW, Hannon GJ. A microRNA polycistron as a potential human oncogene. Nature. 2005;435(7043):828–33.
95. Woods K, Thomson JM, Hammond SM. Direct regulation of an oncogenic micro-RNA cluster by E2F transcription factors. J Biol Chem. 2007;282(4):2130–4.
96. Ciafre S, Galardi S, Mangiola A, Ferracin M, Liu C-G, Sabatino G, Negrini M, Maira G, Croce C, Farace M. Extensive modulation of a set of microRNAs in primary glioblastoma. Biochem Biophys Res Commun. 2005;334(4):1351–8.
97. Gillies JK, Lorimer IA. Regulation of p27Kip1 by miRNA 221/222 in glioblastoma. Cell Cycle. 2007;6(16):2005–9.
98. Roush S, Slack FJ. The let-7 family of microRNAs. Trends Cell Biol. 2008;18(10):505–16.
99. Takamizawa J, Konishi H, Yanagisawa K, Tomida S, Osada H, Endoh H, Harano T, Yatabe Y, Nagino M, Nimura Y. Reduced expression of the let-7 microRNAs in human lung cancers in association with shortened postoperative survival. Cancer Res. 2004;64(11):3753–6.
100. Johnson SM, Grosshans H, Shingara J, Byrom M, Jarvis R, Cheng A, Labourier E, Reinert KL, Brown D, Slack FJ. RAS is regulated by the let-7 microRNA family. Cell. 2005;120(5):635–47.
101. Sampson VB, Rong NH, Han J, Yang Q, Aris V, Soteropoulos P, Petrelli NJ, Dunn SP, Krueger LJ. MicroRNA let-7a down-regulates MYC and reverts MYC-induced growth in Burkitt lymphoma cells. Cancer Res. 2007;67(20):9762–70.
102. Lee YS, Dutta A. The tumor suppressor microRNA let-7 represses the HMGA2 oncogene. Genes Dev. 2007;21(9):1025–30.
103. Mayr C, Hemann MT, Bartel DP. Disrupting the pairing between let-7 and Hmga2 enhances oncogenic transformation. Science. 2007;315(5818):1576–9.
104. Yu F, Yao H, Zhu P, Zhang X, Pan Q, Gong C, Huang Y, Hu X, Su F, Lieberman J. let-7 regulates self renewal and tumorigenicity of breast cancer cells. Cell. 2007;131(6):1109–23.
105. Wang T, Wang G, Hao D, Liu X, Wang D, Ning N, Li X. Aberrant regulation of the LIN28A/LIN28B and let-7 loop in human malignant tumors and its effects on the hallmarks of cancer. Mol Cancer. 2015;14:125. doi:10.1186/s12943-015-0402-5.
106. Navarro F, Lieberman J. miR-34 and p53: new insights into a complex functional relationship. PLoS ONE. 2015;10(7):e0132767.

107. Volinia S, Calin GA, Liu CG, Ambs S, Cimmino A, Petrocca F, Visone R, Iorio M, Roldo C, Ferracin M, Prueitt RL, Yanaihara N, Lanza G, Scarpa A, Vecchione A, Negrini M, Harris CC, Croce CM. A microRNA expression signature of human solid tumors defines cancer gene targets. Proc Natl Acad Sci U S A. 2006;103(7):2257–61. doi:10.1073/pnas.0510565103.
108. Petrocca F, Vecchione A, Croce CM. Emerging role of miR-106b-25/miR-17-92 clusters in the control of transforming growth factor β signaling. Cancer Res. 2008;68(20):8191–4.
109. Volinia S, Calin GA, Liu C-G, Ambs S, Cimmino A, Petrocca F, Visone R, Iorio M, Roldo C, Ferracin M. A microRNA expression signature of human solid tumors defines cancer gene targets. Proc Natl Acad Sci U S A. 2006;103(7):2257–61.
110. Meng F, Henson R, Wehbe–Janek H, Ghoshal K, Jacob ST, Patel T. MicroRNA-21 regulates expression of the PTEN tumor suppressor gene in human hepatocellular cancer. Gastroenterology. 2007;133(2):647–58.
111. Medina PP, Nolde M, Slack FJ. OncomiR addiction in an in vivo model of microRNA-21-induced pre-B-cell lymphoma. Nature. 2010;467(7311):86–90.
112. Hatley ME, Patrick DM, Garcia MR, Richardson JA, Bassel-Duby R, Van Rooij E, Olson EN. Modulation of K-Ras-dependent lung tumorigenesis by MicroRNA-21. Cancer Cell. 2010;18(3):282–93.
113. Cimmino A, Calin GA, Fabbri M, Iorio MV, Ferracin M, Shimizu M, Wojcik SE, Aqeilan RI, Zupo S, Dono M, Rassenti L, Alder H, Volinia S, Liu CG, Kipps TJ, Negrini M, Croce CM. miR-15 and miR-16 induce apoptosis by targeting BCL2. Proc Natl Acad Sci U S A. 2005;102(39):13944–9. doi:10.1073/pnas.0506654102.
114. Salerno E, Scaglione BJ, Coffman FD, Brown BD, Baccarini A, Fernandes H, Marti G, Raveche ES. Correcting miR-15a/16 genetic defect in New Zealand Black mouse model of CLL enhances drug sensitivity. Mol Cancer Ther. 2009;8(9):2684–92.
115. Roccaro AM, Sacco A, Thompson B, Leleu X, Azab AK, Azab F, Runnels J, Jia X, Ngo HT, Melhem MR. MicroRNAs 15a and 16 regulate tumor proliferation in multiple myeloma. Blood. 2009;113(26):6669–80.
116. Musumeci M, Coppola V, Addario A, Patrizii M, Maugeri-Sacca M, Memeo L, Colarossi C, Francescangeli F, Biffoni M, Collura D. Control of tumor and microenvironment cross-talk by miR-15a and miR-16 in prostate cancer. Oncogene. 2011;30(41):4231–42.
117. Chan JA, Krichevsky AM, Kosik KS. MicroRNA-21 is an antiapoptotic factor in human glioblastoma cells. Cancer Res. 2005;65(14):6029–33.
118. Si M, Zhu S, Wu H, Lu Z, Wu F, Mo Y. miR-21-mediated tumor growth. Oncogene. 2007;26(19):2799–803.
119. Huang N, Wu J, Qiu W, Lyu Q, He J, Xie W, Xu N, Zhang Y. MiR-15a and miR-16 induce autophagy and enhance chemosensitivity of Camptothecin. Cancer Biol Ther. 2015;16(6):941–8.
120. Frankel LB, Lund AH. MicroRNA regulation of autophagy. Carcinogenesis. 2012;33(11):2018–25. doi:10.1093/carcin/bgs266.
121. Ruan K, Fang X, Ouyang G. MicroRNAs: novel regulators in the hallmarks of human cancer. Cancer Lett. 2009;285(2):116–26.
122. Benetti R, Gonzalo S, Jaco I, Muñoz P, Gonzalez S, Schoeftner S, Murchison E, Andl T, Chen T, Klatt P. A mammalian microRNA cluster controls DNA methylation and telomere recombination via Rbl2-dependent regulation of DNA methyltransferases. Nat Struct Mol Biol. 2008;15(3):268–79.
123. Dinami R, Ercolani C, Petti E, Piazza S, Ciani Y, Sestito R, Sacconi A, Biagioni F, le Sage C, Agami R. miR-155 drives telomere fragility in human breast cancer by targeting TRF1. Cancer Res. 2014;74(15):4145–56.
124. Tazawa H, Tsuchiya N, Izumiya M, Nakagama H. Tumor-suppressive miR-34a induces senescence-like growth arrest through modulation of the E2F pathway in human colon cancer cells. Proc Natl Acad Sci. 2007;104(39):15472–7.
125. Giglio S, Cirombella R, Amodeo R, Portaro L, Lavra L, Vecchione A. MicroRNA miR-24 promotes cell proliferation by targeting the CDKs inhibitors p27Kip1 and p16INK4a. J Cell Physiol. 2013;228(10):2015–23.

126. Wang S, Aurora AB, Johnson BA, Qi X, McAnally J, Hill JA, Richardson JA, Bassel-Duby R, Olson EN. The endothelial-specific microRNA miR-126 governs vascular integrity and angiogenesis. Dev Cell. 2008;15(2):261–71.
127. Yamakuchi M, Lotterman CD, Bao C, Hruban RH, Karim B, Mendell JT, Huso D, Lowenstein CJ. P53-induced microRNA-107 inhibits HIF-1 and tumor angiogenesis. Proc Natl Acad Sci. 2010;107(14):6334–9.
128. Lee DY, Deng Z, Wang C-H, Yang BB. MicroRNA-378 promotes cell survival, tumor growth, and angiogenesis by targeting SuFu and Fus-1 expression. Proc Natl Acad Sci. 2007;104(51):20350–5.
129. Pecot CV, Rupaimoole R, Yang D, Akbani R, Ivan C, Lu C, Wu S, Han H-D, Shah MY, Rodriguez-Aguayo C. Tumour angiogenesis regulation by the miR-200 family. Nat Commun. 2013;4.
130. Park S-M, Gaur AB, Lengyel E, Peter ME. The miR-200 family determines the epithelial phenotype of cancer cells by targeting the E-cadherin repressors ZEB1 and ZEB2. Genes Dev. 2008;22(7):894–907.
131. Gibbons DL, Lin W, Creighton CJ, Rizvi ZH, Gregory PA, Goodall GJ, Thilaganathan N, Du L, Zhang Y, Pertsemlidis A. Contextual extracellular cues promote tumor cell EMT and metastasis by regulating miR-200 family expression. Genes Dev. 2009;23(18):2140–51.
132. Ma L, Young J, Prabhala H, Pan E, Mestdagh P, Muth D, Teruya-Feldstein J, Reinhardt F, Onder TT, Valastyan S. miR-9, a MYC/MYCN-activated microRNA, regulates E-cadherin and cancer metastasis. Nat Cell Biol. 2010;12(3):247–56.
133. Frankel LB, Christoffersen NR, Jacobsen A, Lindow M, Krogh A, Lund AH. Programmed cell death 4 (PDCD4) is an important functional target of the microRNA miR-21 in breast cancer cells. J Biol Chem. 2008;283(2):1026–33.
134. Pencheva N, Tavazoie SF. Control of metastatic progression by microRNA regulatory networks. Nat Cell Biol. 2013;15(6):546–54.
135. Ryan BM, Robles AI, Harris CC. Genetic variation in microRNA networks: the implications for cancer research. Nat Rev Cancer. 2010;10(6):389–402.
136. Schetter AJ, Nguyen GH, Bowman ED, Mathé EA, Yuen ST, Hawkes JE, Croce CM, Leung SY, Harris CC. Association of inflammation-related and microRNA gene expression with cancer-specific mortality of colon adenocarcinoma. Clin Cancer Res. 2009;15(18):5878–87.
137. Schetter AJ, Heegaard NH, Harris CC. Inflammation and cancer: interweaving microRNA, free radical, cytokine and p53 pathways. Carcinogenesis. 2010;31(1):37–49.
138. Costinean S, Zanesi N, Pekarsky Y, Tili E, Volinia S, Heerema N, Croce CM. Pre-B cell proliferation and lymphoblastic leukemia/high-grade lymphoma in Eμ-miR155 transgenic mice. Proc Natl Acad Sci. 2006;103(18):7024–9.
139. Gironella M, Seux M, Xie M-J, Cano C, Tomasini R, Gommeaux J, Garcia S, Nowak J, Yeung ML, Jeang K-T. Tumor protein 53-induced nuclear protein 1 expression is repressed by miR-155, and its restoration inhibits pancreatic tumor development. Proc Natl Acad Sci. 2007;104(41):16170–5.
140. Lin X, Chen L, Yao Y, Zhao R, Cui X, Chen J, Hou K, Zhang M, Su F, Chen J. CCL18-mediated down-regulation of miR98 and miR27b promotes breast cancer metastasis. Oncotarget. 2015;6(24):20485.
141. Gao P, Tchernyshyov I, Chang T-C, Lee Y-S, Kita K, Ochi T, Zeller KI, De Marzo AM, Van Eyk JE, Mendell JT. c-Myc suppression of miR-23a/b enhances mitochondrial glutaminase expression and glutamine metabolism. Nature. 2009;458(7239):762–5.
142. Okada H, Kohanbash G, Lotze MT. MicroRNAs in immune regulation—opportunities for cancer immunotherapy. Int J Biochem Cell Biol. 2010;42(8):1256–61.
143. Ueda R, Kohanbash G, Sasaki K, Fujita M, Zhu X, Kastenhuber ER, McDonald HA, Potter DM, Hamilton RL, Lotze MT. Dicer-regulated microRNAs 222 and 339 promote resistance of cancer cells to cytotoxic T-lymphocytes by down-regulation of ICAM-1. Proc Natl Acad Sci. 2009;106(26):10746–51.
144. Banerjee A, Schambach F, DeJong CS, Hammond SM, Reiner SL. Micro-RNA-155 inhibits IFN-γ signaling in CD4+ T cells. Eur J Immunol. 2010;40(1):225–31.

145. Lin S-L, Chiang A, Chang D, Ying S-Y. Loss of mir-146a function in hormone-refractory prostate cancer. RNA. 2008;14(3):417–24.
146. Mott JL, Kobayashi S, Bronk SF, Gores GJ. mir-29 regulates Mcl-1 protein expression and apoptosis. Oncogene. 2007;26(42):6133–40.
147. Tagawa H, Seto M. A microRNA cluster as a target of genomic amplification in malignant lymphoma. Leukemia. 2005;19(11):2013–6.
148. Hayashita Y, Osada H, Tatematsu Y, Yamada H, Yanagisawa K, Tomida S, Yatabe Y, Kawahara K, Sekido Y, Takahashi T. A polycistronic microRNA cluster, miR-17-92, is overexpressed in human lung cancers and enhances cell proliferation. Cancer Res. 2005;65(21):9628–32.
149. Yanaihara N, Caplen N, Bowman E, Seike M, Kumamoto K, Yi M, Stephens RM, Okamoto A, Yokota J, Tanaka T. Unique microRNA molecular profiles in lung cancer diagnosis and prognosis. Cancer Cell. 2006;9(3):189–98.
150. Shell S, Park S-M, Radjabi AR, Schickel R, Kistner EO, Jewell DA, Feig C, Lengyel E, Peter ME. Let-7 expression defines two differentiation stages of cancer. Proc Natl Acad Sci. 2007;104(27):11400–5.
151. Yu SL, Chen HY, Chang GC, Chen CY, Chen HW, Singh S, Cheng CL, Yu CJ, Lee YC, Chen HS, Su TJ, Chiang CC, Li HN, Hong QS, Su HY, Chen CC, Chen WJ, Liu CC, Chan WK, Chen WJ, Li KC, Chen JJ, Yang PC. MicroRNA signature predicts survival and relapse in lung cancer. Cancer Cell. 2008;13(1):48–57. doi:10.1016/j.ccr.2007.12.008.
152. Schetter AJ, Leung SY, Sohn JJ, Zanetti KA, Bowman ED, Yanaihara N, Yuen ST, Chan TL, Kwong DL, Au GK, Liu CG, Calin GA, Croce CM, Harris CC. MicroRNA expression profiles associated with prognosis and therapeutic outcome in colon adenocarcinoma. JAMA. 2008;299(4):425–36. doi:10.1001/jama.299.4.425.
153. Cortez MA, Bueso-Ramos C, Ferdin J, Lopez-Berestein G, Sood AK, Calin GA. MicroRNAs in body fluids-the mix of hormones and biomarkers. Nat Rev Clin Oncol. 2011;8(8):467–77. doi:10.1038/nrclinonc.2011.76.
154. Coulouarn C, Factor VM, Andersen JB, Durkin ME, Thorgeirsson SS. Loss of miR-122 expression in liver cancer correlates with suppression of the hepatic phenotype and gain of metastatic properties. Oncogene. 2009;28(40):3526–36. doi:10.1038/onc.2009.211.
155. Valadi H, Ekstrom K, Bossios A, Sjostrand M, Lee JJ, Lotvall JO. Exosome-mediated transfer of mRNAs and microRNAs is a novel mechanism of genetic exchange between cells. Nat Cell Biol. 2007;9(6):654–U672. doi:10.1038/ncb1596.
156. Ohshima K, Inoue K, Fujiwara A, Hatakeyama K, Kanto K, Watanabe Y, Muramatsu K, Fukuda Y, Ogura S, Yamaguchi K, Mochizuki T. Let-7 microRNA family is selectively secreted into the extracellular environment via exosomes in a metastatic gastric cancer cell line. PLoS ONE. 2010;5(10):e13247. doi:10.1371/journal.pone.0013247.
157. Pichler M, Calin GA. MicroRNAs in cancer: from developmental genes in worms to their clinical application in patients. Brit J Cancer. 2015;113(4):569–73. doi:10.1038/bjc.2015.253.
158. Trang P, Weidhaas J, Slack F. MicroRNAs as potential cancer therapeutics. Oncogene. 2008;27:S52–7.
159. Lanford RE, Hildebrandt-Eriksen ES, Petri A, Persson R, Lindow M, Munk ME, Kauppinen S, Ørum H. Therapeutic silencing of microRNA-122 in primates with chronic hepatitis C virus infection. Science. 2010;327(5962):198–201.
160. Elmen J, Lindow M, Silahtaroglu A, Bak M, Christensen M, Lind-Thomsen A, Hedtjärn M, Hansen JB, Hansen HF, Straarup EM. Antagonism of microRNA-122 in mice by systemically administered LNA-antimiR leads to up-regulation of a large set of predicted target mRNAs in the liver. Nucleic Acids Res. 2008;36(4):1153–62.
161. Gaur A, Jewell DA, Liang Y, Ridzon D, Moore JH, Chen C, Ambros VR, Israel MA. Characterization of microRNA expression levels and their biological correlates in human cancer cell lines. Cancer Res. 2007;67(6):2456–68.
162. Kota J, Chivukula RR, O'Donnell KA, Wentzel EA, Montgomery CL, Hwang H-W, Chang T-C, Vivekanandan P, Torbenson M, Clark KR. Therapeutic microRNA delivery suppresses tumorigenesis in a murine liver cancer model. Cell. 2009;137(6):1005–17.

163. Brown CJ, Ballabio A, Rupert JL, Lafreniere RG, Grompe M, Tonlorenzi R, Willard HF. A gene from the region of the human X inactivation centre is expressed exclusively from the inactive X chromosome. Nature. 1991;349(6304):38–44.
164. Lee J, Davidow LS, Warshawsky D. Tsix, a gene antisense to Xist at the X-inactivation centre. Nat Genet. 1999;21(4):400–4.
165. Sotomaru Y, Katsuzawa Y, Hatada I, Obata Y, Sasaki H, Kono T. Unregulated expression of the imprinted genes H19 and Igf2r in mouse uniparental fetuses. J Biol Chem. 2002;277(14):12474–8.
166. Rinn JL, Kertesz M, Wang JK, Squazzo SL, Xu X, Brugmann SA, Goodnough LH, Helms JA, Farnham PJ, Segal E. Functional demarcation of active and silent chromatin domains in human HOX loci by noncoding RNAs. Cell. 2007;129(7):1311–23.
167. Willingham A, Orth A, Batalov S, Peters E, Wen B, Aza-Blanc P, Hogenesch J, Schultz P. A strategy for probing the function of noncoding RNAs finds a repressor of NFAT. Science. 2005;309(5740):1570–3.
168. Lipovich L, Johnson R, Lin C-Y. MacroRNA underdogs in a microRNA world: evolutionary, regulatory, and biomedical significance of mammalian long non-protein-coding RNA. Biochim Biophys Acta. 2010;1799(9):597–615.
169. Guttman M, Amit I, Garber M, French C, Lin MF, Feldser D, Huarte M, Zuk O, Carey BW, Cassady JP. Chromatin signature reveals over a thousand highly conserved large non-coding RNAs in mammals. Nature. 2009;458(7235):223–7.
170. Kapranov P, Cheng J, Dike S, Nix DA, Duttagupta R, Willingham AT, Stadler PF, Hertel J, Hackermüller J, Hofacker IL. RNA maps reveal new RNA classes and a possible function for pervasive transcription. Science. 2007;316(5830):1484–8.
171. Dieci G, Fiorino G, Castelnuovo M, Teichmann M, Pagano A. The expanding RNA polymerase III transcriptome. Trends Genet. 2007;23(12):614–22.
172. Yin Q-F, Yang L, Zhang Y, Xiang J-F, Wu Y-W, Carmichael GG, Chen L-L. Long noncoding RNAs with snoRNA ends. Mol Cell. 2012;48(2):219–30.
173. Hung T, Wang Y, Lin MF, Koegel AK, Kotake Y, Grant GD, Horlings HM, Shah N, Umbricht C, Wang P. Extensive and coordinated transcription of noncoding RNAs within cell-cycle promoters. Nat Genet. 2011;43(7):621–9.
174. Ørom UA, Derrien T, Beringer M, Gumireddy K, Gardini A, Bussotti G, Lai F, Zytnicki M, Notredame C, Huang Q. Long noncoding RNAs with enhancer-like function in human cells. Cell. 2010;143(1):46–58.
175. Salzman J, Gawad C, Wang PL, Lacayo N, Brown PO. Circular RNAs are the predominant transcript isoform from hundreds of human genes in diverse cell types. PLoS ONE. 2012;7(2):e30733.
176. Yang L, Duff MO, Graveley BR, Carmichael GG, Chen L-L. Genomewide characterization of non-polyadenylated RNAs. Genome Biol. 2011;12(2):R16.
177. Huarte M, Guttman M, Feldser D, Garber M, Koziol MJ, Kenzelmann-Broz D, Khalil AM, Zuk O, Amit I, Rabani M. A large intergenic noncoding RNA induced by p53 mediates global gene repression in the p53 response. Cell. 2010;142(3):409–19.
178. Heo JB, Sung S. Vernalization-mediated epigenetic silencing by a long intronic noncoding RNA. Science. 2011;331(6013):76–9.
179. Wang D, Garcia-Bassets I, Benner C, Li W, Su X, Zhou Y, Qiu J, Liu W, Kaikkonen MU, Ohgi KA. Reprogramming transcription by distinct classes of enhancers functionally defined by eRNA. Nature. 2011;474(7351):390–4.
180. Kim T-K, Hemberg M, Gray JM, Costa AM, Bear DM, Wu J, Harmin DA, Laptewicz M, Barbara-Haley K, Kuersten S. Widespread transcription at neuronal activity-regulated enhancers. Nature. 2010;465(7295):182–7.
181. Kanhere A, Viiri K, Araújo CC, Rasaiyaah J, Bouwman RD, Whyte WA, Pereira CF, Brookes E, Walker K, Bell GW. Short RNAs are transcribed from repressed polycomb target genes and interact with polycomb repressive complex-2. Mol Cell. 2010;38(5):675–88.
182. Kino T, Hurt DE, Ichijo T, Nader N, Chrousos GP. Noncoding RNA Gas5 is a growth arrest and starvation-associated repressor of the glucocorticoid receptor. Sci Signal. 2010;3(107):ra8.

183. Tripathi V, Ellis JD, Shen Z, Song DY, Pan Q, Watt AT, Freier SM, Bennett CF, Sharma A, Bubulya PA. The nuclear-retained noncoding RNA MALAT1 regulates alternative splicing by modulating SR splicing factor phosphorylation. Mol Cell. 2010;39(6):925–38.
184. Jia H, Osak M, Bogu GK, Stanton LW, Johnson R, Lipovich L. Genome-wide computational identification and manual annotation of human long noncoding RNA genes. RNA. 2010;16(8):1478–87.
185. Han D, Wang M, Ma N, Xu Y, Jiang Y, Gao X. Long noncoding RNAs: novel players in colorectal cancer. Cancer Lett. 2015;361(1):13–21.
186. Di Gesualdo F, Capaccioli S, Lulli M. A pathophysiological view of the long non-coding RNA world. Oncotarget. 2014;5(22):10976.
187. Pasmant E, Sabbagh A, Vidaud M, Bièche I. ANRIL, a long, noncoding RNA, is an unexpected major hotspot in GWAS. FASEB J. 2011;25(2):444–8.
188. Burd CE, Jeck WR, Liu Y, Sanoff HK, Wang Z, Sharpless NE. Expression of linear and novel circular forms of an INK4/ARF-associated non-coding RNA correlates with atherosclerosis risk. PLoS Genet. 2010;6(12):e1001233.
189. Laurent GS, Wahlestedt C, Kapranov P. The Landscape of long noncoding RNA classification. Trends Genet. 2015;31(5):239–51.
190. Su F, Li D, Zeng M, Song E. A cytoplasmic NF-κB interacting long noncoding RNA blocks IκB phosphorylation and suppresses breast cancer metastasis. Cancer Cell. 2015;27:370–81.
191. Wang KC, Chang HY. Molecular mechanisms of long noncoding RNAs. Mol Cell. 2011;43(6):904–14.
192. Nagano T, Mitchell JA, Sanz LA, Pauler FM, Ferguson-Smith AC, Feil R, Fraser P. The Air noncoding RNA epigenetically silences transcription by targeting G9a to chromatin. Science. 2008;322(5908):1717–20.
193. Lee JT. The X as model for RNA's niche in epigenomic regulation. Cold Spring Harb Perspect Biol. 2010;2(9):a003749.
194. Yap KL, Li S, Muñoz-Cabello AM, Raguz S, Zeng L, Mujtaba S, Gil J, Walsh MJ, Zhou M-M. Molecular interplay of the noncoding RNA ANRIL and methylated histone H3 lysine 27 by polycomb CBX7 in transcriptional silencing of INK4a. Mol Cell. 2010;38(5):662–74.
195. Kotake Y, Nakagawa T, Kitagawa K, Suzuki S, Liu N, Kitagawa M, Xiong Y. Long noncoding RNA ANRIL is required for the PRC2 recruitment to and silencing of p15INK4B tumor suppressor gene. Oncogene. 2011;30(16):1956–62.
196. Gutschner T, Diederichs S. The hallmarks of cancer: a long non-coding RNA point of view. RNA Biol. 2012;9(6):703–19.
197. Mercer TR, Mattick JS. Structure and function of long noncoding RNAs in epigenetic regulation. Nat Struct Mol Biol. 2013;20(3):300–7.
198. Bjornsson HT, Fallin MD, Feinberg AP. An integrated epigenetic and genetic approach to common human disease. TRENDS in Genetics. 2004;20(8):350–8.
199. Brookes E, Shi Y. Diverse epigenetic mechanisms of human disease. Annu Rev Genet. 2014;48:237–68.
200. Feinberg AP, Vogelstein B. Hypomethylation distinguishes genes of some human cancers from their normal counterparts. Nature. 1983;301(5895):89–92.
201. Feinberg AP, Tycko B. The history of cancer epigenetics. Nat Rev Cancer. 2004;4(2):143–53.
202. Khalil AM, Guttman M, Huarte M, Garber M, Raj A, Morales DR, Thomas K, Presser A, Bernstein BE, van Oudenaarden A. Many human large intergenic noncoding RNAs associate with chromatin-modifying complexes and affect gene expression. Proc Natl Acad Sci. 2009;106(28):11667–72.
203. Schwartz YB, Pirrotta V. Polycomb silencing mechanisms and the management of genomic programmes. Nat Rev Genet. 2007;8(1):9–22.
204. Chu C, Qu K, Zhong FL, Artandi SE, Chang HY. Genomic maps of long noncoding RNA occupancy reveal principles of RNA-chromatin interactions. Mol Cell. 2011;44(4):667–78.
205. Tsai M-C, Manor O, Wan Y, Mosammaparast N, Wang JK, Lan F, Shi Y, Segal E, Chang HY. Long noncoding RNA as modular scaffold of histone modification complexes. Science. 2010;329(5992):689–93.

206. Sharma S, Mandal P, Sadhukhan T, Chowdhury RR, Mondal NR, Chakravarty B, Chatterjee T, Roy S, Sengupta S. Bridging links between long noncoding RNA HOTAIR and HPV oncoprotein E7 in cervical cancer pathogenesis. Sci Rep. 2015;5.
207. Zeng X, Chen S, Huang H. Phosphorylation of EZH2 by CDK1 and CDK2: a possible regulatory mechanism of transmission of the H3K27me3 epigenetic mark through cell divisions. Cell Cycle. 2011;10(4):579–83.
208. E-b Z, Kong R, D-d Y, L-h Y, Sun M, Han L, Xu T-p, Xia R, Yang J-s, De W. Long noncoding RNA ANRIL indicates a poor prognosis of gastric cancer and promotes tumor growth by epigenetically silencing of miR-99a/miR-449a. Oncotarget. 2014;5(8):2276.
209. Law JA, Jacobsen SE. Establishing, maintaining and modifying DNA methylation patterns in plants and animals. Nat Rev Genet. 2010;11(3):204–20.
210. Steinhoff C, Paulsen M, Kielbasa S, Walter J, Vingron M. Expression profile and transcription factor binding site exploration of imprinted genes in human and mouse. BMC Genomics. 2009;10(1):144.
211. Ogawa O, Eccles MR, Szeto J, McNoe LA, Yun K, Maw MA, Smith PJ, Reeve AE. Relaxation of insulin-like growth factor II gene imprinting implicated in Wilms' tumour. Nature. 1993;362(6422):749–51.
212. Rainier S, Johnson LA, Dobry CJ, Ping AJ, Grundy PE, Feinberg AP. Relaxation of imprinted genes in human cancer. Nature. 1993;362:747–9.
213. Kaneda A, Feinberg AP. Loss of imprinting of IGF2: a common epigenetic modifier of intestinal tumor risk. Cancer Res. 2005;65(24):11236–40.
214. Nakagawa H, Chadwick RB, Peltomäki P, Plass C, Nakamura Y, de La Chapelle A. Loss of imprinting of the insulin-like growth factor II gene occurs by biallelic methylation in a core region of H19-associated CTCF-binding sites in colorectal cancer. Proc Natl Acad Sci. 2001;98(2):591–6.
215. Spatz A, Borg C, Feunteun J. X-chromosome genetics and human cancer. Nat Rev Cancer. 2004;4(8):617–29.
216. Zhao J, Sun BK, Erwin JA, Song J-J, Lee JT. Polycomb proteins targeted by a short repeat RNA to the mouse X chromosome. Science. 2008;322(5902):750–6.
217. Sun BK, Deaton AM, Lee JT. A transient heterochromatic state in Xist preempts X inactivation choice without RNA stabilization. Mol Cell. 2006;21(5):617–28.
218. Schmitz K-M, Mayer C, Postepska A, Grummt I. Interaction of noncoding RNA with the rDNA promoter mediates recruitment of DNMT3b and silencing of rRNA genes. Genes Dev. 2010;24(20):2264–9.
219. Malek E, Jagannathan S, Driscoll JJ. Correlation of long non-coding RNA expression with metastasis, drug resistance and clinical outcome in cancer. Oncotarget. 2014;5(18):8027.
220. Yang L, Froberg JE, Lee JT. Long noncoding RNAs: fresh perspectives into the RNA world. Trends Biochem Sci. 2014;39(1):35–43.
221. Tano K, Mizuno R, Okada T, Rakwal R, Shibato J, Masuo Y, Ijiri K, Akimitsu N. MALAT-1 enhances cell motility of lung adenocarcinoma cells by influencing the expression of motility-related genes. FEBS Lett. 2010;584(22):4575–80.
222. Matsui K, Nishizawa M, Ozaki T, Kimura T, Hashimoto I, Yamada M, Kaibori M, Kamiyama Y, Ito S, Okumura T. Natural antisense transcript stabilizes inducible nitric oxide synthase messenger RNA in rat hepatocytes. Hepatology. 2008;47(2):686–97.
223. Yoon J-H, Abdelmohsen K, Srikantan S, Yang X, Martindale JL, De S, Huarte M, Zhan M, Becker KG, Gorospe M. LincRNA-p21 suppresses target mRNA translation. Mol Cell. 2012;47(4):648–55.
224. Poliseno L, Salmena L, Zhang J, Carver B, Haveman WJ, Pandolfi PP. A coding-independent function of gene and pseudogene mRNAs regulates tumour biology. Nature. 2010;465(7301):1033–8.
225. Cai X, Cullen BR. The imprinted H19 noncoding RNA is a primary microRNA precursor. RNA. 2007;13(3):313–6.

226. Keniry A, Oxley D, Monnier P, Kyba M, Dandolo L, Smits G, Reik W. The H19 lincRNA is a developmental reservoir of miR-675 that suppresses growth and Igf1r. Nat Cell Biol. 2012;14(7):659–65.
227. Smith CM, Steitz JA. Classification of gas5 as a multi-small-nucleolar-RNA (snoRNA) host gene and a member of the 5′-terminal oligopyrimidine gene family reveals common features of snoRNA host genes. Mol Cell Biol. 1998;18(12):6897–909.
228. Gupta RA, Shah N, Wang KC, Kim J, Horlings HM, Wong DJ, Tsai M-C, Hung T, Argani P, Rinn JL. Long non-coding RNA HOTAIR reprograms chromatin state to promote cancer metastasis. Nature. 2010;464(7291):1071–6.
229. Yang Z, Zhou L, Wu L-M, Lai M-C, Xie H-Y, Zhang F, Zheng S-S. Overexpression of long non-coding RNA HOTAIR predicts tumor recurrence in hepatocellular carcinoma patients following liver transplantation. Ann Surg Oncol. 2011;18(5):1243–50.
230. Niinuma T, Suzuki H, Nojima M, Nosho K, Yamamoto H, Takamaru H, Yamamoto E, Maruyama R, Nobuoka T, Miyazaki Y. Upregulation of miR-196a and HOTAIR drive malignant character in gastrointestinal stromal tumors. Cancer Res. 2012;72(5):1126–36.
231. Kim K, Jutooru I, Chadalapaka G, Johnson G, Frank J, Burghardt R, Kim S, Safe S. HOTAIR is a negative prognostic factor and exhibits pro-oncogenic activity in pancreatic cancer. Oncogene. 2013;32(13):1616–25.
232. X-h L, Z-l L, Sun M, Liu J, Wang Z-x, De W. The long non-coding RNA HOTAIR indicates a poor prognosis and promotes metastasis in non-small cell lung cancer. BMC Cancer. 2013;13(1):464.
233. Nie Y, Liu X, Qu S, Song E, Zou H, Gong C. Long non-coding RNA HOTAIR is an independent prognostic marker for nasopharyngeal carcinoma progression and survival. Cancer Sci. 2013;104(4):458–64.
234. Li D, Feng J, Wu T, Wang Y, Sun Y, Ren J, Liu M. Long intergenic noncoding RNA HOTAIR is overexpressed and regulates PTEN methylation in laryngeal squamous cell carcinoma. Am J Pathol. 2013;182(1):64–70.
235. Chen FJ, Sun M, Li SQ, Wu QQ, Ji L, Liu ZL, Zhou GZ, Cao G, Jin L, Xie HW. Upregulation of the long non-coding rna hotair promotes esophageal squamous cell carcinoma metastasis and poor prognosis. Mol Carcinog. 2013;52(11):908–15.
236. Heubach J, Monsior J, Deenen R, Niegisch G, Szarvas T, Niedworok C, Schulz WA, Hoffmann MJ. The long noncoding RNA HOTAIR has tissue and cell type-dependent effects on HOX gene expression and phenotype of urothelial cancer cells. Mol Cancer. 2015;14(1):108.
237. Nakayama I, Shibazaki M, Yashima-Abo A, Miura F, Sugiyama T, Masuda T, Maesawa C. Loss of HOXD10 expression induced by upregulation of miR-10b accelerates the migration and invasion activities of ovarian cancer cells. Int J Oncol. 2013;43(1):63–71.
238. Wu Y, Liu J, Zheng Y, You L, Kuang D, Liu T. Suppressed expression of long non-coding RNA HOTAIR inhibits proliferation and tumourigenicity of renal carcinoma cells. Tumor Biol. 2014;35(12):11887–94.
239. Ma M, Li C, Zhang Y, Weng M, Zhang M, Qin Y, Gong W, Quan Z. Long non-coding RNA HOTAIR, a c-Myc activated driver of malignancy, negatively regulates miRNA-130a in gallbladder cancer. Mol Cancer. 2014;13:156.
240. Li Z, Li C, Liu C, Yu S, Zhang Y. Expression of the long non-coding RNAs MEG3, HOTAIR, and MALAT-1 in non-functioning pituitary adenomas and their relationship to tumor behavior. Pituitary. 2014;18(1):42–7.
241. Chiyomaru T, Yamamura S, Fukuhara S, Yoshino H, Kinoshita T, Majid S, Saini S, Chang I, Tanaka Y, Enokida H. Genistein inhibits prostate cancer cell growth by targeting miR-34a and oncogenic HOTAIR. PLoS ONE. 2013;8(8):e70372.
242. Tang L, Zhang W, Su B, Yu B. Long noncoding RNA HOTAIR is associated with motility, invasion, and metastatic potential of metastatic melanoma. BioMed Res Int. 2013;2013:251098.
243. He X, Bao W, Li X, Chen Z, Che Q, Wang H, Wan X-P. The long non-coding RNA HOTAIR is upregulated in endometrial carcinoma and correlates with poor prognosis. Int J Mol Med. 2014;33(2):325–32.

244. Wang F, Yuan JH, Wang SB, Yang F, Yuan SX, Ye C, Yang N, Zhou WP, Li WL, Li W. Oncofetal long noncoding RNA PVT1 promotes proliferation and stem cell-like property of hepatocellular carcinoma cells by stabilizing NOP2. Hepatology. 2014;60(4):1278–90.
245. Ding J, Li D, Gong M, Wang J, Huang X, Wu T, Wang C. Expression and clinical significance of the long non-coding RNA PVT1 in human gastric cancer. OncoTargets Ther. 2014;7:1625.
246. Prensner JR, Iyer MK, Balbin OA, Dhanasekaran SM, Cao Q, Brenner JC, Laxman B, Asangani IA, Grasso CS, Kominsky HD. Transcriptome sequencing across a prostate cancer cohort identifies PCAT-1, an unannotated lincRNA implicated in disease progression. Nat Biotechnol. 2011;29(8):742–9.
247. Zhai H, Fesler A, Schee K, Fodstad Ø, Flatmark K, Ju J. Clinical significance of long intergenic noncoding RNA-p21 in colorectal cancer. Clin Colorectal Cancer. 2013;12(4):261–6.
248. Mourtada-Maarabouni M, Pickard M, Hedge V, Farzaneh F, Williams G. GAS5, a non-protein-coding RNA, controls apoptosis and is downregulated in breast cancer. Oncogene. 2009;28(2):195–208.
249. Sun M, F-y J, Xia R, Kong R, J-h L, Xu T-p, Y-w L, E-b Z, X-h L, De W. Decreased expression of long noncoding RNA GAS5 indicates a poor prognosis and promotes cell proliferation in gastric cancer. BMC Cancer. 2014;14(1):319.
250. Martens-Uzunova E, Jalava S, Dits N, van Leenders G, Møller S, Trapman J, Bangma C, Litman T, Visakorpi T, Jenster G. Diagnostic and prognostic signatures from the small non-coding RNA transcriptome in prostate cancer. Oncogene. 2012;31(8):978–91.
251. Qiao H-P, Gao W-S, Huo J-X, Yang Z-S. Long non-coding RNA GAS5 functions as a tumor suppressor in renal cell carcinoma. Asian Pac J Cancer Prev. 2013;14(2):1077–82.
252. Svoboda M, Slyskova J, Schneiderova M, Makovicky P, Bielik L, Levy M, Lipska L, Hemmelova B, Kala Z, Protivankova M. HOTAIR long non-coding RNA is a negative prognostic factor not only in primary tumors, but also in the blood of colorectal cancer patients. Carcinogenesis. 2014;35(7):1510–5.
253. Zhang J, Zhang P, Wang L, H-l P, Ma L. Long non-coding RNA HOTAIR in carcinogenesis and metastasis. Acta Biochim Biophys Sin. 2013;46:1–5.
254. JiP D, Wang W. mALAT-1, anovelnoncoding RnA, and thymos in beta4 predict metastasis and survival in early-stage non-small cell lung cancer. Oncogene. 2003;22(39):8031.
255. Zhu Z, Gao X, He Y, Zhao H, Yu Q, Jiang D, Zhang P, Ma X, Huang H, Dong D. An insertion/deletion polymorphism within RERT-lncRNA modulates hepatocellular carcinoma risk. Cancer Res. 2012;72(23):6163–72.
256. Han Y, Liu Y, Nie L, Gui Y, Cai Z. Inducing cell proliferation inhibition, apoptosis, and motility reduction by silencing long noncoding ribonucleic acid metastasis-associated lung adenocarcinoma transcript 1 in urothelial carcinoma of the bladder. Urology. 2013;81(1):209.e201–7.
257. Davis IJ, Hsi B-L, Arroyo JD, Vargas SO, Yeh YA, Motyckova G, Valencia P, Perez-Atayde AR, Argani P, Ladanyi M. Cloning of an Alpha-TFEB fusion in renal tumors harboring the t (6; 11)(p21; q13) chromosome translocation. Proc Natl Acad Sci. 2003;100(10):6051–6.
258. Walsh AL, Tuzova AV, Bolton EM, Lynch TH, Perry AS. Long noncoding RNAs and prostate carcinogenesis: the missing 'linc'? Trends Mol Med. 2014;20(8):428–36.
259. Zhang XW, Bu P, Liu L, Zhang XZ, Li J. Overexpression of long non-coding RNA PVT1 in gastric cancer cells promotes the development of multidrug resistance. Biochem Biophys Res Commun. 2015;462(3):227–32. doi:10.1016/j.bbrc.2015.04.121.
260. Yang Y-R, Zang S-Z, Zhong C-L, Li Y-X, Zhao S-S, Feng X-J. Increased expression of the lncRNA PVT1 promotes tumorigenesis in non-small cell lung cancer. Int J Clin Exp Pathol. 2014;7(10):6929.
261. Shi Y, Wang Y, Luan W, Wang P, Tao T, Zhang J, Qian J, Liu N, You Y. Long non-coding RNA H19 promotes glioma cell invasion by deriving miR-675. PLoS ONE. 2014;9(1):e86295.
262. Zhu M, Chen Q, Liu X, Sun Q, Zhao X, Deng R, Wang Y, Huang J, Xu M, Yan J. lncRNA H19/miR-675 axis represses prostate cancer metastasis by targeting TGFBI. FEBS J. 2014;281(16):3766–75.

263. Peng W, Fan H. Long non-coding RNA PANDAR correlates with poor prognosis and promotes tumorigenesis in hepatocellular carcinoma. Biomed Pharmacother. 2015;72:113–8.
264. Cao Y, Bryan TM, Reddel RR. Increased copy number of the TERT and TERC telomerase subunit genes in cancer cells. Cancer Sci. 2008;99(6):1092–9.
265. Sugita M, Tanaka N, Davidson S, Sekiya S, Varella-Garcia M, West J, Drabkin HA, Gemmill RM. Molecular definition of a small amplification domain within 3q26 in tumors of cervix, ovary, and lung. Cancer Genet Cytogenet. 2000;117(1):9–18.
266. Oliva J, Bardag-Gorce F, French BA, Li J, French SW. The regulation of non-coding RNA expression in the liver of mice fed DDC. Exp Mol Pathol. 2009;87(1):12–9.
267. Panzitt K, Tschernatsch MM, Guelly C, Moustafa T, Stradner M, Strohmaier HM, Buck CR, Denk H, Schroeder R, Trauner M. Characterization of HULC, a novel gene with striking up-regulation in hepatocellular carcinoma, as noncoding RNA. Gastroenterology. 2007;132(1):330–42.
268. Cui M, Xiao Z, Wang Y, Zheng M, Song T, Cai X, Sun B, Ye L, Zhang X. Long noncoding RNA HULC modulates abnormal lipid metabolism in Hepatoma cells through an miR-9–mediated RXRA signaling pathway. Cancer Res. 2015;75(5):846–57.
269. Rossignol F, Vaché C, Clottes E. Natural antisense transcripts of hypoxia-inducible factor 1alpha are detected in different normal and tumour human tissues. Gene. 2002;299(1):135–40.
270. Thrash-Bingham CA, Tartof KD. aHIF: a natural antisense transcript overexpressed in human renal cancer and during hypoxia. J Natl Cancer Inst. 1999;91(2):143–51.
271. Cayre A, Rossignol F, Clottes E, Penault-Llorca F. aHIF but not HIF-1alpha transcript is a poor prognostic marker in human breast cancer. Breast Cancer Res. 2003;5(6):R223–30.
272. Kawakami T, Okamoto K, Sugihara H, Hattori T, Reeve AE, Ogawa O, Okada Y. The roles of supernumerical X chromosomes and XIST expression in testicular germ cell tumors. J Urol. 2003;169(4):1546–52.
273. Weakley SM, Wang H, Yao Q, Chen C. Expression and function of a large non-coding RNA gene XIST in human cancer. World J Surg. 2011;35(8):1751–6.
274. Kawakami T, Zhang C, Taniguchi T, Kim CJ, Okada Y, Sugihara H, Hattori T, Reeve AE, Ogawa O, Okamoto K. Characterization of loss-of-inactive X in Klinefelter syndrome and female-derived cancer cells. Oncogene. 2004;23(36):6163–9.
275. Benoît M-H, Hudson TJ, Maire G, Squire JA, Arcand SL, Provencher D, Mes-Masson A-M, Tonin PN. Global analysis of chromosome X gene expression in primary cultures of normal ovarian surface epithelial cells and epithelial ovarian cancer cell lines. Int J Oncol. 2007;30(1):5–17.
276. Luo M, Li Z, Wang W, Zeng Y, Liu Z, Qiu J. Long non-coding RNA H19 increases bladder cancer metastasis by associating with EZH2 and inhibiting E-cadherin expression. Cancer Lett. 2013;333(2):213–21.
277. Li H, Yu B, Li J, Su L, Yan M, Zhu Z, Liu B. Overexpression of lncRNA H19 enhances carcinogenesis and metastasis of gastric cancer. Oncotarget. 2014;5(8):2318.
278. Laner T, Schulz WA, Engers R, Müller M, Florl AR. Hypomethylation of the XIST gene promoter in prostate cancer. Oncol Res. 2005;15(5):257–64.
279. Song M, Park JH, Jeong KS, Park DS, Kang MS, Lee S. Quantification of CpG methylation at the 5′-region of XIST by pyrosequencing from human serum. Electrophoresis. 2007;28(14):2379–84.
280. Bhan A, Mandal SS. LncRNA HOTAIR: a master regulator of chromatin dynamics and cancer. Biochim Biophys Acta. 2015;1856(1):151–64.
281. Ding C, Yang Z, Lv Z, Du C, Xiao H, Peng C, Cheng S, Xie H, Zhou L, Wu J. Long non-coding RNA PVT1 is associated with tumor progression and predicts recurrence in hepatocellular carcinoma patients. Oncol Lett. 2015;9(2):955–63.
282. W-h S, Wu Q-q, S-q L, Yang T-x, Z-h L, Y-s T, Tuo L, Wang S, Cao X-F. Upregulation of the long noncoding RNA PCAT-1 correlates with advanced clinical stage and poor prognosis in esophageal squamous carcinoma. Tumor Biol. 2015;36(4):2501–7.

283. Tsai M-C, Spitale RC, Chang HY. Long intergenic noncoding RNAs: new links in cancer progression. Cancer Res. 2011;71(1):3–7.
284. Scaiewicz V, Sorin V, Fellig Y, Birman T, Mizrahi A, Galula J, Abu-Lail R, Shneider T, Ohana P, Buscail L. Use of H19 gene regulatory sequences in DNA-based therapy for pancreatic cancer. J Oncol. 2010;2010:178174.
285. Smaldone MC, Davies BJ. BC-819, a plasmid comprising the H19 gene regulatory sequences and diphtheria toxin A, for the potential targeted therapy of cancers. Curr Opin Mol Ther. 2010;12(5):607–16.
286. Jeck WR, Sorrentino JA, Wang K, Slevin MK, Burd CE, Liu J, Marzluff WF, Sharpless NE. Circular RNAs are abundant, conserved, and associated with ALU repeats. RNA. 2013;19(2):141–57.
287. Guo JU, Agarwal V, Guo H, Bartel DP. Expanded identification and characterization of mammalian circular RNAs. Genome Biol. 2014;15(409):10.1186.
288. Jeck WR, Sharpless NE. Detecting and characterizing circular RNAs. Nat Biotechnol. 2014;32(5):453–61.
289. Zhang X-O, Wang H-B, Zhang Y, Lu X, Chen L-L, Yang L. Complementary sequence-mediated exon circularization. Cell. 2014;159(1):134–47.
290. Graveley BR. Molecular biology: power sequencing. Nature. 2008;453(7199):1197–8.
291. Wilhelm BT, Marguerat S, Watt S, Schubert F, Wood V, Goodhead I, Penkett CJ, Rogers J, Bähler J. Dynamic repertoire of a eukaryotic transcriptome surveyed at single-nucleotide resolution. Nature. 2008;453(7199):1239–43.
292. Qu S, Yang X, Li X, Wang J, Gao Y, Shang R, Sun W, Dou K, Li H. Circular RNA: a new star of noncoding RNAs. Cancer Lett. 2015;365(2):141–8.
293. Westholm JO, Miura P, Olson S, Shenker S, Joseph B, Sanfilippo P, Celniker SE, Graveley BR, Lai EC. Genome-wide analysis of Drosophila circular RNAs reveals their structural and sequence properties and age-dependent neural accumulation. Cell Rep. 2014;9(5):1966–80.
294. Ivanov A, Memczak S, Wyler E, Torti F, Porath HT, Orejuela MR, Piechotta M, Levanon EY, Landthaler M, Dieterich C. Analysis of intron sequences reveals hallmarks of circular RNA biogenesis in animals. Cell Rep. 2015;10(2):170–7.
295. Danan M, Schwartz S, Edelheit S, Sorek R. Transcriptome-wide discovery of circular RNAs in Archaea. Nucleic Acids Res. 2011;40:3131–42.
296. Kelly S, Greenman C, Cook PR, Papantonis A. Exon skipping is correlated with Exon circularization. J Mol Biol. 2015;427:2414–7.
297. Li Z, Huang C, Bao C, Chen L, Lin M, Wang X, Zhong G, Yu B, Hu W, Dai L. Exon-intron circular RNAs regulate transcription in the nucleus. Nat Struct Mol Biol. 2015;22(3):256–64.
298. Ashwal-Fluss R, Meyer M, Pamudurti NR, Ivanov A, Bartok O, Hanan M, Evantal N, Memczak S, Rajewsky N, Kadener S. circRNA biogenesis competes with pre-mRNA splicing. Mol Cell. 2014;56(1):55–66.
299. Chen I, Chen CY, Chuang TJ. Biogenesis, identification, and function of exonic circular RNAs. Wiley Interdiscip Rev RNA. 2015;6(5):563–79.
300. Salzman J, Chen RE, Olsen MN, Wang PL, Brown PO. Cell-type specific features of circular RNA expression. PLoS Genet. 2013;9(9):e1003777.
301. Lasda E, Parker R. Circular RNAs: diversity of form and function. RNA. 2014;20(12):1829–42.
302. Liang D, Wilusz JE. Short intronic repeat sequences facilitate circular RNA production. Genes Dev. 2014;28(20):2233–47.
303. Chen L-L, Carmichael GG. Gene regulation by SINES and inosines: biological consequences of A-to-I editing of Alu element inverted repeats. Cell Cycle. 2008;7(21):3294–301.
304. Lovci MT, Ghanem D, Marr H, Arnold J, Gee S, Parra M, Liang TY, Stark TJ, Gehman LT, Hoon S. Rbfox proteins regulate alternative mRNA splicing through evolutionarily conserved RNA bridges. Nat Struct Mol Biol. 2013;20(12):1434–42.
305. Conn SJ, Pillman KA, Toubia J, Conn VM, Salmanidis M, Phillips CA, Roslan S, Schreiber AW, Gregory PA, Goodall GJ. The RNA binding protein Quaking regulates formation of circRNAs. Cell. 2015;160(6):1125–34.

306. Athanasiadis A, Rich A, Maas S. Widespread A-to-I RNA editing of Alu-containing mRNAs in the human transcriptome. PLoS Biol. 2004;2(12):e391.
307. Chen T, Xiang J-F, Zhu S, Chen S, Yin Q-F, Zhang X-O, Zhang J, Feng H, Dong R, Li X-J. ADAR1 is required for differentiation and neural induction by regulating microRNA processing in a catalytically independent manner. Cell Res. 2015;25(4):459–76.
308. Li Y, Zheng Q, Bao C, Li S, Guo W, Zhao J, Chen D, Gu J, He X, Huang S. Circular RNA is enriched and stable in exosomes: a promising biomarker for cancer diagnosis. Cell Res. 2015;25(8):981.
309. Suzuki H, Tsukahara T. A view of pre-mRNA splicing from RNase R resistant RNAs. Int J Mol Sci. 2014;15(6):9331–42.
310. Bahn JH, Zhang Q, Li F, Chan T-M, Lin X, Kim Y, Wong DT, Xiao X. The landscape of microRNA, Piwi-interacting RNA, and circular RNA in human saliva. Clin Chem. 2015;61(1):221–30.
311. Wang PL, Bao Y, Yee M-C, Barrett SP, Hogan GJ, Olsen MN, Dinneny JR, Brown PO, Salzman J. Circular RNA is expressed across the eukaryotic tree of life. PLoS ONE. 2014;9(3):e90859.
312. C-y C, Sarnow P. Initiation of protein synthesis by the eukaryotic translational apparatus on circular RNAs. Science. 1995;268(5209):415–7.
313. Wang Y, Wang Z. Efficient backsplicing produces translatable circular mRNAs. RNA. 2015;21(2):172–9.
314. Hansen TB, Kjems J, Damgaard CK. Circular RNA and miR-7 in cancer. Cancer Res. 2013;73(18):5609–12.
315. Li F, Zhang L, Li W, Deng J, Zheng J, An M, Lu J, Zhou Y. Circular RNA ITCH has inhibitory effect on ESCC by suppressing the Wnt/β-catenin pathway. Oncotarget. 2015;6(8):6001–13.
316. Bachmayr-Heyda A, Reiner AT, Auer K, Sukhbaatar N, Aust S, Bachleitner-Hofmann T, Mesteri I, Grunt TW, Zeillinger R, Pils D. Correlation of circular RNA abundance with proliferation-exemplified with colorectal and ovarian cancer, idiopathic lung fibrosis, and normal human tissues. Sci Rep. 2015;5:8057.
317. Li P, Chen S, Chen H, Mo X, Li T, Shao Y, Xiao B, Guo J. Using circular RNA as a novel type of biomarker in the screening of gastric cancer. Clin Chim Acta. 2015;444:132–6.
318. Taulli R, Loretelli C, Pandolfi PP. From pseudo-ceRNAs to circ-ceRNAs: a tale of cross-talk and competition. Nat Struct Mol Biol. 2013;20(5):541–3.
319. Petkovic S, Müller S. RNA circularization strategies in vivo and in vitro. Nucleic Acids Res. 2015;43(4):2454–65.
320. Hansen TB, Wiklund ED, Bramsen JB, Villadsen SB, Statham AL, Clark SJ, Kjems J. miRNA-dependent gene silencing involving Ago2-mediated cleavage of a circular antisense RNA. EMBO J. 2011;30(21):4414–22.
321. Chou Y-T, Lin H-H, Lien Y-C, Wang Y-H, Hong C-F, Kao Y-R, Lin S-C, Chang Y-C, Lin S-Y, Chen S-J. EGFR promotes lung tumorigenesis by activating miR-7 through a Ras/ERK/Myc pathway that targets the Ets2 transcriptional repressor ERF. Cancer Res. 2010;70(21):8822–31.
322. Yin G, Chen R, Alvero AB, Fu H-H, Holmberg J, Glackin C, Rutherford T, Mor G. TWISTing stemness, inflammation and proliferation of epithelial ovarian cancer cells through MIR199A2/214. Oncogene. 2010;29(24):3545–53.
323. Ueda T, Volinia S, Okumura H, Shimizu M, Taccioli C, Rossi S, Alder H, C-g L, Oue N, Yasui W. Relation between microRNA expression and progression and prognosis of gastric cancer: a microRNA expression analysis. Lancet Oncol. 2010;11(2):136–46.
324. Xu X-L, Jiang Y-H, Feng J-G, Su D, Chen P-C, Mao W-M. MicroRNA-17, microRNA-18a, and microRNA-19a are prognostic indicators in esophageal squamous cell carcinoma. Ann Thorac Surg. 2014;97(3):1037–45.
325. Kong D, Piao Y-S, Yamashita S, Oshima H, Oguma K, Fushida S, Fujimura T, Minamoto T, Seno H, Yamada Y. Inflammation-induced repression of tumor suppressor miR-7 in gastric tumor cells. Oncogene. 2012;31(35):3949–60.

326. Chao CW, Chan DC, Kuo A, Leder P. The mouse formin (Fmn) gene: abundant circular RNA transcripts and gene-targeted deletion analysis. Mol Med. 1998;4(9):614.
327. Zhang Y, Yang L, Chen L-L. Life without A tail: new formats of long noncoding RNAs. Int J Biochem Cell Biol. 2014;54:338–49.
328. Hentze MW, Preiss T. Circular RNAs: splicing's enigma variations. EMBO J. 2013;32(7):923–5.
329. Elbashir SM, Harborth J, Lendeckel W, Yalcin A, Weber K, Tuschl T. Duplexes of 21-nucleotide RNAs mediate RNA interference in cultured mammalian cells. Nature. 2001;411(6836):494–8.
330. Xia H, Mao Q, Paulson HL, Davidson BL. siRNA-mediated gene silencing in vitro and in vivo. Nat Biotechnol. 2002;20(10):1006–10.
331. Petrocca F, Lieberman J. Promise and challenge of RNA interference–based therapy for cancer. J Clin Oncol. 2011;29(6):747–54.
332. Siomi H, Siomi MC. On the road to reading the RNA-interference code. Nature. 2009;457(7228):396–404.
333. Aravin AA, Lagos-Quintana M, Yalcin A, Zavolan M, Marks D, Snyder B, Gaasterland T, Meyer J, Tuschl T. The small RNA profile during Drosophila melanogaster development. Dev Cell. 2003;5(2):337–50.
334. Lau NC, Seto AG, Kim J, Kuramochi-Miyagawa S, Nakano T, Bartel DP, Kingston RE. Characterization of the piRNA complex from rat testes. Science. 2006;313(5785):363–7.
335. Kim VN. Small RNAs just got bigger: piwi-interacting RNAs (piRNAs) in mammalian testes. Genes Dev. 2006;20(15):1993–7.
336. Siddiqi S, Matushansky I. Piwis and piwi-interacting RNAs in the epigenetics of cancer. J Cell Biochem. 2012;113(2):373–80.
337. Brennecke J, Aravin AA, Stark A, Dus M, Kellis M, Sachidanandam R, Hannon GJ. Discrete small RNA-generating loci as master regulators of transposon activity in Drosophila. Cell. 2007;128(6):1089–103.
338. Houwing S, Kamminga LM, Berezikov E, Cronembold D, Girard A, Van Den Elst H, Filippov DV, Blaser H, Raz E, Moens CB. A role for Piwi and piRNAs in germ cell maintenance and transposon silencing in Zebrafish. Cell. 2007;129(1):69–82.
339. Cheng J, Guo JM, Xiao BX, Miao Y, Jiang Z, Zhou H, Li QN. piRNA, the new non-coding RNA, is aberrantly expressed in human cancer cells. Clin Chim Acta Int J Clin Chem. 2011;412(17–18):1621–5. doi:10.1016/j.cca.2011.05.015.
340. Cheng J, Deng H, Xiao B, Zhou H, Zhou F, Shen Z, Guo J. piR-823, a novel non-coding small RNA, demonstrates in vitro and in vivo tumor suppressive activity in human gastric cancer cells. Cancer Lett. 2012;315(1):12–7. doi:10.1016/j.canlet.2011.10.004.
341. Yan H, Wu QL, Sun CY, Ai LS, Deng J, Zhang L, Chen L, Chu ZB, Tang B, Wang K, Wu XF, Xu J, Hu Y. piRNA-823 contributes to tumorigenesis by regulating de novo DNA methylation and angiogenesis in multiple myeloma. Leukemia. 2015;29(1):196–206. doi:10.1038/leu.2014.135.
342. Chu H, Hui G, Yuan L, Shi D, Wang Y, Du M, Zhong D, Ma L, Tong N, Qin C, Yin C, Zhang Z, Wang M. Identification of novel piRNAs in bladder cancer. Cancer Lett. 2015;356(2 Pt B):561–7. doi:10.1016/j.canlet.2014.10.004.
343. Law PT, Qin H, Ching AK, Lai KP, Co NN, He M, Lung RW, Chan AW, Chan TF, Wong N. Deep sequencing of small RNA transcriptome reveals novel non-coding RNAs in hepatocellular carcinoma. J Hepatol. 2013;58(6):1165–73. doi:10.1016/j.jhep.2013.01.032.
344. Kwon C, Tak H, Rho M, Chang HR, Kim YH, Kim KT, Balch C, Lee EK, Nam S. Detection of PIWI and piRNAs in the mitochondria of mammalian cancer cells. Biochem Biophys Res Commun. 2014;446(1):218–23. doi:10.1016/j.bbrc.2014.02.112.
345. Ferreira HJ, Heyn H, Garcia del Muro X, Vidal A, Larriba S, Munoz C, Villanueva A, Esteller M. Epigenetic loss of the PIWI/piRNA machinery in human testicular tumorigenesis. Epigenetics. 2014;9(1):113–8. doi:10.4161/epi.27237.

346. Zhang H, Ren Y, Xu H, Pang D, Duan C, Liu C. The expression of stem cell protein Piwil2 and piR-932 in breast cancer. Surg Oncol. 2013;22(4):217–23. doi:10.1016/j.suronc.2013.07.001.
347. Hashim A, Rizzo F, Marchese G, Ravo M, Tarallo R, Nassa G, Giurato G, Santamaria G, Cordella A, Cantarella C, Weisz A. RNA sequencing identifies specific PIWI-interacting small non-coding RNA expression patterns in breast cancer. Oncotarget. 2014;5(20):9901–10.
348. Grewal SS. Why should cancer biologists care about tRNAs? tRNA synthesis, mRNA translation and the control of growth. Biochim Biophys Acta. 2015;1849(7):898–907. doi:10.1016/j.bbagrm.2014.12.005.
349. Kiss-László Z, Henry Y, Bachellerie J-P, Caizergues-Ferrer M, Kiss T. Site-specific ribose methylation of preribosomal RNA: a novel function for small nucleolar RNAs. Cell. 1996;85(7):1077–88.
350. Ganot P, Bortolin M-L, Kiss T. Site-specific pseudouridine formation in preribosomal RNA is guided by small nucleolar RNAs. Cell. 1997;89(5):799–809.
351. Kishore S, Stamm S. The snoRNA HBII-52 regulates alternative splicing of the serotonin receptor 2C. Science. 2006;311(5758):230–2.
352. Kawaji H, Nakamura M, Takahashi Y, Sandelin A, Katayama S, Fukuda S, Daub CO, Kai C, Kawai J, Yasuda J. Hidden layers of human small RNAs. BMC Genomics. 2008;9(1):157.
353. Thorenoor N, Slaby O. Small nucleolar RNAs functioning and potential roles in cancer. Tumor Biol. 2015;36(1):41–53.
354. Siprashvili Z, Webster DE, Johnston D, Shenoy RM, Ungewickell AJ, Bhaduri A, Flockhart R, Zarnegar BJ, Che Y, Meschi F, Puglisi JD, Khavari PA. The noncoding RNAs SNORD50A and SNORD50B bind K-Ras and are recurrently deleted in human cancer. Nat Genet. 2015. doi:10.1038/ng.3452.
355. Wan Y, Kertesz M, Spitale RC, Segal E, Chang HY. Understanding the transcriptome through RNA structure. Nat Rev Genet. 2011;12(9):641–55.
356. Maenner S, Blaud M, Fouillen L, Savoye A, Marchand V, Dubois A, Sanglier-Cianferani S, Van Dorsselaer A, Clerc P, Avner P. 2-D structure of the A region of Xist RNA and its implication for PRC2 association. PLoS Biol. 2010;8(1):301.
357. Bertani S, Sauer S, Bolotin E, Sauer F. The noncoding RNA Mistral activates Hoxa6 and Hoxa7 expression and stem cell differentiation by recruiting MLL1 to chromatin. Mol Cell. 2011;43(6):1040 6.
358. Wang KC, Yang YW, Liu B, Sanyal A, Corces-Zimmerman R, Chen Y, Lajoie BR, Protacio A, Flynn RA, Gupta RA. A long noncoding RNA maintains active chromatin to coordinate homeotic gene expression. Nature. 2011;472(7341):120–4.

Chapter 2
The Working Modules of Long Noncoding RNAs in Cancer Cells

Ling Li and Xu Song

Abstract It is clear that RNA is more than just a messenger between gene and protein. The mammalian genome is pervasively transcribed, giving rise to tens of thousands of noncoding transcripts, especially long noncoding RNAs (lncRNAs). Whether all of these large transcripts are functional remains to be elucidated, but it is evident that there are many lncRNAs that seem not to be the "noise" of the transcriptome. Recent studies have set out to decode the regulatory role and functional diversity of lncRNAs in human physiological and pathological processes, and accumulating evidence suggests that most of the functional lncRNAs achieve their biological functions by controlling gene expression. In this chapter, we will organize these studies to provide a detailed description of the involvement of lncRNAs in the major steps of gene expression that include epigenetic regulation, RNA transcription, posttranscriptional RNA processing, protein translation, and posttranslational protein modification and highlight the molecular mechanisms through which lncRNAs function, involving the interactions between lncRNAs and other biological macromolecules.

Keywords Long noncoding RNAs • Epigenome • Transcription • RNA spicing • Protein modification

2.1 Introduction

A large range of biological processes involved in cancer progression, such as cell differentiation, proliferation, apoptosis, and metastasis, are widely reported to be associated with long noncoding RNAs (lncRNAs), which are thought to work in cis on neighboring genes or in trans to regulate distantly located genes or molecular targets in the nucleus and cytoplasm. It has been clear that lncRNAs can function through quite diverse mechanisms and that their interaction with DNA, RNA, or

L. Li • X. Song (✉)
Sichuan University, 29 Wangjiang Road, Chengdu, Sichuan, China
e-mail: lingli1980@scu.edu.cn; xusong@scu.edu.cn

E. Song (ed.), *The Long and Short Non-coding RNAs in Cancer Biology*,
Advances in Experimental Medicine and Biology 927,
DOI 10.1007/978-981-10-1498-7_2

protein is a well-established action mode [1]. On the basis of the intermolecular interactions, most of the characterized lncRNAs are shown to function in gene expression control by acting as decoys, guides, or scaffolds [2, 3]. The "guiding" lncRNAs, like Kcnq1ot1 [4] and lincRNA-p21 [5], are associated with chromatin regulatory protein complexes or transcriptional co-regulators and recruit them to specific genomic DNA regions to regulate transcription; "decoying" lncRNAs, such as GAS5 [6], Lethe [7], and PANDA [8], mimic and compete with their consensus DNA-binding motifs for binding nuclear receptors or transcriptional factors in the nuclei; "scaffolding" lncRNAs, including HOTAIR [9], XIST [10], and NRON [11], bring specific regulatory proteins into proximity with each other to function as a unique complex. Furthermore, many lncRNAs are exclusively expressed in specific stages of tissue differentiation and development or present apparent cell-type-specific expression patterns and distinct subcellular localization [12, 13]. Although some of such lncRNAs are merely by-products of transcription that don't possess any regulatory function, they can faithfully reflect the action of gene expression or activation of signaling pathway and therefore act as "signaling" molecules [3].

Altogether, the current studies suggest that lncRNAs could be involved in almost each step of gene expression, such as epigenetic regulation, RNA transcription, posttranscriptional RNA processing, protein translation, and posttranslational protein modification, through in cis or in trans manner, and the dysregulation of lncRNAs could cause broad changes in cell signaling pathways.

2.2 Epigenetic Regulation

2.2.1 lncRNAs Involved in Histone Modification

Epigenetic regulation of gene expression, which is characterized as the altered transcription without any change in gene sequence, is generally reported to play roles in organism development as well as in tumorigenesis. Recently, the involvement of lncRNAs in epigenetic regulation has been widely documented. Kcnq1ot1, a nuclear-retained lncRNA with the length of 91.5 kb, is transcribed by RNA polymerase II (RNAP II) from the intron 10 of Kcnq1 gene in an antisense orientation to Kcnq1 [14]. Kcnq1ot1 was first found to be associated with the lineage-specific silencing of dozens of genes within the Kcnq1 locus [14, 15], and the subsequent studies further indicated that its silencing effect was achieved by the interactions with chromatin and with the H3K9- and H3K27-specific histone methyltransferases G9a and polycomb repressive complex 2 (PRC2) [4]. HOTAIR and Air are the other two well-characterized lncRNAs involved in chromatin remodeling, which exert function through a similar fashion, including accumulation at the chromatin regions of silenced genes and the subsequent mediation of repressive histone modification through recruiting specific histone modifiers such as G9a, PRC1, and PRC2 [9, 16–18].

2.2.2 lncRNAs Involved in DNA Methylation

Although interaction with histone modifiers is a major mechanism through which lncRNAs function in epigenetic regulation, some lncRNAs are also implicated in establishment and maintenance of DNA methylation patterns. DNA methylation is mediated by the members of the DNA methyltransferase (DNMT) family, conventionally classified as de novo (DNMT3a and DNMT3b) and maintenance (DNMT1) DNMTs. The ecCEBPA represents one of the lncRNAs that modulate DNA methylation by interacting with DNMT1, which functions as a decoying transcript to sequestrate DNMT1, resulting in prevention of CEBPA gene locus methylation [19]. The best characterized DNMT3a- and DNMT3b-binding lncRNA is Dum, which recruits both de novo methylation and maintenance DNMTs to silence its neighboring gene in cis [20].

Interestingly, some lncRNAs, especially the antisense lncRNAs, have been identified to simultaneously mediate DNA and histone modifications at the loci of silenced genes [21]. TMS1/ASC is a tumor suppressor gene that encodes a pro-apoptotic signaling factor operating in the intrinsic and extrinsic cell death pathways [22]. TMS1/ASC was originally identified as a downstream target of DNA methyltransferase-1 (DNMT1), and subsequent studies further showed that it was subjected to the hypermethylation-mediated epigenetic silencing in a wide range of human tumors [23–27]. In addition to DNA methylation, the inactivation of TMS1/ASC is also controlled by certain other epigenetic events such as the G9a-mediated histone H3K9 methylation [28], which is commonly linked to the methylation of nearby CpG sites [29, 30]. Biochemical interactions between DNA and histone methyltransferases were thought to provide a molecular explanation, at least in part, for the combinatorial pattern of DNA and histone modifications in chromatin [29–32]. In a recent study, an antisense lncRNA of TMS1/ASC, termed TMS1AS, was further revealed to regulate outputs of its sense counterpart by interacting with the DNMT1/G9a complex and aiding in recruitment of the complex to the sense promoter (see Fig. 2.1). This interesting finding not only highlighted the significant involvement of lncRNAs in epigenetic regulation of gene expression but also revealed a potential crosstalk between DNA methylation and histone modification established by lncRNA. Kcnq1ot1 is another example of such antisense lncRNAs. In addition to its role in regulating histone modification, Kcnq1ot1 was also reported to be required for the silencing of ubiquitously imprinted genes (UIGs) through guiding and maintaining the CpG methylation at methylated regions flanking the UIGs [33].

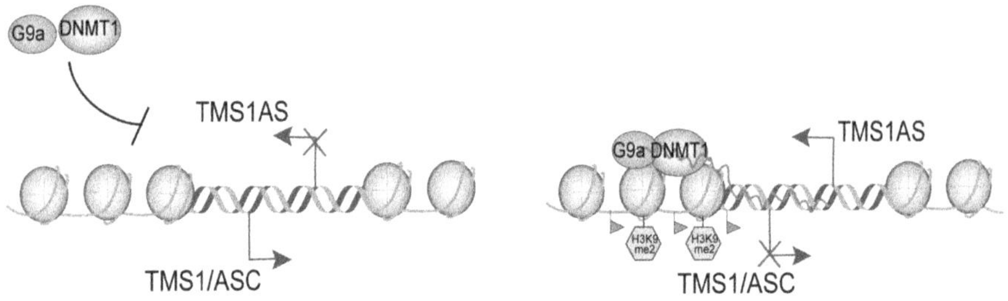

Fig. 2.1 TMS1AS regulates TMS1/ASC at the epigenetic level. Upon being transcribed, TMS1AS can act in cis to recruit the chromatin repressor proteins DNMT1 and G9a to TMS1/ASC promoter

2.2.3 Correlation Between the Sense and Antisense Transcripts

The antisense lncRNAs described above represent guiding transcripts that modulate their sense counterparts negatively. However, sense-antisense pairs are frequently revealed to express in a concordant manner [34, 35]. TARID is one of the antisense lncRNAs that positively correlate with their sense counterparts. The underlying mechanism implicates its simultaneous association with sense promoter and the regulator of DNA demethylation GADD45A, which in turn recruits thymine-DNA glycosylase along with members of TET family to induce sense promoter demethylation [36]. Thus, it can be speculated that although many antisense lncRNAs regulate their sense counterparts through a general mechanism serving as a genomic address label for specific epigenetic modification enzymes, they may exert completely opposite effect on gene expression on the basis of property of the associated enzyme partner.

Certainly, certain lncRNAs involved in epigenetic regulation also possess the ability to control multiple targets in addition to their sense counterparts. It has been speculated that lncRNAs recruit epigenetic modification complexes by binding to target sites through three mechanisms: tethering to its nascent transcription locus, directly hybridizing to genomic targets, or interacting with a DNA-binding protein [37].

2.3 RNA Transcription

Transcription is a tightly regulated process in eukaryotes. In addition to the well-known protein factors such as RNAP, general transcriptional factors, and gene-specific transcriptional factors, it is also suggested that ncRNAs, which include small ncRNAs and lncRNAs, exert regulatory functions in the complicated network to make gene expression more symphonic. Thus, a current central issue is to obtain a full understanding of the potential role of lncRNAs in regulated gene transcription programs, possibly through diverse mechanisms.

2.3.1 Role of NcRNAs in Controlling Initiation and Elongation of Transcription

The expression of protein-coding genes in mammalian genomes begins with the assembly of the preinitiation complex (PIC) that brings RNAP II to gene promoters. Following this step, U1 snRNA can induce transcriptional initiation by specifically binding to and stimulating TFIIH to phosphorylate the C-terminal domain of RNAP II [38]. However, after transcriptional initiation and promoter clearance, RNAP II is frequently paused near the transcription start site on numerous genes, and the regulation of RNAP II pause release has been recognized as a critical step in activation of gene transcription [39].

The transition of RNAP II to productive elongation requires active recruitment of P-TEFb, a cyclin-dependent kinase responsible for phosphorylation of the C-terminal domain of RNAP II and other key transcription elongation factors. The 7SK RNA is able to repress transcription elongation by, in combination with HEXIM1/2, forming an inactive complex that sequesters P-TEFb and then prevents its active recruitment [40, 41]. The SR-splicing factor SRSF2 is a newly identified component of 7SK complex assembled at gene promoters. Upon its binding to promoter-associated nascent RNA, SRSF2 can mediate the switch of P-TEFb from the 7SK complex to RNAP II, making the paused transcription elongation reactive [42].

2.3.2 lncRNAs Regulate Transcription of Specific Genes

Through recruiting and modulating the activities of co-regulators, lncRNAs may act as selective ligands to prevent the transcription of target genes. DNA damage has been reported to induce the production of several lncRNAs from the 5' regulatory region of cyclin D1 (CCND1). These induced lncRNAs specifically bind and allosterically modify TLS, a regulatory sensor of DNA damage, leading to the interaction of the modified TLS with CREB-binding protein (CBP) that thereby inhibits the transcription of CCND1 [43]. Martianov et al. have reported another example of inhibitory lncRNA that is induced by serum starvation and functions as a promoter-specific transcriptional repressor. Upon being transcribed from the upstream minor promoter of DHFR gene, this lncRNA not only bonds transcriptional factor II B (TFIIB) to prevent its association with the major promoter but also formed a stable complex with the major promoter that interfered with the promoter-directed recruitment of TFIIB [44].

In addition to the inhibitory effects on transcription, lncRNAs can also act as activators or coactivators to upregulate gene expression. For instance, steroid receptor RNA activator (SRA) is an lncRNA that acts as a eukaryotic transcriptional coactivator for steroid hormone receptors [45]. Another lncRNA, called Evf-2, which is transcribed from one of the two Dlx-5/6 conserved intergenic regions,

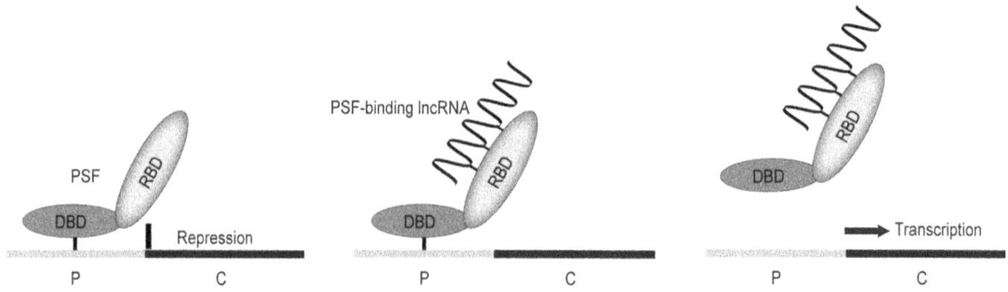

Fig. 2.2 Regulation of gene transcription by PSF protein and PSF-binding lncRNAs. The first diagram on the *left* shows the PSF binding to the promoter (*P*) of a gene via the DNA-binding domain (*DBD*), which represses transcription of the coding region (*C*). The second diagram in the *center* shows the binding of an lncRNA to the RNA-binding domain (*RBD*) of PSF. The third diagram on the *right* shows the release of PSF from the promoter and initiation of transcription

activates transcriptional activity of homeodomain proteins by directly influencing Dlx-2 activity [46]. The functional importance of gene enhancers in regulated gene expression has been well established, and the subsequent identification of bidirectional lncRNAs transcribed from enhancers, termed as enhancer RNAs (eRNAs), adds another functional layer to the transcriptional regulatory elements [47–49]. A subtype of eRNAs, which are derived from enhancers adjacent to E2-upregulated coding genes, has been observed to contribute to E2-dependent gene activation by stabilizing E2-/ERα-/eRNA-induced enhancer-promoter looping, suggesting that eRNAs are not merely a reflection of enhancer activation but a functional player [50].

Based on the ability to repress proto-oncogenes, PSF functions as a tumor suppressor protein, whose oncogenesis suppression activity, however, is impaired by the enhancement of its RNA-binding ability [51, 52]. The subsequent studies identify several PSF-binding lncRNAs from mice and humans, which include retroposon-derived lncRNAs such as VL30 as well as frame-disrupted and tumorigenesis-related noncoding transcripts such as MALAT1. All the lncRNAs promote tumorigenesis in mice and humans through a mechanism of reversible regulation on proto-oncogene transcription, including the protein PSF that binds to the regulatory region of a proto-oncogene and represses transcription and PSF-binding lncRNA that binds to PSF, forming a PSF-lncRNA complex that dissociates from a proto-oncogene, activating transcription [53–57] (see Fig. 2.2).

2.3.3 lncRNAs Join P53 Network by Regulating Transcription

P53 is considered to be one of the most common denominators in human cancer that plays a central role in regulatory networks responsible for cancer-related stress [58–60]. It seems controversial to explain p53 pathway only from the perspective of protein-coding genes [61–63]. As expected, a large number of miRNAs and conserved miRNA families are found to be direct transcriptional targets of p53 that mediate the p53 function or serve as the upstream regulators of p53 [64]. In addition to the small ncRNAs, several studies further illustrate the linkage between lncRNAs and p53 pathway from the perspective of molecular biology. Among these lncRNAs, p53-activated lincRNA-p21 serves as a repressor, through interacting with heterogeneous nuclear ribonucleoprotein K (hnRNP-K), in p53-dependent transcriptional response. Inhibition of lincRNA-p21 results in a global change in expression of hundreds of gene targets, a majority of which are downstream targets repressed by p53 and responsible for p53-mediated apoptosis [5]. PANDA RNA is another downstream target of p53. Upon being induced by p53, PANDA RNA acts as a decoy through an interaction with the transcriptional factor NF-YA that prevents NF-YA from activating expression of pro-apoptosis genes, leading to cell cycle arrest [8].

2.3.4 Circular Intronic RNA (ciRNA)

During the process of pre-mRNA splicing, the 5' splice site within intron sequence undergoes a nucleophilic attack by the downstream branch point to form a circular structure called the lariat, and the resultant free 5' exon then attacks the 3' splice site within intron, leading to the joint of two exons and the release of the intron lariat. In general, the intron lariats are debranched and degraded rapidly. However, a debranching failure, which seems to be determined by the presence of specific elements near the 5' splice site and the branch point site, can result in the formation of the circular intronic RNAs, referred to as ciRNAs [65–67].

ciRNAs are prominently found in the nucleus and are often more stable than their parent linear mRNAs [67]. Once generated, ciRNAs seem to function in cis to regulate the transcription of their parent genes. For example, a relatively abundant ciRNA called ci-ankrd52 was found to interact with elongating RNAP II and facilitate transcription. Many ciRNAs remain at their "sites of synthesis" in the nucleus. However, a portion of ciRNAs do localize to additional sites in the nucleus, suggesting that they may have certain transacting effects other than regulating the transcription of their parent genes [67].

2.4 Posttranscriptional RNA Processing

2.4.1 Alternative Splicing

For all eukaryotes, one of the most versatile processing steps in the life of an mRNA molecule is pre-mRNA splicing, through which the noncoding introns are removed and the neighboring exons are ligated together before protein translation. Although the spliceosome functions in splicing catalysis, some additional transacting protein factors, such as the serine-/arginine-rich (SR) nuclear proteins (SR proteins), the SR-related proteins, the small nuclear ribonucleoproteins (snRNPs), and the heterogeneous nuclear ribonucleoproteins (hnRNPs), are well characterized to be required for the choice of splice sites, where spliceosome assembles and the following splicing occurs [68–70]. More than 90 % of human multi-exon-containing genes are speculated to be alternatively spliced in tissue- and cell-specific manners [71, 72], and in some spectacular examples, thousands of distinct gene products (protein isoforms) can be generated from a single gene [73], pointing to alternative pre-mRNA splicing as a common mechanism in regulation and diversification of gene function. Interestingly, several recent studies have highlighted the significant implication of lncRNAs in the regulation of pre-mRNA splicing.

2.4.1.1 MALAT1

Metastasis-associated lung adenocarcinoma transcript 1 (MALAT1), also known as nuclear-enriched autosomal transcript 2 (NEAT2), was originally identified as a prognostic marker in non-small cell lung cancer [74], and the subsequent studies showed that it was also overexpressed in many other human cancers [75–79].

This nuclear-enriched noncoding transcript predominantly localizes to nuclear speckles, the highly dynamic nuclear subdomains that are enriched with pre-mRNA splicing factors and are thought to serve as sites for the assembly, modification, and/or storage of the pre-mRNA processing machinery [80, 81]. Another nuclear-enriched autosomal noncoding transcript, referred to as NEAT1, has been shown to be significantly implicated in the structural maintenance of paraspeckles, the nuclear subdomains that control mRNA nuclear export [82–85]. MALAT1, however, does not possess a similar capacity to establish or maintain the functional nuclear subdomains since its depletion does not disrupt the architecture of nuclear speckles [86]. Nevertheless, MALAT1 is required for proper localization of several SR proteins to nuclear speckles [86], and its depletion has been reported to compromise the recruitment of SR proteins from nuclear speckles to the sites of transcription, where splicing occurs [87]. Furthermore, it has been demonstrated that MALAT1 can act as a "molecular sponge" by interacting with SR proteins, especially SRSF1, and thereby modulate the concentration of "splicing-competent" SR proteins in cells [86]. Altogether, these studies suggest that MALAT1 regulates pre-mRNA splicing by influencing the localization and/or activity of SR proteins. This "sponge" mechanism is also applicable to the untranslated region (UTR) of certain protein-coding genes. For instance, the repeat-containing RNA from the CTG expansion in 3'UTR of DMPK gene accumulates in the nucleus and affects the activity of splicing factors; similarly, CGG repeats occurred in FMR1 gene can recruit a set of splicing regulators into nuclear inclusions [88].

It has been suggested that the MALAT1enrichment in nuclear speckles occurs only when RNAP II-dependent transcription is active [87]. Since nuclear speckles do not represent the major sites of transcription or splicing, this finding seems contrary to the recent studies showing that MALAT1 binds many nascent pre-mRNAs derived from actively transcribed gene loci [89, 90]. Although the subcellular location of MALAT1 remains puzzling, these studies highlight the significant involvement of lncRNAs in the control of pre-mRNA splicing and the potential contribution of lncRNAs to the links between gene transcription and pre-mRNA splicing.

MALAT1 has a strong influence on pre-mRNA splicing patterns in human HeLa and fibroblast cells [86, 91, 92]. However, significant splicing changes were not observed when MALAT1 gene was knocked out in human lung cancer cells or in MALAT1 knockout mice [93, 94]. Although some other aspects of gene expression such as transcription have been shown to be affected in these cases, these studies still implicate that human and mouse genomes may encode certain functionally redundant products that possess the ability to regulate pre-mRNA splicing and compensate the loss of MALAT1. Certainly, we cannot exclude the possibility that MALAT1 only exerts the regulatory function in pre-mRNA splicing under a particular cell state or in specific cellular contexts.

2.4.1.2 sno-lncRNAs

The sno-lncRNAs (snoRNA-related long noncoding RNAs), a class of newly identified nuclear-enriched lncRNAs, are derived from the intron sequences and terminate in either box C/D or box H/ACA small nucleolar RNA (snoRNA) structures at their 5' and 3' ends [95]. At least 19 endogenous sno-lncRNAs have been identified in human, rhesus monkey, and mouse [96]. Most of them have been demonstrated to present apparent tissue- and species-specific expression patterns. Furthermore, their terminal snoRNA structures, but not the internal sequences, are highly conserved across species evolution.

The sno-lncRNAs encoded by the q11-q13 region of human chromosome 15 have been well characterized [95]. During exonucleolytic trimming, the terminal snoRNA structures protect the internal sequences between them from degradation, leading to the accumulation of these sno-lncRNAs to high levels (expression similar to that of some histone mRNAs). Although sno-lncRNAs and snoRNAs are processed by the same machinery and both of them are generally originated from the introns of protein-coding genes, sno-lncRNAs mainly accumulate near their sites of synthesis but do not co-localize with nucleoli or Cajal bodies, where snoRNAs reside [95, 97]. The difference in their cellular distribution strongly indicates that these sno-lncRNAs do not have a similar function as snoRNAs, which play roles in the modification of other noncoding RNAs including rRNAs and snRNAs. Indeed, these sno-lncRNAs strongly associate with RbFox2, a member of Fox family splicing regulators that is known to regulate many posttranscriptional events including alternative splicing. Moreover, these sno-lncRNAs can significantly influence the RbFox2-regulated splicing events through acting, at least in part, as molecular sinks that prevent RbFox2 from targeting its mRNA targets [95].

2.4.2 lncRNAs Function as miRNA Sponges

On the basis of direct base pairing to target sites within untranslated regions of mRNAs, the small regulatory ncRNAs such as miRNAs and endo-siRNAs can function as important posttranscriptional regulators that trigger mRNA degradation or translational inhibition. Since lncRNAs achieve their functions through employing quite diverse strategies, they are generally considered to be a class of regulatory RNAs fundamentally distinct from the small RNAs. Nevertheless, current studies have revealed that lncRNAs can serve as the precursors of certain small ncRNAs, such as miR-675 derived from the H19 lncRNA [98, 99]. Furthermore, the link between lncRNA and miRNA has been presented by the finding that several lncRNAs possess the ability to affect the miRNA activity, which is attributed to the internal miRNA-matching sequence. For instance, in human hepatocellular carcinoma metastases, the lncRNA activated by TGF-β, termed as lncRNA-ATB, was reported to act as a sponge for miR-200 family. The sponge effect was mediated by the selectively conserved miRNA target sites contained in lncRNA-ATB, which strongly bond miR-200 s to suppress their activities, resulting in increased levels of ZEB1 and ZEB2, two miR-200 s targets, and the ultimate induction of epithelial-mesenchymal transition (EMT) and invasion [100]. Moreover, identification of other sponge lncRNAs, such as tumor-suppressive PTEN competitive endogenous RNAs (ceRNAs) and muscle-specific linc-MD1 that targets miR-133 and miR-135, suggests that the lncRNA-mediated miRNA sponge effects can be a general phenomenon in diseases as well as in normal physiology [101–106].

In addition to the canonical linear mRNAs, it is now clear that thousands of protein-coding genes can also be processed to generate circular RNAs (circRNAs), which are resistant to exonuclease-mediated degradation due to the fact that they do not have 5' and 3' ends and the two ends have been joined together [107–113]. Distinct from ciRNAs that are derived from introns and reside in the nuclei, circRNAs are almost exclusively encoded by exons and predominantly localize in the cytoplasm. circRNAs are produced by noncanonical splicing events called "backsplicing," in which a splice donor site is jointed to a splice acceptor site further upstream in the primary transcript [111]. The strategy employed by circRNAs to exit the nucleus remains to be elucidated. Since the nuclear envelope breaks down during mitosis, it seems reasonable to hypothesize that circRNAs exit the nucleus during this phase of the cell cycle. However, expression of certain circRNAs such as ciRS-7 is also detected in neuronal tissues, where mitotic division occurs at a low frequency [110, 114].

CiRS-7 and Sry are the two well-characterized circRNAs [110, 114]. These particular circRNAs contain many binding sites for specific miRNAs (miR-7 and miR-138, respectively). Following their accumulation in the cytoplasm, they can act as sponges that titrate the miRNAs from their other RNA targets, thereby modulating the miRNA activities. Although current attention has been focused on circRNA's sponge effects, several lines of evidence also suggest their other functional possibilities. For instance, expression of ciRS-7 but not miR-7 has been detected in some areas of the mouse adult hippocampus, suggesting that ciRS-7 may have certain roles other than interacting with the miRNA [110]. Furthermore, most other circRNAs have been identified to contain few miRNA binding sites, indicating that they may achieve their functions through other strategies such as binding to RNA-binding proteins to form RNA-protein complexes [108, 113].

2.5 Protein Translation

lincRNA-RoR is one of the few lncRNAs whose detailed function in the regulation of p53 pathway has been characterized. Distinct from the p53-regulated lincRNA-p21 and PANDA, lincRNA-RoR acts as an upstream regulator of p53 in response to DNA damage. MDM2 is a well-known upstream negative p53 regulator, which causes p53 degradation through the ubiquitin-proteasome pathway. LincRNA-RoR, however, has been demonstrated to function via a different mechanism, involving its direct interaction with hnRNP-I that suppresses p53 translation and inhibits p53-mediated cell cycle arrest and apoptosis [115].

As described above, lincRNA-p21 was initially revealed to be induced by p53 during DNA damage and recruit hnRNP-K via physical interaction to facilitate p53-mediated repression of gene transcription [5]. Interestingly, the subsequent study showed that it possessed an additional role in translational control, the output of which was determined by the presence or absence of RNA-binding protein HuR [116]. In the presence of HuR, the association of HuR with lincRNA-p21 facilitates the recruitment of let-7/Ago2 to lincRNA-p21, leading to lower lincRNA-p21 stability. On the contrary, lincRNA-p21 is stable and accumulates when HuR is absent. It then interacts with the mRNAs CTNNB1 and JUNB and translational repressor Rck, repressing the translation of the targeted mRNAs.

The lncRNA-mRNA gene pairs have been found to be prevalent in mammalian genomes, and the cluster of natural antisense transcripts (NATs) is revealed to constitute a surprisingly large fraction of lncRNAs [34, 35]. Unlike the nuclear-retained NATs such as Tsix, Airn, and HOTAIR that guide the epigenetic modification complexes to the target loci, some NATs can form RNA duplexes with the mRNA of paired gene, leading to the change in mRNA translation. Zeb2/Sip1 is a transcriptional repressor of E-cadherin. After the Snail1-induced EMT, expression of the Zeb2/Sip1 NAT leads to an increase in Zeb2/Sip1 protein level without any change at the mRNA level. The Zeb2/Sip1 NAT is complementary to the 5' splice site of an intron in the 5' UTR of Zeb2 mRNA. Expression of the Zeb2/Sip1 NAT upon EMT can mask the spice site, preventing deletion of the intron. As a result, the translation machinery can then recognize and bind to an internal ribosome entry site (IRES) in the retained intron, resulting in more efficient Zeb2/Sip1 translation [117]. The Uchl1 NAT is another antisense lncRNA that promotes translation of the paired mRNA. The Uchl1 contains a region that overlaps with the first 73 nucleotides of Uchl1 mRNA. Under stress conditions in which cap-dependent translation is inhibited, the Uchl1 NAT moves from the nucleus to the cytoplasm and hybridizes

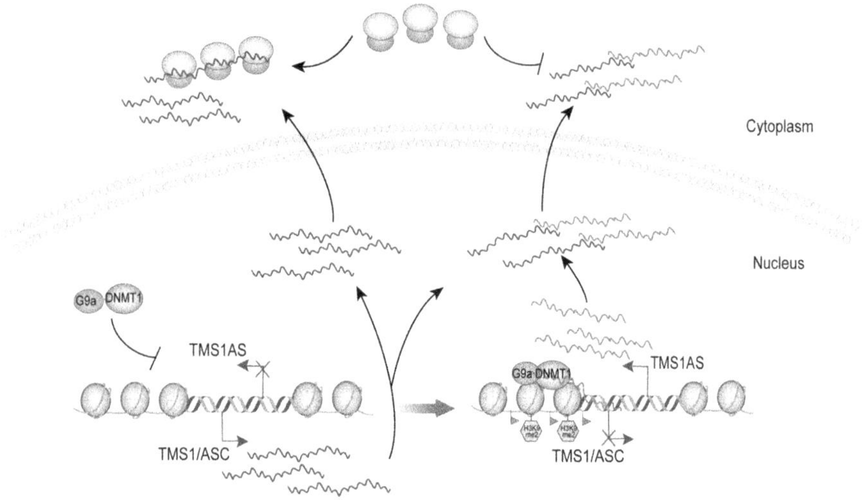

Fig. 2.3 Proposed model of TMS1AS-mediated regulation of TMS1/ASC expression. TMS1AS is transcribed from the opposite strand of TMS1/ASC gene. Following transcription, a fraction of TMS1AS transcripts act in cis by recruiting the chromatin repressor proteins DNMT1 and G9a to TMS1/ASC promoter, and other transcripts interact with TMS1/ASC mRNA through direct base pairing, forming intermolecular duplexes that suppress TMS1/ASC translation by interfering with the ribosome assembly on TMS1/ASC mRNA. This novel mechanism results in a target-specific regulation of TMS1/ASC at both epigenetic and translational levels by its antisense counterpart

with Uchl1 mRNA to switch on its cap-independent translation; that is, the Uchl1 NAT acts as an internal ribosomal entry element to promote selective translation [118]. Although the above RNA duplexes work in promoting mRNA translation, the RNA-RNA pairing between TMS1AS and TMS1/ASC mRNA has been demonstrated to suppress TMS1/ASC translation by interfering with the ribosome assembly on TMS1/ASC mRNA. As described above, TMS1AS also regulates TMS1/ASC expression at the epigenetic level. Thus, this NAT links different effector mechanisms to simultaneously operate in the different aspects of TMS1/ASC regulation, contributing to establishing a more strict fashion of gene expression and enhancing the efficacy of gene expression control (see Fig. 2.3).

2.6 Posttranslational Protein Modification

Upon being synthesized by ribosomes, the nascent polypeptide chains undergo posttranslational modification (PTM) to form the mature protein product. In addition, PTM is also an important strategy employed to control protein activity. The initial evidence for a function of lncRNAs in PTM comes from the research on the lncRNA-mediated epigenetic gene silencing. As described above, lncRNAs have been associated with gene silencing through guiding enzymes involved in chromatin remodeling, such as PRC2 and G9a, to specific genomic DNA regions. In other words, lncRNAs can function in the recruitment of histone posttranslational modification machinery.

The function of most lncRNAs depends on their ability to interact with proteins, implying that lncRNAs may also directly interact with functional domains of signaling proteins and thus regulate signal transduction. NF-κB is a critical link between inflammation and cancer, and its aberrant activation has been observed in many tumors. Despite the classical negative regulators that include IκB, an lncRNA termed NKILA has been recently reported to play a significant role in regulating NF-κB signaling and repressing cancer-associated inflammation. NKILA binds to NF-κB/ IκB complex and inhibits NF-κB signaling by masking the phosphorylation sites of IκB and stabilizing the complex. Importantly, NKILA expression is significantly decreased in many breast cancers and is associated with cancer metastasis and poor patient prognosis [119]. This action mode is also applicable to lnc-DC. This transcript has been reported to regulate STAT3 signal transduction by interacting with STAT3 in the cytoplasm of dendritic cells and modulating its phosphorylation [120].

2.7 Summary and Perspectives

The discovery that lncRNAs are central to numerous pivotal biological processes may reflect ancient connections between lncRNAs and the regulation of developmental and physiological decisions, whose disruption can lead to physiological disorders as well as many types of diseases. Comparison of gene expression profiles of tumor and normal cells has revealed a linkage of lncRNAs with tumorigenesis. Most of the oncogenic and tumor suppressor lncRNAs are characterized to function in gene expression control, and their interaction with DNA, RNA, or protein is a well-established action mode [1]. However, there are only limited studies decoding the detailed molecular mechanism of these cancer-related lncRNAs. It has long been acknowledged that lncRNAs contain functionally redundant sequences. Nevertheless, their core functionality relies heavily on the cooperative action of their dispersed functional domains [121]. Similar to miRNAs, lncRNAs can also serve as the potential molecular targets for diagnosis and treatment of cancer. Thus, it is important to decode the molecular features of lncRNAs, which include the consensus motif and structural element that determine the intermolecular interactions, to help understand the involvement and significance of lncRNAs in tumor biology.

References

1. Guttman M, Rinn JL. Modular regulatory principles of large non-coding RNAs. Nature. 2012;482:339–46.
2. Ulitsky I, Bartel DP. LincRNAs: genomics, evolution, and mechanisms. Cell. 2013;154:26–46.
3. Wang KC, Chang HY. Molecular mechanisms of long non-coding RNAs. Mol Cell. 2011;43:904–14.
4. Pandey RR, Mondal T, Katayama F, et al. Kcnq1ot1 antisense non-coding RNA mediates lineage-specific transcriptional silencing through chromatin-level regulation. Mol Cell. 2008;32:232–46.
5. Huarte M, Guttman M, Feldser D, et al. A large intergenic non-coding RNA induced by p53 mediates global gene repression in the p53 response. Cell. 2010;142:409–19.
6. Kino T, Hurt DE, Ichijo T, et al. Non-coding RNA gas5 is a growth arrest- and starvation-associated repressor of the glucocorticoid receptor. Sci Signal. 2010;3:ra8.
7. Rapicavoli NA, Qu K, Zhang J, et al. A mammalian pseudogene lncRNA at the interface of inflammation and anti-inflammatory therapeutics. eLife. 2013;2:e00762.
8. Hung T, Wang Y, Lin MF, et al. Extensive and coordinated transcription of non-coding RNAs within cell-cycle promoters. Nat Genet. 2011;43:621–9.
9. Tsai MC, Manor O, Wan Y, et al. Long non-coding RNA as modular scaffold of histone modification complexes. Science. 2010;329:689–93.
10. Jeon Y, Lee JT. YY1 tethers Xist RNA to the inactive X nucleation center. Cell. 2011;146:119–33.
11. Sharma S, Findlay GM, Bandukwala HS, et al. Dephosphorylation of the nuclear factor of activated T cells (NFAT) transcription factor is regulated by an RNA-protein scaffold complex. Proc Natl Acad Sci U S A. 2011;108:11381–6.
12. Guttman M, Amit I, Garber M, et al. Chromatin signature reveals over a thousand highly conserved large non-coding RNAs in mammals. Nature. 2009;458:223–7.
13. Ulitsky I, Shkumatava A, Jan CH, et al. Conserved function of lincRNAs in vertebrate embryonic development despite rapid sequence evolution. Cell. 2011;147:1537–50.
14. Mancini-Dinardo D, Steele SJ, Levorse JM, et al. Elongation of the Kcnq1ot1 transcript is required for genomic imprinting of neighboring genes. Genes Dev. 2006;20(10):1268–82.
15. Fitzpatrick GV, Soloway PD, Higgins MJ. Regional loss of imprinting and growth deficiency in mice with a targeted deletion of KvDMR1. Nat Genet. 2002;32(3):426–31.
16. Gupta RA, Shah N, Wang KC, et al. Long non-coding RNA HOTAIR reprograms chromatin state to promote cancer metastasis. Nature. 2010;464(7291):1071–6.
17. Nagano T, Mitchell JA, Sanz LA, et al. The Air non-coding RNA epigenetically silences transcription by targeting G9a to chromatin. Science. 2008;322(5908):1717–20.
18. Rinn JL, Kertesz M, Wang JK, et al. Functional demarcation of active and silent chromatin domains in human HOX loci by non-coding RNAs. Cell. 2007;129(7):1311–23.
19. Di Ruscio A, Ebralidze AK, Benoukraf T, et al. DNMT1-interacting RNAs block gene-specific DNA methylation. Nature. 2013;503(7476):371–6.
20. Wang L, Zhao Y, Bao X, et al. LncRNA Dum interacts with Dnmts to regulate Dppa2 expression during myogenic differentiation and muscle regeneration. Cell Res. 2015;25(3):335–50.
21. Wutz A. RNA-mediated silencing mechanisms in mammalian cells. Prog Mol Biol Transl Sci. 2011;101:351–76.
22. McConnell BB, Vertino PM. TMS1/ASC: the cancer connection. Apoptosis. 2004;9:5–18.
23. Conway KE, McConnell BB, Bowring CE, et al. TMS1, a novel proapoptotic caspase recruitment domain protein, is a target of methylation-induced gene silencing in human breast cancers. Cancer Res. 2000;60(22):6236–42.

24. Moriai R, Tsuji N, Kobayashi D, et al. A proapoptotic caspase recruitment domain protein gene, TMS1, is hypermethylated in human breast and gastric cancers. Anticancer Res. 2002;22(6C):4163–8.
25. Stone AR, Bobo W, Brat DJ, et al. Aberrant methylation and down-regulation of TMS1/ASC in human glioblastoma. Am J Pathol. 2004;165(4):1151–61.
26. Virmani A, Rathi A, Sugio K, et al. Aberrant methylation of TMS1 in small cell, non small cell lung cancer and breast cancer. Int J Cancer. 2003;106(2):198–204.
27. Yokoyama T, Sagara J, Guan X, et al. Methylation of ASC/TMS1, a proapoptotic gene responsible for activating procaspase-1, in human colorectal cancer. Cancer Lett. 2003;202(1):101–8.
28. Kapoor-Vazirani P, Kagey JD, Powell DR, Vertino PM. Role of hMOF-dependent histone H4 lysine 16 acetylation in the maintenance of TMS1/ASC gene activity. Cancer Res. 2008;68(16):6810–21.
29. Chang Y, Sun L, Kokura K, et al. MPP8 mediates the interactions between DNA methyltransferase Dnmt3a and H3K9 methyltransferase GLP/G9a. Nat Commun. 2011;2:533.
30. Hashimoto H, Vertino PM, Cheng X. Molecular coupling of DNA methylation and histone methylation. Epigenomics. 2010;2(5):657–69.
31. Cedar H, Bergman Y. Linking DNA methylation and histone modification: patterns and paradigms. Nat Rev Genet. 2009;10(5):295–304.
32. Estève PO, Chin HG, Smallwood A, et al. Direct interaction between DNMT1 and G9a coordinates DNA and histone methylation during replication. Genes Dev. 2006;20(22):3089–103.
33. Mohammad F, Pandey GK, Mondal T, et al. Long non-coding RNA-mediated maintenance of DNA methylation and transcriptional gene silencing. Development. 2012;139(15):2792–803.
34. Katayama S, Tomaru Y, Kasukawa T, et al. Antisense transcription in the mammalian transcriptome. Science. 2005;309(5740):1564–6.
35. Sigova AA, Mullen AC, Molinie B, et al. Divergent transcription of long non-coding RNA/mRNA gene pairs in embryonic stem cells. Proc Nat Acad Sci USA. 2013;110(8):2876–81.
36. Arab K, Park YJ, Lindroth AM, et al. Long non-coding RNA TARID directs demethylation and activation of the tumor suppressor TCF21 via GADD45A. Mol Cell. 2014;55(4):604–14.
37. Quinodoz S, Guttman M. Long non-coding RNAs: an emerging link between gene regulation and nuclear organization. Trends Cell Biol. 2014;24(11):651–63.
38. Kwek KY, Murphy S, Furger A, et al. U1 snRNA associates with TFIIH and regulates transcriptional initiation. Nat Struct Biol. 2002;9(11):800–5.
39. Adelman K, Lis JT. Promoter-proximal pausing of RNA polymerase II: emerging roles in metazoans. Nat Rev Genet. 2012;13(10):720–31.
40. Peterlin BM, Price DH. Controlling the elongation phase of transcription with P-TEFb. Mol Cell. 2006;23(3):297–305.
41. Yik JH, Chen R, Nishimura R, et al. Inhibition of P-TEFb (CDK9/Cyclin T) kinase and RNA polymerase II transcription by the coordinated actions of HEXIM1 and 7SK snRNA. Mol Cell. 2003;12(4):971–82.
42. Ji X, Zhou Y, Pandit S, et al. SR proteins collaborate with 7SK and promoter-associated nascent RNA to release paused polymerase. Cell. 2013;153(4):855–68.
43. Wang ET, Sandberg R, Luo S, et al. Alternative isoform regulation in human tissue transcriptomes. Nature. 2008;456(7221):470–6.
44. Martianov I, Ramadass A, Barros S, et al. Repression of the human dihydrofolate reductase gene by a non-coding interfering transcript. Nature. 2007;445:666–70.
45. Lanz RB, Razani B, Goldberg AD, O'Malley BW. Distinct RNA motifs are important for coactivation of steroid hormone receptors by steroid receptor RNA activator (SRA). Proc Natl Acad Sci U S A. 2002;99:16081–6.

46. Feng J, Bi C, Clark BS, et al. The Evf-2 non-coding RNA is transcribed from the Dlx-5/6 ultraconserved region and functions as a Dlx-2 transcriptional coactivator. Genes Dev. 2006;20:1470–84.
47. Hah N, Danko CG, Core L, et al. A rapid, extensive, and transient transcriptional response to estrogen signaling in breast cancer cells. Cell. 2011;145:622–34.
48. Kim TK, Hemberg M, Gray JM, et al. Widespread transcription at neuronal activity-regulated enhancers. Nature. 2010;465:182–7.
49. Wang D, Garcia-Bassets I, Benner C, et al. Reprogramming transcription by distinct classes of enhancers functionally defined by eRNA. Nature. 2011;474:390–4.
50. Li W, Notani D, Ma Q, et al. Functional roles of enhancer RNAs for oestrogen-dependent transcriptional activation. Nature. 2013;498:516–20.
51. Figueroa A, Fujita Y, Gorospe M. Hacking RNA: Hakai promotes tumorigenesis by enhancing the RNA-binding function of PSF. Cell Cycle. 2009;8:3648–51.
52. Galietta A, Gunby RH, Redaelli S, et al. NPM/ALK binds and phosphorylates the RNA/DNA-binding protein PSF in anaplastic large-cell lymphoma. Blood. 2007;100:2600–9.
53. Song X, Wang B, Bromberg M, et al. Retroviral-mediated transmission of a mouse VL30 RNA to human melanoma cells promotes metastasis in an immunodeficient mouse model. Proc Natl Acad Sci U S A. 2002;99:6269–73.
54. Song X, Sui A, Garen A. Binding of mouse VL30 retrotransposon RNA to PSF protein induces genes repressed by PSF: effects on steroidogenesis and oncogenesis. Proc Natl Acad Sci U S A. 2004;101:621–6.
55. Song X, Sun Y, Garen A. Roles of PSF protein and VL30 RNA in reversible gene regulation. Proc Natl Acad Sci U S A. 2005;102:12189–93.
56. Li L, Feng T, Lian Y. Role of human non-coding RNAs in the control of tumorigenesis. Proc Natl Acad Sci U S A. 2009;106:16794–8.
57. Wang G, Cui Y, Zhang G, et al. Regulation of proto-oncogene transcription, cell proliferation, and tumorigenesis in mice by PSF protein and a VL30 non-coding RNA. Proc Natl Acad Sci U S A. 2009;106:16794–8.
58. Lane DP, Fischer PM. Turning the key on p53. Nature. 2004;427:789–90.
59. Lowe SW, Cepero E, Evan G. Intrinsic tumour suppression. Nature. 2004;432:307–15.
60. Vogelstein BB, Lane DD, Levine AA. Surfing the p53 network. Nature. 2000;408:307–10.
61. Brugarolas J, Chandrasekaran C, Gordon JI, et al. Radiation-induced cell cycle arrest compromised by p21 deficiency. Nature. 1995;337:552–7.
62. Yu J, Zhang L, Hwang PM, et al. Identification and classification of p53-regulated genes. Proc Natl Acad Sci U S A. 1996;96:14517–22.
63. Zhao R, Gish K, Mruphy M, et al. Analysis of p53-regualted gene expression patterns using oligonucleotide arrays. Genes Dev. 2000;14:981–93.
64. Suzuki HI, Miyazono K. Dynamics of microRNA biogenesis: crosstalk between p53 network and microRNA processing pathway. J Mol Med (Berl). 2010;88:1085–94.
65. Gardner EJ, Nizami ZF, Talbot Jr CC, Gall JG. Stable intronic sequence RNA (sisRNA), a new class of non-coding RNA from the oocyte nucleus of Xenopus tropicalis. Genes Dev. 2012;26:2550–9.
66. Qian L, Vu MN, Carter M, Wilkinson MF. A spliced intron accumulates as a lariat in the nucleus of T cells. Nucleic Acids Res. 1992;20:5345–50.
67. Zhang Y, Zhang XO, Chen T, et al. Circular intronic long non-coding RNAs. Mol Cell. 2013;51:792–806.
68. Blencowe BJ. Alternative splicing: new insights from global analyses. Cell. 2006;126(1):37–47.
69. Lin S, Fu XD. SR proteins and related factors in alternative splicing. Adv Exp Med Biol. 2007;623:107–22.
70. Long JC, Caceres JF. The SR protein family of splicing factors: master regulators of gene expression. Biochem J. 2009;417(1):15–27.

71. Pan Q, Shai O, Lee LJ, et al. Deep surveying of alternative splicing complexity in the human transcriptome by high-throughput sequencing. Nat Genet. 2008;40(12):1413–5.
72. Wang X, Arai S, Song X, et al. Induced ncRNAs allosterically modify RNA-binding proteins in cis to inhibit transcription. Nature. 2008;454:126–30.
73. Schmucker D, Clemens JC, Shu H, et al. Drosophila Dscam is an axon guidance receptor exhibiting extraordinary molecular diversity. Cell. 2000;101(6):671–84.
74. Ji P, Diederichs S, Wang W, et al. MALAT-1, a novel non-coding RNA, and thymosin beta4 predict metastasis and survival in early-stage non-small cell lung cancer. Oncogene. 2003;22:8031–41.
75. Guffanti A, Iacono M, Pelucchi P, et al. A transcriptional sketch of a primary human breast cancer by 454 deep sequencing. BMC Genomics. 2009;10:163.
76. Lai MC, Yang Z, Zhou L, et al. Long non-coding RNA MALAT-1 overexpression predicts tumor recurrence of hepatocellular carcinoma after liver transplantation. Med Oncol. 2011;29:1810–6.
77. Lin R, Maeda S, Liu C, et al. A large non-coding RNA is a marker for murine hepatocellular carcinomas and a spectrum of human carcinomas. Oncogene. 2007;26:851–8.
78. Luo JH, Ren B, Keryanov S, et al. Transcriptomic and genomic analysis of human hepatocellular carcinomas and hepatoblastomas. Hepatology. 2006;44:1012–24.
79. Schmidt LH, Spieker T, Koschmieder S, et al. The long non-coding MALAT-1 RNA indicates a poor prognosis in non-small cell lung cancer and induces migration and tumor growth. J Thorac Oncol. 2011;6:1984–92.
80. Hall LL, Smith KP, Byron M, Lawrence JB. Molecular anatomy of a speckle. Anat Rec A: Discov Mol Cell Evol Biol. 2006;288:664–75.
81. Lamond AI, Spector DL. Nuclear speckles: a model for nuclear organelles. Nat Rev Mol Cell Biol. 2003;4:605–12.
82. Chen LL, Carmichael GG. Altered nuclear retention of mRNAs containing inverted repeats in human embryonic stem cells: functional role of a nuclear non-coding RNA. Mol Cell. 2009;35:467–78.
83. Clemson CM, Hutchinson JN, Sara SA, et al. An architectural role for a nuclear non-coding RNA: NEAT1 RNA is essential for the structure of paraspeckles. Mol Cell. 2009;33:717–26.
84. Sasaki YT, Ideue T, Sano M, et al. MENepsilon/beta non-coding RNAs are essential for structural integrity of nuclear paraspeckles. Proc Natl Acad Sci U S A. 2009;106:2525–30.
85. Sunwoo H, Dinger ME, Wilusz JE, et al. MEN varepsilon/beta nuclear-retained non-coding RNAs are up-regulated upon muscle differentiation and are essential components of paraspeckles. Genome Res. 2009;19:347–59.
86. Tripathi V, Ellis JD, Shen Z, et al. The nuclear-retained non-coding RNA MALAT1 regulates alternative splicing by modulating SR splicing factor phosphorylation. Mol Cell. 2010;39:925–38.
87. Bernard D, Prasanth KV, Tripathi V, et al. A long nuclear-retained non-coding RNA regulates synaptogenesis by modulating gene expression. EMBO J. 2010;29:3082–93.
88. Sellier C, Rau F, Liu Y, et al. Sam68 sequestration and partial loss of function are associated with splicing alterations in FXTAS patients. EMBO J. 2010;29:1248–61.
89. Engreitz JM, Sirokman K, McDonel P, et al. RNA-RNA interactions enable specific targeting of non-coding RNAs to nascent pre-mRNAs and chromatin sites. Cell. 2014;159:188–99.
90. West JA, Davis CP, Sunwoo H, et al. The long non-coding RNAs NEAT1 and MALAT1 bind active chromatin sites. Mol Cell. 2014;55:791–802.
91. Lin R, Roychowdhury-Saha M, Black C, et al. Control of RNA processing by a large non-coding RNA over-expressed in carcinomas. FEBS Lett. 2011;585:671–6.
92. Tripathi V, Shen Z, Chakraborty A, et al. Long non-coding RNA MALAT1 controls cell cycle progression by regulating the expression of oncogenic transcription factor B-MYB. PLoS Genet. 2013;9:e1003368.

93. Gutschner T, Hämmerle M, Eissmann M, et al. The non-coding RNA MALAT1 is a critical regulator of the metastasis phenotype of lung cancer cells. Cancer Res. 2013;73:1180–9.
94. Zhang B, Arun G, Mao YS, et al. The lncRNA Malat1 is dispensable for mouse development but its transcription plays a cis-regulatory role in the adult. Cell Rep. 2012;2:111–23.
95. Yin QF, Yang L, Zhang Y, et al. Long non-coding RNAs with snoRNA ends. Mol Cell. 2012;48:219–30.
96. Zhang XO, Yin QF, Wang HB, et al. Species-specific alternative splicing leads to unique expression of sno-lncRNAs. BMC Genomics. 2014;15:287.
97. Matera AG, Terns RM, Terns MP. Non-coding RNAs: lessons from the small nuclear and small nucleolar RNAs. Nat Rev Mol Cell Biol. 2007;8:209–20.
98. Kapranov P, Cheng J, Dike S, et al. RNA maps reveal new RNA classes and a possible function for pervasive transcription. Science. 2007;316:1484–8.
99. Keniry A, Oxley D, Monnier P, et al. The H19 lincRNA is a developmental reservoir of miR-675 that suppresses growth and Igf1r. Nat Cell Biol. 2012;14:659–65.
100. Yuan JH, Yang F, Wang F, et al. A long non-coding RNA activated by TGF-β promotes the invasion-metastasis cascade in hepatocellular carcinoma. Cancer Cell. 2014;25:666–81.
101. Cesana M, Cacchiarelli D, Legnini I, et al. A long non-coding RNA controls muscle differentiation by functioning as a competing endogenous RNA. Cell. 2011;147:358–69.
102. Franco-Zorrilla JM, Valli A, Todesco M, et al. Target mimicry provides a new mechanism for regulation of microRNA activity. Nat Genet. 2007;39:1033–7.
103. Karreth FA, Tay Y, Perna D, et al. In vivo identification of tumor-suppressive PTEN ceRNAs in an oncogenic BRAF-induced mouse model of melanoma. Cell. 2011;147:382–95.
104. Poliseno L, Salmena L, Zhang J, et al. A coding-independent function of gene and pseudogene mRNAs regulates tumour biology. Nature. 2010;465:1033–8.
105. Sumazin P, Yang X, Chiu HS, et al. An extensive microRNA-mediated network of RNA-RNA interactions regulates established oncogenic pathways in glioblastoma. Cell. 2011;147:370–81.
106. Tay Y, Kats L, Salmena L, et al. Coding-independent regulation of the tumor suppressor PTEN by competing endogenous mRNAs. Cell. 2011;147:344–57.
107. Conn SJ, Pillman KA, Toubia J, et al. The RNA binding protein quaking regulates formation of circRNAs. Cell. 2015;160:1125–34.
108. Guo JU, Agarwal V, Guo H, Bartel DP. Expanded identification and characterization of mammalian circular RNAs. Genome Biol. 2014;15:409.
109. Jeck WR, Sorrentino JA, Wang K, et al. Circular RNAs are abundant, conserved, and associated with ALU repeats. RNA. 2013;19:141–57.
110. Memczak S, Jens M, Elefsinioti A, et al. Circular RNAs are a large class of animal RNAs with regulatory potency. Nature. 2013;495:333–8.
111. Salzman J, Gawad C, Wang PL, et al. Circular RNAs are the predominant transcript isoform from hundreds of human genes in diverse cell types. PLoS ONE. 2012;7:e30733.
112. Salzman J, Chen RE, Olsen MN, et al. Cell-type specific features of circular RNA expression. PLoS Genet. 2013;9:e1003777.
113. Wilusz JE, Sharp PA. Molecular biology. A circuitous route to non-coding RNA. Science. 2013;340:440–1.
114. Hansen TB, Jensen TI, Clausen BH, et al. Natural RNA circles function as efficient microRNA sponges. Nature. 2013;495(7441):384–8.
115. Zhang A, Zhou N, Huang J, et al. The human long non-coding RNA-RoR is a p53 repressor in response to DNA damage. Cell Res. 2013;23:340–50.
116. Yoon JH, Abdelmohsen K, Srikantan S, et al. LincRNA-p21 suppresses target mRNA translation. Mol Cell. 2012;47(4):648–55.
117. Beltran M, Puig I, Peña C, et al. A natural antisense transcript regulates Zeb2/Sip1 gene expression during Snail1-induced epithelial-mesenchymal transition. Genes Dev. 2008;22(6):756–69.

118. Carrieri C, Cimatti L, Biagioli M, et al. Long non-coding antisense RNA controls Uchl1 translation through an embedded SINEB2 repeat. Nature. 2012;491(7424):454–7.
119. Liu B, Sun L, Liu Q, et al. A cytoplasmic NF-κB interacting long non-coding RNA blocks IκB phosphorylation and suppresses breast cancer metastasis. Cancer Cell. 2015;27(3):370–81.
120. Wang P, Xue Y, Han Y, et al. The STAT3-binding long non-coding RNA lncc-DC controls human dendritic cell differentiation. Science. 2014;344(6181):310–3.
121. Wutz A, Rasmussen TP, Jaenisch R. Chromosomal silencing and localization are mediated by different domains of Xist RNA. Nat Genet. 2002;30:167–74.

Chapter 3
Methods to Study Long Noncoding RNA Biology in Cancer

Man-Li Luo

Abstract Thousands of long noncoding RNAs (lncRNAs) have been discovered in recent years. The functions of lncRNAs range broadly from regulating chromatin structure and gene expression in the nucleus to controlling messenger RNA (mRNA) processing, mRNA posttranscriptional regulation, cellular signaling, and protein activity in the cytoplasm. Experimental and computational techniques have been developed to characterize lncRNAs in high-throughput scale, to study the lncRNA function in vitro and in vivo, to map lncRNA binding sites on the genome, and to capture lncRNA–protein interactions with the identification of lncRNA-binding partners, binding sites, and interaction determinants. In this chapter, we will discuss these technologies and their applications in decoding the functions of lncRNAs. Understanding these techniques including their advantages and disadvantages and developing them in the future will be essential to elaborate the roles of lncRNAs in cancer and other diseases.

Keywords Long noncoding RNA • lncRNA function • lncRNA structure • RNA-protein interaction • RNA–chromatin interaction

3.1 Introduction

Long noncoding RNAs (lncRNAs) are transcripts greater than 200 nucleotides in length but appear to lack protein-coding potential. Thousands of long noncoding RNAs (lncRNAs) have been discovered in recent years, but their molecular and cellular functions as well as the mechanisms by which they act are largely unexplored. With the relatively few lncRNAs that have been well studied, it is known that they are involved in various physiological and pathological processes at

---

M.-L. Luo (✉)
Medical Research Center, Sun Yat-sen Memorial Hospital, Sun Yat-sen University, 107 Yanjiang West Road, Guangzhou 510120, China
e-mail: luomli@mail.sysu.edu.cn

E. Song (ed.), *The Long and Short Non-coding RNAs in Cancer Biology*, Advances in Experimental Medicine and Biology 927,
DOI 10.1007/978-981-10-1498-7_3

epigenetic, transcriptional, or posttranscriptional level to regulate the expression of related genes. In the nucleus, lncRNAs control the epigenetic state of particular genes, participate in transcriptional regulation, involve in alternative splicing, and constitute subnuclear compartments. In the cytoplasm, lncRNAs can facilitate mRNA decay, stabilize mRNAs, and promote or inhibit the translation of target mRNAs through extended base pairing or function as the precursor of microRNAs or compete for microRNA-mediated inhibition. With the complexity of the lncRNA functions and mechanisms of action, it is fortunate that recent years have witnessed an explosion of technologies that make it possible to identify lncRNA and dissect the RNA domains, sequences, structures, and characteristics. In this chapter, we will review the emerging RNA-centric approaches to screen for functional lncRNAs, to identify the mechanism of action, and to dissect the function for specific lncRNAs. With these technologies, biologists have made substantial progress in discovering lncRNA functions. We will discuss the use of these technologies and address current limitations as well as future directions.

3.2 Screening for Functional lncRNAs

The explosion in sequencing technologies has led to the identification of diverse classes of noncoding RNAs (ncRNAs), including lncRNAs. As there are thousands of lncRNAs whose functions we know very little about, high-throughput methods are needed to screen for the functional lncRNAs in different conditions. Microarray, RNA sequencing (RNA-seq), and other genome-wide methods have been used in a large-scale screening of lncRNAs. However, these systematic methods should be followed by functional validation and mechanism studies in certain working models. Integrating the lncRNA functional screening and mechanistic study will reduce the blindness of lncRNA researches.

3.2.1 Microarray and RNA Sequencing

The microarray is a powerful tool to study the transcriptome. However, comparing to the well-annotated protein-coding genes, the annotations of lncRNAs remain at a preliminary stage. In recent years, there have emerged many new databases for the repository of lncRNA transcripts and annotation, such as lncRNAdb [1], NONCODE [2], and LNCipedia [3]. They are now being organized through RNAcentral, a comprehensive and consistent collection containing over 8.1 million ncRNA sequences [4]. Several important consortia and platforms are also attempting to develop a more unified system for lncRNA annotation. The GENCODE consortium within the ENCODE project has been annotating a comprehensive set of human lncRNAs for several years [5]. To better understand the human lncRNA expression, GENCODE has developed a custom expression microarray targeting more than

20,000 lncRNA transcripts and 17,000 randomly selected protein-coding genes. It is encouraging that the progress in transcriptome annotations and lncRNA microarrays will foster the analysis of lncRNA expressions.

Recent advances in RNA sequencing (RNA-seq) have led to the identification of many novel lncRNAs. RNA-seq, based on next-generation sequencing (NGS) of complementary DNAs (cDNAs), provides both the sequence and frequency of RNA that are present at any particular time in a specific cell type, tissue, or organ [6]. Unlike microarray, RNA-seq accurately reveals entire transcriptomes with high sensitivity and detects expression levels over a wide range [7–9]. RNA-seq studies can also distinguish several classes of lncRNAs, such as antisense transcripts to protein-coding genes, transcripts associated with enhancers or repetitive regions, bidirectional promoter transcripts, and transcripts originating in intergenic regions [10]. RNA-seq is suitable for the transcript quantification. Expression levels of lncRNA can be easily compared across different conditions, developmental stages, and tissues, which will provide dynamic, comprehensive, and sensitive data of lncRNA expression. Another advantage of RNA-seq is that only very few cells can be studied by single-cell RNA-seq, which enables transcriptome analysis of tiny amounts of materials.

Sequencing of transcriptomes by RNA-seq is one of the most powerful methodologies for de novo discovery and expression analyses of lncRNAs. Improving RNA-seq methods is an active area of research both in terms of experimental and computational methods [11]. Biases and artifacts in RNA-seq data are often introduced during reverse transcription of RNA into cDNA. Direct RNA sequencing technologies, which use RNA as a template for sequencing, offer the potential to mitigate some current problems and are under development. Progress in algorithms for analyzing sequence data will increase the power of RNA-seq to unravel transcriptomes as well.

3.2.2 Screening of lncRNAs Based on RNA–Protein Interaction

lncRNAs play widespread roles in gene regulation by diverse mechanisms. lncRNAs interact with protein partners such as chromatin modification proteins and transcription factors. These lncRNA–protein interactions increase the flexibility and complexity of gene regulation. Here we will review representative methods for lncRNA identification from lncRNA–protein interactions.

3.2.2.1 RIP-Chip and RIP-Seq

RNA immunoprecipitation (RIP) is a method used to study RNA interactions with some specific proteins (see Sect. 6.2.1). In the RIP method, cells are lysed and the immunoprecipitation is performed with an antibody that targets the protein of interest. By isolating the protein, the RNA interacting with the protein will also be precipitated. The purified RNA–protein complexes can be separated by performing an RNA extraction. Then RIP can be coupled to microarray (RIP-chip) or sequencing (RIP-seq) to identify RNAs bound by a protein of interest [10]. The disadvantage is that there will be nonspecific RNA interaction and high signal-to-noise ratio in the data [12]. Zhao et al. have utilized RIP-seq technology to capture a genome-wide pool of long transcripts (>200 nt) associated with PRC2 and identified classes of medically significant targets, including dozens of imprinted loci, hundreds of oncogene and tumor suppressor loci, and multiple stem-cell-related domains [13]. Because chromatin modifiers such as PRC2 play a central role in maintaining stem cell pluripotency and in cancer, a genome-wide profile of regulatory RNAs will be a valuable resource to elucidate potential mechanisms by which RNA regulates Polycomb.

3.2.2.2 CLIP-Seq

Cross-linking immunoprecipitation and high-throughput sequencing (CLIP-seq) combine UV cross-linking, immunoprecipitation, and high-throughput sequencing to identify binding sites of RNA-binding proteins. Both RIP and CLIP rely on the specific interaction of RNA and protein for immunoprecipitation. CLIP depends on cross-linking induced mutation sites (CIMS) to localize protein–RNA binding sites [14]. Compared with DNA–protein cross-linking by formaldehyde, CLIP uses UV cross-linking which only links proteins to RNAs and avoids protein–protein cross-links. The RNA–protein complexes are then immunoprecipitated. As the UV cross-linking is irreversible, the next step is digestion with proteinase K. This digestion leaves a peptide at the cross-link site that modifies the chemical structure of the nucleotide. The isolated RNA is reverse transcribed to cDNA and then analyzed with sequencing (CLIP-seq) [14]. As reverse transcription of the fragments through the cross-link site introduces mutations that are specific to each separate CLIP, this confers an advantage to determine the protein–RNA binding sites.

3.3 Identifying lncRNAs

3.3.1 Cloning the Full-Length Sequence of lncRNA

Rapid amplification of cDNA ends (RACE) is a technique used in molecular biology to obtain the full-length sequence of an RNA transcript found within a cell. RACE results in the production of a cDNA copy of the RNA sequence of interest, produced through reverse transcription, followed by polymerase chain reaction (PCR) amplification of the cDNA copies. The amplified cDNA copies are then sequenced and, if long enough, should map to a unique genomic region. RACE can provide the sequence of an RNA transcript from a small known sequence within the transcript to the 5′ end (5′ RACE-PCR) or 3′ end (3′ RACE-PCR) of the RNA.

RACE involves 3′ RACE and 5′ RACE. 3′ end amplification of messenger RNA (mRNA) by the PCR synthesizes the first strands of cDNAs. 5′ end amplification of RNA by the nest PCR synthesizes the first strand of cDNA using a gene-specific primer 1. Then RNA is reverse transcribed as above, but the gene-specific primer 1 is substituted for (dT) adaptor. The tailed cDNA is then amplified by PCR using a mixture of a nested gene-specific primer 2 and a combination of a complementary homopolymer-containing anchor primer and corresponding adapter primer, which allow for the amplification of unknown sequences between the gene-specific primer 2 and the 5′ end of the mRNA.

Taking advantage of the latest sequencing technology, Olivarius et al. have developed a high-throughput deep-RACE method which combines RACE with next-generation sequencing for investigating the transcriptional starting sites of genes of interest. It allows for the parallel analysis of multiple genes and is free of time-consuming cloning steps. In comparison to the sequencing of RACE-PCR products, this approach is more precise and more cost-effective even for batches [15]. Pastori et al. have used this method to systematically interrogate the FMR1 gene locus for the occurrence of novel lncRNAs. They found this technique was highly sensitive and enabled the detection of very low-abundance transcripts [16].

3.3.2 Predicting the Coding Potential of lncRNA

As the lncRNAs lack protein-coding potential, assessing the protein-coding status for a transcript is critical for the identification of lncRNAs. The principles for distinguishing coding potential of transcripts include (1) a lack of evolutionary conservation of the identified open reading frames (ORFs), (2) a lack of homology in known protein domains, and (3) a lack of the ability to template significant protein production, etc. [10].

Accordingly, there are various computational tools to calculate the protein-coding potential. The ORF Finder (http://www.ncbi.nlm.nih.gov/gorf/gorf.html) is a graphical analysis tool which finds all open reading frames of a selectable minimum size in a sequence [17]. This tool identifies all open reading frames using the standard or alternative genetic codes. Other tools include Coding Substitution Frequency (CSF) [18], Coding Potential Calculator (CPC) [19], Coding-Potential Assessment Tool (CPAT) [20], Coding–Non-coding Index (CNCI) [21], etc. CSF develops algorithms to score conserved ORFs across several organisms and provides a general strategy for determining the coding potential. However, such a conservation-based method may fail to detect new proteins which do not contain a conserved ORF [18]. CPC (http://cpc.cbi.pku.edu.cn/) searches for the putative ORFs and homologies based on the identified ORFs, as well as the output of parsing protein database [19]. Not like the alignment-based approaches, CPAT (http://lilab.research.bcm.edu/cpat/) uses a logistic regression model and employs the information of ORF embedded in transcripts to develop the classifier [20]. CNCI (http://www.bioinfo.org/software/cnci/) classifies protein-coding and lncRNA transcripts by exploiting the intrinsic components contained in sequences instead of predicting the ORF [21].

3.3.3 *Ribosome Enrichment Assay*

Ribosome profiling, or Ribo-Seq, is a technique that uses specialized mRNA sequencing to determine which mRNAs are being actively translated [22]. It is based on the deep sequencing of ribosome-protected mRNA fragments and enables genome-wide investigation of translation with subcodon resolution. This method involves digestion of RNA followed by separation of 80S ribosomes based on their size and density through a sucrose cushion. The associated RNAs are then sequenced to assess the occupancy of ribosomes on RNAs. Ribosome profiling can be used to identify mRNAs associated with 80S ribosomes, distinguish ribosome-free three prime untranslated regions (3′ UTRs) on these messages, and assess the quantitative dynamics of translation within the translated regions [23]. Ribosome profiling involves similar sequencing library preparation and data analysis to RNA-seq, but unlike RNA-seq, which sequences all of the mRNA of a given sequence present in a sample, it targets only mRNA sequences protected by the ribosome during the process of decoding by translation [22].

Since its invention, the ribosome profiling technique has been utilized for global analysis of translation in both prokaryotic and eukaryotic organisms. Studies using this approach have already provided new insights into the identity and the amount of proteins that are produced by cells, as well as detailed views into the mechanism of protein synthesis itself. Notably, the detection of 80S ribosomes on RNA does not alone provide evidence that the RNA encodes a functional protein. It is important to use the 80S footprint data provided by ribosome profilings to distinguish true messengers that encode functional proteins from those that are noncoding. Ribosome profiling has proven to be a highly useful tool for exploring the peptide-coding potential of the lncRNAs, which have higher ribosome occupancy than 3′ UTRs and possibly can be translated into functional proteins [24]. Guttman et al. have performed a more complete analysis of the ribosome profiling data and studied the pattern of ribosome occupancy on transcripts to evaluate whether long intergenic noncoding RNAs (lincRNAs) show evidence of encoding functional proteins [25]. In particular, they analyzed the existing ribosome profiling data across other non-coding regions, including 5′ UTRs and classical noncoding RNAs. They found that the known noncoding controls also showed ribosome profiles that differed from those of the 3′ UTRs of protein-coding genes. Moreover, 5′ UTRs resembled coding regions of protein-coding genes more closely than lincRNAs did. Their findings clearly demonstrate that lincRNAs are likely to function directly as RNA molecules rather than through encoded protein products [25]. Taken together, ribosome profiling bridges the gap between global measurements of steady-state mRNA and protein levels and is a powerful method to study coding potential of the lncRNAs.

3.3.4 *Subcellular Location of lncRNAs*

3.3.4.1 Fractionation qPCR

lncRNAs play a variety of roles in different subcellular domains. In the nucleus, lncRNAs control the epigenetic state of particular genes, participate in transcriptional regulation, involve in alternative splicing, and constitute subnuclear compartments. In the cytoplasm, lncRNAs regulate gene expression by affecting the stability of mRNAs and altering the translation efficacy of target mRNAs. They function as precursor of microRNAs (miRNAs) or compete for miRNA-mediated inhibition, leading to increased expression of the mRNA [26]. Therefore, lncRNAs must localize themselves to their particular site of action within the important cell. To identify specific subcellular distribution of lncRNAs will be helpful to study the functions and mechanisms of action of lncRNAs.

Successful isolation of pure nuclei and cytoplasmic fractions is critical in this assay. Commercial kits have been developed for the isolation of both RNA and native protein from the same experimental sample. Tissue or cultured cells are first homogenized in ice-cold cell disruption buffer to prepare a total cell lysate. Since the homogenization is performed quickly on ice and in the presence of detergent, both protein and RNA can be purified directly from this lysate. For RNA isolation, a part of the total cell lysate is immediately mixed with an equal volume of lysis/binding solution. RNA is then purified from the mixture using an RNA-binding glass fiber filter. The RNA isolated from total, nuclear, or cytoplasmic fractions can be subjected to cDNA synthesis and reverse transcription polymerase chain reaction (RT-PCR). The effectiveness of cell fractionation can be easily checked by Western blotting with antibodies for proteins found predominantly in the nucleus or cytoplasm of cells. Once the fractionation is confirmed, quantitative RT-PCR can be performed with specific primers for lncRNA, and the distribution and abundance of lncRNA in the nucleus and cytoplasm will be determined.

3.3.4.2 FISH

Fluorescence in situ hybridization (FISH) is a method to localize nucleic acid targets in fixed cells for subcellular cytogenetic or gene expression studies. It relies on fluorophore-labeled DNA or RNA probes to count and localize specific genes or regions along chromosomes, to detect mutations, or to analyze temporal and spatial gene expression. FISH is often used to detect and localize specific RNA targets (mRNA, lncRNA, and miRNA) in cells. Compared with other methods of DNA or RNA detection, FISH can provide the exact subcellular localization and expression of the target DNA or RNA with a visible method by using the specific probe. Indeed, direct observation of lncRNA localization by RNA FISH has led to many of the early hypotheses about lncRNA function that now serve as paradigms in the field, such as X-inactive specific transcript (XIST), metastasis-associated lung adenocarcinoma transcript 1 (MALAT1), nuclear-enriched abundant transcript 1 (NEAT1), myocardial infarction-associated transcript (MIAT), etc. [27].

RNA FISH begins with the specific probe construction and involves in bathing the sample in a high concentration of probes that are complementary in some way to the target RNA species. After the hybridization, excess unbound probe is washed away, theoretically leaving only those specifically bound to the target molecule. Differences among the variants of RNA FISH typically revolve around the type of nucleic acid used for the probe and the type of labeling scheme used to detect the probe via microscopy.

Highly abundant RNAs in the cell are easily detected by regular RNA FISH, whereas the vast majority of lncRNAs are considerably less abundant. More recently, researchers have developed and applied single-molecule RNA FISH techniques based on hybridization of multiple short, fluorescently labeled, oligonucleotide probes to estimate the absolute level and subcellular localization of even low-abundance lncRNAs [27, 28]. The single-molecule RNA FISH falls into two categories: one that uses some form of signal amplification and the other that relies on direct detection of signal. Direct detection involves labeling the probes themselves with fluorophores. In order to achieve single-molecule sensitivity, the probes must have enough fluorescence to be detectable above background autofluorescence. One technique is to use a set of short single-stranded DNA oligonucleotides complementary to various regions of the target RNA, each labeled with one or more fluorescent moieties [28]. The binding of multiple probes localizes enough fluorophores to the target RNA so that the RNA is easily visible as a fluorescent spot via fluorescence microscopy. The benefit of using several oligonucleotide probes at the same time is that the off-target binding of a single oligonucleotide in the probe pool will either be undetectable or readily distinguishable to the much brighter spots corresponding to the true RNA, thus reducing the chances of false positives. False negatives are similarly unlikely, for even if a single probe out of the pool fails to bind, the rest are likely to bind [29]. Collectively, RNA FISH is an indispensable tool for the detection and localization of lncRNA, and the single-molecule RNA FISH enables quantification and spatial resolution of single lncRNA molecules within cells.

3.4 Detecting lncRNA Expression

As lncRNAs are a novel class of RNA transcripts, it is important to characterize the expression of lncRNAs in various systems. There are several methods to analyze their expression, including quantitative reverse transcription polymerase chain reaction (qRT-PCR), in situ hybridization (ISH), FISH, and Northern blotting. qRT-PCR and Northern blotting are able to detect small changes in gene expression both in cells and tissues, while Northern blotting often faces the problem with sample degradation by ribonucleases (RNases). Comparing to qRT-PCR, ISH, and FISH,

Northern blotting has a low sensitivity. Moreover, ISH and FISH can both localize the gene sites and measure the gene expression level at the same time.

3.4.1 Northern Blot

Northern blot is a technique used to study gene expression by RNA detection (or isolated mRNA) in a sample. Northern blotting involves the use of electrophoresis to separate RNA samples by size and detection with a hybridization probe complementary to part of or the entire target sequence. Northern blot can be used to determine lncRNA abundance and to identify different splicing variants of a given lncRNA. Compared to RT-PCR, Northern blotting has a high specificity, which is important to reduce false-positive results. The advantages of using Northern blotting include the detection of RNA size, the observation of alternate splice products, the use of probes with partial homology, the measurement of the quality and quantity of RNA on the gel prior to blotting, and the feasibility of storing and re-probing the membranes for years after blotting [30].

Firstly, RNA is isolated and followed by gel electrophoresis for separation. Next, the negatively charged RNA is transferred onto a nylon membrane driven by capillary forces. The membrane is exposed to intense UV light to induce RNA cross-linking. Subsequently, the RNA fixed in the membrane is hybridized with the labeled probe, forming a double-stranded RNA–DNA or RNA–RNA structure. Hybridization is usually carried out overnight and detected by autoradiography or with the use of a phosphor-imager apparatus. So far, Northern blot analysis is still a standard technique used in the detection and quantification of mRNA levels because it allows for a direct comparison of the mRNA abundance between samples on a single membrane.

3.4.2 qRT-PCR

qRT-PCR is a technique based on the polymerase chain reaction (PCR), which is used to amplify and simultaneously detect or quantify a targeted DNA or RNA expression. Quantitative PCR is carried out in a thermal cycler by detecting the fluorescence emitted by the excited fluorophore, while qRT-PCR begins with the specific primer designation and synthesis, after which the RNA is isolated and reverse transcribed from the samples or cells. A marked fluorophore is then added to this mixture in a thermal cycler, where the fluorescence intensity is detected and analyzed to measure the gene expression [31, 32].

3.4.3 ISH and FISH

Comparing to FISH, in situ hybridization (ISH) is mainly used in formalin-fixed paraffin-embedded (FFPE) or frozen tissue section samples to detect and analyze the specific gene expression without fluorescence coupled to the probe. The application of ISH is to localize and measure the targeted RNA expression, dependent on the specific probe designation and construction (see Sect. 2.4.2).

3.5 Manipulating lncRNA Expression

Although techniques to systematically investigate lncRNAs are limited, methods to comprehensively study cellular roles of protein-coding genes stay available. The typical way to discern cellular function is by manipulating lncRNA expression, including selective knockdown, knockout or overexpression of a specific lncRNA, and examination of the resulting phenotype.

3.5.1 Loss of Function

3.5.1.1 siRNA and shRNA

Small interfering RNA (siRNA) and short hairpin RNA (shRNA) are powerful gene knockdown approaches which imitate the endogenous natural process of RNA interference (RNAi) mechanisms to silence the targeted gene expression with high specificity and selectivity. Both the designation and synthesis of siRNA and shRNA are based on the bioinformatics analysis and optimization. Additionally, optimized shRNA constructs endow high potency and sustainable effects resulting in less off-target effects.

siRNA begins with the designation based on the target gene sequence alignment and selection. Then paired double-stranded antisense RNA is synthesized to be delivered to the cells via transfection. shRNA method relies on the high efficient interfering vector construction by molecular clone technology, which contains the interfering fragment and forms a short hairpin stem-loop structure to block the gene expression [33–35]. Although siRNA and shRNA utilize a similar cellular mechanism that involves the RNA-induced silencing complex (RISC), which method to use depends on several factors such as cell type, time demands, and the need for transient versus stable integration. The siRNA-mediated knockdown is not permanent, and the introduction of shRNA allows for stable integration and long-term knockdown of the targeted gene.

RNAi is an invaluable tool to knock down a target gene of interest, including the lncRNAs. However, compared with protein-coding mRNAs or miRNAs, lncRNAs are more difficult to target by RNAi. Moreover, RNAi has its limitations: (1) siRNAs not only silence their specific target gene but also influence the expression of other genes (off-target effects). (2) The efficiency of siRNA is not predictable, and finding a highly efficient specific siRNA can be time-consuming and cost intensive. (3) Some transcripts can be hard to target due to their strong secondary structure and incorporation into large protein complexes or their intracellular localization. Thus, a knockdown by RNAi to activating long noncoding RNA (alncRNA) sometimes might not be sufficient to evoke a phenotype and uncover its physiological function.

3.5.1.2 Antisense Oligonucleotides (ASOs)

Antisense oligonucleotides (ASOs) are 15–20-mer single-stranded deoxyribonucleotides that are artificially synthesized to hybridize with the complementary mRNA in a sequence-specific manner. The formation of the ASO–mRNA heteroduplex either triggers RNase H activity, leading to mRNA degradation, inducing translational arrest by steric hindrance of ribosomal activity, and interfering with mRNA maturation by inhibiting splicing, or destabilizes pre-mRNA in the nucleus, resulting in downregulation of target mRNA expression.

The first generation of ASO is 2′-DNA oligomers uniformly modified with the phosphorothioate backbone substitution and worked predominantly through an RNase H-dependent mechanism. In order to improve upon the first-generation ASO drugs, many different modifications to the core nucleoside monomer unit of the ASO have been evaluated for their effects on affinity for complementary RNA, nuclease resistance, and ASO potency. The most advanced second-generation antisense designs are 2′-O-methoxyethylribose (MOE) gapmer oligonucleotides. MOE-modified ASOs show increased affinity toward a complementary RNA and are highly resistant toward degradation by nucleases [36].

Although MOE provides a substantial improvement in affinity, bicyclic nucleoside modifications such as 2′,4′-methylene bridged nucleic acids also called as "locked nucleic acid" (LNA) have been shown to provide a further increase in affinity [37]. LNA oligonucleotides are synthesized in different formats, such as all-LNA, LNA/DNA mixmers, or LNA/DNA gapmers. Essentially, all aspects of antisense technology have profited from LNA due to its unprecedented affinity, good or even improved mismatch discrimination, low toxicity, and increased metabolic stability. LNA nucleotides are used to increase the sensitivity and specificity of expression in FISH probes, quantitative PCR probes, and other techniques based on oligonucleotides. The combination of antisense DNA and LNA can enhance the inhibition efficiency of gene expression [37]. LNA is particularly attractive for in vivo applications that are inaccessible to RNAi technology [38].

Achieving consistent knockdown has proven to be more challenging for lncRNAs than for mRNAs or miRNAs, because lncRNAs are often localized to the nucleus. It has been reported that nuclear lncRNAs are more effectively suppressed using ASOs and cytoplasmic lncRNAs are more effectively suppressed using RNAi.

Using both methods, dual-localized lncRNAs are suppressed. A mixed-modality approach combining ASOs and RNAi reagents improves knockdown efficacy [39].

3.5.1.3 CRISPR/Cas-Based Methods

RNAi-mediated knockdown is the most common strategy for ablating gene function. However, knockdown by RNAi is incomplete and unpredictable with off-target effects and provides only temporary inhibition of gene function. Recently, complete loss-of-function approaches have become available with the discovery of zinc-finger nucleases (ZFNs), transcription activator-like effector nucleases (TALENs), and, later, the more powerful clustered regularly interspaced short palindromic repeats (CRISPR/Cas) system. Currently, the type II CRISPR/Cas9 system from *Streptococcus pyogenes* is the most widely used CRISPR system and has been successfully applied to edit human genomes [40].

CRISPR/Cas9 is a RNA-guided gene-editing technique that makes use of a bacterially derived protein (Cas9) and a synthetic guide RNA to introduce a double-strand break at a specific location within the genome, resulting in adding or deleting base pairs at specifically targeted DNA loci. Editing is achieved by transfecting a cell with the Cas9 protein along with a specially designed guide RNA (gRNA) that directs the cut through hybridization with its matching genomic sequence. When Cas9 cuts the target sequence, the cell repairs the damage by replacing the original sequence with an altered version [41, 42]. CRISPR simplifies the process of deleting, adding, or modifying genes and has successfully been tested in cells of 20 species, including humans [43].

The CRISPR/Cas9 system has been applied as an efficient tool for engineering site-specific mutations of protein-coding genes in the genome. The knockout of ncRNA, especially lncRNA, is relatively more difficult than that of coding genes with this system. One of the challenges is that a small deletion or insertion generated by the standard CRISPR/Cas system may not necessarily lead to functional loss of a given noncoding gene because of lacking an open reading frame, especially in polyploidy human cells. To overcome this challenge, Ho et al. have adopted a selection system that allows for marker genes to integrate into the genome through homologous recombination (HR). Moreover, they constructed a dual-guide RNA vector that could make two cuts simultaneously at designated sites such that a large fragment could be deleted. With these approaches, they were able to successfully generate knockouts for miRNAs and lncRNAs in various human cell lines [44]. For the in vivo model, Han et al. have explored the potential of using the CRISPR/Cas9 system to generate large genomic deletions of lncRNAs in mice. They developed an efficient one-step strategy to target the maternally expressed lncRNA, Rian, on chromosome 12 in mice and showed that paired single-guide RNAs (sgRNAs) could precisely generate large deletions up to 23 kb. The deletions are heritable, and the efficiency can be further improved up to 33 % by combining multiple sgRNAs [45]. Taken together, with the improvement of the CRISPR/Cas9 system, it will largely contribute to generating better in vivo and in vitro models for lncRNA research.

3.5.2 *Gain of Function*

Upregulating a target lncRNA expression in cells is a key method to investigate the lncRNA function. The overexpression of a specific lncRNA always begins with performing the rapid amplification of cDNA ends (RACE) to acquire the whole sequences, which are then constructed to a vector to express the lncRNA, the function of which is to be analyzed.

Previously, regular ectopic expression vectors were always used for the above purpose until two recent studies showed single endogenous gene activation using dCas9-based activators [46, 47]. Coexpression of this transactivator and combinations of gRNAs in human cells induce specific expression of endogenous target genes, demonstrating a simple and versatile approach for RNA-guided lncRNA overexpression. However, the change in gene expression achieved by this approach is small for many applications, which is less than or around fivefold activation. Most recently Konermann et al. have overcome this low efficiency of gene activation by turning the CRISPR sgRNA into a modular platform for assembling multiple different transcriptional activators. Applications of this dCas9-based transcription modulator can induce more than 100-fold activation and enable the dissection of many types of genetic elements, including lncRNAs. Moreover, this system allows for multiplexed activation of ten genes simultaneously [48]. Thus it may be possible to deliver defined combinations of lncRNAs and various effector proteins to the same genomic location using one sgRNA molecule. However, CRISPR targeting may have off-target effects, and additional validation experiments are needed to confirm any effects of altered lncRNA expression using this approach.

3.6 Mapping lncRNA Binding Sites on the Genome

lncRNAs that regulate gene expression are often found to be localized to chromatin, which suggests their direct involvement with the accompanying protein machineries. They are thought to be important in dosage compensation. To investigate the exact roles of the RNAs at chromatin, determining their binding sites on the genome is necessary. Currently the occupancy sites of most lncRNAs are not known. There are three methods to address this matter, which include chromatin isolation by RNA purification (ChIRP), capture hybridization analysis of RNA targets (CHART), and RNA antisense purification (RAP). ChIRP uses antisense DNA oligonucleotides to capture and purify specific lncRNA–chromatin complexes from the cell, while CHART and RAP differ in methods of cross-linking, chromatin shearing, and the density and length of the oligonucleotide probes. The three methods share some similarities in the procedure, but have one major difference in the oligonucleotide probe design. ChIRP and RAP do not require a priori knowledge of the lncRNA domains involved in chromatin interaction and tile oligonucleotides across the entire target RNA so that all potential hybridization spots are fully used. In contrast, CHART takes a different approach and uses the RNase H assay to narrow down the search space for effective probe design.

3.6.1 Chromatin Isolation by RNA Purification (ChIRP)-Seq

Chromatin isolation by RNA purification (ChIRP) is based on affinity capture of target lncRNA–chromatin complex by tiling antisense oligos, which then generates a map of genomic binding sites at a resolution of several hundred bases with high sensitivity and low background. It is applicable to many lncRNAs because the design of affinity probes is straightforward given the RNA sequence and requires no knowledge of the RNA's structure or functional domains [49]. Oligonucleotide probes are designed to be complementary to the RNA of interest and labeled with biotin. Cells are collected and cross-linked with glutaraldehyde to preserve RNA–chromatin interactions. Then cross-linked cells are lysed and DNA is sheared by sonication. The DNA is later isolated from an aliquot of the bound complex by treatment with proteinase followed by RNA-seq to digest associated protein and RNA. The purified DNA is then sequenced on a next-generation DNA sequencing system, where a pileup of reads at specific locations on the genome indicates that the RNA of interest has bound to that region of the genome. This helps delineate specific genomic regions that interact with RNA [50].

ChIRP-seq has greatly improved our understanding of RNA–DNA interactions at high resolution on a genomic scale. As with all experiments, proper controls are required to interpret the results. Like ChIRP-seq, not all binding events are necessarily functional, and additional studies are required to ascertain the biological consequences of RNA occupancy on chromatin. In addition, RNA–RNA interactions can also be assayed by ChIRP. In this case, instead of proteins, RNAs are eluted and subjected to qRT-PCR or high-throughput sequencing analysis.

3.6.2 Capture Hybridization Analysis of RNA Targets (CHART)-Seq

Capture hybridization analysis of RNA targets (CHART) is a method to analyze RNA targets that are analogous to chromatin immunoprecipitation (ChIP) for proteins [51]. To identify the genomic targets of those large noncoding RNAs which can act on chromatin at sites distant from where they are transcribed, this endogenous RNA is enriched from cross-linked chromatin extracts using short biotinylated complementary oligodeoxyribonucleotides. The proteins and DNA enriched under these conditions can then be examined to determine the targets of RNA.

The CHART protocol enables the affinity capture and study of endogenous RNAs from cross-linked nuclear extracts. After cross-linking and fragmenting chromatin, capture oligonucleotides are hybridized to an RNA bound to specific genomic regions and proteins and isolated using streptavidin resin. After stringent washing, captured material is eluted with RNase H, which cleaves RNA–DNA heteroduplexes created by hybridization of the capture oligonucleotide to the target RNA. DNA and proteins can be isolated from the CHART-enriched material and subjected to next-generation sequencing or Western blot analysis, respectively [52]. These procedures can be applied to the study of DNA that co-purifies with alncRNA and provide mechanistic insights into lncRNA function by identifying putative trans genomic binding sites for endogenous RNAs.

3.6.3 RNA Antisense Purification (RAP)-Seq

RNA antisense purification (RAP) is a method for selective purification of endogenous RNA complexes from cell extracts that enables mapping of RNA interactions with chromatin [53]. In RAP, cells are cross-linked to fix endogenous RNA complexes, which will be purified through hybrid capture with biotinylated antisense oligos. DNA loci that interact with the target RNA are identified using high-throughput DNA sequencing. By cross-linking endogenous macromolecular complexes prior to RNA capture, RAP allows for identification of proteins, RNA, and DNA loci that cross-link to and co-purify with the target RNA. Compared with previous approaches for examining RNA–chromatin interactions such as FISH, RNA-centric biochemical purification enables genome-wide mapping of RNA–DNA interactions by coupling the protocol with high-throughput DNA sequencing (RAP-DNA). Thus, RAP provides an important tool for systematic interrogation of lncRNA function and mechanism.

In the RAP protocol, antisense nucleotide oligo probes are designed for selective retrieval of RNA target by RAP. The cells are collected and cross-linked with disuccinimidyl glutarate and formaldehyde to preserve RNA–chromatin interactions. Then the cross-linked cells are lysed and DNA is sheared by sonication. Next, biotinylated DNA probes are hybridized to RNA. Finally RNA and DNA fraction from RAP samples are extracted and assayed by quantitative RT-PCR and by sequencing, respectively [54]. Compared to similar protocols, the most distinctive and important feature of RAP is its use of long (120-nucleotide) capture probes tiled across the entire target RNA. This probe design strategy robustly captures any lncRNA and enables the use of stringent hybridization to wash conditions that dramatically reduce nonspecific interactions of off-target nucleic acids or proteins [54].

Taken together, ChIRP, CHART, and RAP are all based on the same idea—using biotinylated oligonucleotides complementary to the RNA of interest as a handle with which to pull down associated proteins or, more accurately, chromatin. The next-generation sequencing and/or mass spectrometry will then be used to identify the proteins associated with the RNA and the genomic locations at which those interactions occurred.

3.7 Discovering RNA–Protein Interactions

As one major mechanism for lncRNA to exert its function is to serve as a scaffold via RNA–protein interaction, it is important to investigate which lncRNAs are binding to a protein of interest. Moreover, no matter what the lncRNAs serve, such as molecular signals, decoys, or guides, they recruit protein accomplices. Thus, the identification of the lncRNA-bound proteome is essential for understanding lncRNA functions.

3.7.1 Identifying the lncRNA-Bound Proteome

RNA chromatography combined with mass spectrometry represents a widely used experimental approach to identify RNA-binding proteins that recognize specific RNA targets [55]. RNA chromatography method relies on the use of a cocktail of RNases in the elution step. Such results in the release of proteins specifically associated with the RNA ligand and almost eliminates the background noise, allowing for a more sensitive and thorough detection of RNA-binding proteins to recognize a specific RNA transcript. The process goes as follows: firstly, the RNA transcript of interest is covalently attached to the agarose beads. Then, the coupled RNA and beads are incubated with protein extracts, followed by stringent washes. Next, the RNA–protein complexes are denatured, and the bead-coupled RNP complex is released from the beads by RNase treatment, followed by SDS electrophoresis and mass spectrometry of specific protein bands [56].

Although this method has been carried out to study a large number of functional RNA molecules, there are major downsides. Firstly, the RNA baits are often not physiologically concentrated or correctly structured, thus leading to artifactual protein interactions that occur in in vitro conditions. Secondly, post-lysis association between RNA and proteins contributes to false-positive hits, which can in part be circumvented by genetically tagging RNAs with affinity sequences. The tagged RNAs and their binding factors can then be purified with affinity proteins fixed on a solid support. The advantage over RNA chromatography is that the tagged transcripts are transcribed in vivo and are thus more likely to be exposed to the same biochemical environment as their endogenous counterparts [57]. In the future, improved RNA chromatography method may be developed to capture protein-binding partners of individual RNA motifs, embedded in full-length transcripts, by using sequence-specific cleaving reagents such as RNAzymes, DNAzymes, or oligonucleotides coupled with RNase H.

Direct capture of lncRNA-binding proteins by antisense oligonucleotide probes, as with the aforementioned ChIRP and RAP methods, can sidestep the complications introduced by RNA chromatography or genetic tagging. Comprehensive identification of RNA-binding proteins by mass spectrometry (ChIRP-MS) and RNA antisense purification mass spectrometry (RAP-MS) have been used to study RNA–protein interaction. Chu et al. have performed ChIRP-MS to systematically identify Xist RNA-binding proteins [57]. Xist is tightly associated with the nuclear matrix, which makes extraction under native conditions very inefficient. Physical disruption by sonication could solubilize the Xist RNA, but given its long size (17–19 kb), shearing would break the RNA into fragments. These problems can be resolved by ChIRP, in which cross-linking and subsequent disruption solubilize the RNA while preserving its interactions, and tiling antisense oligonucleotide probes ensure capture of all possible fragments of the RNA.

McHugh et al. have identified a set of highly specific and reproducible proteins that directly interact with Xist using the RAP-MS method [58]. Briefly, RAP-MS uses UV cross-linking to create covalent linkages between directly interacting RNA and protein and purifies lncRNAs in denaturing conditions to disrupt non-covalent interactions. This UV cross-linking and denaturing approach, which is utilized by methods such as CLIP, is known to identify only direct RNA–protein interactions and to separate interactions that are cross-linked in the cell from those that merely associate in solution. Thus, this RAP-MS experiment identifies direct Xist-interacting proteins. Given the generality of the RAP-MS approach, it is expected that this method will be broadly applicable for defining the proteins that directly interact with lncRNAs.

The other way to run the MS analysis is to use quantitative MS to simultaneously compare the proteins in the sample and control. In one popular method used for RNA–protein analysis, cells are metabolically labeled to generate differentially tagged protein pools for MS analysis, in which the isotopes of the proteins are compared to provide direct quantification [59]. The advantage to this approach is that the ratios of peptides from the experimental and control samples can be directly compared to allow for discrimination of true binding partners from nonspecific interactors. This method can account for some of the issues associated with abundant protein association.

3.7.2 Validating the RNA–Protein Interaction

RNA–protein interaction data obtained by RIP-seq, CLIP-seq, ChIRP-MS, RAP-MS, or other high-throughput methods need to be verified by techniques such as RIP, RNA pulldown, or RNA-electrophoretic mobility shift assay (EMSA). RIP has been developed to identify lncRNA species that bind to a protein of interest. On the other hand, if the research focus is to identify the proteins that are bound to a

Table 3.1 Methods for identifying lncRNA–protein–DNA interactions

Method	Application	Characteristics
RIP	Detect in vivo RNA–protein interactions	RIP involves immunoprecipitation of a protein of interest using an antibody. RNAs bound to the protein will be isolated and detected by real-time PCR, microarrays, or sequencing
RNA pulldown	Detect in vitro RNA–protein interactions	RNA pulldown uses transcribed RNAs or synthetic RNAs in vitro for labeling and selectively capture proteins with labeled RNA
CLIP	Capture in vivo RNA–protein interactions	CLIP is similar to RIP, except it adds a UV cross-linking step before immunoprecipitation, which specifically and irreversibly links proteins to RNAs that are in very close proximity
ChIRP	Map the genomic binding sites of chromatin-associated lncRNAs	ChIRP, CHART, and RAP all uses biotinylated oligonucleotides complementary to the RNA of interest to pull down associated proteins and chromatin. ChIRP uses relatively short antisense DNA oligonucleotides to capture and purify specific lncRNA–chromatin complexes from the cell
CHART	Identify the genomic targets of lncRNAs	CHART targets the RNA with a few short DNA-based oligonucleotides and identifies genomic DNA and proteins cross-linked to the RNA of interest. CHART uses RNase H to cleave RNA–DNA heteroduplexes created by hybridization of the capture oligonucleotide to the target RNA
RAP	Identify the genomic targets of lncRNAs	RAP uses relatively long capture probes tiled across the entire target RNA and identifies proteins, RNA, and DNA loci that cross-link to and co-purify with the target RNA

given lncRNA, lncRNA pulldown will help to identify the protein molecules that interact with a specific lncRNA. As the methods for the study of RNA–protein or RNA–protein–DNA have similar characteristics and are confusing, we have summarized those methods in Table 3.1

3.7.2.1 RNA Immunoprecipitation (RIP)

As aforementioned in Sect. 1.2.1, RIP is an antibody-based technique used to map in vivo RNA–protein interactions. There are two main classes of these methods, native RIP and cross-linked RIP [60]. Native RIP begins with immunoprecipitation of endogenous complexes of RNA-binding proteins and co-isolation of RNAs associated with the immunoprecipitated complex. For the validation of RNA–protein interaction, the RNA in the complex can be isolated and identified by quantitative reverse transcription PCR, while the protein can be subjected to Western blot analysis. The limitation of native RIP is due to the nonphysiological formation of RNA–protein interactions in solution, which can be overcome by the cross-linked RIP. The cross-linked RIP is to treat cells with cross-linking methods, such as formaldehyde cross-linking. By cross-linking RNA–protein complexes and purifying the complex under denaturing conditions, one can distinguish in vivo interactions that are cross-linked in the cell from interactions formed subsequently in solution [60].

Huarte et al. have discovered multiple lincRNAs in the p53 pathway. The mass spectrometry analysis demonstrated the interaction between lincRNA-p21 and hnRNP-K. To validate this interaction, they performed RIP with an antibody against hnRNP-K from nuclear extracts. Then they further performed cross-linked RIP with formaldehyde cross-linked cells followed by stringent washing conditions to rule out potential nonspecific interactions and observed a greater and very significant enrichment of lincRNA-p21 in the hnRNP-K RIP [61]. Thus RIP is a powerful technique that can be used to detect the in vivo interaction between individual proteins and specific lncRNA. The drawback of this method is that RNAs of high abundance are easily purified, while specific interactions that occur with low-abundance transcripts may be masked by nonspecific interactions that occur with highly abundant transcripts.

3.7.2.2 RNA Pulldown

Proteins interact with RNA through electrostatic interactions, hydrogen bonding, hydrophobic interactions, and base stacking. Protein–RNA interactions are significantly influenced by the tertiary structure on the RNA molecules. Therefore, in assays to identify protein–RNA interactions, both the RNA and protein(s) must be correctly folded to allow proper binding.

The RNA pulldown assay uses high-affinity tags such as biotin to selectively extract a protein–RNA complex from a sample in vitro. In this assay, RNAs are biotin labeled and transcribed with T7 RNA polymerase and then purified. Biotin-HRP Northern blot is performed to demonstrate that all the RNAs are biotinylated and transcribed at the right size. Then biotinylated RNA is prepared to allow for proper secondary structure formation, which is a critical step for the function where end labeling is preferred. Folded RNA is then incubated with cell lysates, after which streptavidin agarose beads are added to each binding reaction and further incubated. Lastly the beads are washed and boiled in SDS buffer, and the retrieved protein can be detected by Western blot [62]. The advantage of RNA pulldown assay is the enrichment of low abundant targets. Huarte et al. have carried out RNA pulldown to identify proteins possibly associated with lincRNA-p21 and further used this experiment to detect the interaction of truncated versions of lincRNA-p21 with heterogeneous nuclear ribonucleoprotein K (hnRNP-K). These analyses successfully identified a 780 nt region at the 5′ end of lincRNA-p21 required for the interaction with hnRNP-K [61].

3.7.2.3 RNA-EMSA

An electrophoretic mobility shift assay (EMSA) or mobility shift electrophoresis, also referred as a gel shift assay, gel mobility shift assay, band shift assay, or gel retardation assay, is a common affinity electrophoresis technique used to study protein–DNA or protein–RNA interactions. This procedure can determine if a protein or mixture of proteins is capable of binding to a given DNA or RNA sequence and can sometimes indicate if more than one protein molecule is involved in the binding complex. Gel shift assays are often performed in vitro concurrently with DNase footprinting, primer extension, and promoter-probe experiments when studying transcription initiation, DNA replication, DNA repair or RNA processing, and maturation. Although precursors can be found in earlier literature, most current assays are based on methods described by Garner and Revzin and Fried and Crothers [63, 64].

The RNA-EMSA detects protein–RNA interactions through changes in migration speed during gel electrophoresis. In this essay, a labeled RNA probe is incubated with a protein sample (typically from a cell lysate) to initiate binding and formation of the interaction complex. The binding reaction is then separated via nondenaturing polyacrylamide gel electrophoresis. Like protein–DNA complexes, a protein–RNA complex migrates more slowly than a free RNA probe through a gel matrix, which causes a migration shift relative to the nonbound RNA probe. Hereby, specificity is determined through a competition reaction, where excess unlabeled RNA is incubated in the binding reaction, resulting in a decrease in the shifted signal if the labeled and unlabeled RNA sequences compete for binding of the same protein. Alternatively, the protein–RNA complex may be cross-linked and the reaction runs on a denaturing gel. In this case, specificity is determined through visualization of a single shifted band. Traditionally, RNA probes are radioactively labeled for detection, although fluorescent and chemiluminescent detection is also possible [65]. Nonradioactive RNA end-labeling techniques are limited, but more versatile biotin and fluorescent labeling methods are now available. With the RNA-EMSA method, it is easy to screen RNA mutants for binding efficiency.

3.7.2.4 In Situ Visualization of RNA and Proteins

Protein immunofluorescence (IF)/fluorescence in situ hybridization (FISH) double labeling (IF/FISH) is a combining method to simultaneously detect RNA and multiple protein targets in fixed cells. It relies on specific antibody and fluorophore-labeled RNA probes to localize specific genes or regions along chromosomes, analyzing temporal and spatial gene expression. IF/FISH is often used to detect and co-localize the interaction between specific RNA (mRNA, lncRNA, and miRNA) and associated protein in cells and tissue samples.

To simultaneously preserve antigens for protein detection and enable RNA probe penetration for FISH, IF/FISH begins with protein IF, followed by RNA FISH. Briefly, the cells or tissue section is fixed and incubated with the specific primary antibody, and then the secondary antibody is applied. The cells or tissue section is then permeabilized and rehydrated for FISH. The specific fluorophore-labeled probe is synthesized and used to hybridize to the target RNA, where the signal is detected under a fluorescence microscope [66]. The advantage of IF/FISH is that the readout is visual, and a co-localized signal for both the RNA and protein of interest indicates possible complex formation. The main challenges of a combined IF and FISH analysis are to preserve nuclear organization and the epitope detected by the antibody (IF) as far as possible and to allow for the penetration of the FISH probe on detection of nuclear primary transcripts (RNA FISH). Achieving these goals depends on several factors: the copy number of the endogenous molecules, the length and GC content of the RNA probe, the structural features of the tissues and cells, and the sensitivity of complexes to denaturing chemicals. Many investigators have developed protocols for IF/FISH in various tissues. One needs to choose the optimized permeabilization and fixation methods for different tissues and RNA targets for the best result of combining protein IF with RNA FISH experiments [67].

3.8 Probing and Analyzing lncRNA Structure

While functional roles of several lncRNAs have been determined, the molecular mechanisms are not well understood. Because RNA functional performance is typically driven by its secondary and tertiary organization, determining the structural and functional domains of lncRNAs, as well as sequence-specific requirements, will lay the foundation for a detailed mechanistic understanding of lncRNAs.

3.8.1 Bioinformatics to Predict RNA Structure: Mfold, RNAfold, and RNA Structure

Structural architecture plays a key role in understanding the mechanism of functional RNAs. Before the mechanisms of many functional RNAs are understood, it's fundamentally important to have an extensive knowledge on secondary structure and comparative sequence analysis. Although RNA is in general single stranded, the bases have a strong propensity to interact in two principal ways, either perpendicular to their planes (stacking) or hydrogen bonded within the base planes (pairing) [68]. Paired residues are indicated by connecting lines in the RNA secondary structure diagrams composed of stems, bulges, and loops [69]. The base pairs in RNA are A–U and G–C, but the non-Watson–Crick G–U wobble pair has approximately the same stability as an A–U is commonly observed in all the medium-sized and larger RNA molecules [70]. It is reported that single-stranded regions often serve as landing pads for proteins. Loops have been shown to be critical in the recognition of small molecules as well as controlling the formation of long-distance interactions within complex RNAs. Extended stem structures have been implicated in RNA-based diseases [71].

Computational calculations which determine the secondary structure (the base-paired helical regions) of an RNA molecule using energy calculations are readily available. The Mfold web server (http://unafold.rna.albany.edu/?q=mfold) uses minimum free energy (MFE) RNA structure prediction algorithm and provides easy access to RNA and DNA folding and hybridization softwares [72]. The RNAfold web server (http://rna.tbi.univie.ac.at/cgi-bin/RNAfold.cgi) also uses MFE RNA structure prediction algorithm and predicts secondary structures of single-stranded RNA or DNA sequences [73]. The RNA structure (http://rna.urmc.rochester.edu/RNAstructureWeb/) is a program used to predict lowest free energy structures and base pair probabilities for RNA or DNA sequences [74]. It is also available to predict maximum expected accuracy structures which include pseudoknots. Structure prediction can be constrained using experimental data, including RNA-selective 2′-hydroxyl acylation and primer extension (SHAPE), enzymatic cleavage, and chemical modification accessibility [72].

3.8.2 RNA Secondary Structure Probing and RNA Footprinting

Chemical and enzymatic footprinting methods are used to map RNA secondary and tertiary structure, to monitor ligand interactions and conformational changes, and to study protein–RNA interactions. The binding of a protein to an RNA sequence protects the region of the RNA from ribonuclease (RNase) digestion, and this protected region is known as the protein's "footprint" [75]. Such methods provide data at single-nucleotide resolution that nicely complements the structural information available from X-ray diffraction, nuclear magnetic resonance (NMR) spectroscopy, or cryo-electron microscopy. Conventional footprinting methods are among the most powerful ones to obtain first valuable insight at nucleotide resolution into RNA–protein and RNA–ligand interactions. They have the advantage of requiring little material and not having technical constraints like the need for good-quality crystals for X-ray diffraction or high solubility samples for NMR.

In this protocol, end-labeled RNAs with and without bound protein are digested with RNase, and the products of digestion are analyzed by gel electrophoresis on denaturing polyacrylamide gels. If the experiment is performed properly, a comparison of the banding patterns from the two samples will reveal the binding site of the protein or "footprint," which will be detected as a region without bands in the protein-bound sample. In the sample without bound protein, the bands should cover the entirety of the RNA molecule. To establish appropriate digestion conditions for the procedure (i.e., ≤1 cleavage event per molecule), it is necessary to titrate the amount of RNase under a range of time and temperature conditions. RNase I cleaves after every nucleotide of RNA and works well under many assay conditions, but other enzymes with different cleavage specificities can also be used. RNase VI is preferable for analyzing structured RNA, while RNase A is for using pyrimidine-rich RNAs and RNase T1 for G-rich RNAs [76]. Choosing enzymes with preference for double-stranded (such as RNase VI) versus single-stranded (such as RNase I) RNA may also be helpful. Often, a combination of nucleases is advantageous.

3.8.3 RNA-Selective 2′-Hydroxyl Acylation and Primer Extension (SHAPE)

RNA has the unique ability to base-pair with itself and other nucleic acids to form complex structures. Hence the information content in RNA is encoded not only in its linear sequence of bases but also in the complex folding of RNA molecules. Chemical probing is a powerful technique for RNA structure analysis, where the RNA of interest is "modified" by a chemical reagent in some way. The reagent can be a small organic molecule, a metal ion, or an RNase enzyme. The modification can be performed using purified components in vitro or in a complex biological environment in situ or in vivo, where the reaction with the RNA is relatively limited and any two modification events are uncorrelated. It can result in either the cleavage of the RNA or the formation of a covalent chemical adduct between the RNA and probe molecule. Chemical cleavage is usually detected by resolving end-labeled RNA fragments by size. Both classes of modification, cleavage, and adduct formation can be detected as a stop to primer extension mediated by a reverse transcriptase enzyme with sites of modification inferred from the length of the resulting cDNA fragments [77].

A general chemical functionality that all RNAs have is a 2′-hydroxyl group in the ribose ring, and the reactivity of the 2′-hydroxyl in RNA is gated by local nucleotide flexibility. In other words, the 2′-hydroxyl is reactive at single-stranded and conformationally flexible positions, but unreactive at nucleotides constrained by base pairing. Recent efforts have been focused on developing reagents that modify RNA as a function of RNA 2′-hydroxyl group reactivity. Such RNA structure probing techniques can be read out by primer extension in experiments termed RNA SHAPE [71].

SHAPE chemistry interrogates local nucleotide dynamics in RNAs of virtually any size in a single experiment. A SHAPE experiment employs hydroxyl-selective electrophiles that react with the 2′-hydroxyl group to form a 2′-O-ester adduct over periods of a few minutes, but that can be as short as 1 s. SHAPE measures local nucleotide dynamics because flexible nucleotides are better able to adopt conformations conducive to electrophilic attack by the 2′-hydroxyl group [71]. Mechanistic studies show that SHAPE reactivities are largely independent of nucleotide type and correlate strongly with the model-free order parameter calculated from NMR relaxation measurements on RNA, suggesting that this chemistry provides a quantitative measure of local nucleotide flexibility.

RNA structure probing has continued to evolve from using blunt measurements through RNA cleavage to high-resolution analysis with chemical probes [77]. The 2′-hydroxyl acylation has emerged as a leading chemical probing technique owing to its generality and quantitative output. The extension of this technology to in vivo measurements presents a mandate that researchers who focus on understanding RNA structure and function should begin to study RNA structure within the native environment of the cell. Enabling in vivo SHAPE measurements transcriptome-wide to yield an RNA structure-ome that measures all four nucleobases will be a major milestone in the near future.

3.8.4 Parallel Analysis of RNA Structure (PARS)

To increase the throughput of obtaining experimental RNA structure data independent of RNA size, Chang's team developed the parallel analysis of RNA structure (PARS) method which used structure-specific enzymes for in vitro profiling of the secondary structure of RNA species at single-nucleotide resolution [78, 79]. PARS couples traditional RNA footprinting with high-throughput sequencing to identify double- and single-stranded regions of many RNAs in solution simultaneously. The method provides genome-wide RNA structural information at single-nucleotide resolution, greatly expanding the number of RNA secondary structures probed. Briefly, double- and single-stranded bases that are present in a pool of isolated cellular RNAs are cleaved using RNase V1 and S1 nuclease, which are double- and single-strand specific, respectively. Instead of detecting the cleavage sites by running a sequencing gel, the cleavage sites are identified by converting the RNA into a cDNA library and performing high-throughput sequencing on this library. The secondary structures of thousands of RNAs can be determined simultaneously because the deep sequencing reads can be mapped specifically to the yeast transcriptome, thus enabling researchers to identify where the cleavage event has occurred. The extent of enzymatic cleavage at a base is estimated by counting the number of reads that start at that particular base. Depending on the sequencing depth of the libraries, structural information on thousands of RNAs can be obtained [78]. This information can provide insight into the structural properties of most mRNAs in the cell, as well as how structural organization of groups of genes can affect cellular biology.

In addition to yeast, the PARS approach should be applicable to probing in vitro the structures of RNAs isolated from many other organisms [79]. It should also be useful for studying structural RNA changes under different solution conditions. However, performing PARS on organisms with highly repetitive genomes may be challenging because of the difficulty in mapping the sequencing reads to the individual transcripts accurately. Similarly to an RNA sequencing library, paired-end sequencing of a PARS library can enhance mapping of sequencing reads, as information from the two ends of the read can extend beyond the repetitive region and reside in a region with unique sequence content. This enables the accurate assignment of the read to a specific transcript.

3.9 Identifying RNA–Protein Interfaces

RNAs are flexible molecules that display complex secondary and tertiary structures almost as diverse as their function. Although they are commonly single stranded, their structures include short lengths of double helices, hairpin loops, bulges, and pseudoknots. Proteins tend to interact with RNA, which forms complex secondary structure elements such as stem-loops and bulges. In addition, non-Watson–Crick base pairing can occur in loop regions of RNA structures which can also be preferentially identified by proteins. Protein–RNA interactions are essential for many biological processes. However, the structural mechanisms underlying these interactions are not fully understood.

3.9.1 RNA Mechanically Induced Trapping of Molecular Interactions (RNA-MITOMI)

MITOMI was originally developed to map the binding affinities of proteins to various DNA target sequences, but MITOMI applications have since been expanded to measure a broad range of molecular interactions, including protein–RNA, protein–protein, antibody–protein, and protein–small molecule [80]. The MITOMI platform combines two different techniques, microarrays and a new microfluidic detection mechanism. A MITOMI microarray consists of a planar substrate, usually an epoxy-coated glass slide, onto which minute amounts of biological solutions are printed using standard microarray robots. Slides can contain thousands of spots of biological solutions, including DNA, RNA, and small molecules. Once spotted, a microfluidic chip is aligned and bonded to the microarray. MITOMI chips generally consist of two layers: a flow layer, which is in contact with the substrate and contains a network of microfluidic channels to perform the biological assays, and a control layer, which comprises MITOMI button membranes and valves to control fluid flow [81].

Aiming at identifying the contribution of RNA structural and sequence features to affinity for a given ligand, Martin et al. have developed a microfluidic platform that allows for the integrated synthesis and functional assays for programmable RNA libraries. They arrayed DNA oligos, which served as transcription templates for the RNA library, and overlaid the MITOMI chip onto the array such that each spot was compartmentalized in a unique microchamber. They then flowed an in vitro transcription mix onto the chip, transcribing RNA from each DNA spot. The transcribed poly(A)-tailed RNA molecules hybridized to the surface-immobilized capture probe. After the RNA synthesis, they used a quencher probe to quantify RNA capture in each detection chamber. Such design allowed for co-transcriptional RNA folding and quantification of RNA capture without incorporation of fluorophore-modified nucleotides or use of intercalating dyes, which can alter RNA structure or function. Next they incubated N-terminal glutathione S-transferase (GST)-tagged stem-loop-binding protein (SLBP) with Texas Red dye conjugated to a GST antibody and flowed the protein across the immobilized RNA library. Both RNA and protein captured in each chamber after equilibration using a microarray scanner were quantified. Thus the interaction of a comprehensive library of RNA mutants with stem-loop-binding protein precisely defined the RNA structural and sequence features that govern affinity [80].

RNA-MITOMI analysis is complementary with sequencing-based methods for RNA interactomics, which can identify thousands of putative RNA–ligand interactions that need to be validated. Furthermore, motifs may be recombined and assayed to support engineering of ncRNAs or untranslated regions with new functions.

3.9.2 Domain-Specific ChIRP (dChIRP)

In 2011, Chang and colleagues reported chromatin isolation by RNA purification (ChIRP), a genome-wide RNA-centric approach to identify RNA–chromatin interactions [49]. To obtain deeper insights into lncRNA function, they recently reported a variation on ChIRP, called dChIRP (domain-specific ChIRP), which dissects the functions of individual domains of a lncRNA within its natural environment, the cell [82]. For an RNA of interest, dChIRP can identify domain-level intramolecular and intermolecular RNA–RNA, RNA–protein, and RNA–DNA interactions and map the RNA's genomic binding sites with higher precision than domain-agnostic methods.

The dChIRP approach involves first designing biotinylated, antisense oligonucleotide pools that target specific domains (defined by functional evidence or simply by dividing up the lncRNA by length) of alncRNA. Next, whole cells are cross-linked to preserve lncRNA interactions. The chromatin fraction is sonicated to fragment the lncRNA into domain-sized lengths. The pooled, biotinylated oligonucleotides are then added to the divided chromatin samples and allowed to hybridize. Finally, the lncRNA regions of interest are purified along with any binding partners, on magnetic streptavidin beads [82, 83]. Each purification is then subjected to different analyses: immunoblotting, reverse transcription quantitative PCR, and quantitative PCR or sequencing to identify RNA–protein, RNA–RNA, and RNA–chromatin interactions, respectively.

Chang's team applied dChIRP to study putative functional domains in the roX1 lncRNA of *Drosophila melanogaster* [82]. Their analysis revealed the functional domain architecture of roX1, suggesting "three-fingered hand" structure with three distinct domains extending from a "palm." The team found that these finger domains interacted with chromatin and the male-specific lethal riboprotein complex and behaved as independent functional RNA subunits. Importantly, by focusing only on the domains that strongly associate with DNA, dChIRP improved the signal-to-noise ratio by >20-fold over traditional ChIRP, which interrogated interactions across the whole of roX1. This method enables lncRNAs to be characterized at the domain level, thus providing insights into how the modularity of specific RNA domains could contribute to the diverse functions of lncRNAs in gene expression regulation.

3.9.3 RNase Protection Assay

The RNase protection assay (RPA) was first used to analyze RNA structures and then has been adapted to study RNA–protein complexes [84]. Subsequently, oligonucleotide-targeted RNase H protection assay had been developed and became a powerful approach to analyze protein-binding domains in ribonucleoprotein particles (RNPs). In such an assay, the RNA component of a RNP and, in an essential control reaction, the corresponding deproteinized RNA are targeted with an antisense DNA oligonucleotide and RNase H. If the oligonucleotide is able to anneal to the complementary sequence of the RNA, RNase H will cleave the RNA within the double-stranded DNA/RNA region. However, protein binding to a specific RNA sequence may prevent hybridization of the DNA oligonucleotide, thereby protecting the RNA molecule from endonucleolytic cleavage, indicating a site of interaction between protein and RNA [85].

RPA assays allow greater flexibility in the integrity of target RNA, requiring very short segments for hybridization and detection. RNase H requires only a four-base-pair hybrid with a DNA probe in order to cleave the RNA molecule of interest. Using many small probes allows the entire sequence of RNA to be mapped for sites of interaction. On the one hand, an RNase H protection analysis can usually be carried out with crude cell extract and does not require further RNP purification. On the other hand, purified RNP fractions are preferable when a crude extract contains RNase activity or a heterogeneous RNP population of a specific RNA. The cleavage pattern of RNase H digestion can be analyzed by Northern blotting or RT-PCR. In addition, the investigation of RNP fragments, for example, by native gel electrophoresis, may reveal important structural information about a RNP.

3.9.4 Structural Analysis of Protein–RNA Interactions with Biotin-Labeling Assay and Mass Spectrometry

Large protein–RNA structures are often less amenable to crystallographic and/or NMR analysis. To enable rapid and accurate mapping of protein–RNA contacts, Kvaratskhelia et al. have described a footprinting methodology which used biotin-labeling assay and mass spectrometry for the identification of amino acids in the protein of interest that interact with cognate RNA [86]. The method exploits differential accessibility of the primary amine modifying reagent N-hydroxysuccinimide

(NHS)–biotin to lysine residues in the free protein versus the protein–RNA complex. Subsequent mass spectrometric analysis enables accurate identification of these residues [87]. This method is a combination of the protein-footprinting steps comprising nucleoprotein modification by NHS–biotin, SDS-PAGE fractionation of different protein subunits, in-gel proteolysis, and comparative MS analysis of peptide fragments.

Monitoring lysine accessibility is a logical choice, as lysine–phosphate backbone contacts play a key role in the formation of many nucleoprotein complexes. Introducing SDS polyacrylamide gel electrophoresis (PAGE) and in-gel proteolysis prior to mass spectrometry is important for the following reasons. SDS-PAGE allows separation of individual protein subunits based on their molecular weight differences. Thereafter, contact lysines can be accurately assigned to individual components of a multi-subunit complex. The methodology can be expanded to probe other RNA-interacting amino acids such as Arg, Trp, Tyr, His, and Cys using corresponding commercially available reagents and represents a powerful tool for high-resolution solution structural studies of protein–nucleic acid complexes [88].

3.10 Detecting Regulation of lncRNA Expression

Recently increased attention has been focused on the regulation of the expression of lncRNAs, which are transcribed by RNA polymerase II. Many lncRNAs are expressed in a tissue- or development-specific pattern and have the hallmarks typical of pol II-transcribed gene products, 7-methylguanosine capping, and polyadenylation. The expression of lncRNA is tightly controlled. At the transcriptional level, DNA methylation at CpG dinucleotides and histone modification is the main epigenetic mechanism to regulate lncRNA expression. Meanwhile, lncRNA has been shown to be the target of well-known transcription factors. At the posttranscriptional level, lncRNA can be targeted by miRNA- and RNA-binding proteins, which causes lncRNA degradation. Here we will discuss the methylation-specific PCR and bisulfate sequencing PCR methods to detect DNA methylation, the chromatin immunoprecipitation assay to detect histone modification and transcription factor binding, the reporter assay to characterize putative promoter or regulatory factors, and the DNA-EMSA to determine the DNA–protein interaction.

3.10.1 Methylation-Specific PCR (MSP) and Bisulfate Sequencing PCR (BSP)

DNA methylation, a stable and heritable epigenetic mark, is one of the most important mechanisms to directly modulate differential gene expression profiles depending on the cell type. It is known that a large proportion of lncRNAs are transcribed from promoters with low CpG dinucleotide contents, and methylation-specific PCR (MSP) has become a rapid method for the analysis of DNA methylation patterns in CpG islands, independent of the use of methylation-sensitive restriction enzymes [89]. This assay entails initial modification of DNA by sodium bisulfite, converting all unmethylated, but not methylated, cytosines to uracil, and subsequent amplification with two primer pairs, which are detectable methylated and unmethylated DNA, respectively. MSP, which requires only small quantities of DNA, is sensitive to 0.1 % methylated alleles of a given CpG island locus and can be performed on DNA extracted from paraffin-embedded samples. It also eliminates the false-positive results inherent to previous PCR-based approaches which rely on differential restriction enzyme cleavage to distinguish methylated from unmethylated DNA. In addition, by incorporating some basic automation, samples can be prepared and analyzed in a 96-well plate format. Therefore, the method can be used either quantitatively (qRT-PCR based) or qualitatively (using agarose gels) to detect changes in DNA methylation.

Meanwhile, bisulfate sequencing PCR (BSP) is the gold standard for methylation detection and can quantitate the detection of gene methylation level. This method is based on the finding that the amination reactions of cytosine and 5-methylcytosine (5mC) proceed with very different consequences after the treatment of sodium bisulfate [90]. In this regard, cytosines in single-stranded DNA will be converted into uracil residues and recognized as thymine in subsequent PCR amplification and sequencing, while 5mCs are immune to this conversion and remain as cytosines to be distinguished from unmethylated cytosines. A subsequent PCR process is necessary to determine the methylation status in the loci of interest by using specific methylation primers after the bisulfite treatment. The actual methylation status can be determined either through direct PCR product sequencing (detection of average methylation status) or subcloning sequencing (detection of single-molecule distribution of methylation patterns). Moreover, bisulfite sequencing analysis cannot only indentify DNA methylation status along the DNA single strand but also detect the DNA methylation patterns of DNA double strands since the converted DNA strands are no longer self-complementary and the amplification products can be measured individually [89].

Taken together, MS-PCR is a more rapid way to detect changes in DNA methylation than bisulfite sequencing, with the former providing only a relative difference in the CpG density within a genomic region and the latter the single-nucleotide resolution of CpG methylation. However, bisulfite-based DNA methylation analysis has more quantitative accuracy, detection sensitivity, high efficiency, and a wide spectrum for sample analysis.

3.10.2 Chromatin Immunoprecipitation (ChIP)

The ChIP method can be used to monitor DNA-binding factors, such as transcription factors, posttranscriptionally modified histones, and members of the transcription complex, to assess regulatory input, epigenetic modifications, and transcriptional activity [91]. The ChIP assay method allows for the analysis of DNA–protein interactions in living cells by treating the cells with formaldehyde or other cross-linking reagents in order to stabilize the interactions for downstream purification and detection. In this approach, an antibody to the transcription factor of interest is used to immunoprecipitate the protein–DNA complex that has been cross-linked in whole cells. The immunoprecipitation is then subjected to PCR using primers that flank the DNA element of interest, after which the products are quantified using electrophoresis or qPCR [92]. In order to visualize these complexes, it is essential to have a tissue or cell line that expresses the protein of interest, as the transcription factors bound to a gene may vary depending on the cell type and level of gene expression, and designing an oligonucleotide sequence that will yield an adequate PCR product may be challenging at times. Despite the challenges, particularly in characterizing novel transcription factors, ChIP assay is still a powerful approach to demonstrate protein–DNA interactions in vivo.

3.10.3 Reporter Assay

Since many lncRNA promoters have been found to have putative transcription factor binding sites, the reporter assay can be used to test such putative promoter and regulatory sequence or even the enhancer and repressor regions. The reporter genes are fusions of a target promoter DNA sequence and a reporter gene DNA sequence. The promoter DNA sequence is customized by the researcher, and the reporter gene DNA sequence is coded for a protein with detectable properties such as firefly luciferase, *Renilla luciferase*, or alkaline phosphatase. These genes produce enzymes only when the promoter of interest is activated. The enzyme, in turn, catalyzes a substrate to produce either light, color change or other reactions that can be detected by spectroscopic instrumentation. With the signal from the reporter gene being used as an indirect determinant for the translation of endogenous proteins driven from the same promoter, reporter assay provides a sensitive readout of the transcriptional activity of a DNA fragment and can be used to test the effect of specific treatments on the transcriptional activity of a promoter.

3.10.4 DNA Electrophoretic Mobility Shift Assay (DNA-EMSA)

The EMSA is used to study proteins binding to known DNA oligonucleotide probes and to assess the degree of affinity or specificity of the interaction [93]. The technique is based on the observation that protein–DNA complexes migrate more slowly than free DNA molecules when subjected to nondenaturing polyacrylamide or agarose gel electrophoresis. Because the rate of DNA migration is shifted or retarded upon protein binding, the assay is also referred to as a gel shift or gel retardation assay. Adding a protein-specific antibody to the binding components creates an even larger complex (antibody–protein–DNA) which migrates even slower during electrophoresis. This is known as a supershift and can be used to confirm protein identities.

The gel shift assay is performed by incubating a purified recombinant protein, nuclear, or cell extract with [32P]-end-labeled DNA fragment containing the putative protein binding site. The reaction products are then analyzed on a nondenaturing polyacrylamide gel. The specificity of this interaction is confirmed by competition experiment using unlabeled DNA fragment of interest [93]. Now nonradioactive EMSA using biotinylated or fluorescently labeled DNA probes has become available. EMSA can be used to study the gene regulation and determine the DNA–protein interaction, especially in detecting low-abundance DNA-binding proteins from lysates. Moreover, this method is good for testing binding site mutations using many probe configurations with the same lysate and binding affinity through DNA probe mutational analysis.

3.11 Animal Models of lncRNA

Numerous murine models have been developed to study human cancer. One of the most widely used models is the human tumor xenograft, where human tumor cells are transplanted either subcutaneously or orthotopically into immunocompromised mice. Another model is the genetically engineered mouse (GEM) model, where the genetic profile of the mice is altered such that one or several genes thought to be involved in the transformation or malignancy are overexpressed, deleted, or mutated. Each model has its use in studying lncRNA function in cancer initiation, development, and metastasis.

3.11.1 Genetically Engineered Mouse (GEM) Models

The xenograft models that use human tumor cell lines or primary human tumors have been used for decades to increase our understanding of factors affecting tumor growth. However, recent information regarding the key influence of the tumor microenvironment on tumor progression and growth has led to greater reliance on GEM tumor models using immunocompetent mice. In the GEM model, the tumor microenvironment can be mirrored as much as possible in a murine tumor. Moreover, specific genetic abnormalities can be induced at specific ages in the tissue-type of origin, and the stages of tumor progression can be studied over time. To probe the in vivo function of lncRNAs in cancer, one can generate transgenic overexpression or knockout mice and either observe the spontaneous tumor formation or cross the mice with relevant tumor models to examine tumor onset or metastasis.

3.11.1.1 Transgenic Mouse Model

In 1974, Rudolf Jaenisch injected SV40 virus DNA into a mouse blastocyst, which created the first mouse carrying the exogenous gene. Later, some researchers injected the murine leukemia virus into mouse embryos, producing the mouse that was stably genetic through reproductive system and could pass along stable exogenous gene expression to its offspring. These mice are called transgenic mice. The basic principle of constructing transgenic mouse is that the reconstructed target genes (or the genome fragments) are injected into the donor mouse's fertilized eggs (or embryo cells before implantation) by microscopic injections, and then this fertilized egg (or embryo cells before implantation) is implanted into recipient animals' oviduct or uterus in order to develop into transgenic animals with foreign genes [94]. This technique not only reveals the function of exogenous gene but also serves for the mass production of engineered animals through analysis of the relationship between the transgene and the phenotype of experimental mice.

Currently, transgenic mice for lncRNA overexpression have been generated for experiments. Embryonic stem (ES) cell gene targeting technology is the most common method. Guo et al. have generated bone marrow-expressing lncRNA-BGL3 transgenic mice by the microinjection and found expression of lncRNA-BGL3 in the transgenic mice significantly inhibits Bcr–Abl-mediated primary bone marrow transformation [95]. Gomez et al. have generated mice that expressed NeST RNA transgenically by pronuclear microinjections. NeST cDNA was cloned downstream of a CD4 + and CD8+ T cell-specific promoter, and mice that expressed NeST RNA, either in its natural chromosomal environment or by transgenic delivery, displayed increased resistance to *Salmonella*-induced pathogenesis but increased susceptibility to Theiler's virus persistence [96]. The generation of transgenic mouse model can determine the role of lncRNAs in normal development and pathogenesis, while the implication of lncRNAs in tumorigenesis, metastasis, and progression remains to be further investigated in transgenic mouse models.

3.11.1.2 Knockout Mouse Model

For the in vivo loss-of-function analysis, knockout (KO) mouse model of lncRNA can be used. Most of the KO strategies of the protein-coding genes need the translation of the RNA transcript to achieve the disruptive effect and are not feasible for lncRNAs. Thus, lncRNA targeting strategies must prevent the whole transcript from being made [97].

The first strategy to knock out the lncRNAs requires the deletion of the complete gene sequence which relies on homologous recombination and can be used to generate constitutive or conditional knockout animals. The lncRNA MALAT1 knockout mice were constitutively generated by this method [98].

The second strategy to knock out the lncRNAs replaces the lncRNA sequence with a reporter gene (monoallelic or biallelic), e.g., LacZ. If the endogenous lncRNA promoter is kept intact during the replacement, it can drive reporter gene expression to reveal lncRNA expression patterns in vivo. A large-scale knockout project to investigate the phenotypes of 18 lncRNA knockouts in mice has used this method [99].

The third strategy for the lncRNA KO is to remove the endogenous promoter. The deletion can be small (only a few hundred base pairs) and might only minimally perturb the genomic locus (in contrast to the deletion of the complete gene) [97]. However, many protein-coding and noncoding genes have alternative promoters and hence will retain expression of one or more isoforms, if only one promoter is targeted. Moreover, promoter deletions may affect both lncRNA and protein-coding genes in close proximity that share the promoter sequence.

The fourth strategy for targeting lncRNAs is the integration of strong transcriptional stop signals at the very 5′ end of the noncoding transcript. The integration of polyadenylation (poly A) signals at the beginning of the transcript causes premature cleavage and polyadenylation of the lncRNA and finally its degradation [97]. This approach can yield strong, but also incomplete, lncRNA depletion, depending on the lncRNA abundance.

A novel strategy to target lncRNAs is the use of CRISPR/Cas9 system for engineering large genomic deletions of lncRNAs. The CRISPR/Cas9 system is an efficient genome-editing tool for deletion of protein-coding genes. However, conventional gene targeting by homologous recombination is not efficient for the generation of large fragment deletions. Han et al. have used the CRISPR/Cas9 system and paired sgRNAs to generate large fragment deletions of up to 23 kb of the maternally expressed lncRNA gene, Rian, in mice, and demonstrated that the deletions are precise and heritable. They used the CRISPR/Cas9 system with two sgRNAs, N-sgRNA1 and C-sgRNA1, to induce a 23 kb deletion in the Rian gene, and found targeting efficiency could increase with more sgRNAs [45]. Although there are challenges in using the CRISPR/Cas9 system to delete lncRNA in vivo, future development of this technology will largely contribute to generating better animal models for lncRNA research.

3.11.2 *Tumor Xenografts*

Given that not every human lncRNA has the orthologous mouse counterpart, it is important to use the human tumor xenograft models to study the in vivo function of cancer-related lncRNA. First the expression of lncRNAs is manipulated by overexpression or knockdown in cancer cells. Then the cancer cells with stable expression of lncRNA or shRNA are injected into athymic nude mice, severe combined immunodeficiency (SCID) mice, or nonobese diabetic (NOD)/SCID mice, under the skin or into the organ type in which the tumor originated or from the tail vein. Depending on the number of cells injected, the tumor will develop over 1–8 weeks (or in some instances 1–4 months or longer). One can observe the tumor formation by measuring the tumor volume and weight and examine metastasis after euthanizing the animal.

Many investigators have used human tumor xenografts to evaluate the lncRNA function in tumor growth and metastasis. To quantify metastatic potential of HOX transcript antisense RNA (HOTAIR) in vivo, Gupta et al. have performed tail vein xenografts and compared the rates of lung colonization. HOTAIR expression allowed SK-BR3 cells to colonize the lung in 80 % of animals, whereas vector expression in the nonmetastatic cell line SK-BR3 never showed lung colonization [100]. NF-κB interacting long noncoding RNA (NKILA) increased apoptosis and reduced migration in breast cancer cells. To further study the function of NKILA in vivo, Liu et al. have overexpressed or knocked down NKILA expression in breast cancer cells. They found ectopic NKILA expression in MBA-MD-231-inhibited cancer metastasis to the lungs, liver, and lymph nodes and prolonged mice survival, whereas silencing NKILA in MCF7 significantly promoted cancer metastasis and reduced mice survival [101].

3.12 Conclusion

In summary, both the human tumor xenograft and the GEM models are useful for enhancing our understanding of lncRNA function in cancer development. Each has its strengths and limitations, with the GEM model being best for examining the role of specific lncRNAs in tumor development and progression and the human tumor xenograft being irreplaceable for human lncRNA that does not have mouse ortholog. In the future, these models may be used not only for determining the consequences of alterations in individual lncRNA but also for determining the mechanisms of cooperation between multiple genes in tumorigenesis and for evaluating novel approaches that may target specific lncRNA or pathways for cancer therapeutics.

References

1. Quek XC, Thomson DW, Maag J-LV, et al. LncRNAdb v2.0: expanding the reference database for functional long non-coding RNAs. Nucleic Acids Res. 2015;43(Database issue):D168–73.
2. Bu D, Yu KT, Sun SL, et al. NONCODE v3.0: integrative annotation of long non-coding RNAs. Nucleic Acids Res. 2012;40(Database issue):D210–5.
3. Volders PJ, Verheggen K, Menschaert G, et al. An update on LNCipedia: a database for annotated human lncRNA sequences. Nucleic Acids Res. 2015;43(Database issue):D174–80.
4. RNAcentral Consortium. RNAcentral: an international database of ncRNA sequences. Nucleic Acids Res. 2015;43(Database issue):D123–9.
5. Fatica A, Bozzoni I. Long non-coding RNAs: new players in cell differentiation and development. Nat Rev Genet. 2014;15(1):7–21.
6. Derrien T, Johnson R, Bussotti G, et al. The GENCODE v7 catalog of human long noncoding RNAs: analysis of their gene structure, evolution, and expression. Genome Res. 2012;22(9):1775–89.
7. Marguerat S, Bahler J. RNA-seq: from technology to biology. Cell Mol Life Sci. 2010;67(4):569–79.
8. Ozsolak F, Milos PM. RNA sequencing: advances, challenges and opportunities. Nat Rev Genet. 2011;12(2):87–98.
9. Wang Z, Gerstein M, Snyder M. RNA-seq: a revolutionary tool for transcriptomics. Nat Rev Genet. 2009;10(1):57–63.
10. Atkinson SR, Marguerat S, Bahler J. Exploring long non-coding RNAs through sequencing. Semin Cell Dev Biol. 2012;23(2):200–5.
11. Heyer EE, Ozadam H, Ricci EP, et al. An optimized kit-free method for making strand-specific deep sequencing libraries from RNA fragments. Nucleic Acids Res. 2015;43(1):e2.
12. Mili S, Steitz JA. Evidence for reassociation of RNA-binding proteins after cell lysis: implications for the interpretation of immunoprecipitation analyses. RNA. 2004;10(11):1692–4.
13. Zhao J, Ohsumi TK, Kung JT, et al. Genome-wide identification of polycomb-associated RNAs by RIP-seq. Mol Cell. 2010;40(6):939–53.
14. Zhang C, Darnell RB. Mapping in vivo protein-RNA interactions at single-nucleotide resolution from HITS-CLIP data. Nat Biotechnol. 2011;29(7):607–14.
15. Olivarius S, Plessy C, Carninci P. High-throughput verification of transcriptional starting sites by Deep-RACE. Biotechniques. 2009;46(2):130–2.
16. Pastori C, Peschansky VJ, Barbouth D, et al. Comprehensive analysis of the transcriptional landscape of the human FMR1 gene reveals two new long non-coding RNAs differentially expressed in Fragile X syndrome and Fragile X-associated tremor/ataxia syndrome. Hum Genet. 2014;133(1):59–67.
17. Wheeler DL, Church DM, Federhen S, et al. Database resources of the National Center for Biotechnology. Nucleic Acids Res. 2003;31(1):28–33.
18. Lin MF, Jungreis I, Kellis M. PhyloCSF: a comparative genomics method to distinguish protein coding and non-coding regions. Bioinformatics. 2011;27(13):i275–82.
19. Kong L, Zhang Y, Ye ZQ, et al. CPC: assess the protein-coding potential of transcripts using sequence features and support vector machine. Nucleic Acids Res. 2007;35(Web Server issue):W345–9.
20. Wang L, Park HJ, Dasari S, et al. CPAT: Coding-Potential Assessment Tool using an alignment-free logistic regression model. Nucleic Acids Res. 2013;41(6):e74.
21. Sun L, Luo H, Bu D, et al. Utilizing sequence intrinsic composition to classify protein-coding and long non-coding transcripts. Nucleic Acids Res. 2013;41(17):e166.
22. Ingolia NT. Ribosome profiling: new views of translation, from single codons to genome scale. Nat Rev Genet. 2014;15(3):205–13.

23. Ingolia NT, Ghaemmaghami S, Newman JR, et al. Genome-wide analysis in vivo of translation with nucleotide resolution using ribosome profiling. Science. 2009;324(5924):218–23.
24. Ingolia NT, Lareau LF, Weissman JS. Ribosome profiling of mouse embryonic stem cells reveals the complexity and dynamics of mammalian proteomes. Cell. 2011;147(4):789–802.
25. Guttman M, Russell P, Ingolia NT, et al. Ribosome profiling provides evidence that large non-coding RNAs do not encode proteins. Cell. 2013;154(1):240–51.
26. Zhang K, Shi ZM, Chang YN, et al. The ways of action of long non-coding RNAs in cytoplasm and nucleus. Gene. 2014;547(1):1–9.
27. Cabili MN, Dunagin MC, McClanahan PD, et al. Localization and abundance analysis of human lncRNAs at single-cell and single-molecule resolution. Genome Biol. 2015;16:20.
28. Pinaud R, Mello CV, Velho TA, et al. Detection of two mRNA species at single-cell resolution by double-fluorescence in situ hybridization. Nat Protoc. 2008;3(8):1370–9.
29. Dunagin M, Cabili MN, Rinn J, et al. Visualization of lncRNA by single-molecule fluorescence in situ hybridization. Methods Mol Biol. 2015;1262:3–19.
30. Streit S, Michalski CW, Erkan M, et al. Northern blot analysis for detection and quantification of RNA in pancreatic cancer cells and tissues. Nat Protoc. 2009;4(1):37–43.
31. Nolan T, Hands RE, Bustin SA. Quantification of mRNA using real-time RT-PCR. Nat Protoc. 2006;1(3):1559–82.
32. Schmittgen TD, Livak KJ. Analyzing real-time PCR data by the comparative C(T) method. Nat Protoc. 2008;3(6):1101–8.
33. Rosner M, Siegel I, Fuchs C, et al. Efficient siRNA-mediated prolonged gene silencing in human amniotic fluid stem cells. Nat Protoc. 2010;5(6):1081–95.
34. Schaniel C, Li F, Schafer XL, et al. Delivery of short hairpin RNAs – triggers of gene silencing – into mouse embryonic stem cells. Nat Methods. 2006;3(5):397–400.
35. Moore CB, Guthrie EH, Huang MT, et al. Short hairpin RNA (shRNA): design, delivery, and assessment of gene knockdown. Methods Mol Biol. 2010;629:141–58.
36. Teplova M, et al. Crystal structure and improved antisense properties of 2′-O-(2-methoxyethyl)-RNA. Nat Struct Biol. 1999;6(6):535–9.
37. Obernosterer G, Martinez J, Alenius M. Locked nucleic acid-based in situ detection of microRNAs in mouse tissue sections. Nat Protoc. 2007;2(6):1508–14.
38. Grunweller A, Hartmann RK. Locked nucleic acid oligonucleotides: the next generation of antisense agents? BioDrugs. 2007;21(4):235–43.
39. Lennox KA, Behlke MA. Cellular localization of long non-coding RNAs affects silencing by RNAi more than by antisense oligonucleotides. Nucleic Acids Res. 2016;44(2):863–77. doi:10.1093/nar/gkv1206.
40. Boettcher M, McManus MT. Choosing the right tool for the job: RNAi, TALEN, or CRISPR. Mol Cell. 2015;58(4):575–85.
41. Cong L, Ran FA, Cox D, et al. Multiplex genome engineering using CRISPR/Cas systems. Science. 2013;339(6121):819–23.
42. Mali P, Esvelt KM, Church GM. Cas9 as a versatile tool for engineering biology. Nat Methods. 2013;10(10):957–63.
43. Esvelt KM, Smidler AL, Catteruccia F, et al. Concerning RNA-guided gene drives for the alteration of wild populations. Elife. 2014;3(1), e03401. doi:10.7554/eLife.03401.
44. Ho TT, Zhou N, Huang J, et al. Targeting non-coding RNAs with the CRISPR/Cas9 system in human cell lines. Nucleic Acids Res. 2015;43(3):e17.
45. Han J, Zhang J, Chen L, et al. Efficient in vivo deletion of a large imprinted lncRNA by CRISPR/Cas9. RNA Biol. 2014;11(7):829–35.
46. Maeder ML, Linder SJ, Reyon D, et al. Robust, synergistic regulation of human gene expression using TALE activators. Nat Methods. 2013;10(3):243–5.
47. Perez-Pinera P, Ousterout DG, Brunger JM, et al. Synergistic and tunable human gene activation by combinations of synthetic transcription factors. Nat Methods. 2013;10(3):239–42.

48. Konermann S, Brigham MD, Trevino AE, et al. Genome-scale transcriptional activation by an engineered CRISPR-Cas9 complex. Nature. 2015;517(7536):583–8.
49. Chu C, Qu K, Zhong FL, et al. Genomic maps of long non-coding RNA occupancy reveal principles of RNA-chromatin interactions. Mol Cell. 2011;44(4):667–78.
50. Chu C, Quinn J, Chang HY. Chromatin isolation by RNA purification (ChIRP). J Vis Exp. 2012;61:3912. doi:10.3791/3912.
51. Davis CP, West JA. Purification of specific chromatin regions using oligonucleotides: capture hybridization analysis of RNA targets (CHART). Methods Mol Biol. 2015;1262:167–82.
52. Simon MD. Capture hybridization analysis of RNA targets (CHART). Curr Protoc Mol Biol. 2013; Chapter 21:Unit 21 25. doi:10.1002/0471142727.mb2125s101.
53. Engreitz JM, Pandya-Jones A, McDonel P, et al. The Xist lncRNA exploits three-dimensional genome architecture to spread across the X chromosome. Science. 2013;341(6147):1237973.
54. Engreitz J, Lander ES, Guttman M. RNA antisense purification (RAP) for mapping RNA interactions with chromatin. Methods Mol Biol. 2015;1262:183–97.
55. Chu C, Spitale RC, Chang HY. Technologies to probe functions and mechanisms of long non-coding RNAs. Nat Struct Mol Biol. 2015;22(1):29–35.
56. Michlewski G, Caceres JF. RNase-assisted RNA chromatography. RNA. 2010;16(8):1673–8.
57. Chu C, Zhang QC, da Rocha ST, et al. Systematic discovery of Xist RNA binding proteins. Cell. 2015;161(2):404–16.
58. McHugh CA, Chen CK, Chow A, et al. The Xist lncRNA interacts directly with SHARP to silence transcription through HDAC3. Nature. 2015;521(7551):232–6.
59. Bantscheff M, Schirle M, Sweetman G, et al. Quantitative mass spectrometry in proteomics: a critical review. Anal Bioanal Chem. 2007;389(4):1017–31.
60. McHugh CA, Russell P, Guttman M. Methods for comprehensive experimental identification of RNA-protein interactions. Genome Biol. 2014;15(1):203.
61. Huarte M, Guttman M, Feldser D, et al. A large intergenic non-coding RNA induced by p53 mediates global gene repression in the p53 response. Cell. 2010;142(3):409–19.
62. Bonifacino JS, Dell'Angelica ED, Springer TA. Immunoprecipitation. Curr Protoc Neurosci. 2006; Chapter 5:Unit 5 24.
63. Garner MM, Revzin A. A gel electrophoresis method for quantifying the binding of proteins to specific DNA regions: application to components of the Escherichia coli lactose operon regulatory system. Nucleic Acids Res. 1981;9(13):3047–60.
64. Fried M, Crothers DM. Equilibria and kinetics of lac repressor-operator interactions by polyacrylamide gel electrophoresis. Nucleic Acids Res. 1981;9(23):6505–25.
65. Wang K, Gao Y, Peng X, et al. Using FAM labeled DNA oligos to do RNA electrophoretic mobility shift assay. Mol Biol Rep. 2010;37(6):2871–5.
66. Zimmerman SG, Peters NC, Altaras AE, et al. Optimized RNA ISH, RNA FISH and protein-RNA double labeling (IF/FISH) in Drosophila ovaries. Nat Protoc. 2013;8(11):2158–79.
67. Chaumeil J, Augui S, Chow JC, et al. Combined immunofluorescence, RNA fluorescent in situ hybridization, and DNA fluorescent in situ hybridization to study chromatin changes, transcriptional activity, nuclear organization, and X-chromosome inactivation. Methods Mol Biol. 2008;463:297–308.
68. Turner DH, Sugimoto N, Freier SM. RNA structure prediction. Annu Rev Biophys Biophys Chem. 1988;17:167–92.
69. Zwieb C. The principles of RNA structure architecture. Methods Mol Biol. 2014;1097:33–43.
70. Giese MR, Betschart K, Dale T, et al. Stability of RNA hairpins closed by wobble base pairs. Biochemistry. 1998;37(4):1094–100.
71. Spitale RC, Flynn RA, Torre EA, et al. RNA structural analysis by evolving SHAPE chemistry. Wiley Interdiscip Rev RNA. 2014;5(6):867–81.
72. Zuker M. Mfold web server for nucleic acid folding and hybridization prediction. Nucleic Acids Res. 2003;31(13):3406–15.

73. Denman RB. Using RNAFOLD to predict the activity of small catalytic RNAs. Biotechniques. 1993;15(6):1090–5.
74. Mathews DH. Using an RNA secondary structure partition function to determine confidence in base pairs predicted by free energy minimization. RNA. 2004;10(8):1178–90.
75. Fourmy D, Yoshizawa S. Protein-RNA footprinting: an evolving tool. Wiley Interdiscip Rev RNA. 2012;3(4):557–66.
76. Peng Y, Soper TJ, Woodson SA. RNase footprinting of protein binding sites on an mRNA target of small RNAs. Methods Mol Biol. 2012;905:213–24.
77. Weeks KM. Advances in RNA structure analysis by chemical probing. Curr Opin Struct Biol. 2010;20(3):295–304.
78. Kertesz M, Wan Y, Mazor E, et al. Genome-wide measurement of RNA secondary structure in yeast. Nature. 2010;467(7311):103–7.
79. Wan Y, Qu K, Ouyang Z, et al. Genome-wide mapping of RNA structure using nuclease digestion and high-throughput sequencing. Nat Protoc. 2013;8(5):849–69.
80. Martin L, Meier M, Lyons SM, et al. Systematic reconstruction of RNA functional motifs with high-throughput microfluidics. Nat Methods. 2012;9(12):1192–4.
81. Rockel S, Geertz M, Maerkl SJ. MITOMI: a microfluidic platform for in vitro characterization of transcription factor-DNA interaction. Methods Mol Biol. 2012;786:97–114.
82. Quinn JJ, Ilik IA, Qu K, et al. Revealing long non-coding RNA architecture and functions using domain-specific chromatin isolation by RNA purification. Nat Biotechnol. 2014;32(9):933–40.
83. Quinn JJ, Chang HY. In situ dissection of RNA functional subunits by domain-specific chromatin isolation by RNA purification (dChIRP). Methods Mol Biol. 2015;1262: 199–213.
84. Carey MF, Peterson CL, Smale ST. The RNase protection assay. Cold Spring Harb Protoc. 2013;2013(3). doi:10.1101/pdb.prot071910.
85. Gunzl A, Palfi Z, Bindereif A. Analysis of RNA-protein complexes by oligonucleotide-targeted RNase H digestion. Methods. 2002;26(2):162–9.
86. Kvaratskhelia M, Miller JT, Budihas SR, et al. Identification of specific HIV-1 reverse transcriptase contacts to the viral RNA: tRNA complex by mass spectrometry and a primary amine selective reagent. Proc Natl Acad Sci U S A. 2002;99(25):15988–93.
87. Kvaratskhelia M, Grice SF. Structural analysis of protein-RNA interactions with mass spectrometry. Methods Mol Biol. 2008;488:213–9.
88. Mendoza VL, Vachet RW. Probing protein structure by amino acid-specific covalent labeling and mass spectrometry. Mass Spectrom Rev. 2009;28(5):785–815.
89. Li Y, Tollefsbol TO. DNA methylation detection: bisulfite genomic sequencing analysis. Methods Mol Biol. 2011;791:11–21.
90. Frommer M, McDonald LE, Millar DS, et al. A genomic sequencing protocol that yields a positive display of 5-methylcytosine residues in individual DNA strands. Proc Natl Acad Sci U S A. 1992;89(5):1827–31.
91. Furey TS. ChIP-seq and beyond: new and improved methodologies to detect and characterize protein-DNA interactions. Nat Rev Genet. 2012;13(12):840–52.
92. Bhan A, Mandal SS. Estradiol-induced transcriptional regulation of long non-coding RNA, HOTAIR. Methods Mol Biol. 2016;1366:395–412.
93. Evertts AG, Zee BM, Garcia BA. Modern approaches for investigating epigenetic signaling pathways. J Appl Physiol (1985). 2010;109(3):927–33.
94. Doyle A, McGarry MP, Lee NA, et al. The construction of transgenic and gene knockout/knockin mouse models of human disease. Transgenic Res. 2012;21(2):327–49.
95. Guo G, Kang Q, Zhu X, et al. A long non-coding RNA critically regulates Bcr-Abl-mediated cellular transformation by acting as a competitive endogenous RNA. Oncogene. 2015;34(14): 768–79.
96. Gomez JA, Wapinski OL, Yang YW, et al. The NeST long ncRNA controls microbial susceptibility and epigenetic activation of the interferon-gamma locus. Cell. 2013;152(4):743–54.

97. Bassett AR, Akhtar A, Barlow DP, et al. Considerations when investigating lncRNA function in vivo. Elife. 2014;3:e03058.
98. Eissmann M, Gutschner T, Hämmerle M, et al. Loss of the abundant nuclear non-coding RNA MALAT1 is compatible with life and development. RNA Biol. 2012;9(8):1076–87.
99. Sauvageau M, Goff LA, Lodato S, et al. Multiple knockout mouse models reveal lincRNAs are required for life and brain development. Elife. 2013;2:e01749.
100. Gupta RA, Shah N, Wang KC, et al. Long non-coding RNA HOTAIR reprograms chromatin state to promote cancer metastasis. Nature. 2010;464(7291):1071–6.
101. Liu B, Sun L, Liu Q, et al. A cytoplasmic NF-kappaB interacting long non-coding RNA blocks IkappaB phosphorylation and suppresses breast cancer metastasis. Cancer Cell. 2015;27(3):370–81.

Chapter 4
Relationship Between Noncoding RNA Dysregulation and Epigenetic Mechanisms in Cancer

Hiromu Suzuki, Reo Maruyama, Eiichiro Yamamoto, Takeshi Niinuma, and Masahiro Kai

Abstract Epigenetic alterations, including aberrant DNA methylation and histone modification, play key roles in the dysregulation of tumor-related genes, thereby affecting numerous cellular processes, including cell proliferation, cell adhesion, apoptosis, and metastasis. In recent years, studies have demonstrated that short and long noncoding RNAs (ncRNAs) are key players in the initiation and progression of cancer, and epigenetic mechanisms are deeply involved in their dysregulation. Indeed, the growing list of microRNA (miRNA) genes aberrantly methylated in cancer suggests that a large number of miRNAs act as tumor suppressors or oncogenes. In addition, emerging evidence suggests that dysregulation of long ncRNAs (lncRNAs) plays critical roles in tumorigenesis. And because ncRNAs are involved in regulating gene expression through interaction with epigenetic modifiers, their dysregulation appears causally related to epigenetic alterations in cancer. Dissection of the interrelationships between ncRNAs and epigenetic alterations has the potential to reveal novel approaches to the diagnosis and treatment of cancer.

Keywords Epigenetic mechanism • RNA dysregulation • ncRNA • Tumor-related genes

---

H. Suzuki (✉) • R. Maruyama • T. Niinuma • M. Kai
Department of Molecular Biology, School of Medicine, Sapporo Medical University, S1, S17, Chuo-ku, Sapporo 060-8556, Japan
e-mail: hsuzuki@sapmed.ac.jp

E. Yamamoto
Department of Molecular Biology, School of Medicine, Sapporo Medical University, S1, S17, Chuo-ku, Sapporo 060-8556, Japan

Department of Gastroenterology, Rheumatology and Clinical Immunology, School of Medicine, Sapporo Medical University, S1, S17, Chuo-ku, Sapporo 060-8556, Japan

E. Song (ed.), *The Long and Short Non-coding RNAs in Cancer Biology*,
Advances in Experimental Medicine and Biology 927,
DOI 10.1007/978-981-10-1498-7_4

4.1 Introduction

Advances in genome analysis have revealed that a much larger portion of the human genome is transcribed into RNA than previously recognized. Evidence emerging in recent years has highlighted the biological and pathological importance of RNA molecules that lack protein-coding potential, which are collectively referred to as noncoding RNAs (ncRNAs) [1]. Generally, ncRNAs are classified into two categories, small (<200 nt) and large (>200 nt), though there are many subclasses of ncRNAs within these categories. Among them, microRNAs (miRNAs) are the most intensively studied small ncRNAs in cancer.

miRNAs are endogenous, small (~22 nt) ncRNAs that function at the posttranscriptional level as negative regulators of gene expression [1, 2]. Annotation of their genomic locations suggests many miRNA genes are located within intergenic regions, though they are also found within both exons and introns in either the sense or antisense orientation. Like genes encoding proteins, miRNA genes are mainly transcribed by RNA polymerase II. They are initially transcribed as large RNA precursors called pri-miRNAs, which are processed within the nucleus by Drosha, a ribonuclease III enzyme, in collaboration with Pasha (also known as DGCR8). Evidence from numerous studies indicates that subsets of miRNAs act as tumor suppressor genes or oncogenes, and their dysregulation is a common feature of human cancers [1–3]. More specifically, expression of miRNAs is generally downregulated in tumor tissues, as compared to the corresponding normal tissues, which suggests many miRNAs may act as tumor suppressors. Although the mechanism underlying the altered miRNA expression in cancer is not yet fully understood, recent studies have shown that epigenetic mechanisms play important roles in the regulation of miRNA expression.

Epigenetic gene silencing associated with promoter CpG island hypermethylation is a common mechanism by which tumor suppressor genes are inactivated during tumorigenesis (see Fig. 4.1). It is now apparent that many miRNA genes are also targets of methylation-associated silencing in cancer, and the list of miRNAs silenced by methylation in cancer is rapidly growing (see Table 4.1) [74, 75]. Epigenetic gene regulation is also tightly linked to histone modification. For instance, promoters of transcriptionally active miRNA genes are characterized by active histone marks, such as acetylated histone H3 and trimethylated histone H3 lysine 4 (H3K4me3), while those of silenced genes are marked by repressive modifications, including dimethylated histone H3 lysine 9 (H3K9me2), trimethylated histone H3 lysine 9 (H3K9me3), and trimethylated histone H3 lysine 27 (H3K27me3) (see Fig. 4.1) [75, 76]. What's more, recent evidence indicates that miRNAs can modulate epigenetic regulation by targeting epigenetic modifiers and that miRNA dysregulation leads to aberrant DNA methylation or histone modification in a subset of tumors.

Long ncRNAs (lncRNAs) are broadly defined as transcribed RNA molecules greater than 200 nt in length and lacking an open reading frame of significant length (less than 100 amino acids) [77, 78]. Although there are no specific definitions,

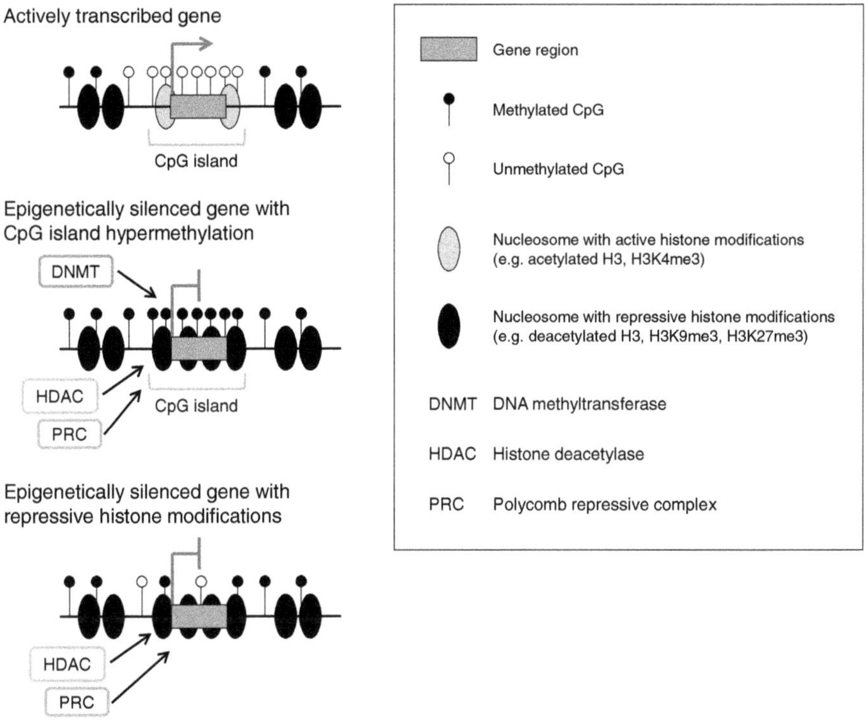

Fig. 4.1 Epigenetic regulation of gene transcription

lncRNAs can be categorized into several subgroups based on their locations and characteristics. For instance, antisense RNAs are transcribed from the opposite strand of a protein-coding gene, while lncRNAs transcribed from intergenic regions are referred to as large intergenic ncRNAs (lincRNAs) [36]. Although the specific functions of the majority of lncRNAs remain unknown, recent studies have begun to shed light on the critical roles played by these molecules in a variety of cellular processes, including differentiation, development, and tumorigenesis. Overexpression or repression of several lncRNAs in cancer is reportedly mediated by aberrant DNA methylation or histone modification. Notably, recent studies have revealed that a subset of lncRNAs act as guides or scaffolds for epigenetic modifiers and recruit them to specific genomic loci. Thus, aberrant expression of such lncRNAs could contribute to tumorigenesis by inducing aberrant epigenetic gene regulation. In this section, we will review the interrelationships between ncRNA dysregulation and epigenetic alterations in cancer and highlight their biological and clinical implications.

Table 4.1 miRNAs silenced by DNA hypermethylation in cancer

Gene	Tumor type	Target gene	References
miR-1-1	Liver, colon	FOXP1, MET, HDAC4, ANXA2, BDNF	[4, 5]
miR-9 family	Multiple types	FGFR1, CDK6, CDX2, CDH1	[6–14]
miR-10b	Stomach	MAPRE1	[15]
miR-31	Prostate, breast	AR	[16, 17]
miR-34 family	Multiple types	MET, CDK4, CCNE2, MYC, CDK6, E2F3, NOTCH4	[6, 18–28]
miR-124 family	Multiple types	CDK6, VIM, SMYD3, E2F6, IQGAP1, IGFBP7	[7, 19, 29–34]
miR-125b	Breast	ETS1	[35]
miR-127	Prostate, bladder, colon	BCL6	[36]
miR-129-2	Endometrium, colon, esophagus, stomach, liver	SOX4	[9, 24, 37–39]
miR-132	Pancreas, prostate	TALIN2	[40, 41]
miR-137	Head and neck, stomach, colon, breast	CDK6, CDC42, LSD1	[9, 42]
miR-143	Leukemia	MLL-AF4	[43]
miR-145	Prostate, lung	MYC, FSCN1, EGFR, MUC1	[44–46]
miR-148a	Colon, head and neck, lung, breast, pancreas, brain	TGIF2	[6, 47]
miR-152	Endometrium, bladder, lung	DNMT1, E2F3, MET, RICTOR	[7, 48]
miR-181c	Stomach	NOTCH4, KRAS	[49]
miR-193a	Liver, leukemia, bladder	SRSF2, KIT	[50–52]
miR-200 family	Colon, breast, bladder, lung, head and neck	ZEB1, ZEB2	[42, 53–58]
miR-203	Liver, leukemia, myeloma, MALToma, cervix	ABL1, ABCE1, CDK6	[32, 34, 59–61]
miR-205	Bladder, lung	ZEB1, ZEB2	[56, 58]
miR-218	Head and neck	RICTOR	[62]
miR-219-2	Stomach		[63]
miR-335	Breast	SOX4, TNC	[64]
miR-345	Colon	BAG3	[65]
miR-375	Esophagus, head and neck, melanoma, cervix	IGF1R, PDK1	[34, 42, 66–68]
miR-487b	Lung	SUZ12, BMI1, WNT5A, MYC, KRAS	[69]
miR-512	Stomach	MCL1	[70]
miR-708	Leukemia	IKBKB	[71]
miR-941	Colon, liver	KDM6B	[72, 73]
miR-1237	Colon		[72]
miR-1247	Colon		[72]

4.2 Identification of Epigenetically Dysregulated miRNAs

The first evidence of an epigenetic mechanism involved in silencing miRNA genes in cancer came from a pharmacological unmasking experiment. Using miRNA microarray assays, Saito et al. have analyzed the expression profiles of miRNAs in T24 bladder cancer cells treated with or without the DNA methyltransferase (DNMT) inhibitor 5-aza-2'-deoxycytidine and the histone deacetylase (HDAC) inhibitor 4-phenylbutyric acid [29]. The drug treatment led to upregulation of several miRNAs, including miR-127, which were encoded within the CpG island. Upregulation of miR-127 was associated with DNA demethylation as well as increases in acetylated histone H3 and H3K4me3, suggesting that miR-127 is epigenetically silenced in cancer cells. Experimental evidence also confirmed that the proto-oncogene BCL6 is a target of miR-127, indicating the miRNA can act as a tumor suppressor [29]. Thereafter, comparisons of miRNA expression profiles between the HCT116 colorectal cancer (CRC) cell line and the same cell line in which both DNMT1 and DNMT3B were genetically disrupted revealed the epigenetic silencing of miR-124 family genes and miR-34b/c in CRC [18, 79]. Since then, similar screenings have identified a number of other methylated miRNA genes in various human malignancies [6, 49, 70, 80, 81].

In addition to pharmacological or genetic unmasking experiments, genome-wide DNA methylation analysis has also been used to identify epigenetically silenced miRNA genes. For example, through a combination of methylated CpG island amplification (MCA) and CpG island microarray analysis, methylation of the miR-9-1 gene was detected in pancreatic cancer [15]. In addition, microarray analysis using the Infinium BeadChip revealed miR-10b methylation in gastric cancer (GC) [72]. Yan et al. have performed a deep sequencing of methylated DNA-binding domain (MBD)-isolated DNA in CRC cells and identified several methylated miRNA genes, including miR-941, miR-1237, and miR-1247 [82].

A number of miRNA genes epigenetically silenced in cancer have been detected using genome-wide histone modification analyses. The combination of chromatin immunoprecipitation (ChIP)-on-chip and miRNA microarray analyses with prostate cancer cells revealed that miRNA gene expression correlates positively with H3K4me3 and inversely with H3K27me3 at their promoter regions [19]. Analysis of histone modification in acute lymphoblastic leukemia (ALL) using ChIP-on-chip revealed that the CpG islands of 13 miRNA genes are associated with high levels of H3K9me2 and low levels of H3K4me3, suggesting these miRNAs are epigenetically silenced in ALL [83]. Subsequent analysis confirmed the hypermethylation of the CpG islands of selected miRNA genes, including miR-9 family, miR-34 family, and miR-124 family genes. Identification of epigenetically dysregulated miRNA genes in cancer is sometimes impeded by the limited annotation of the primary transcripts of miRNA genes, but earlier studies showed that H3K4me3 could be a useful mark with which to identify the active promoter regions of miRNA genes [83, 84]. Using ChIP-seq experiments with CRC cells, we assessed genome-wide histone modification and identified the putative promoter regions for 174 primary miRNA genes, among which 37 were predicted targets of epigenetic silencing [4]. Similarly, integrative analysis of genome-wide DNA methylation and histone modification in chronic lymphocytic leukemia (CLL) revealed 128 miRNA genes with aberrant DNA methylation at their promoters [85]. Interestingly, among these genes, 38 exhibited hypermethylation while 90 showed hypomethylation, which is indicative of epigenetically silenced and activated miRNAs, respectively. Consistent with these findings, the hypermethylated regions included a number of well-documented epigenetically silenced miRNA genes, including miR-9-2, miR-124-2, and miR-129-2, while the hypomethylation was associated with the upregulation of oncogenic miRNAs (oncomirs), including miR-21 and miR-155 [85].

Epigenetic silencing of several tumor-suppressive miRNAs was discovered through functional screening. For instance, assessment of the antiproliferative effects of a panel of 327 synthetic miRNAs in oral and endometrial cancer cell lines revealed nearly 100 miRNAs that suppressed cancer cell growth [48, 62]. Approximately half of those were associated with CpG islands, and subsequent analyses identified methylation of miR-218 and miR-152 in oral and endometrial cancers, respectively [48, 62].

4.3 miRNA Dysregulation via DNA Methylation

4.3.1 miR-124 Family

Epigenetic silencing of miR-124 family genes was first identified in CRC cells [29]. MiR-124 is thought to exert tumor suppressor effects by targeting cyclin-dependent kinase 6 (CDK6), and epigenetic silencing of miR-124 leads to CDK6 activation and Rb phosphorylation [29, 30]. Within the human genome, three independent loci (miR-124-1, miR-124-2, and miR-124-3) encode the identical mature miR-124, and all of them are associated with CpG islands, which are frequently methylated in multiple types of solid tumors, including colorectal, gastric, lung, liver, pancreatic, and cervical cancers [7, 29, 31–34]. Methylation of miR-124 genes is also reported in hematological malignancies, including ALL [30] and non-Hodgkin's lymphoma [86]. In all, miR-124 methylation is associated with higher recurrence and mortality rates and may be an independent prognostic factor for disease-free and overall survival [30].

Methylation of the miR-124 genes is also observed in the gastric mucosa of *Helicobacter pylori* (*H. pylori*)-positive healthy individuals, which suggests the methylation may be induced by the *H. pylori* infection [31]. Importantly, among *H. pylori*-negative individuals, miR-124 genes are more highly methylated in the noncancerous gastric mucosae of GC patients than in those of healthy individuals, indicating that miR-124 methylation may be involved in an epigenetic field defect in the stomach [31]. Methylation of the miR-124 genes is also acquired during human papillomavirus-mediated transformation in keratinocytes, suggesting that miR-124 methylation may be a useful marker for detection of cervical cancer and high-grade precursor lesions [34].

4.3.2 *miR-34 Family*

Members of the miR-34 gene family (miR-34a, miR-34b, and miR-34c) are known to be direct targets of p53, and their ectopic expression in cancer cells induces cell cycle arrest and apoptosis [87, 88]. Forced expression of miR-34 in cancer cells leads to the downregulation of candidate target genes, including MET, CDK4, cyclin E2, and MYC [6, 18, 87]. Within the human genome, miR-34a is located on chromosome 1p36, while miR-34b and miR-34c are co-transcribed from a single transcription unit on chromosome 11q23. CpG islands in the promoter regions of these genes are targets of aberrant hypermethylation in oral, esophageal, gastric, colorectal, pancreatic, breast, lung, and renal cancers; malignant mesothelioma; melanoma; and hematological malignancies [18–27, 79]. Methylation of miR-34b/c has also been linked to cancer metastasis [6] and invasion [28]. Similar to miR-124, methylation of miR-34b/c in the gastric mucosa is associated with *H. pylori* infection, and the noncancerous gastric mucosae of patients with multiple GCs show higher levels of miR-34b/c methylation than those of patients with a single GC, again indicating involvement in an epigenetic field defect [21]. In addition, silencing miR-34c through DNA methylation promotes self-renewal and epithelial–mesenchymal transition (EMT) in breast tumor-initiating cells [89]. These findings, as well as their contribution to the p53 network, strongly imply that miR-34 family members act as tumor suppressors in cancer.

4.3.3 *miR-9 Family*

Methylation of the promoter CpG island of the miR-9-1 gene was first reported in breast and pancreatic cancers [8, 81]. Thereafter, methylation of all miR-9 family genes (miR-9-1, miR-9-2, and miR-9-3) was identified through screening for methylated miRNA genes in metastatic cancer cell lines [6]. Consistent with that report, methylation of miR-9-1 is associated with lymph node metastasis in CRC [9], and methylation of miR-9-1 and miR-9-3 correlates with metastatic recurrence of renal cell carcinoma [10]. All of the miR-9 family genes are simultaneously methylated in GC, and ectopic expression of miR-9 inhibits proliferation, migration, and invasion by GC cells [11]. MiR-9 targets FGFR1 and CDK6 in ALL [12] and CDX2 in GC cells [90], suggesting it acts as a tumor suppressor. Xenoestrogen, which may increase the risk of breast cancer, can induce methylation-associated silencing of miR-9-3 in breast epithelial cells, indicating that methylation of this gene could be a hallmark of early breast cancer development [13]. Methylation of miR-9-3 is also a frequent event in CLL, and loss of miR-9 expression may be associated with activation of the NF-κB pathway [14]. In contrast to these findings, however, another study showed that miR-9 is activated by MYC and MYCN in breast cancer and that miR-9 promotes metastasis through downregulation of E-cadherin (CDH1) [91]. These results are indicative of the functional complexity of miRNAs in cancer cells and suggest that miRNAs may exert opposite effects in different tissues or settings.

4.3.4 miR-200 Family and miR-205

The miR-200 gene family (miR-200a, miR-200b, miR-200c, miR-141, and miR-429) and miR-205 encode key regulators of EMT by directly targeting ZEB1 and ZEB2, which are transcriptional repressors that downregulate CDH1 [92–94]. Within the human genome, the miR-200 family genes are grouped into two polycistronic units, miR-200b/200a/429 and miR-200c/141, located on chromosomes 1 and 12, respectively [53]. Expression of miR-200c/141 is inversely associated with methylation of the promoter CpG island in both normal and tumor cells [95]. Methylation of miR-200c/141 correlates with the invasiveness of breast cancer cells, and induction of EMT through ectopic expression of the transcription factor Twist in immortalized human mammary epithelial cells is accompanied by increased methylation of miR-200c/141 [54]. Similarly, in non-small cell lung cancer (NSCLC), promoter methylation is associated with loss of miR-200c expression, which is in turn associated with poor differentiation, lymph node metastasis, and weaker CDH1 expression [55]. Another study has demonstrated that the CpG islands of both units (miR-200b/200a/429 and miR-200c/141) are unmethylated in cancer cells with epithelial features, but are methylated and silenced in transformed cells with mesenchymal characteristics [53]. In bladder cancer, both units of the miR-200 family as well as miR-205 are coordinately silenced in association with promoter methylation [56]. In CRC patients, liver metastatic lesions exhibit lower methylation and higher expression of miR-200c/141 than the primary tumors [57]. Experimental evidence suggests that establishment of EMT in primary CRCs and the reverse phenotype of mesenchymal–epithelial transition (MET) are modulated in the metastatic lesions through epigenetic regulation of miR-200c [57]. Epigenetic silencing of the miR-200 family and miR-205 has also been reported in carcinogen-treated lung epithelial cells, suggesting induction of EMT through miRNA dysregulation is an early event during lung carcinogenesis [58].

4.3.5 miR-137

Methylation of miR-137 was first reported in oral cancer [79] and was subsequently reported in colorectal [9, 96], gastric [97], and bladder cancers [98]. MiR-137 methylation is associated with a poorer survival rate among patients with head and neck squamous cell carcinoma (HNSCC) [99] and could be detected in oral rinses collected from HNSCC patients, suggesting its utility as a cancer biomarker [100]. Within cancer cells, miR-137 targets CDK6, LSD1, CDC42, and KDM1A, which is indicative of its tumor suppressor function [79, 96, 101, 102], while in normal cells, miR-137 regulates neuronal differentiation by targeting EZH2 and MIB1 [103, 104].

4.3.6 *miR-145*

MiR-145 is known to act as a tumor suppressor and is downregulated in colorectal [105], breast [106], ovarian [107], and prostate cancers [44]. Moreover, an association between promoter methylation and miR-145 silencing has been confirmed in prostate cancer [44]. It has also been shown that both DNA methylation and p53 mutation are major causes of miR-145 downregulation in prostate cancer [45]. It appears p53 represses c-MYC by transcriptionally activating miR-145, which directly targets c-MYC [108], and miR-145 also suppresses cancer cell proliferation, invasion, and metastasis by targeting MUC1, IRS1, FSCN1, and BNIP3 [109–112]. In prostate cancer, downregulation of miR-145 expression and increased BNIP3 expression are both associated with a poor prognosis [112]. And one recent study has showed that epigenetic silencing of miR-145 correlates with brain metastasis in lung cancer [46]. Loss of miR-145 in cancer cells leads to upregulation of its target genes, including OCT4, EGFR, MUC1, and c-MYC, which contribute to malignant progression and metastasis [46].

4.3.7 *miR-129-2*

Methylation of miR-129-2 has been reported in colorectal, gastric, esophageal, hepatic, and endometrial cancers [9, 24, 37, 38]. MiR-129 targets SOX4, an oncogene frequently upregulated in malignancies, and an association between miR-129-2 methylation and SOX4 overexpression had been detected in both endometrial and gastric cancers [37, 38]. In endometrial cancer, miR-129-2 methylation is associated with microsatellite instability, MLH1 methylation, and poor overall survival [59]. In GC, miR-129 targets ABC transporters related to multidrug resistance (ABCB1, ABCC5, and AGCG1), suggesting methylation of miR-129-2 may modulate chemoresistance in GC [113]. Consistent with that idea, downregulated expression of miR-129 is associated with poor clinicopathological features in GC [114].

Table 4.2 miRNAs activated by DNA hypomethylation in cancer

Gene	Tumor type	Target gene	References
let7a-3	Lung		[115]
miR-191	Liver	TIMP3	[116]
miR-196b	Stomach		[117]
miR-200a/200b	Pancreas	ZEB2	[118]
miR-375	Breast	RASD1	[119]
miR-519d	Liver	CDKN1A, PTEN, AKT3, TIMP2	[120]
miR-663	Breast	HSPG2	[121]

4.3.8 Hypomethylation of miRNA Genes

DNA hypomethylation in the promoter region is one of the mechanisms by which miRNA genes are activated in cancer (see Table 4.2). For instance, the CpG island of let-7a-3 is methylated in normal cells but is hypomethylated in lung adenocarcinoma, leading to the gene's elevated expression [115]. In lung cancer cells, let-7a-3 acts as an oncogene through its regulation of genes involved in cell proliferation, adhesion, and differentiation [115]. In addition, members of the miR-196 family (miR-196a and miR-196b) are located within the HOX gene clusters and are often overexpressed in tumors, which is consistent with their oncogenic functionality [122]. The miR-196b gene is embedded within a CpG island, and hypomethylation in this region is associated with its overexpression in GC [117]. In hepatocellular carcinoma (HCC), two other miRNAs, miR-191 and miR-519d, are both upregulated in association with DNA hypomethylation [116, 120]. Experimental evidence indicates that miR-191 directly targets TIMP3, while miR-519d targets CDKN1A (p21), PTEN, AKT3, and TIMP2, which is suggestive of the oncogenic potential of these miRNAs [116, 120].

4.3.9 Clinical Implications of DNA Methylation

Numerous studies have demonstrated the usefulness of aberrantly expressed miRNAs as cancer biomarkers. Likewise, aberrant methylation of miRNA genes may also be a useful marker for detecting cancers and predicting outcomes. For example, methylation of miR-124 family genes is an independent prognostic factor for disease-free and overall survival in ALL [30]. A more comprehensive analysis of epigenetically silenced miRNA genes in ALL has identified methylation of 13 miRNA genes (miR-9-1, miR-9-2, miR-9-3, miR-10b, miR-34b, miR-34c, miR124-1, miR-124-2, miR-124-3, miR-132, miR-212, miR-196b, and miR-203), and ALLs with at least one methylated miRNA gene show significantly poorer disease-free and overall survival than the unmethylated group [19]. Similarly, methylation of miR-9-3, miR-124-2, and miR124-3 is associated with larger tumor size and poorer progression-free survival in NSCLC [7], and methylation of miR-124-3 is associated with disease recurrence in clear cell renal carcinoma [123]. Methylation of miR-34b/c is also associated with a high probability of recurrence and poor overall survival in NSCLC [22]. Finally, miR-34a methylation in combination with high c-MET and β-catenin expression could be a useful marker predictive of distant metastasis of colon cancer [124].

Aberrant DNA methylation of miRNA genes detected in body fluids is also a potentially useful cancer biomarker. We have shown that methylation of miRNA genes can be detected in urine specimens from bladder cancer patients, and a panel of four miRNA genes (miR-137, miR-124-2, miR-124-3, and miR-9-3) could be a useful biomarker for cancer detection [98]. Methylation of miR-34b/c is reportedly detected in 75 % of fecal specimens from CRC patients and in 16 % of specimens from high-grade dysplasia patients, suggesting it could be a useful feces-based screening marker [125]. In addition, we have found that methylation of a panel of genes, including miR-34b/c, in mucosal wash fluid collected during colonoscopy is potentially an effective biomarker for predicting the invasiveness of CRCs [126]. The degree of methylation of miR-34b/c detected in the serum-circulating DNA is also associated with malignant pleural mesothelioma [127]. Moreover, a recent large case-control study has showed that aberrantly methylated miRNA genes may be present in the blood before cancer diagnosis. Cordero et al. have reported that altered methylation of a series of miRNA genes (miR-328, miR-675, miR-1307, miR-1286, miR-1275, miR-1910, miR-24-1, and miR-548a-1) is detectable in peripheral blood cells collected from individuals who go on to develop breast cancer, suggesting it may represent a biomarker for early detection or risk of cancer [128].

As mentioned, levels of miR-124 and miR-34b/c methylation are elevated in GC and in the background gastric mucosa, suggesting that methylation is involved in an epigenetic field defect in the stomach. To assess the clinical utility of miR-34b/c methylation as a GC risk marker, we carried out a prospective study in a cohort of early GC patients who underwent curative endoscopic resection [129]. We analyzed the levels of miR-34b/c methylation in the background noncancerous gastric mucosa at the first detection of the cancer. During the follow-up period after resection, the cumulative incidence of metachronous GC was much higher among patients with elevated methylation levels. In a similar study using a large cohort of early GC patients, Asada et al. have found that miR-124-3 methylation could be a predictive marker of metachronous GC [130].

4.4 miRNA Dysregulation Due to Histone Modification

It is now evident that histone modification also plays an important role in the dysregulation of miRNAs in cancer. In HCC, for example, overexpression of the oncomir miR-224 is mediated by the histone acetylation [131]. In addition, Chang et al. have showed that wild-type BRCA1 epigenetically represses the oncomir

miR-155 by recruiting HDAC2 to the miR-155 promoter, whereas a BRCA1 R1699Q mutant relieves the repression and causes miR-155 overexpression [132].

By contrast, overexpression of HDACs (HDAC1, HDAC2, and HDAC3) leads to downregulation of the tumor-suppressive miRNAs miR-15a, miR-16, and miR-29b in CLL [133]. Inhibition of HDACs induces robust accumulation of active histone marks at the promoters of miRNAs and their increased expression, which in turn leads to downregulation of their target genes, BCL2 and MCL1. Another study has showed that in lymphoma, MYC represses miR-15a/miR-16-1 by recruiting HDAC3 to their promoters [134]. In HCC, upregulation of HDACs (HDAC1-3) is associated with repression of miR-449, which leads to activation of the putative miR-449 target gene c-MET [135].

Some miRNA genes are silenced in association with repressive histone marks without DNA methylation. For instance, downregulation of miR-212 in lung cancer cells is reportedly associated with H3K9me2 and H3K27me3 but not DNA hypermethylation [136]. MiR-212 exerts a pro-apoptotic effect in lung cancer cells by targeting the anti-apoptotic gene PED, and inhibition of HDAC and the histone methyltransferase EZH2 strongly reactivates miR-212 expression in lung cancer cells. It has also been found that miR-708 is repressed by H3K27me3 in metastatic breast cancer [137]. MiR-708 targets NNAT, a regulator of intracellular Ca^{2+}, and silencing miR-708 leads to elevated intracellular Ca^{2+} levels and increased cell migration and metastasis. Upregulation of EZH2 in HCC leads to the silencing of several tumor-suppressive miRNAs, including miR-139-5p, miR-125b, miR-101, let-7c, and miR-200b [138]. Moreover, pathway analysis has showed that EZH2-silenced miRNAs play a regulatory role in modulating cell motility and metastasis-related pathways [138].

4.5 Epigenetic Alterations Induced by miRNAs

4.5.1 DNA Methylation

Several lines of evidence indicate that dysregulation of miRNAs can lead to aberrant DNA methylation in cancer (see Fig. 4.2a). For example, the miR-29 family (miR-29a, miR-29b, and miR-29c), which is downregulated in lung cancer, directly targets two DNA methyltransferases, DNMT3A and DNMT3B [139]. Ectopic expression of the miR-29 family in lung cancer cells suppresses DNMT3A/B and restores expression of the methylation-silenced tumor suppressor genes FHIT and WWOX [139]. In addition, miR-143 and miR-199a-3p target DNMT3A in CRC and testicular cancer, respectively [140, 141]. Several miRNAs, including miR-152, miR-148a, miR-185, miR-343, and miR-126, reportedly target DNMT1, and downregulation of these miRNAs is associated with methylation-associated silencing of tumor suppressor genes in cancer [48, 142–146]. For instance, downregulation of miR-152 correlates with increased DNMT1 expression in

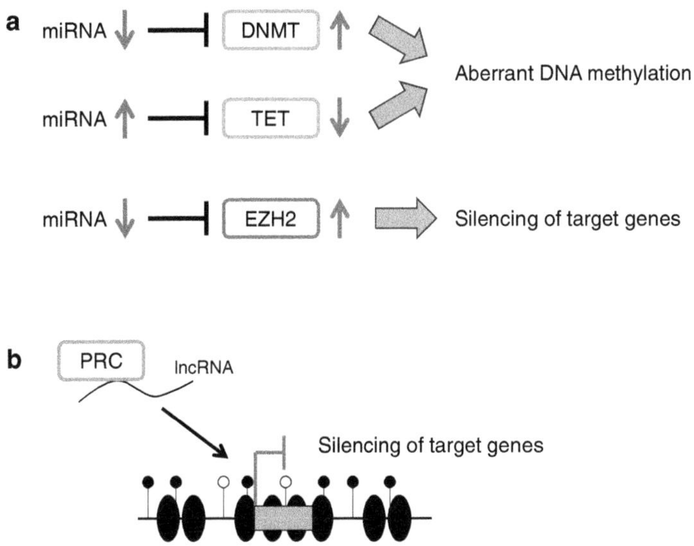

Fig. 4.2 (**a**) Epigenetic alterations induced by miRNAs. (**b**) Epigenetic alterations induced by lncRNAs

hepatitis B virus-related HCC tissues, and forced expression of miR-152 in HCC cells markedly reduces DNMT1 expression and global DNA methylation levels [142]. MiR-342 is also downregulated in CRC, and restoration of miR-342 in CRC cells leads to reactivation of the tumor suppressor genes ADAM23 and RASSF1A via promoter demethylation [144]. Recently, miR-34b has been shown to target DNMTs and HDACs in prostate cancer cells [147]. As in other malignancies, miR-34b is silenced via CpG island methylation in prostate cancer, and low miR-34b expression is strongly associated with poor survival. Ectopic expression of miR-34b in prostate cancer cells suppresses DNMTs and HDACs and activates the endogenous miR-34b gene, which is indicative of a positive feedback loop [147].

The ten eleven translocation (TET) family, which includes TET1, TET2, and TET3, has been recently shown to act as DNA demethylases by converting 5-methylcytosine to 5-hydroxymethylcytosine (5-hmC) [148]. Reduced TET activity and lower genome-wide levels of 5-hmC have been reported in several cancer types, and recent studies indicate that miRNA dysregulation could be involved. In HCC, miR-494 reportedly acts as an epigenetic regulator that directly targets TET1 [149]. Upregulation of miR-494 in HCC cells leads to epigenetic silencing of invasion suppressing miRNA genes through inhibition of DNA demethylation in promoter CpG islands. In addition, miR-767 targets TET1 and TET3, and overexpression of miR-767 in cancer cells represses TET1 and TET3 and reduces genomic 5-hmC levels [150]. Collectively, these results indicate that dysregulation of specific miRNAs may be causally related to aberrant DNA methylation in cancer.

4.5.2 *Histone Modifications*

Like the relationship between miRNAs and DNMTs, dysregulation of miRNAs that regulate histone modifiers results in aberrant histone modifications in cancer (see Fig. 4.2a). EZH2 functions as a catalytic subunit of the polycomb repressive complex 2 (PRC2), which methylates H3K27. Evidence suggests that EZH2 acts as an oncogene, and its overexpression in prostate and breast cancers promotes tumorigenesis, invasiveness, and metastasis [151, 152]. EZH2 is a target of several miRNAs, including miR-101, and genomic loss of miR-101 is an important factor underlying EZH2 overexpression in prostate cancer [153]. Reduced expression of miR-101 and upregulation of EZH2 occur in parallel during the progression of prostate cancer, and genomic loss of miR-101 is more frequently seen in metastatic disease than localized cancers [153]. The inverse association between miR-101 and EZH2 has also been reported in bladder, gastric, lung, and renal cancers [154–157]. In addition, other miRNAs, including miR-26a [158], miR-98 [159], miR-124 [160], miR-144 [161], miR-214 [162], and let-7 [163], are also reported to negatively regulate EZH2.

Several studies have shown an intriguing interrelationship between miRNA dysregulation and epigenetic modifiers in cancer. In prostate and breast cancer cells, EZH2 represses a set of miRNAs (miR-181c, miR-181b, miR-200b, miR-200c, and miR-203) that negatively regulate the PRC1 subcomponents BMI1 and RING2, and the relationship between miRNA and PRC protein levels has been confirmed in primary prostate cancer tissue [164]. These results suggest that key miRNAs link PRC2 to PRC1 to form a regulatory axis of epigenetic silencing machinery. Another study has demonstrated that MYC, HDAC3, and EZH2 coordinately repress miR-29 in B-cell lymphoma [165]. Interestingly, MYC contributes to EZH2 upregulation through repression of EZH2-targeting miR-26a, while EZH2 induces MYC via inhibition of MYC-targeting miR-494. These results suggest the existence of a positive MYC-miRNA-EZH2 feedback loop that mediates persistent overexpression of MYC and EZH2 in lymphoma. Combined inhibition of HDAC3 and EZH2 restores miR-29 expression and suppresses lymphoma cell growth, suggesting the MYC-EZH2-miR-29 axis could be a promising target for epigenetic therapy in aggressive B-cell lymphomas [165].

4.6 Epigenetic Dysregulation of lncRNA Genes

Although the function of most lncRNAs remains unknown, a number of biologically functional lncRNAs have been reported [77]. Recent studies have also begun to shed light on epigenetic mechanisms that play essential roles in the dysregulation of lncRNAs in cancer. For instance, genome-wide DNA methylation analysis using deep sequencing with Barrett's esophagus (BE) and esophageal adenocarcinoma (EAC) reveals hypomethylation in intragenic and repetitive genomic elements as well as in noncoding regions [166]. Among these, AFAP1-AS1 is significantly hypomethylated in both BE and EAC. AFAP1-AS1 knockdown in EAC cells induces apoptosis and suppresses proliferation, migration, and invasion, which is suggestive of its oncogenic function. Similarly, hypomethylation of the promoter region of ZEB1-AS1 is associated with its overexpression and a poor prognosis in HCC [167]. Functionally, ZEB1-AS1 upregulates ZEB1 expression and induces EMT, and ectopic expression of ZEB1-AS1 in HCC cells promotes proliferation, invasion, and metastasis. By contrast, expression of MEG3 is downregulated in association with promoter CpG island methylation in HCC [168]. Overexpression of MEG3 in HCC cells suppresses proliferation and induces apoptosis, which is indicative of its tumor suppressor function. Notably, reduced expression of miR-29a, which can modulate DNMT1 and DNMT3B, contributes to the methylation-associated silencing of MEG3 in HCC cells [168]. These results indicate an interrelationship between miRNAs, lncRNAs, and epigenetic gene regulation in cancer.

Two lncRNA genes, DLEU1 and DLEU2, have been mapped to a region at chromosomal band 13q14.3, which is frequently deleted in hematopoietic and solid tumors [169]. DLEU2 is also known to be a primary transcript of miR-15a and miR-16-1, which act as tumor suppressors in CLL [170]. More than 50 % of CLL cases harbor a deletion at 13q14.3, indicating the tumor suppressor function of this region. Another recent study has showed that, in CLL, DNA demethylation at the transcription start sites of DLEU1 and DLEU2 is associated with upregulation of these lncRNAs, which leads to downregulation of neighboring tumor suppressor genes in a cis-regulatory manner [169]. Notably, all genes in the cluster at 13q14.3 are modulators of the NF-κB pathway. This suggests the epigenetic dysregulation of these lncRNAs attenuates the tumor suppressor function at 13q14.3, which targets NF-κB in CLL cells.

To unravel the DNA methylation profiles of lincRNA genes in cancer, Zhi et al. have re-annotated Infinium HumanMethylation450 BeadChip array data obtained from 4629 tumor samples and 705 samples of normal tissue [171]. They have classified the 2461 lincRNA genes as prone to methylation (PM) or resistant to methylation (RM) based on their promoter methylation patterns in tumors. Promoters of RM lincRNA genes are evolutionally conserved among species and ubiquitously expressed in normal tissues. Moreover, patterns of lincRNA promoter methylation are associated with cancer status, subtype or prognosis. These results suggest that a large number of lncRNA genes are potential targets of aberrant DNA methylation, and they could be useful biomarkers and potential therapeutic targets [171].

Histone modifications also play an essential role in the regulation of lncRNA expression. For example, the lncRNA SPRY4-IT1 is silenced via H3K27me3 in NSCLC [172]. SPRY4-IT1 is derived from an intron within the SPRY4 gene, and its ectopic expression in NSCLC cells exerts significant antitumor effects. Knocking down EZH2 restores SPRY4-IT4 expression, and inhibition of SPRY4-IT1 partially rescues the impaired oncogenic phenotype resulting from EZH2 knockdown. This suggests SPRY1-IT1 repression has an important role in EZH2-induced oncogenesis.

4.7 Epigenetic Alterations Induced by lncRNAs

LncRNAs are involved in all aspects of gene regulation, including chromosome dosage compensation, imprinting, epigenetic regulation, transcription, mRNA splicing, and translation [77]. Recent studies have demonstrated that a number of lncRNAs dysregulated in cancer are deeply involved in epigenetic regulation (see Fig. 4.2b). One of the best documented examples is HOTAIR, which is frequently overexpressed in multiple types of cancer. The oncogenic function of HOTAIR was first reported in breast cancer, where HOTAIR binds to PRC2 to silence genes encoding invasion and metastasis suppressors [173]. Upregulation of HOTAIR is associated with a poor prognosis in breast, liver, colon, and pancreatic cancers, as well as in gastrointestinal stromal tumors [173–177]. In neuroblastoma, weak expression of another lncRNA, NBAT-1, is a strong indicator of a poor clinical outcome [178]. The promoter region of NBAT-1 is hypermethylated in high-risk neuroblastomas, and loss of NBAT-1 increases cellular proliferation and invasion. NBAT-1 controls tumor progression by interacting with EZH2 to repress its target genes, and it promotes differentiation of neuroblastoma cells.

Several lncRNAs reportedly modulate DNA methylation in cancer. linc-POU3F3, which is overexpressed in esophageal squamous cell carcinoma, induces hypermethylation of the promoter CpG island of a neighboring POU3F3 gene, which encodes a transcription factor [179]. linc-POU3F3 physically interacts with EZH2, and it is thought that EZH2 recruits DNMTs to the promoter region of POU3F3. Functional evidence suggests that overexpression of linc-POU3F3 promotes tumorigenesis by epigenetically silencing POU3F3.

Highly upregulated in liver cancer (HULC) is a lncRNA specifically overexpressed in HCC, where it influences abnormal lipid metabolism by activating the acyl-CoA synthetase subunit ASCL1 [180]. Functional analysis has revealed that HULC upregulates ASCL1 by activating the transcription factor PPARA, which is a target of miR-9. HULC activates PPARA by inducing CpG island methylation and thus silencing the miR-9 family genes.

Merry et al. have screened for lncRNAs that directly interact with DNMT1 and identified DNMT1-associated colon cancer repressed lncRNA (DACOR1), which is expressed in normal colon and repressed in colon cancer [181]. Induction of DACOR1 in colon cancer cells suppresses colony formation and downregulates cystathionine β-synthase, which leads to increased levels of S-adenosyl methionine, a key methyl donor for DNA methylation. Repression of DACOR1 may thus promote colon tumorigenesis by inducing global DNA hypomethylation [181].

DNA repair machinery is known to play an important role in DNA demethylation involving activation-induced cytidine deaminase and thymine-DNA glycosylase [182]. The stress response protein GADD45A recruits the DNA repair machinery to specific genomic sites, where it catalyzes the replacement of methylated cytosines with unmethylated ones. It has been recently reported that a lncRNA, TARID, modulates DNA demethylation using the same mechanism [182]. The tumor suppressor gene TCF21 and its neighboring lncRNA gene, TARID, are concordantly methylated in multiple malignancies, including NSCLC, HNSCC, and ovarian cancer. In normal cells, TARID interacts with both the TCF21 promoter and GADD45A and modulates DNA demethylation at the TCF21 promoter. TARID is thus required to maintain the TCF21 promoter open, and reduced expression of TARID is thought to induce methylation-associated silencing of TCF21 and to promote tumorigenesis. Supporting that hypothesis, downregulation of TARID and methylation of TCF21 are also seen in clear cell sarcoma of the kidney [183].

4.8 Concluding Remarks

In this review, we have highlighted the relationship between epigenetic alterations and dysregulation of short and long ncRNAs in cancer. Aberrant DNA methylation and histone modification are the major mechanisms underlying ncRNA dysregulation in cancer, and methylation of a subset of miRNA genes may be useful biomarkers for detecting cancer or predicting clinical outcome. Replacement of silenced tumor-suppressive miRNAs in cancer cells could be an effective strategy for cancer treatment. And cross talk between lncRNA dysregulation and epigenetic mechanisms could be an important driver promoting tumorigenesis. We anticipate that further understanding of the causes and consequences of ncRNA dysregulation will open the door to novel biomarkers and therapeutic targets in cancer.

Acknowledgments Dr. William Goldman has helped edit this manuscript.

References

1. Esteller M. Non-coding RNAs in human disease. Nat Rev Genet. 2011;12(12):861–74. doi:10.1038/nrg3074.
2. Esquela-Kerscher A, Slack FJ. Oncomirs – microRNAs with a role in cancer. Nat Rev Cancer. 2006;6(4):259–69. doi:10.1038/nrc1840.
3. Croce CM. Causes and consequences of microRNA dysregulation in cancer. Nat Rev Genet. 2009;10(10):704–14. doi:10.1038/nrg2634.
4. Suzuki H, Takatsuka S, Akashi H, et al. Genome-wide profiling of chromatin signatures reveals epigenetic regulation of MicroRNA genes in colorectal cancer. Cancer Res. 2011;71(17):5646–58. doi:10.1158/0008-5472.CAN-11-1076.
5. Datta J, Kutay H, Nasser MW, et al. Methylation mediated silencing of MicroRNA-1 gene and its role in hepatocellular carcinogenesis. Cancer Res. 2008;68(13):5049–58. doi:10.1158/0008-5472.CAN-07-6655.
6. Lujambio A, Calin GA, Villanueva A, et al. A microRNA DNA methylation signature for human cancer metastasis. Proc Natl Acad Sci U S A. 2008;105(36):13556–61. doi:10.1073/pnas.0803055105.
7. Kitano K, Watanabe K, Emoto N, et al. CpG island methylation of microRNAs is associated with tumor size and recurrence of non-small-cell lung cancer. Cancer Sci. 2011;102(12):2126–31. doi:10.1111/j.1349-7006.2011.02101.x.

8. Lehmann U, Hasemeier B, Christgen M, et al. Epigenetic inactivation of microRNA gene hsa-mir-9-1 in human breast cancer. J Pathol. 2008;214(1):17–24. doi:10.1002/path.2251.
9. Bandres E, Agirre X, Bitarte N, et al. Epigenetic regulation of microRNA expression in colorectal cancer. Int J Cancer. 2009;125(11):2737–43. doi:10.1002/ijc.24638.
10. Hildebrandt MA, Gu J, Lin J, et al. Hsa-miR-9 methylation status is associated with cancer development and metastatic recurrence in patients with clear cell renal cell carcinoma. Oncogene. 2010;29(42):5724–8. doi:10.1038/onc.2010.305.
11. Tsai KW, Liao YL, Wu CW, et al. Aberrant hypermethylation of miR-9 genes in gastric cancer. Epigenetics. 2011;6(10):1189–97. doi:10.4161/epi.6.10.16535.
12. Rodriguez-Otero P, Roman-Gomez J, Vilas-Zornoza A, et al. Deregulation of FGFR1 and CDK6 oncogenic pathways in acute lymphoblastic leukaemia harbouring epigenetic modifications of the MIR9 family. Br J Haematol. 2011;155(1):73–83. doi:10.1111/j.1365-2141.2011.08812.x.
13. Hsu PY, Deatherage DE, Rodriguez BA, et al. Xenoestrogen-induced epigenetic repression of microRNA-9-3 in breast epithelial cells. Cancer Res. 2009;69(14):5936–45. doi:10.1158/0008-5472.CAN-08-4914.
14. Wang LQ, Kwong YL, Kho CS, et al. Epigenetic inactivation of miR-9 family microRNAs in chronic lymphocytic leukemia-implications on constitutive activation of NFkappaB pathway. Mol Cancer. 2013;12:173. doi:10.1186/1476-4598-12-173.
15. Kim K, Lee HC, Park JL, et al. Epigenetic regulation of microRNA-10b and targeting of oncogenic MAPRE1 in gastric cancer. Epigenetics. 2011;6(6):740–51.
16. Lin PC, Chiu YL, Banerjee S, et al. Epigenetic repression of miR-31 disrupts androgen receptor homeostasis and contributes to prostate cancer progression. Cancer Res. 2013;73(3):1232–44. doi:10.1158/0008-5472.CAN-12-2968.
17. Augoff K, McCue B, Plow EF, Sossey-Alaoui K. miR-31 and its host gene lncRNA LOC554202 are regulated by promoter hypermethylation in triple-negative breast cancer. Mol Cancer. 2012;11:5. doi:10.1186/1476-4598-11-5.
18. Toyota M, Suzuki H, Sasaki Y, et al. Epigenetic silencing of microRNA-34b/c and B-cell translocation gene 4 is associated with CpG island methylation in colorectal cancer. Cancer Res. 2008;68(11):4123–32. doi:10.1158/0008-5472.CAN-08-0325.
19. Roman-Gomez J, Agirre X, Jimenez-Velasco A, et al. Epigenetic regulation of microRNAs in acute lymphoblastic leukemia. J Clin Oncol. 2009;27(8):1316–22. doi:10.1200/JCO.2008.19.3441.
20. Lodygin D, Tarasov V, Epanchintsev A, et al. Inactivation of miR-34a by aberrant CpG methylation in multiple types of cancer. Cell Cycle. 2008;7(16):2591–600.
21. Suzuki H, Yamamoto E, Nojima M, et al. Methylation-associated silencing of microRNA-34b/c in gastric cancer and its involvement in an epigenetic field defect. Carcinogenesis. 2010;31(12):2066–73. doi:10.1093/carcin/bgq203.
22. Wang Z, Chen Z, Gao Y, et al. DNA hypermethylation of microRNA-34b/c has prognostic value for stage non-small cell lung cancer. Cancer Biol Ther. 2011;11(5):490–6. doi:10.4161/cbt.11.5.14550.
23. Vogt M, Munding J, Gruner M, et al. Frequent concomitant inactivation of miR-34a and miR-34b/c by CpG methylation in colorectal, pancreatic, mammary, ovarian, urothelial, and renal cell carcinomas and soft tissue sarcomas. Virchows Arch. 2011;458(3):313–22. doi:10.1007/s00428-010-1030-5.
24. Chen X, Hu H, Guan X, et al. CpG island methylation status of miRNAs in esophageal squamous cell carcinoma. Int J Cancer. 2012;130(7):1607–13. doi:10.1002/ijc.26171.
25. Kubo T, Toyooka S, Tsukuda K, et al. Epigenetic silencing of microRNA-34b/c plays an important role in the pathogenesis of malignant pleural mesothelioma. Clin Cancer Res. 2011;17(15):4965–74. doi:10.1158/1078-0432.CCR-10-3040.
26. Mazar J, Khaitan D, DeBlasio D, et al. Epigenetic regulation of microRNA genes and the role of miR-34b in cell invasion and motility in human melanoma. PLoS One. 2011;6(9):e24922. doi:10.1371/journal.pone.0024922.

27. Wong KY, Yim RL, So CC, et al. Epigenetic inactivation of the MIR34B/C in multiple myeloma. Blood. 2011;118(22):5901–4. doi:10.1182/blood-2011-06-361022.
28. Watanabe K, Emoto N, Hamano E, et al. Genome structure-based screening identified epigenetically silenced microRNA associated with invasiveness in non-small-cell lung cancer. Int J Cancer. 2012;130(11):2580–90. doi:10.1002/ijc.26254.
29. Lujambio A, Ropero S, Ballestar E, et al. Genetic unmasking of an epigenetically silenced microRNA in human cancer cells. Cancer Res. 2007;67(4):1424–9. doi:10.1158/0008-5472.CAN-06-4218.
30. Agirre X, Vilas-Zornoza A, Jimenez-Velasco A, et al. Epigenetic silencing of the tumor suppressor microRNA Hsa-miR-124a regulates CDK6 expression and confers a poor prognosis in acute lymphoblastic leukemia. Cancer Res. 2009;69(10):4443–53. doi:10.1158/0008-5472.CAN-08-4025.
31. Ando T, Yoshida T, Enomoto S, et al. DNA methylation of microRNA genes in gastric mucosae of gastric cancer patients: its possible involvement in the formation of epigenetic field defect. Int J Cancer. 2009;124(10):2367–74. doi:10.1002/ijc.24219.
32. Furuta M, Kozaki KI, Tanaka S, et al. miR-124 and miR-203 are epigenetically silenced tumor-suppressive microRNAs in hepatocellular carcinoma. Carcinogenesis. 2010;31(5):766–76. doi:10.1093/carcin/bgp250.
33. Wang P, Chen L, Zhang J, et al. Methylation-mediated silencing of the miR-124 genes facilitates pancreatic cancer progression and metastasis by targeting Rac1. Oncogene. 2014;33(4):514–24. doi:10.1038/onc.2012.598.
34. Wilting SM, van Boerdonk RA, Henken FE, et al. Methylation-mediated silencing and tumour suppressive function of hsa-miR-124 in cervical cancer. Mol Cancer. 2010;9:167. doi:10.1186/1476-4598-9-167.
35. Zhang Y, Yan LX, Wu QN, et al. miR-125b is methylated and functions as a tumor suppressor by regulating the ETS1 proto-oncogene in human invasive breast cancer. Cancer Res. 2011;71(10):3552–62. doi:10.1158/0008-5472.CAN-10-2435.
36. Saito Y, Liang G, Egger G, et al. Specific activation of microRNA-127 with down-regulation of the proto-oncogene BCL6 by chromatin-modifying drugs in human cancer cells. Cancer Cell. 2006;9(6):435–43. doi:10.1016/j.ccr.2006.04.020.
37. Huang YW, Liu JC, Deatherage DE, et al. Epigenetic repression of microRNA-129-2 leads to overexpression of SOX4 oncogene in endometrial cancer. Cancer Res. 2009;69(23):9038–46. doi:10.1158/0008 5472.CAN-09-1499.
38. Shen R, Pan S, Qi S, et al. Epigenetic repression of microRNA-129-2 leads to overexpression of SOX4 in gastric cancer. Biochem Biophys Res Commun. 2010;394(4):1047–52. doi:10.1016/j.bbrc.2010.03.121.
39. Lu CY, Lin KY, Tien MT, et al. Frequent DNA methylation of MiR-129-2 and its potential clinical implication in hepatocellular carcinoma. Genes Chromosomes Cancer. 2013;52(7):636–43. doi:10.1002/gcc.22059.
40. Zhang S, Hao J, Xie F, et al. Down-regulation of miR-132 by promoter methylation contributes to pancreatic cancer development. Carcinogenesis. 2011;32(8):1183–9. doi:10.1093/carcin/bgr105.
41. Formosa A, Lena AM, Markert EK, et al. DNA methylation silences miR-132 in prostate cancer. Oncogene. 2013;32(1):127–34. doi:10.1038/onc.2012.14.
42. Wiklund ED, Gao S, Hulf T, et al. MicroRNA alterations and associated aberrant DNA methylation patterns across multiple sample types in oral squamous cell carcinoma. PLoS ONE. 2011;6(11):e27840. doi:10.1371/journal.pone.0027840.
43. Dou L, Zheng D, Li J, et al. Methylation-mediated repression of microRNA-143 enhances MLL-AF4 oncogene expression. Oncogene. 2012;31(4):507–17. doi:10.1038/onc.2011.248.
44. Zaman MS, Chen Y, Deng G, et al. The functional significance of microRNA-145 in prostate cancer. Br J Cancer. 2010;103(2):256–64. doi:10.1038/sj.bjc.6605742.
45. Suh SO, Chen Y, Zaman MS, et al. MicroRNA-145 is regulated by DNA methylation and p53 gene mutation in prostate cancer. Carcinogenesis. 2011;32(5):772–8. doi:10.1093/carcin/bgr036.

46. Donzelli S, Mori F, Bellissimo T, et al. Epigenetic silencing of miR-145-5p contributes to brain metastasis. Oncotarget. 2015;6(34):35183–201. doi:10.18632.
47. Li S, Chowdhury R, Liu F, et al. Tumor-suppressive miR148a is silenced by CpG island hypermethylation in IDH1-mutant gliomas. Clin Cancer Res. 2014;20(22):5808–22. doi:10.1158/1078-0432.CCR-14-0234.
48. Tsuruta T, Kozaki K, Uesugi A, et al. miR-152 is a tumor suppressor microRNA that is silenced by DNA hypermethylation in endometrial cancer. Cancer Res. 2011;71(20):6450–62. doi:10.1158/0008-5472.CAN-11-0364.
49. Hashimoto Y, Akiyama Y, Otsubo T, et al. Involvement of epigenetically silenced microRNA-181c in gastric carcinogenesis. Carcinogenesis. 2010;31(5):777–84. doi:10.1093/carcin/bgq013.
50. Ma K, He Y, Zhang H, et al. DNA methylation-regulated miR-193a-3p dictates resistance of hepatocellular carcinoma to 5-fluorouracil via repression of SRSF2 expression. J Biol Chem. 2012;287(8):5639–49. doi:10.1074/jbc.M111.291229.
51. Gao XN, Lin J, Li YH, et al. MicroRNA-193a represses c-kit expression and functions as a methylation-silenced tumor suppressor in acute myeloid leukemia. Oncogene. 2011;30(31):3416–28. doi:10.1038/onc.2011.62.
52. Lv L, Deng H, Li Y, et al. The DNA methylation-regulated miR-193a-3p dictates the multi-chemoresistance of bladder cancer via repression of SRSF2/PLAU/HIC2 expression. Cell Death Dis. 2014;5:e1402. doi:10.1038/cddis.2014.367.
53. Davalos V, Moutinho C, Villanueva A, et al. Dynamic epigenetic regulation of the microRNA-200 family mediates epithelial and mesenchymal transitions in human tumorigenesis. Oncogene. 2012;31(16):2062–74. doi:10.1038/onc.2011.383.
54. Neves R, Scheel C, Weinhold S, et al. Role of DNA methylation in miR-200c/141 cluster silencing in invasive breast cancer cells. BMC Res Notes. 2010;3:219. doi:10.1186/1756-0500-3-219.
55. Ceppi P, Mudduluru G, Kumarswamy R, et al. Loss of miR-200c expression induces an aggressive, invasive, and chemoresistant phenotype in non-small cell lung cancer. Mol Cancer Res. 2010;8(9):1207–16. doi:10.1158/1541-7786.MCR-10-0052.
56. Wiklund ED, Bramsen JB, Hulf T, et al. Coordinated epigenetic repression of the miR-200 family and miR-205 in invasive bladder cancer. Int J Cancer. 2011;128(6):1327–34. doi:10.1002/ijc.25461.
57. Hur K, Toiyama Y, Takahashi M, et al. MicroRNA-200c modulates epithelial-to-mesenchymal transition (EMT) in human colorectal cancer metastasis. Gut. 2013;62(9):1315–26. doi:10.1136/gutjnl-2011-301846.
58. Tellez CS, Juri DE, Do K, et al. EMT and stem cell-like properties associated with miR-205 and miR-200 epigenetic silencing are early manifestations during carcinogen-induced transformation of human lung epithelial cells. Cancer Res. 2011;71(8):3087–97. doi:10.1158/0008-5472.CAN-10-3035.
59. Craig VJ, Cogliatti SB, Rehrauer H, et al. Epigenetic silencing of microRNA-203 dysregulates ABL1 expression and drives Helicobacter-associated gastric lymphomagenesis. Cancer Res. 2011;71(10):3616–24. doi:10.1158/0008-5472.CAN-10-3907.
60. Bueno MJ, Perez de Castro I, Gomez de Cedron M, et al. Genetic and epigenetic silencing of microRNA-203 enhances ABL1 and BCR-ABL1 oncogene expression. Cancer Cell. 2008;13(6):496–506. doi:10.1016/j.ccr.2008.04.018.
61. Wong KY, Liang R, So CC, et al. Epigenetic silencing of MIR203 in multiple myeloma. Br J Haematol. 2011;154(5):569–78. doi:10.1111/j.1365-2141.2011.08782.x.
62. Uesugi A, Kozaki K, Tsuruta T, et al. The tumor suppressive microRNA miR-218 targets the mTOR component Rictor and inhibits AKT phosphorylation in oral cancer. Cancer Res. 2011;71(17):5765–78. doi:10.1158/0008-5472.CAN-11-0368.
63. Lei H, Zou D, Li Z, et al. MicroRNA-219-2-3p functions as a tumor suppressor in gastric cancer and is regulated by DNA methylation. PLoS One. 2013;8(4):e60369. doi:10.1371/journal.pone.0060369.

64. Png KJ, Yoshida M, Zhang XH, et al. MicroRNA-335 inhibits tumor reinitiation and is silenced through genetic and epigenetic mechanisms in human breast cancer. Genes Dev. 2011;25(3):226–31. doi:10.1101/gad.1974211.
65. Tang JT, Wang JL, Du W, et al. MicroRNA 345, a methylation-sensitive microRNA is involved in cell proliferation and invasion in human colorectal cancer. Carcinogenesis. 2011;32(8):1207–15. doi:10.1093/carcin/bgr114.
66. Kong KL, Kwong DL, Chan TH, et al. MicroRNA-375 inhibits tumour growth and metastasis in oesophageal squamous cell carcinoma through repressing insulin-like growth factor 1 receptor. Gut. 2012;61(1):33–42. doi:10.1136/gutjnl-2011-300178.
67. Mazar J, DeBlasio D, Govindarajan SS, et al. Epigenetic regulation of microRNA-375 and its role in melanoma development in humans. FEBS Lett. 2011;585(15):2467–76. doi:10.1016/j.febslet.2011.06.025.
68. Li X, Lin R, Li J. Epigenetic silencing of microRNA-375 regulates PDK1 expression in esophageal cancer. Dig Dis Sci. 2011;56(10):2849–56. doi:10.1007/s10620-011-1711-1.
69. Xi S, Xu H, Shan J, et al. Cigarette smoke mediates epigenetic repression of miR-487b during pulmonary carcinogenesis. J Clin Invest. 2013;123(3):1241–61. doi:10.1172/JCI61271.
70. Saito Y, Suzuki H, Tsugawa H, et al. Chromatin remodeling at Alu repeats by epigenetic treatment activates silenced microRNA-512-5p with down-regulation of Mcl-1 in human gastric cancer cells. Oncogene. 2009;28(30):2738–44. doi:10.1038/onc.2009.140.
71. Baer C, Oakes CC, Ruppert AS, et al. Epigenetic silencing of miR-708 enhances NF-kappaB signaling in chronic lymphocytic leukemia. Int J Cancer. 2015;137(6):1352–61. doi:10.1002/ijc.29491.
72. Yan H, Choi AJ, Lee BH, Ting AH. Identification and functional analysis of epigenetically silenced microRNAs in colorectal cancer cells. PLoS ONE. 2011;6(6):e20628. doi:10.1371/journal.pone.0020628.
73. Zhang PP, Wang XL, Zhao W, et al. DNA methylation-mediated repression of miR-941 enhances lysine (K)-specific demethylase 6B expression in hepatoma cells. J Biol Chem. 2014;289(35):24724–35. doi:10.1074/jbc.M114.567818.
74. Lopez-Serra P, Esteller M. DNA methylation-associated silencing of tumor-suppressor microRNAs in cancer. Oncogene. 2012;31(13):1609–22. doi:10.1038/onc.2011.354.
75. Suzuki H, Maruyama R, Yamamoto E, Kai M. DNA methylation and microRNA dysregulation in cancer. Mol Oncol. 2012;6(6):567–78. doi:10.1016/j.molonc.2012.07.007.
76. Prensner JR, Chinnaiyan AM. The emergence of lncRNAs in cancer biology. Cancer Discov. 2011;1(5):391–407. doi:10.1158/2159-8290.CD-11-0209.
77. Maruyama R, Suzuki H. Long non-coding RNA involvement in cancer. BMB Rep. 2012;45(11):604–11.
78. Cabili MN, Trapnell C, Goff L, Koziol M, et al. Integrative annotation of human large intergenic non-coding RNAs reveals global properties and specific subclasses. Genes Dev. 2011;25(18):1915–27. doi:10.1101/gad.17446611.
79. Kozaki K, Imoto I, Mogi S, et al. Exploration of tumor-suppressive microRNAs silenced by DNA hypermethylation in oral cancer. Cancer Res. 2008;68(7):2094–105. doi:10.1158/0008-5472.CAN-07-5194.
80. Bi C, Chung TH, Huang G et al. Genome-wide pharmacologic unmasking identifies tumor suppressive microRNAs in multiple myeloma. Oncotarget, 2015;6(28):26508–18. doi:10.18632/oncotarget.4769
81. Omura N, Li CP, Li A, et al. Genome-wide profiling of methylated promoters in pancreatic adenocarcinoma. Cancer Biol Ther. 2008;7(7):1146–56. doi:6208 [pii].
82. Ke XS, Qu Y, Rostad K, et al. Genome-wide profiling of histone h3 lysine 4 and lysine 27 trimethylation reveals an epigenetic signature in prostate carcinogenesis. PLoS One. 2009;4(3):e4687. doi:10.1371/journal.pone.0004687.
83. Marson A, Levine SS, Cole MF, et al. Connecting microRNA genes to the core transcriptional regulatory circuitry of embryonic stem cells. Cell. 2008;134(3):521–33. doi:10.1016/j.cell.2008.07.020.

84. Ozsolak F, Poling LL, Wang Z, et al. Chromatin structure analyses identify miRNA promoters. Genes Dev. 2008;22(22):3172–83. doi:10.1101/gad.1706508.
85. Baer C, Claus R, Frenzel LP, et al. Extensive promoter DNA hypermethylation and hypomethylation is associated with aberrant microRNA expression in chronic lymphocytic leukemia. Cancer Res. 2012;72(15):3775–85. doi:10.1158/0008-5472.CAN-12-0803.
86. Wong KY, So CC, Loong F, et al. Epigenetic inactivation of the miR-124-1 in haematological malignancies. PLoS ONE. 2011;6(4):e19027. doi:10.1371/journal.pone.0019027.
87. He L, He X, Lim LP, et al. A microRNA component of the p53 tumour suppressor network. Nature. 2007;447(7148):1130–4. doi:10.1038/nature05939.
88. Bommer GT, Gerin I, Feng Y, et al. P53-mediated activation of miRNA34 candidate tumor-suppressor genes. Curr Biol. 2007;17(15):1298–307. doi:10.1016/j.cub.2007.06.068.
89. Yu F, Jiao Y, Zhu Y, et al. MicroRNA 34c gene down-regulation via DNA methylation promotes self-renewal and epithelial-mesenchymal transition in breast tumor-initiating cells. J Biol Chem. 2012;287(1):465–73. doi:10.1074/jbc.M111.280768.
90. Rotkrua P, Akiyama Y, Hashimoto Y, et al. MiR-9 down-regulates CDX2 expression in gastric cancer cells. Int J Cancer. 2011;129(11):2611–20. doi:10.1002/ijc.25923.
91. Ma L, Young J, Prabhala H, et al. miR-9, a MYC/MYCN-activated microRNA, regulates E-cadherin and cancer metastasis. Nat Cell Biol. 2010;12(3):247–56. doi:10.1038/ncb2024.
92. Gregory PA, Bert AG, Paterson EL, et al. The miR-200 family and miR-205 regulate epithelial to mesenchymal transition by targeting ZEB1 and SIP1. Nat Cell Biol. 2008;10(5):593–601. doi:10.1038/ncb1722.
93. Korpal M, Lee ES, Hu G, Kang Y. The miR-200 family inhibits epithelial-mesenchymal transition and cancer cell migration by direct targeting of E-cadherin transcriptional repressors ZEB1 and ZEB2. J Biol Chem. 2008;283(22):14910–4. doi:10.1074/jbc.C800074200.
94. Park SM, Gaur AB, Lengyel E, Peter ME. The miR-200 family determines the epithelial phenotype of cancer cells by targeting the E-cadherin repressors ZEB1 and ZEB2. Genes Dev. 2008;22(7):894–907. doi:10.1101/gad.1640608.
95. Vrba L, Jensen TJ, Garbe JC, et al. Role for DNA methylation in the regulation of miR-200c and miR-141 expression in normal and cancer cells. PLoS One. 2010;5(1):e8697. doi:10.1371/journal.pone.0008697.
96. Balaguer F, Link A, Lozano JJ, et al. Epigenetic silencing of miR-137 is an early event in colorectal carcinogenesis. Cancer Res. 2010;70(16):6609–18. doi:0008–5472.CAN-10-0622 [pii].
97. Chen Q, Chen X, Zhang M, et al. miR-137 is frequently down-regulated in gastric cancer and is a negative regulator of Cdc42. Dig Dis Sci. 2011;56(7):2009–16. doi:10.1007/s10620-010-1536-3.
98. Shimizu T, Suzuki H, Nojima M, et al. Methylation of a panel of microRNA genes is a novel biomarker for detection of bladder cancer. Eur Urol. 2013;63(6):1091–100. doi:10.1016/j.eururo.2012.11.030.
99. Langevin SM, Stone RA, Bunker CH, et al. MicroRNA-137 promoter methylation is associated with poorer overall survival in patients with squamous cell carcinoma of the head and neck. Cancer. 2011;117(7):1454–62. doi:10.1002/cncr.25689.
100. Langevin SM, Stone RA, Bunker CH, et al. MicroRNA-137 promoter methylation in oral rinses from patients with squamous cell carcinoma of the head and neck is associated with gender and body mass index. Carcinogenesis. 2010;31(5):864–70. doi:10.1093/carcin/bgq051.
101. Liu M, Lang N, Qiu M, et al. miR-137 targets Cdc42 expression, induces cell cycle G1 arrest and inhibits invasion in colorectal cancer cells. Int J Cancer. 2011;128(6):1269–79. doi:10.1002/ijc.25452.
102. Althoff K, Beckers A, Odersky A, et al. MiR-137 functions as a tumor suppressor in neuroblastoma by down-regulating KDM1A. Int J Cancer. 2013;133(5):1064–73. doi:10.1002/ijc.28091.
103. Szulwach KE, Li X, Smrt RD, et al. Cross talk between microRNA and epigenetic regulation in adult neurogenesis. J Cell Biol. 2010;189(1):127–41. doi:10.1083/jcb.200908151.

104. Smrt RD, Szulwach KE, Pfeiffer RL, et al. MicroRNA miR-137 regulates neuronal maturation by targeting ubiquitin ligase mind bomb-1. Stem Cells. 2010;28(6):1060–70. doi:10.1002/stem.431.
105. Michael MZ, O' Connor SM, van Holst Pellekaan NG, et al. Reduced accumulation of specific microRNAs in colorectal neoplasia. Mol Cancer Res. 2003;1(12):882–91.
106. Iorio MV, Ferracin M, Liu CG, et al. MicroRNA gene expression deregulation in human breast cancer. Cancer Res. 2005;65(16):7065–70. doi:10.1158/0008-5472.CAN-05-1783.
107. Iorio MV, Visone R, Di Leva G, et al. MicroRNA signatures in human ovarian cancer. Cancer Res. 2007;67(18):8699–707. doi:10.1158/0008-5472.CAN-07-1936.
108. Sachdeva M, Zhu S, Wu F, et al. p53 represses c-Myc through induction of the tumor suppressor miR-145. Proc Natl Acad Sci U S A. 2009;106(9):3207–12. doi:10.1073/pnas.0808042106.
109. Sachdeva M, Mo YY. MicroRNA-145 suppresses cell invasion and metastasis by directly targeting mucin 1. Cancer Res. 2010;70(1):378–87. doi:10.1158/0008-5472.CAN-09-2021.
110. Shi B, Sepp-Lorenzino L, Prisco M, et al. Micro RNA 145 targets the insulin receptor substrate-1 and inhibits the growth of colon cancer cells. J Biol Chem. 2007;282(45):32582–90. doi:10.1074/jbc.M702806200.
111. Fuse M, Nohata N, Kojima S, et al. Restoration of miR-145 expression suppresses cell proliferation, migration and invasion in prostate cancer by targeting FSCN1. Int J Oncol. 2011;38(4):1093–101. doi:10.3892/ijo.2011.919.
112. Chen X, Gong J, Zeng H, et al. MicroRNA145 targets BNIP3 and suppresses prostate cancer progression. Cancer Res. 2010;70(7):2728–38. doi:10.1158/0008-5472.CAN-09-3718.
113. Wu Q, Yang Z, Xia L, et al. Methylation of miR-129-5p CpG island modulates multi-drug resistance in gastric cancer by targeting ABC transporters. Oncotarget. 2014;5(22):11552–63. doi:2594 [pii].
114. Tsai KW, Wu CW, Hu LY, et al. Epigenetic regulation of miR-34b and miR-129 expression in gastric cancer. Int J Cancer. 2011;129(11):2600–10. doi:10.1002/ijc.25919.
115. Brueckner B, Stresemann C, Kuner R, et al. The human let-7a-3 locus contains an epigenetically regulated microRNA gene with oncogenic function. Cancer Res. 2007;67(4):1419–23. doi:10.1158/0008-5472.CAN-06-4074.
116. He Y, Cui Y, Wang W, et al. Hypomethylation of the hsa-miR-191 locus causes high expression of hsa-mir-191 and promotes the epithelial-to-mesenchymal transition in hepatocellular carcinoma. Neoplasia. 2011;13(9):841–53.
117. Tsai KW, Hu LY, Wu CW, et al. Epigenetic regulation of miR-196b expression in gastric cancer. Genes Chromosomes Cancer. 2010;49(11):969–80. doi:10.1002/gcc.20804.
118. Li A, Omura N, Hong SM, et al. Pancreatic cancers epigenetically silence SIP1 and hypomethylate and overexpress miR-200a/200b in association with elevated circulating miR-200a and miR-200b levels. Cancer Res. 2010;70(13):5226–37.
119. de Souza Rocha Simonini P, Breiling A, et al. Epigenetically deregulated microRNA-375 is involved in a positive feedback loop with estrogen receptor alpha in breast cancer cells. Cancer Res. 2010;70(22):9175–84. doi:10.1158/0008-5472.CAN-10-1318.
120. Fornari F, Milazzo M, Chieco P, et al. In hepatocellular carcinoma miR-519d is up-regulated by p53 and DNA hypomethylation and targets CDKN1A/p21, PTEN, AKT3 and TIMP2. J Pathol. 2012;227(3):275–85. doi:10.1002/path.3995.
121. Hu H, Li S, Cui X, et al. The overexpression of hypomethylated miR-663 induces chemotherapy resistance in human breast cancer cells by targeting heparin sulfate proteoglycan 2 (HSPG2). J Biol Chem. 2013;288(16):10973–85. doi:10.1074/jbc.M112.434340.
122. Popovic R, Riesbeck LE, Velu CS, et al. Regulation of mir-196b by MLL and its overexpression by MLL fusions contributes to immortalization. Blood. 2009;113(14):3314–22. doi:10.1182/blood-2008-04-154310.
123. Gebauer K, Peters I, Dubrowinskaja N, et al. Hsa-mir-124-3 CpG island methylation is associated with advanced tumours and disease recurrence of patients with clear cell renal cell carcinoma. Br J Cancer. 2013;108(1):131–8. doi:10.1038/bjc.2012.537.

124. Siemens H, Neumann J, Jackstadt R, et al. Detection of miR-34a promoter methylation in combination with elevated expression of c-Met and beta-catenin predicts distant metastasis of colon cancer. Clin Cancer Res. 2013;19(3):710–20. doi:10.1158/1078-0432.CCR-12-1703.
125. Kalimutho M, Di Cecilia S, Del Vecchio BG, et al. Epigenetically silenced miR-34b/c as a novel faecal-based screening marker for colorectal cancer. Br J Cancer. 2011;104(11):1770–8. doi:10.1038/bjc.2011.82.
126. Kamimae S, Yamamoto E, Yamano HO, et al. Epigenetic alteration of DNA in mucosal wash fluid predicts invasiveness of colorectal tumors. Cancer Prev Res (Phila). 2011;4(5):674–83. doi:10.1158/1940-6207.CAPR-10-0214.
127. Muraoka T, Soh J, Toyooka S, et al. The degree of microRNA-34b/c methylation in serum-circulating DNA is associated with malignant pleural mesothelioma. Lung Cancer. 2013;82(3):485–90. doi:10.1016/j.lungcan.2013.09.017.
128. Cordero F, Ferrero G, Polidoro S, et al. Differentially methylated microRNAs in prediagnostic samples of subjects who developed breast cancer in the European Prospective Investigation into Nutrition and Cancer (EPIC-Italy) cohort. Carcinogenesis. 2015;36(10):1144–53. doi:10.1093/carcin/bgv102.
129. Suzuki R, Yamamoto E, Nojima M, et al. Aberrant methylation of microRNA-34b/c is a predictive marker of metachronous gastric cancer risk. J Gastroenterol. 2014;49(7):1135–44. doi:10.1007/s00535-013-0861-7.
130. Asada K, Nakajima T, Shimazu T, et al. Demonstration of the usefulness of epigenetic cancer risk prediction by a multicentre prospective cohort study. Gut. 2015;64(3):388–96. doi:10.1136/gutjnl-2014-307094.
131. Wang Y, Toh HC, Chow P, et al. MicroRNA-224 is up-regulated in hepatocellular carcinoma through epigenetic mechanisms. FABES J. 2012;26(7):3032–41. doi:10.1096/fj.11-201855.
132. Chang S, Wang RH, Akagi K, et al. Tumor suppressor BRCA1 epigenetically controls oncogenic microRNA-155. Nat Med. 2011;17(10):1275–82. doi:10.1038/nm.2459.
133. Sampath D, Liu C, Vasan K, et al. Histone deacetylases mediate the silencing of miR-15a, miR-16, and miR-29b in chronic lymphocytic leukemia. Blood. 2012;119(5):1162–72. doi:10.1182/blood-2011-05-351510.
134. Zhang X, Chen X, Lin J, et al. Myc represses miR-15a/miR-16-1 expression through recruitment of HDAC3 in mantle cell and other non-Hodgkin B-cell lymphomas. Oncogene. 2012;31(24):3002–8. doi:10.1038/onc.2011.470.
135. Buurman R, Gurlevik E, Schaffer V, et al. Histone deacetylases activate hepatocyte growth factor signaling by repressing microRNA-449 in hepatocellular carcinoma cells. Gastroenterology. 2012;143(3):811–20 e815. doi:10.1053/j.gastro.2012.05.033
136. Incoronato M, Urso L, Portela A, et al. Epigenetic regulation of miR-212 expression in lung cancer. PLoS one. 2011;6(11):e27722. doi:10.1371/journal.pone.0027722.
137. Ryu S, McDonnell K, Choi H, et al. Suppression of miRNA-708 by polycomb group promotes metastases by calcium-induced cell migration. Cancer Cell. 2013;23(1):63–76. doi:10.1016/j.ccr.2012.11.019.
138. Au SL, Wong CC, Lee JM, et al. Enhancer of zeste homolog 2 epigenetically silences multiple tumor suppressor microRNAs to promote liver cancer metastasis. Hepatology. 2012;56(2):622–31. doi:10.1002/hep.25679.
139. Fabbri M, Garzon R, Cimmino A, et al. MicroRNA-29 family reverts aberrant methylation in lung cancer by targeting DNA methyltransferases 3A and 3B. Proc Natl Acad Sci U S A. 2007;104(40):15805–10. doi:10.1073/pnas.0707628104.
140. Ng EK, Tsang WP, Ng SS, et al. MicroRNA-143 targets DNA methyltransferases 3A in colorectal cancer. Br J Cancer. 2009;101(4):699–706. doi:10.1038/sj.bjc.6605195.
141. Chen BF, Gu S, Suen YK, et al. microRNA-199a-3p, DNMT3A, and aberrant DNA methylation in testicular cancer. Epigenetics. 2014;9(1):119–28. doi:10.4161/epi.25799.
142. Huang J, Wang Y, Guo Y, Sun S. Down-regulated microRNA-152 induces aberrant DNA methylation in hepatitis B virus-related hepatocellular carcinoma by targeting DNA methyltransferase 1. Hepatology. 2010;52(1):60–70. doi:10.1002/hep.23660.

143. Braconi C, Huang N, Patel T. MicroRNA-dependent regulation of DNA methyltransferase-1 and tumor suppressor gene expression by interleukin-6 in human malignant cholangiocytes. Hepatology. 2010;51(3):881–90. doi:10.1002/hep.23381.
144. Wang H, Wu J, Meng X, et al. MicroRNA-342 inhibits colorectal cancer cell proliferation and invasion by directly targeting DNA methyltransferase 1. Carcinogenesis. 2011;32(7):1033–42. doi:10.1093/carcin/bgr081.
145. Zhang Z, Tang H, Wang Z, et al. MiR-185 targets the DNA methyltransferases 1 and regulates global DNA methylation in human glioma. Mol Cancer. 2011;10:124. doi:10.1186/1476-4598-10-124.
146. Liu R, Gu J, Jiang P, et al. DNMT1-microRNA126 epigenetic circuit contributes to esophageal squamous cell carcinoma growth via ADAM9-EGFR-AKT signaling. Clin Cancer Res. 2015;21(4):854–63. doi:10.1158/1078-0432.CCR-14-1740.
147. Majid S, Dar AA, Saini S, et al. miRNA-34b inhibits prostate cancer through demethylation, active chromatin modifications, and AKT pathways. Clin Cancer Res. 2013;19(1):73–84. doi:10.1158/1078-0432.CCR-12-2952.
148. Branco MR, Ficz G, Reik W. Uncovering the role of 5-hydroxymethylcytosine in the epigenome. Nat Rev Genet. 2012;13(1):7–13. doi:10.1038/nrg3080.
149. Chuang KH, Whitney-Miller CL, Chu CY, et al. MicroRNA-494 is a master epigenetic regulator of multiple invasion-suppressor microRNAs by targeting ten eleven translocation 1 in invasive human hepatocellular carcinoma tumors. Hepatology. 2015;62(2):466–80. doi:10.1002/hep.27816.
150. Loriot A, Van Tongelen A, Blanco J, et al. A novel cancer-germline transcript carrying pro-metastatic miR-105 and TET-targeting miR-767 induced by DNA hypomethylation in tumors. Epigenetics. 2014;9(8):1163–71. doi:10.4161/epi.29628.
151. Varambally S, Dhanasekaran SM, Zhou M, et al. The polycomb group protein EZH2 is involved in progression of prostate cancer. Nature. 2002;419(6907):624–9. doi:10.1038/nature01075.
152. Kleer CG, Cao Q, Varambally S, et al. EZH2 is a marker of aggressive breast cancer and promotes neoplastic transformation of breast epithelial cells. Proc Natl Acad Sci U S A. 2003;100(20):11606–11. doi:10.1073/pnas.1933744100.
153. Varambally S, Cao Q, Mani RS, et al. Genomic loss of microRNA-101 leads to overexpression of histone methyltransferase EZH2 in cancer. Science. 2008;322(5908):1695–9. doi:10.1126/science.1165395.
154. Friedman JM, Liang G, Liu CC, et al. The putative tumor suppressor microRNA-101 modulates the cancer epigenome by repressing the polycomb group protein EZH2. Cancer Res. 2009;69(6):2623–9. doi:10.1158/0008-5472.CAN-08-3114.
155. Wang HJ, Ruan HJ, He XJ, et al. MicroRNA-101 is down-regulated in gastric cancer and involved in cell migration and invasion. Eur J Cancer. 2010;46(12):2295–303.
156. Zhang JG, Guo JF, Liu DL, et al. MicroRNA-101 exerts tumor-suppressive functions in non-small cell lung cancer through directly targeting enhancer of zeste homolog 2. J Thorac Oncol. 2011;6(4):671–8. doi:10.1097/JTO.0b013e318208eb35.
157. Sakurai T, Bilim VN, Ugolkov AV, et al. The enhancer of zeste homolog 2 (EZH2), a potential therapeutic target, is regulated by miR-101 in renal cancer cells. Biochem Biophys Res Commun. 2012;422(4):607–14. doi:10.1016/j.bbrc.2012.05.035.
158. Wong CF, Tellam RL. MicroRNA-26a targets the histone methyltransferase Enhancer of Zeste homolog 2 during myogenesis. J Biol Chem. 2008;283(15):9836–43. doi:10.1074/jbc.M709614200.
159. Alajez NM, Shi W, Hui AB, et al. Enhancer of Zeste homolog 2 (EZH2) is overexpressed in recurrent nasopharyngeal carcinoma and is regulated by miR-26a, miR-101, and miR-98. Cell Death Dis. 2010;1:e85. doi:10.1038/cddis.2010.64.
160. Zheng F, Liao YJ, Cai MY, et al. The putative tumour suppressor microRNA-124 modulates hepatocellular carcinoma cell aggressiveness by repressing ROCK2 and EZH2. Gut. 2012;61(2):278–89. doi:10.1136/gut.2011.239145.

161. Guo Y, Ying L, Tian Y, et al. miR-144 down-regulation increases bladder cancer cell proliferation by targeting EZH2 and regulating Wnt signaling. FEBS J. 2013;280(18):4531–8. doi:10.1111/febs.12417.
162. Derfoul A, Juan AH, Difilippantonio MJ, et al. Decreased microRNA-214 levels in breast cancer cells coincides with increased cell proliferation, invasion and accumulation of the Polycomb Ezh2 methyltransferase. Carcinogenesis. 2011;32(11):1607–14. doi:10.1093/carcin/bgr184.
163. Kong D, Heath E, Chen W, et al. Loss of let-7 up-regulates EZH2 in prostate cancer consistent with the acquisition of cancer stem cell signatures that are attenuated by BR-DIM. PLoS ONE. 2012;7(3):e33729. doi:10.1371/journal.pone.0033729.
164. Cao Q, Mani RS, Ateeq B, et al. Coordinated regulation of polycomb group complexes through microRNAs in cancer. Cancer Cell. 2011;20(2):187–99. doi:10.1016/j.ccr.2011.06.016.
165. Zhang X, Zhao X, Fiskus W, et al. Coordinated silencing of MYC-mediated miR-29 by HDAC3 and EZH2 as a therapeutic target of histone modification in aggressive B-Cell lymphomas. Cancer Cell. 2012;22(4):506–23. doi:10.1016/j.ccr.2012.09.003.
166. Wu W, Bhagat TD, Yang X et al. Hypomethylation of non-coding DNA regions and overexpression of the long non-coding RNA, AFAP1-AS1, in Barrett's esophagus and esophageal adenocarcinoma. Gastroenterology, 2013;144(5):956–66 e954. doi:10.1053/j.gastro.2013.01.019
167. Li T, Xie J, Shen C, et al. Up-regulation of long non-coding RNA ZEB1-AS1 promotes tumor metastasis and predicts poor prognosis in hepatocellular carcinoma. Oncogene. 2015. doi:10.1038/onc.2015.223.
168. Braconi C, Kogure T, Valeri N, et al. microRNA-29 can regulate expression of the long non-coding RNA gene MEG3 in hepatocellular cancer. Oncogene. 2011;30(47):4750–6. doi:10.1038/onc.2011.193.
169. Garding A, Bhattacharya N, Claus R, et al. Epigenetic up-regulation of lncRNAs at 13q14.3 in leukemia is linked to the In Cis down-regulation of a gene cluster that targets NF-κB. PLoS Genet. 2013;9(4):e1003373. doi:10.1371/journal.pgen.1003373.
170. Calin GA, Dumitru CD, Shimizu M, et al. Frequent deletions and down-regulation of micro-RNA genes miR15 and miR16 at 13q14 in chronic lymphocytic leukemia. Proc Natl Acad Sci U S A. 2002;99(24):15524–9.
171. Zhi H, Ning S, Li X, et al. A novel reannotation strategy for dissecting DNA methylation patterns of human long intergenic non-coding RNAs in cancers. Nucleic Acids Res. 2014;42(13):8258–70. doi:10.1093/nar/gku575.
172. Sun M, Liu XH, Lu KH, et al. EZH2-mediated epigenetic suppression of long non-coding RNA SPRY4-IT1 promotes NSCLC cell proliferation and metastasis by affecting the epithelial-mesenchymal transition. Cell Death Dis. 2014;5:e1298. doi:10.1038/cddis.2014.256.
173. Gupta RA, Shah N, Wang KC, et al. Long non-coding RNA HOTAIR reprograms chromatin state to promote cancer metastasis. Nature. 2010;464(7291):1071–6. doi:10.1038/nature08975.
174. Yang Z, Zhou L, Wu LM, et al. Overexpression of long non-coding RNA HOTAIR predicts tumor recurrence in hepatocellular carcinoma patients following liver transplantation. Ann Surg Oncol. 2011;18(5):1243–50. doi:10.1245/s10434-011-1581-y.
175. Kogo R, Shimamura T, Mimori K, et al. Long non-coding RNA HOTAIR regulates polycomb-dependent chromatin modification and is associated with poor prognosis in colorectal cancers. Cancer Res. 2011;71(20):6320–6. doi:10.1158/0008-5472.CAN-11-1021.
176. Kim K, Jutooru I, Chadalapaka G, et al. HOTAIR is a negative prognostic factor and exhibits pro-oncogenic activity in pancreatic cancer. Oncogene. 2012. doi:10.1038/onc.2012.193.
177. Niinuma T, Suzuki H, Nojima M, et al. Up-regulation of miR-196a and HOTAIR drive malignant character in gastrointestinal stromal tumors. Cancer Res. 2012;72(5):1126–36. doi:10.1158/0008-5472.CAN-11-1803.

178. Pandey GK, Mitra S, Subhash S, et al. The risk-associated long non-coding RNA NBAT-1 controls neuroblastoma progression by regulating cell proliferation and neuronal differentiation. Cancer Cell. 2014;26(5):722–37. doi:10.1016/j.ccell.2014.09.014.
179. Li W, Zheng J, Deng J, You Y et al. Increased levels of the long intergenic non-protein coding RNA POU3F3 promote DNA methylation in esophageal squamous cell carcinoma cells. Gastroenterology, 2014;146(7):1714–26 e1715. doi:10.1053/j.gastro.2014.03.002
180. Cui M, Xiao Z, Wang Y, Zheng M, et al. Long non-coding RNA HULC modulates abnormal lipid metabolism in hepatoma cells through an miR-9-mediated RXRA signaling pathway. Cancer Res. 2015;75(5):846–57. doi:10.1158/0008-5472.CAN-14-1192.
181. Merry CR, Forrest ME, Sabers JN, et al. DNMT1-associated long non-coding RNAs regulate global gene expression and DNA methylation in colon cancer. Hum Mol Genet. 2015;24(21):6240–53. doi:10.1093/hmg/ddv343.
182. Arab K, Park YJ, Lindroth AM, et al. Long non-coding RNA TARID directs demethylation and activation of the tumor suppressor TCF21 via GADD45A. Mol Cell. 2014;55(4):604–14. doi:10.1016/j.molcel.2014.06.031.
183. Gooskens SL, Gadd S, Guidry Auvil JM, et al. TCF21 hypermethylation in genetically quiescent clear cell sarcoma of the kidney. Oncotarget. 2015;6(18):15828–41. doi:4682 [pii].

Chapter 5
Noncoding RNAs in Growth and Death of Cancer Cells

Anfei Liu and Shanrong Liu

Abstract The mammalian genomes are mostly comprised of noncoding genes. And mammalian genomes are characterized by pervasive expression of different types of noncoding RNAs (ncRNAs). In sharp contrast to previous collections, these ncRNAs show strong purifying selection evolutionary conservation. Previous studies indicated that only a small fraction of the mammalian genome codes for messenger RNAs destined to be translated into peptides or proteins, and it is generally assumed that a large portion of transcribed sequences—including pseudogenes and several classes of ncRNAs—do not give rise to peptides or proteins. However, ribosome profiling suggests that ribosomes occupy many regions of the transcriptome thought to be noncoding. Moreover, these observations highlight a potentially large and complex set of biologically regulated translational events from transcripts formerly thought to lack coding potential. Furthermore, accumulating evidence from previous studies has suggested that the novel translation products exhibit temporal regulation similar to that of proteins known to be involved in many biological activity processes. In this review, we focus on the coding potential of noncoding genes and ncRNAs. We also sketched the possible mechanisms for their coding activities. Overall, our review provides new insights into the word of central dogma and is an expansive resource of functional annotations for biomedical research. At last, the outcome of the majority of the translation events and their potential biological purpose remain an intriguing topic for future investigation.

Keywords Cancer cells • Noncoding RNA • Proliferation • Cell cycle • Necrosis • Apoptosis • Autophagy

---

A. Liu • S. Liu (✉)
Second Military Medical University, 800 Xiangyin Road, Shanghai, China
e-mail: liushanrong01@126.com

E. Song (ed.), *The Long and Short Non-coding RNAs in Cancer Biology*,
Advances in Experimental Medicine and Biology 927,
DOI 10.1007/978-981-10-1498-7_5

5.1 Introduction

A striking finding about the production of numerous noncoding RNAs (ncRNAs) from mammalian genomes in the past decade has completely shifted our view of gene expression programs, which have historically been based on the assumption that only protein-coding genes could specify cellular functions. Increasing evidence suggests that while it is entirely possible that many of those ncRNAs are transcription noises or by-products of RNA processing, a large fraction of them provide diverse regulatory functions in the cell, and regulatory RNA networks in general represent a crucial interphase between genomics and proteomics [1, 2], impacting both on physiology and disease. The era of ncRNA research has resulted in and benefited from the rapid advance in genomics technologies and informatics approaches [1] that have been developed in recent years and thus has experienced tremendous progress in elucidating the function and mechanism of various ncRNAs.

Classic ncRNA transcripts have been categorized into two groups: housekeeping ncRNAs and regulatory ncRNAs (see Table 5.1). Housekeeping ncRNAs include small ncRNAs such as transfer RNAs (tRNAs) for carrying amino acids, small nucleolus RNAs (snoRNAs) for RNA modifications, and small nuclear RNAs (snRNAs) for RNA splicing, and large ones, such as ribosomal RNAs (rRNAs) for protein synthesis [1]. Regulatory ncRNAs can be divided into two classes based on an arbitrary size cutoff of 200 nt to separate small ncRNAs from long noncoding RNAs (lncRNAs), including the well-annotated microRNAs (miRNAs), which are 18–25 nt noncoding regulatory RNAs that mediate posttranscriptional regulation of gene expression by inhibiting the translation or promoting the degradation of target

Table 5.1 The categorization of ncRNAs

	Abbreviation	Full name	Role
Housekeeping ncRNAs	rRNA	Ribosomal RNA	mRNA reading and decoding
	tRNA	Transfer RNA	Amino acid carriers
	snRNA	Small nuclear RNA	RNA splicing
	snoRNA	Small nucleolar RNA	RNA modification
	TR	Telomere RNA	Chromosome end synthesis
Regulatory ncRNAs	miRNA	Micro RNA	RNA stability and translation control
	siRNA	Endogenous RNA	RNA silencing
	piRNA	Piwi RNA	Gene silencing
	eRNA	Enhancer-derived RNA	Regulation of gene expression
	lncRNA	Long noncoding RNA	Imprinting, epigenetics, nuclear structure
	circRNA	Circular RNA	Regulation of miRNA
	xRNA	X-inactive RNA	RNA silencing
	gRNA	Guide RNA	RNA editing

mRNAs [3], and lncRNAs, which account for approximately 81.8 % of all ncRNAs [4] and have a role in physiologic aspects of cell-type determination and tissue homeostasis [5]. However, miRNA and lncRNA are not alone—there are many other families of ncRNAs that are necessary for regulatory roles in the cell, such as enhancer RNA (eRNA, 50–2000 nt) for transcribing enhancers, guide RNA(gRNA, 60–80 nt) for editing site selection in RNA editing, small interfering RNAs (siRNAs, 19–23 nt) which interfere with the expression of specific genes with complementary nucleotide sequences, and piwi-interacting RNAs (piRNAs, 26–30 nt), which are involved in germline development, silencing of selfish DNA elements, and maintaining germline DNA integrity. piRNAs form complexes with piwi proteins, members of the Argonaute family, and unlike other small RNAs, they are created without RNase Dicer participation [6]. They represent a large group of ncRNAs with many diverse functions, many of which are only just beginning to be understood. Indeed, the number of ncRNAs per genome correlates far better with organism complexity than the number of coding genes, suggesting that RNA-based regulatory mechanisms are critical in the evolution of developmental complexity [2] and tumorigenesis.

The development of human cancer is mediated by both genetic and epigenetic alterations of the cell [7, 8], and ncRNAs play a crucial role in maintaining genomic stability which is essential for cell survival and preventing tumorigenesis [9]. ncRNAs and coding genes cooperatively mediate pathway dysregulation during the development and progression of cancers [10]. It has become increasingly clear that misexpression of ncRNAs is recognized as a hallmark feature in cancers. Studies have shown that tumor tissue and normal tissue exhibit distinct ncRNA expression profiles, and the aberrant ncRNAs can play an important role in cancer development [9], metastasis, prognosis, and patient survival rates [11]. There is mounting evidence of miRNAs acing as either potent oncogenes or tumor suppressor genes, linking to cancer initiation and progression [12]. lncRNAs involved in different biological processes occur in a space- and time-dependent manner [13], and many of them are dysregulated in a wide range of cancers [14]. siRNA-based therapies are known for offering great potential for cancer treatments. RNA interference (RNAi) is a well-conserved, naturally occurring process of posttranscriptional gene silencing initiated by double-strand siRNA. A RNAi-based approach can effectively silence any disease-causing gene in a given cell type or tissue/organ of interest [15]. piRNA pathway is shown active in various cancers. For example, the expression of PIWIL1 and PIWIL2 has been found in a wide range of human cancers such as stomach, breast, gastrointestinal tract, and endometrium [16–18] and recently also in ovarian carcinoma [19]. Among effects of estrogen signaling on the transcriptome of breast cancer cells, induced eRNA contributes to the dynamic generation/stabilization of enhancer: promoter looping between the regulated coding transcription units and these ERα-bound enhancers [20]. In most advanced cancers, telomerase is reactivated and serves to maintain telomere length, and emerging data have also documented the capacity of telomerase to directly regulate cancer-promoting pathways [21]. A circRNA, ciRS-7, is closely coupled to miR-7, and fine-tuning of the miR-7/miR-671/ciRS-7 axis will likely play profound roles in diseases such as

Table 5.2 NcRNAs involved in cancer cell biological process

Cancer cell biological process	ncRNAs	Notes
Apoptosis	lncRNA, miRNA, piRNA, snRNA, snoRNA	1–5
Cell cycle	lncRNA, miRNA, siRNA, piRNA	6–9
Proliferation	lncRNA, miRNA, piRNA, rRNA, tRNA, snoRNA, circRNA	10–16
Autophagy	lncRNA, miRNA, rRNA	17–19
Necrosis	lncRNA, miRNA	20–21

Notes
1. Transforming growth factor-β-induced lncRNA-Smad7 inhibits apoptosis of mouse breast cancer
2. TEL-AML1 regulation of surviving and apoptosis via miRNA-494 and miRNA-320a
3. An lncRNA (GAS5)/SnoRNA-derived piRNA induces activation of TRAIL gene by site-specifically recruiting MLL/COMPASS-like complexes
4. Hyperstable U1snRNA complementary to the K-ras transcripts induces cell death in pancreatic cancer cells
5. GAS5, a noncoding protein, controls apoptosis and is downregulated in breast cancer
Cell cycle
6. Long noncoding RNA UCA1 regulates cell cycle distribution via CREB through PI3-K
7. Inverse correlation of miRNA and cell cycle-associated genes suggests influence of miRNA on benign thyroid nodule tumorigenesis
8. Effects of HMGB1 expression suppressed by siRNA on cell cycle and proliferation of human cervical cancer cell line HeLa
9. Overexpression of Hiwi inhibits the cell growth of chronic myeloid leukemia K562 cells and enhances their chemosensitivity to daunomycin
10. lncRNA-LALR1 accelerates hepatocyte proliferation during liver regeneration by activating Wnt/beta-catenin signaling
11. miRNA-26b inhibits proliferation by targeting PTGS2 in breast cancer
12. piRNA, the new noncoding RNA, is aberrantly expressed in human cancer cells
13. Changes in rRNA transcription influence proliferation and cell fate within a stem cell lineage
14. tRNA-derived microRNA modulates proliferation and the DNA damage response and is downregulated in B-cell lymphoma
15. snoRNA U50 levels are regulated by cell proliferation and rRNA transcription
16. Correlation of circular RNA abundance with proliferation—exemplified with colorectal and ovarian cancer, idiopathic lung fibrosis, and normal human tissues
17. APF lncRNA regulates autophagy and myocardial infarction by targeting miR-188-3p
18. Oncophagy—harnessing miRNA regulation of autophagy in cancer therapy
19. Noncoding rRNA-mediated preferential killing in cancer cells is enhanced by suppression of autophagy in non-transformed counterpart
20. miR-15b/16 protects primary human retinal microvascular endothelial cells against hyperglycemia-induced increases in tumor necrosis factor alpha and suppressor of cytokine signaling
21. Altered expression of long noncoding RNAs during genotoxic stress-induced cell death in human glioma cells

cancer [22]. However, there are still many gaps in our current understanding of the functional roles for the vast majority of these unique ncRNAs.

In this chapter, we pay attention to the relationship between the two mostly studied ncRNAs and related cancer diseases: miRNA and lncRNA (see Table 5.2).

The biogenesis, targeting, and function of these classes of ncRNAs have been extensively studied and reviewed. These ncRNAs are largely involved in biological processes of cancer cells including proliferation, cell cycle, apoptosis, necroptosis, and autophagy. Recent progress suggests that the involvement of ncRNAs in cancers' cellular regulation can be far more prevalent than previously appreciated. Deciphering the role of ncRNAs in cancer benefits not only understanding the molecular basis of this disease but also developing novel clinical tools and treatments. The field of applications for these ncRNAs in cancer harbors great promise.

In this chapter, we will attempt to organize some of the rapidly expanding information, with a focus on roles of ncRNAs involved in the growth and death biological processes of cancer cells, including proliferation, cell cycle, apoptosis, necrosis, and autophagy. Furthermore, we highlight on potential clinical applications of ncRNAs as predictive biomarkers and therapeutic targets for anticancer treatments.

5.2 miRNA in Growth and Death of Cancer Cells

miRNAs comprise a class of endogenously expressed small ncRNAs which are approximately 22 nucleotides in length. The canonical miRNA biogenesis pathway consists of at least four steps: transcription, nuclear and cytoplasmic processing, loading into RNA-induced silencing complex (RISC), and decay. Transcription of miRNAs usually involves RNA polymerase II (Pol II), meaning that miRNA genes share the transcriptional machinery of protein-coding genes, including transcription factors, enhancers, and epigenetic regulation [2]. They exert their function via base pairing with complementary sequences within mRNA molecules. Upon sequence-specific binding of miRNAs, mRNA molecules are destabilized through shortening of their poly(A)tails, or degraded by cleavage of the mRNA strand, or less efficiently translated into proteins by ribosomes [23]. Recent evidences suggest that miRNAs can also recognize specific target sites in gene promoters, modulating gene expression likely by recruiting similar Ago proteins and chromatin-remodeling enzymes in the vicinity of promoters [24]. miRNA regulates diverse biological processes, and bioinformatic data indicates that each miRNA can control hundreds of gene targets, underscoring the potential influence of miRNAs on almost every genetic pathway [25]. In fact, miRNA-mediated regulation is one of the most widespread posttranscriptional regulatory mechanisms in eukaryotes and is estimated to affect the majority of human transcripts [26].

miRNAs have been implicated in cancer diseases by regulating the cellular levels of specific oncogenes or tumor suppressor genes. About half of the annotated human miRNAs map within fragile regions of chromosomes, which are areas of the genome that are associated with various human cancers [25]. Gene therapies that use miRNAs are proven to be an effective approach to blocking tumor progression. Accumulating evidence highlights the crucial role of miRNA-mediated regulation in virtually all cellular processes that modulate malignant transformation of cells including cell growth, differentiation, proliferation, and apoptosis, to name a few. In normal tissues, proper miRNA transcription, processing and binding to complementary sequences on the target mRNA, results in the repression of target gene expression through a block in protein translation or altered mRNA stability, leading to the overall result as normal rates of cellular growth, proliferation, differentiation, and cell death [25]. Alterations in the miRNA balance in the cell can lead to dysregulation of tumor suppressor genes and/or oncogenes controlled by aberrantly expressed miRNAs [26], resulting in the overall outcome involving increased proliferation, promoted cell cycle or cell cycle arrest, decreased levels of apoptosis and necroptosis, or delayed autophagy, ultimately leading to tumor formation [25].

Hereafter, we will summarize the functions and mechanisms of miRNAs in regulating the proliferation, cell cycle, apoptosis, necroptosis, and autophagy of cancer cells.

5.2.1 Proliferation

Proliferation is an important part of cancer development and progression as the cancer cell embodies characteristics that permit survival beyond its normal life span and to proliferate abnormally [27]. The relevance of miRNAs expression correlating with tumor proliferation has been analyzed in detail in the last decade.

miRNAs are thought to function as both tumor suppressors and oncogenes through regulating their target genes and related signal pathway in tumor proliferation. For example, the expression level of miR-31 is significantly higher in cervical cancer patients than in normal individuals (P\0.05). The target genes such as SATB2, TIAM1, and RASA1 have been found to be involved in the biological functions of miR-31 [28]. The miR-34 family is directly transactivated by tumor suppressor p53, which is frequently mutated in various cancers. The MTT assay reveals significant cell proliferation inhibition in miR-34a transfectant compared with the control from HO8910 and SKOV3 cells, which displays lowest expressions of miR-34a [29]. miR-203 is downregulated in renal cancer cell lines ($P<0.05$). Mechanistic investigations confirm FGF2 as a direct target of miR-203, and upregulation of miR-203 can decrease expression of FGF2 [30]. In lung cancer cell line A549 cell line, overexpression of microRNA-126 inhibits the proliferation rate. VEGF is the target gene of microRNA-126, with the latter exerting its function via regulating the former's protein level [31].

Tumor hypoxia is one of the features of tumor microenvironment that contributes to chemoresistance. Early steps in tumor development are associated with a fibrogenic response and the development of a hypoxic environment which favors the survival and proliferation of cancer stem cells [27]. Moreover, hypoxia also regulates the expression of a series of miRNAs. IGF2BP1 is identified as a potential target of miR-196b. miR-196b overexpression decreases IGF2BP1 RNA expression and protein level. The IGF2BP1downregulation by either miR-196b or IGF2BP1 siRNA leads to a decrease in cell viability and proliferation in normal culture conditions [32]. Former studies indicate that c-Myc-mediated repression of miR-15-16 in hypoxia is induced by increased HIF-2α and promotes tumor angiogenesis and metastasis by upregulating FGF2.

In addition, mounting evidence has shown that miRNAs take important roles in self-renewal of cancer cells. One such example is linc-RoR, which has already been shown to control self-renewal and maintain pluripotency of human embryonic stem cells by acting as a miR-145 "sponge" and thus controlling OCT4, NANOG, and SOX2 expression. Similarly, linc-RoR is found to regulate expression of the same transcription factors in endometrial cancer stem cells, and inhibit their differentiation, in a miR-145-dependent manner [26]. In the mouse, members of the let-7 family are shown to be expressed at low levels in self-renewal progenitors (ALDH+/Sca-1+) and induce upon differentiation, suggesting that low levels of let-7 mark the self-renewal compartment and can be used to prospectively isolate normal mammary stem cells (MaSCs). Accordingly, the let-7 family emerges as the most induced group of miRNAs upon estradiol treatment in human luminal cells. Independent studies further suggest that miRNAs of this family are implicated in the self-renewal of cancer stem cells (CSCs) in breast and other cancers, by multiple mechanisms. One of these mechanisms involves the pluripotency gene known as LIN28/LIN28B, which inhibits the function of let-7 by interfering with its biogenesis. In breast cancer, LIN28 expression confers CSC traits and impinges on signaling mechanisms involved in self-renewal of normal and CSCs, such as the Wnt/beta-catenin pathway, NF-κB signaling, and inflammatory cytokine signaling [2].

5.2.2 *Cell Cycle*

Regulation of the cell cycle involves processes crucial to the survival of a cell, including the detection and repair of genetic damage as well as the prevention of uncontrolled cell division. Mitotic cell cycle is accepted as a constantly reproducible sequence of events, which includes four phases: G1 phase, S phase, G2 phase, and M phase. The expression and activation of cyclin-dependent kinases (CDKs) and cyclin play a pivotal role in proceeding through the G1 into S phase (DNA replication) and from the G2 to M phase (mitosis) [33]. These proteins regulate the cell's progression through the stages of the cell cycle and are in turn regulated by numerous proteins, including p53, p21, p16, and cdc25. Downstream targets of cyclin-CDK complexes include pRb and E2F. The cell cycle often is dysregulated in neoplasia due to alterations either in oncogenes that indirectly affect the cell cycle or in tumor suppressor genes or oncogenes that directly impact cell cycle regulation, such as pRb, p53, p16, cyclin D1, or mdm-2. The cell cycle has become an intense subject of research in recent years. The complexity of the regulation of the cell cycle is reflected in the different alterations leading to aberrant cell proliferation and development of cancer. A large number of researchers have supposed that miRNA can regulate cell cycle proteins and induce cell cycle arrest [34].

Tumor-associated cell cycle defects are often mediated by alterations in the CDK activity. Misregulated CDKs induce unscheduled proliferation as well as genomic and chromosomal instability [35]. miRNAs can control cell cycle progression after DNA damage by targeting CHK1, p53, retinoblastoma1 serine phosphates from human chromosome3 (RBSP3), cyclin D, CDC25a, p21, CDK2, WEE1, LK1, and so on [9]. miR-223 affects the G1/S transition of cell cycle, so that the expression of CDK2, CDK4, CDK6, CCND1, CCND2, and CCND3 is significantly downregulated in miR-223 inhibitor. miR-31 is a direct regulator of endogenous expression CDK2 in liver cancer cells [36]. In leukemic cells, CDK2 is found to be a target gene of miR-638 that CDK2 inhibition phenotypically mimics the overexpression of miR-638 [37]. miR-449a is able to regulate the expression of the CDK6 protein, and a lower expression level of miR-449 and a higher expression level of CDK6 may contribute to the occurrence and development of gastric cancer [38].

Cell cycle arrest, also referred to as delay, is produced by a variety of factors that may be intrinsic or extrinsic and may affect several checkpoints [34]. miR-449a inhibits neuroblastoma cell survival and growth through two mechanisms, inducing cell differentiation and cell cycle arrest. Its function in inducing cell cycle arrest involves downregulating its direct targets CDK6 and LEF1 [39]. miR-34a is importantly competent in p53 tumor suppressor network, where p53-dependent activation of miR-34a is widely validated and its upregulation induces cell cycle arrest. miR-34a has many potential target genes, with several of these regular cell cycles, including NMYC, CCND1, CCNE2, CDK4, CDK6, and MET, having been experimentally validated, which lead to a significant reduction in the number of cells in the S phase of the cell cycle and an increase in the percentage of cells in theG0/G1 phase [33]. miR-302/367 cluster can dually regulate cell cycle in gene dose-dependent manner that knockdown of the endogenous miR-302/367 cluster causes cell cycle arrest 7. miR-27a is part of the miR-23a/27a/24-2 cluster. The impact of miR-27a silencing or overexpression on the cell cycle of U251 and U87MG cells is examined in vitro. Treatment with miR-27a-mimics oligonucleotides suppresses U251 cell proliferation, promotes apoptosis by inducing G2/M phase arrest, and impairs the invasive potential of U251 cells in vitro [40].

5.2.3 Apoptosis

Apoptosis is characterized by specific morphological and biochemical changes of dying cells, including cell shrinkage, nuclear condensation and fragmentation, dynamic membrane blebbing, and loss of adhesion to neighbors or to extracellular matrix. Biochemical changes include chromosomal DNA cleavage into internucleosomal fragments, phosphatidylserine externalization, and a number of intracellular substrate cleavages by specific proteolysis [41]. The field of apoptosis research has been moving forward at an alarmingly rapid rate as many of the key apoptotic proteins have been identified. Inappropriate apoptosis is a hallmark in many cancer diseases as the most striking feature of the cancer cells is that they do not undergo apoptosis. This fact is due to the involvement of dynamic interplay between oncogenes and/or inactivated tumor suppressor genes. Both of them have key roles in generation of a tumor [42]. To date, research indicates that there are two main apoptotic pathways in cancer: the extrinsic or death receptor pathway and the intrinsic or mitochondrial pathway. Besides, studies show that the two pathways are linked and that molecules in one pathway can influence the other. Multiple lines of evidence indicate that miRNAs can modulate apoptotic pathways.

The intrinsic (or mitochondrial) apoptotic pathway is triggered by a variety of intracellular stimuli, including DNA damage, cytotoxic drug treatment, growth factor deprivation, and/or oxidative stress. This pathway relies on the formation of a complex termed the apoptosome, which is composed of procaspase-9, apoptotic protease-activating factor 1 (Apaf-1), and cytochrome c. A series of Bcl-2 family members including Bax, Bak, Bcl-2, Bcl-xL, Mcl-1, Bid, and Bim control the release of cytochrome c by regulating mitochondrial membrane permeabilization [43]. It has been reported that miR15-a and miR-16-1 target Bcl-2 to induce apoptosis, while Bcl-2 inhibits mitochondrial-mediated apoptosis, by influencing oligomerization of Bax and Bak at the posttranscriptional level. Additionally, overexpression of miR-15a-miR-16-1 in vitro can decrease Bcl-2 operation and then promote apoptosis [41]. miR-181c can target the 3′ untranslated region of Bcl-2, and the increased level of Bcl-2 caused by the decrease in miR-181c protects mitochondrial morphology from the tumor necrosis factor alpha-induced apoptosis [44]. miR-608 is reported to target Bcl-xL to regulate chordoma malignancy [45]. The tumor-suppressive activity of miR-133a is likely due to the targeted suppression of Bcl-xL and Mcl-1 expression [46]. Besides, the nuclear transcription factor p53 can govern main apoptotic signals that mitochondria receive in the intrinsic pathway of apoptosis. miR-34a is a tie molecule between the p53 and SIRT1 and is composed of a p53/miR-34a/SIRT1 signal feedback loop, which can enhance apoptosis signal and significantly promote cell apoptosis [47].

The extrinsic pathway of apoptosis is initiated by the binding of death ligands such as Fas ligand (FasL), TNF-related apoptosis-inducing ligand (TRAIL), TNF-α, and TNF-like weak inducer of apoptosis (TWEAK) to death receptors in the TNF receptor (TNFR) superfamily [43]. TRAIL is an attractive therapeutic target in cancer because it directly induces tumor cell apoptosis. miR-221 and miR-222 induce TRAIL resistance by directly targeting 30-UTRs of phosphatase and tensin homolog, tissue inhibitor of metalloproteinase-3 (TIMP3), and the cyclin-dependent kinase inhibitor p27kip1 [12]. miR-145, miR-216, miR-182, and miR-96 can interact with DR4/5 and Fas-associated death domain protein. Overexpression of these miRNAs reduce caspase-3 activation through TRAIL-induced apoptosis signaling [12]. miR-20a downregulates Fas expression in osteosarcoma cells, thus enhancing the metastatic capacity of osteosarcoma cells by promoting cell survival in the FasL-positive lung microenvironment [48].

There are also other miRNA-mediated apoptosis pathways. For example, miR-125b-mitochondria-caspase-3 pathway plays a role in doxorubicin resistance in human breast cancer [49]. And previous researches have shown that miRNA-221 and miRNA-222 induce apoptosis via the KIT/AKT signaling pathway in gastrointestinal stromal tumors [50].

5.2.4 Necrosis

Necrosis has always been considered to be almost "accidental" cell death, a random, uncontrolled process. However, emerging evidence has shown that necrosis can be induced and regulated in a similar manner to apoptosis. Regulated necrosis is termed "programmed necrosis" or "necroptosis" to distinguish this process from necrosis induced by physical trauma [43]. Necroptosis can be induced by the activation of the TNF receptor superfamily, T-cell receptors, interferon receptors, toll-like receptors (TLRs), cellular metabolic and genotoxic stresses, or various anticancer agents [51]. Receptor-interacting protein kinases (RIPK), poly(ADPribose) polymerase-1(PARP1), NADPH oxidases, and calpains have also been identified as signaling components of necroptosis pathways in cancer [41]. Ensuing local inflammation induced by necroptosis may promote tumor growth. However, at present, study on how miRNAs regulate necroptotic cell death is still in its infancy.

Tumor necrosis factor-α (TNF-α) triggers necroptotic cell death through an intracellular signaling complex containing RIPK1 and RIPK3 called the necrosome. RIPK1 phosphorylates RIPK3, which phosphorylates the pseudokinase mixed lineage kinase domain-like (MLKL), driving its oligomerization and membrane-disrupting necroptotic activity [52]. miR-155 has been reported to prevent necroptosis in human cardiomyocyte progenitor cells by directly targeting RIP1 [43], and overexpressing miR-155 in cardiomyocyte progenitor cells (CMPCs) reveals that miR-155 attenuates necrotic cell death by $40 \pm 2.3\,\%$ via targeting receptor-interacting protein 1 (RIP1) [53].

Caspase-8 is a cysteine protease that is critically involved in regulating cellular apoptosis and has recently been shown to be important in regulating necrotic cell death. The RIP3 is a key signaling molecule in the necroptosis pathway. Caspase-8 has been demonstrated to be able to cleave and inactivate RIP3 and prevent the pathway of necroptosis. Caspase-8 is a target of miR-874 in the necrotic pathway. Foxo3a regulates caspase-8 expression and the consequent myocardial necrosis through targeting miR-874 [54].

To date, no report has demonstrated how miRNAs regulate other key necroptotic factors, including RIP3, MLKL, and PGAM5 [43].

5.2.5 Autophagy

Autophagy is a catabolic process that allows cellular macromolecules to be broken down and recycled into metabolic precursors. It is a highly conserved, critical process, allowing cells to gain survival advantages under various stress situations due to growth and environmental changes. The process involves enwrapping fractions of the cytoplasm until a double-membrane autophagic vacuole (autophagosome) is formed. The pathways and genes, such as multiple autophagy-related genes (ATGs), involved in autophagy that have been identified in detail in recent years include ATG1, ATG4, LC3/ATG8, and beclin-1. Autophagy serves a dichotomous role in cancer, and recent advances have helped delineate the appropriate settings where inhibiting or promoting autophagy may confer therapeutic efficacy in patients [55]. Silencing Dicer1, an essential processor of miRs, increases levels of ATG protein and formation of autophagosomes in cells, indicating that miRs regulate autophagy [56]. Accumulating evidence indicates that miRNAs contribute significantly to autophagy in cancer mainly through the three following pathways.

miRNAs regulates the expression of key autophagy-related proteins. miR-30a is a potent inhibitor of autophagy by downregulating Beclin-1. miR-30a in human renal cell carcinoma (RCC) interferes with the effectiveness of sorafenib-mediated apoptosis by an autophagy-dependent pathway. Expression of miR-30a is significantly downregulated in several RCC tissues and in RCC cell lines. Accordingly, its targeted gene Beclin-1 is upregulated. Sorafenib activates autophagy in RCC cells (786-0 and A489 lines), evidenced by p62 degradation, Beclin-1/autophagy protein 5 (ATG-5) upregulation, and light chain (LC)3B-I/-II conversion. Exogenously expressing miR-30a in 786-0 or A489 cells inhibits Beclin-1 expression and enhances sorafenib-induced cytotoxicity [57]. The ability of miR106B and miR93 to regulate ATG transcripts in human cancer cell lines (HCT116, SW480, HeLa, and U2OS) is studied using luciferase report assays and bioinformatics analyses. Results show that miR106B and miR93 target ATG16L1 messenger RNA which they reduce levels of ATG16L1. In contrast, miR106B and miR93 antagonists increase formation of autophagosomes [56]. What's more, miR-375 has been reported to inhibit autophagy through its target, ATG7, in hepatocellular carcinoma (HCC) cells [58]. miR-376b expression targets ATG4C and beclin-1, which, in turn, downregulate autophagy induced by nutrition starvation and rapamycin in breast cancer cells [59]. Meanwhile, ectopic expression of miR-7 in human lung cancer and esophageal cancer cell lines enhances autophagy by suppressing epidermal growth factor receptor (EGFR) expression [60].

miRNAs mediate autophagy in genotoxic stress and hypoxia. Hypoxia plays an important role in the tumor microenvironment by allowing the development and maintenance of cancer cells. Studies have shown that hypoxia-inducible factor 1 (HIF-1) regulates a panel of microRNAs, whereas some of microRNAs target HIF-1. For example, miR-210 has emerged as the predominant miRNA regulated by hypoxia. Hypoxia-induced miR-210 can repress glycerol-3-phosphate dehydrogenase 1-like (GPD1L), which, in turn, stabilizes HIF-1α by reducing hyperhydroxylation [61]. miR-96 can promote or inhibit autophagy by principally inhibiting MTOR or ATG7 depending on the expression levels of miR-96. Inhibition of miR-96 abolishes hypoxia-induced autophagy [62]. Under hypoxia, HIF-1α induces miRNA-210 which in turn enhances autophagy and reduces radiosensitivity by downregulating Bcl-2 expression in colon cancer cells [63]. miR-137, which targets the expression of two mitophagy receptors NIX and FUNDC1, markedly inhibits mitochondrial degradation by autophagy without affecting global autophagy.

Impaired mitophagy in response to hypoxia caused by miR-137 is reversed by re-expression of FUNDC1 and NIX expression vectors lacking the miR-137 recognition sites at their 3′ UTR [64].

miRNA is involved in p53-mediated autophagy. A number of previous studies have reported the direct involvement of p53 in autophagy. The cytoplasmic pool of p53 suppresses autophagy under nutrient deprivation [60]. P53 also functions as a key regulator of autophagy, a catabolic pathway for degradation and recycling of proteins and cellular organelles, which has been shown to be dysregulated in cancers [65]. Previous studies suggest that some of the p53-mediated miRNAs may play important roles in autophagy. miR44 induces autophagy in lung cancer cells by targeting the p53-induced glycolysis and apoptosis regulator TIGAR [66]. In colon cancer, miR-502 inhibits autophagy by suppression of Rab1B, a critical mediator of autophagy. The expression of miR-502 is regulated by p53 via a negative feedback regulatory mechanism [65].

5.3 lncRNA in Growth and Death of Cancer Cells

Although studies of small regulatory RNAs, in particular miRNAs, have dominated the field of RNA biology during the past decade, a surprisingly broad spectrum of biological processes is also associated with lncRNAs. Over the past several years, accumulated data have begun to advance the idea that lncRNAs are not just transcriptional noise or cloning artifacts but important supplements to proteins and other effectors in complex regulatory networks. Thus, the focus of scientists is now shifting to one of the most poorly understood yet most common products of transcription from genomes: lncRNAs [13]. The most commonly used definition of lncRNA is an RNA molecule that is longer than 200 nucleotides and that is not translated into a protein. However, this definition may be too simple and does not take into account certain issues. For example, the same RNA can contain both protein-coding genes (PCGs) and noncoding functions. Furthermore, a PCG is usually defined as a transcript that contains an open reading frame (ORF) longer than 100 amino acids, while lncRNAs can contain ORFs longer than 100 amino acids and not necessarily synthesize polypeptides [67]. Although the nomenclature is still evolving, lncRNA typically refers to a polyadenylated long ncRNA that is transcribed by RNA polymerase II and associated with epigenetic signatures common to protein-coding genes, such as trimethylation of histone 3 lysine 4 (H3K4me3) at the transcriptional start site (TSS) and trimethylation of histone 3 lysine 36 (H3K36me3) throughout the gene body [68]. Besides, advances in the depth and quality of transcriptome sequencing have revealed many new classes of lncRNAs. The existing classifications of lncRNAs rest on their descriptive and distinctive properties including their size, their localization, and their function, which have been discussed in previous reviews [69].

lncRNA genes are interspersed in the genome in various possible locations in relation to protein-coding transcripts, such as overlapping, intergenic, or divergent transcripts. Transcription of lncRNAs follows the same rules as for protein-coding genes and is executed by RNA Pol II. Functions of lncRNAs are executed by multiple modes of action and can occur both in the nucleus and in the cytosol. In contrast to miRNAs, which are highly conserved and mainly involved in negative regulation of gene expression at the posttranscriptional level, lncRNAs are poorly conserved and can regulate gene expression (either positively or negatively) at numerous levels by a variety of mechanisms, some of which are yet to be characterized [2]. It is now becoming evident that lncRNAs are important transcriptional outputs of the genome. Generally, lncRNAs have been implicated in gene-regulatory roles, such as chromosome dosage compensation, imprinting, epigenetic regulation, cell cycle control, nuclear and cytoplasmic trafficking, transcription, translation, splicing, cell differentiation, and others. Multiple lines of evidence demonstrate that a number of characterized lncRNAs are implicated in a spectrum of biological processes and that misregulated lncRNA expression can cause various human diseases and cancers [13]. Transcription of individual lncRNAs involved in different biological processes occurs in developmental stage-, tissue-, and organ-specific patterns. Thus, lncRNAs can function as molecular markers that signal the space, time, and expression of gene transcription, specifically reflecting the integrative biological outcome of transcription factors and signaling pathways controlling gene expression in space and time [10].

In this section, we will focus on summarizing the functions and mechanisms of lncRNAs that have been linked to cancer through regulating the proliferation, cell cycle, apoptosis, necroptosis, and autophagy of cancer cells.

5.3.1 Proliferation

As lncRNAs are crucial regulators of gene expression, it is expected that their misregulation will lead to abnormal cellular function and growth defects and cause cancer diseases [14]. lncRNAs have been detected in cancer by various techniques including expression microarrays, tiling arrays, next-generation sequencing, and methylation analysis. These approaches have led to the identification of a wide range of lncRNAs whose expression is significantly associated with cancerous tissues [70]. The abundance of these transcripts in cancer suggests that lncRNA-mediated biology occupies a central place in cancer proliferation. lncRNAs are abundant during embryogenesis and reactivation, or non-suppression of some of these fetal lncRNAs may critically regulate pluripotency and uninhibited cellular growth, thus giving rise to adult or developmental cancers [14]. lncRNAs act through a variety of mechanisms in cancer cell proliferation such as remodeling of chromatin, transcriptional co-activation or co-repression, and protein inhibition, as posttranscriptional modifiers or decoy elements [71].

Accumulating evidence suggests that a major role of lncRNAs is to guide the site specificity of chromatin-modifying complexes to effect epigenetic changes in cancer cell proliferation. The well-characterized lncRNAs ANRIL, XIST, HOTAIR, and KCNQ1OT1 are able to recruit epigenetic modifiers to specific loci to reprogram the chromatin state [71]. ANRIL is significantly upregulated in gastric cancer, and it can promote cell proliferation and inhibit cell apoptosis by silencing of miR99a and miR449a transcription [72]. XIST expression is upregulated in glioma tissues and human glioblastoma stem cells (GSCs). Knockdown of XIST exerts tumor-suppressive functions by reducing cell proliferation, migration, and invasion as well as inducing apoptosis. Mechanistic investigations have defined that XIST and miR-152 are in the same RNA-induced silencing complex (RISC) so that miR-152 mediates the tumor-suppressive effects that the knockdown of XIST exerts [73]. HOTAIR promotes the proliferation of serous ovarian cancer cells by at least partly regulating certain cell cycle- and apoptotic-related genes, including cyclin E, Bcl-2, caspase-9, caspase-3, and BRCA1 [74]. A novel short tandem repeat (STR) polymorphism within KCNQ1OT1 contributes to hepatocellular carcinoma (HCC) cell proliferation, possibly by affecting KCNQ1OT1 and CDKN1C expression through a structure-dependent mechanism, but further functional studies are needed to validate the hypothesis [75].

Other lncRNAs have been found to be key regulators of the protein signaling pathways underlying cancer cell proliferation. The ability of lincRNA-p21 to promote tumor growth is validated in mouse xenograft models. After loss of lincRNA p21, hnRNP-K is inappropriately localized at the promoters of p53-repressed genes, which results in deregulated expression and altered chromatin states of polycomb target genes, a defective G1/S checkpoint, increased proliferation rates, and enhanced reprogramming efficiency [9]. Estrogen plays a critical role in cancer development and progression. H19 is an estrogen-inducible lncRNA and mediates 17β-estradiol-induced cell proliferation in MCF-7 breast cancer cells through the mechanism in which 17β-estradiol produces a dose- and time-dependent induction of H19 expression in MCF-7 cells, which is mediated via ERα as evident by the blockade of this 17β-estradiol effect with ICI 182780, a specific ER antagonist and knockdown of ERα using specific RNAi [76]. Steroid receptor RNA activator (SRA) noncoding RNA and SRA protein together regulates estrogen receptor signaling pathways and plays a critical role in breast cancer cell proliferation [77].

Some lncRNAs are constituents of macromolecular with roles in RNA processing during cancer cell proliferation. MALAT1, a highly conserved long noncoding RNA, which is thought to act at a posttranscriptional level by controlling alternative splicing of pre-mRNAs [71], is deregulated in several types of cancers. Emerging evidences have shown that MALAT1 has been linked to an increase in cancer cell proliferation. MALAT1 contributes to proliferation and metastasis in esophageal squamous cell carcinoma [78], and silencing of MALAT1 by miR-101 and miR-217 inhibits proliferation of esophageal squamous cell carcinoma cells [79]. Other lncRNAs can also act as decoys, sequestering biomolecules and preventing them from fulfilling their cellular functions. An example of this mechanism is represented by the PTENP1 restricting cell proliferation by acting as a microRNA decoy for the tumor suppressor PTEN [71]. miR-21 regulates the potential anticancer effects of icariin on cell proliferation and apoptosis by targeting PTEN, RECK, and Bcl-2 in the ovarian cancer A2780 cells [80]. In HCC, the overexpressed PTENP1 represses the oncogenic PI3K/AKT pathway and inhibits the HCC proliferation in vivo, which is accompanied by enhanced apoptosis, autophagy, and dampened angiogenesis/neovasculature maturation [81].

In addition, there is evidence that lncRNA is involved in the self-renewal maintenance of cancer stem cells (CSCs). Using transcriptome microarray analysis, lncTCF7 is identified as being required for liver CSC self-renewal and tumor propagation through Wnt signaling. Mechanistically, lncTCF7 recruits the SWI/SNF complex to the promoter of TCF7 to regulate its expression, leading to activation of Wnt signaling [82]. MALAT1 is reported to increase the proportion of pancreatic CSCs, maintain self-renewing capacity, decrease the chemosensitivity to anticancer drugs, and accelerate tumor angiogenesis in vitro. The potential mechanism may be that MALAT1 functions as endogenous sponge RNA to interact with miR-200c and miR-145, and upregulates the expression of their target gene Sox2, leading to enhanced stem cell-like phenotypes [83].

5.3.2 Cell Cycle

The connection between the cell cycle and cancer is obvious: cell cycle machinery controls cell proliferation, and cancer is a disease of inappropriate cell proliferation. At least two types of cell cycle control mechanisms are recognized: a cascade of protein phosphorylations that relay a cell from one stage to the next and a set of checkpoints that monitor completion of critical events and delay progression to the next stage if necessary. The first type of control involves a highly regulated kinase family, and the progress through the cycle is accomplished in part by the regulated activity of numerous cyclin-CDK complexes [84]. The cyclin-dependent kinases (CDKs) are a family of serine/threonine kinases controlling progression through the cell cycle. The regulatory subunits of the CDKs, known as cyclins, form complexes with their catalytic partner to function as checkpoint kinases of specific proteins that regulate progression through the cell cycle. The cyclin-CDK complexes govern a linear progression of events that lead cells from a resting state (G0), growth phase (G1), through DNA replication (S), and finally to cell division (M) [85]. The second type of cell cycle regulation is not an essential part of the cycle progression machinery. Cell cycle checkpoints sense flaws in critical events such as DNA replication and chromosome segregation. When checkpoints are activated, for example, by underreplicated or damaged DNA, signals are relayed to the cell cycle progression machinery [84]. Previous work has shown that many lncRNAs have been involved in cell cycle regulation of cancer cells, but many lncRNAs that have a functional role in cell cycle regulation remain to be identified because the functions of only a small percentage of the total lncRNA population are understood. To clarify the roles of lncRNAs in cell cycle regulation, it should be determined how they regulate the target cell cycle regulators and which signaling pathways induce these lncRNAs.

Mutations of checkpoint proteins are frequent in all types of cancer. The tumor suppressor protein p53 is a sequence-specific DNA-binding protein that is able to induce either cell cycle arrest or apoptosis at the cell cycle checkpoints [86]. Although p53 is unstable, it is stabilized and activated via phosphorylation mediated by the ATM/ATR pathway in response to DNA damage. Moreover, p53 is also regulated via phosphorylation at various sites by specific kinases [87]. A prominent p53-induced lncRNA termed lncRNA activator of enhancer domains (LED) is required for p53-induced cell cycle arrest and is involved in the activation of a subset of p53-bound and p53-unbound enhancers by inducing an epigenetic change. Chromatin-binding and eRNA expression analyses show that LED associates with and activates strong enhancers. One prominent target of LED is located at an enhancer region within CDKN1A gene, a potent p53-responsive cell cycle inhibitor. LED knockdown reduces CDKN1A enhancer induction and activity and cell cycle arrest following p53 activation [88]. MALAT1, an abundant nuclear-retained lncRNA, is overexpressed in several cancers, and its elevated expression has been associated with hyper-proliferation and metastasis. Recent studies show that MALAT1 modulates the expression of cell cycle genes and is required for G1/S and mitotic progression. Depletion of MALAT1 leads to activation of p53 and its target genes. The cell cycle defects observed in MALAT1-depleted cells are sensitive to p53 levels, indicating that p53 is a major downstream mediator of MALAT1 activity. Furthermore, MALAT1-depleted cells display reduced expression of B-MYB (Mybl2), an oncogenic transcription factor involved in G2/M progression, due to altered binding of splicing factors on B-MYB pre-mRNA and aberrant alternative splicing [89]. Moreover, there are more lncRNA-regulated cell cycle proteins. The transcription factor MYC is known to regulate lncRNAs and has been implicated in cancer cell proliferation and tumorigenesis. MYC-regulated lncRNAs, named MYCLos, function in cell proliferation and cell cycle by regulating MYC target genes such as CDKN1A (p21) and CDKN2B (p15) [90]. A recently identified lncRNA mapped to chromosome 8q24 is found to be generally upregulated in colon cancer tissues compared to their neighboring normal tissues. CARLo-5 knockdown inhibits GC cell growth through inducing G0/G1 arrest and apoptosis. That p27, p21, and p16 protein expression levels are increased with inhibition of CARLo-5 expression shows CARLo-5 has a role in regulating cell cycle genes, but the precise underlying mechanism still needs to be revealed [91].

lncRNAs also play a role in the cell cycle arrest of cancer development. lncRNAs are induced in a p53-dependent manner, suggesting that they are induced by DNA damage. Also, these reported lncRNAs may participate in cell cycle arrest [87]. There are other lncRNA-related cell cycle arrest pathways. BRAF-activated noncoding RNA (BANCR), an lncRNA, is crucial for cell migration in melanoma cells and non-small cell lung cancer (NSCLC) cells. Downregulation of BANCR contributes to the induction of G0/G1 cell cycle arrest of colorectal cancer cells, at least in part, through the regulation of p21 protein [92]. AK001796, the lncRNA with the most clearly altered expression, is overexpressed in lung cancer tissues and cell lines, but its expression is downregulated in resveratrol-treated lung cancer cells. AK001796 knockdown causes a cell cycle arrest, with significant increases in the percentage of cells in G0/G1 in lung cancer cells. The cell cycle-associated genes up- or downregulated directly or indirectly by lncRNA AK001796 knockdown include (i) TFDP2, which encodes a protein that forms heterodimers with the E2F transcription factors, resulting in the transcriptional activation of cell cycle-regulated genes; (ii) CDC6, which encodes a protein that functions as a regulator of the early steps of DNA replication; (iii) ATR, a cell cycle-checkpoint gene required for cell cycle arrest and DNA damage repair in response to DNA damage; (vi) and CCNB1, which encodes a regulatory protein involved in the cell cycle [93].

5.3.3 Apoptosis

The mechanisms of apoptosis are highly complex and sophisticated, involving an energy-dependent cascade of molecular events. To date, research indicates that there are two main apoptotic pathways: the extrinsic or death receptor pathway and the intrinsic or mitochondrial pathway. However, there is now evidence that the two pathways are linked and that molecules in one pathway can influence the other [94]. While the role of miRNAs in apoptosis control is well established, lncRNAs have received less attention, especially in extrinsic apoptosis pathway. Further studies about the lncRNA-mediated apoptosis pathways are needed.

P53, the gene most commonly mutated in human cancer, is a major regulator of mitochondrial apoptosis. P53 is essential for the cellular response to DNA-damaging stimuli to maintain genomic integrity of cells, mainly by activating a gene expression program that leads to cell cycle arrest or elimination of the damaged cells through programmed cell death, including apoptosis [95]. Of significant interest, recent studies suggest that lncRNAs participate in the p53 apoptosis pathway. P53-regulated lincRNA-p21 expression plays an important role in triggering apoptosis. HuR is a ubiquitous RBP that influences cell proliferation, survival, carcinogenesis, and the stress and immune responses. Association of the RNA-binding protein HuR with lincRNA-p21 favors the recruitment of let-7/Ago2 to lincRNA-p21, leading to lower lincRNA-p21 stability. Under reduced HuR levels, lincRNA-p21 accumulates in human cervical carcinoma HeLa cells, increasing its association with JUNB and CTNNB1 mRNAs and selectively lowering their translation [96], resulting in suppressed apoptosis. The expression of lnc_bc060912 is repressed by p53. In human lung carcinoma cells, researches using a recently developed method for RNA pulldown with formaldehyde cross-linking found that lnc_bc060912 interacted with the two DNA damage repair proteins PARP1 and NPM1. The results suggest that lnc_bc060912, via PARP1 and NPM1, affects cell apoptosis and plays important roles in tumorigenesis and cancer progression [97]. Maternally expressed gene 3(MEG3), an lncRNA, represents as a tumor suppressor gene, and its ectopic expression can inhibit cell proliferation and promote cell apoptosis in human glioma cell line. Accumulation of p53 (TP53) protein and its target gene expression partly contribute to cell growth inhibition induced by MEG3. MEG3 is significantly downregulated in non-small cell lung cancer (NSCLC) tissues that can be affected by DNA methylation and regulates NSCLC cell proliferation and apoptosis, partially via the activation of p53 [98]. DNA damage can induce five lncRNAs from the CDKN1A promoter, and one such lncRNA, named promoter of CDKN1A antisense DNA damage-activated RNA (PANDAR), is induced in a p53-dependent manner. Moreover, PANDAR interacts with the transcription factor NF-YA to limit the expression of pro-apoptotic genes. PANDAR is a direct transcriptional target of p53 in NSCLC cells. PANDAR-mediated growth regulation is in part due to the transcriptional modulation of Bcl-2 by interacting with NF-YA, thus affecting NSCLC cell apoptosis [99].

There are also other mitochondrial apoptosis pathways. BCL-X is a key apoptotic member of the BCL-2 gene family that modulates tumor cell death and growth. Alternative splicing of exon 2 in the BCL-X pre-mRNA produces two isoforms, BCL-XL and BCL-XS, which have been shown to exert antagonistic functions in the apoptotic pathway. A lncRNA named INXS, transcribed from the opposite genomic strand of BCL-X, is five- to ninefold less abundant in tumor cell lines from the kidney, liver, breast, and prostate and in kidney tumor tissues compared with nontumors. Three apoptosis-inducing agents increase INXS lncRNA endogenous expression in the 786-O kidney tumor cell line, increase BCL-XS/BCL-XL mRNA ratio, and activate caspase-3, caspase-7, and caspase-9 in the mitochondrially mediated apoptosis pathway. These effects are abrogated in the presence of INXS knockdown. Similarly, ectopic INXS overexpression causes a shift in splicing toward BCL-XS and activation of caspases, thus leading to apoptosis. The overexpression of INXS lncRNA causes a significant activation of caspase-3, caspase-7, and caspase-9, the major caspases in the mitochondrially mediated apoptosis pathway, but has no effect on initiator caspase-8 of the death receptor extrinsic pathway [100]. Growth arrest-specific 5 (GAS5) encodes multiple snoRNAs within its introns, while exonic sequences produce lncRNA which can act as a riborepressor of the glucocorticoid and related receptors. GAS5 negatively regulates the survival of lymphoid and breast cells and is aberrantly expressed in several cancers. In prostate cancer, high levels of GAS5 lncRNA expression promote basal apoptosis and enhance the action of a range of apoptotic stimuli [101]. The HOX transcript antisense RNA (HOTAIR) gene is located within the homeobox C (HOXC) gene cluster on chromosome 12 and encodes a 2.2 kb lncRNA molecule. In serous ovarian cancer (SOC), HOTAIR is found to regulate apoptosis-related protein such as cyclin E, bcl-2, caspase-9, caspase-3, and BRCA1, but further studies are required to explore the underlying molecular mechanism [74].

There are few studies about lncRNA-mediated extrinsic apoptosis pathway. In gastric cancer, the expression of MRP1 and P-glycoprotein is founded decreased significantly in SGC7901/DDP and BGC823/DDP cells using siRNA to interfere with lncRNA AK022798 expression, and their apoptosis and the expressions of caspase-3 and caspase-8 obviously increase [102]. HOXA cluster antisense RNA 2 (HOXA-AS2) is a long noncoding RNA located between the HOXA3 and HOXA4 genes in the HOXA cluster. Its transcript is expressed in NB4 promyelocytic leukemia cells and human peripheral blood neutrophils, and the expression is increased in NB4 cells treated with all-trans-retinoic acid (ATRA). Knockdown of

HOXA-AS2 expression by transduced shRNA decreases the number of viable cells and increases the proportion of apoptotic cells, measured by annexin V binding and by activity and cleavage of caspase-3, caspase-8, and caspase-9. The increase in death of HOXA-AS2 knockdown cells is accompanied by an elevated TNF-related apoptosis-inducing ligand (TRAIL) levels, but ATRA-induced NB4 cells treated with TRAIL show an increase in HOXA-AS2 expression. These results demonstrate that ATRA induction of HOXA-AS2 suppresses ATRA-induced apoptosis, possibly through a TRAIL-mediated pathway. HOXA-AS2-mediated negative regulation thus contributes to the fine-tuning of apoptosis during ATRA-induced myeloid differentiation in NB4 cells [103].

5.3.4 Necrosis

Regulated necrosis is termed "programmed necrosis" or "necroptosis" to distinguish this process from necrosis induced by physical trauma. Necroptosis can be induced by the engagement of the TNF receptor superfamily, T-cell receptor (TCR), interferon receptors (IFNRs), toll-like receptors (TLRs), cellular metabolic and genotoxic stresses, and a number of anticancer agents [43]. While still lacking enough investigation, emerging studies have shown that lncRNAs participate in controlling necrosis of cancer cells.

The aberrant expression of lncRNAs is detected during genotoxic stress-induced necrosis in human glioma cells. A change in the expression of specific candidate lncRNAs (neat1, GAS5, TUG1, BC200, MALAT1, MEG3, MIR155HG, PAR5, and ST7OT1) is detected during DNA damage-induced apoptosis in human glioma cell lines (U251 and U87) using doxorubicin (DOX) and resveratrol (RES). The expression pattern of these lncRNAs is also detected in human glioma cell lines under necrosis induced using an increased dose of DOX. The results reveal that the lncRNA expression patterns are distinct between genotoxic stress-induced apoptosis and necrosis in human glioma cells. The sets of lncRNA expressed during genotoxic stress-induced apoptosis are DNA-damaging agent specific. Potentially, these lncRNAs are involved in an apoptotic signaling pathway rather than a necrotic signaling pathway. But TUG1 and BC200 are only involved in necrosis rather than apoptosis [104]. Future studies are necessary to elucidate the function and mechanisms of the regulation of individual lncRNAs in necrosis.

5.3.5 Autophagy

The role of autophagy in cancer cells is still under investigation. It seems that autophagy function depends on several factors, for example, step of tumor formation, tissue origin, and gene mutations existing in specific cancer type. Some cancer types like human pancreatic cancers with constitutive Ras activation have elevated levels of autophagy that contributes to their growth and survival. Conversely, other tumor types like human breast, ovarian, and prostate cancers have allelic deletions of the essential autophagy regulator Beclin 1, indicating that decreased autophagy may promote tumor development [105]. In general, autophagy is activated in cancer cells and contributes to tumor cell survival. Autophagy is first linked to human cancer through the identification of Beclin 1, and later a number of studies show that ATG genes and others are also oncogenically associated [106]. lncRNA controls autophagy through regulating a series of autophagy-related proteins in cancer cells.

lncRNA PTENP1 is a pseudogene of the tumor suppressor gene PTEN. The molecular mechanisms of how PTENP1 represses the tumorigenic properties of hepatocellular carcinoma (HCC) cells are related with lncRNA-mediated autophagy pathway. A recent study shows that PTENP1 and PTEN are downregulated in several HCC cells. Sleeping Beauty (SB)-based hybrid baculovirus (BV) vectors are constructed for sustained PTENP1 lncRNA expression. Cotransduction of HCC cells with the SB-BV vector expressing PTENP1 elevates the levels of PTENP1 and PTEN, which suppresses the oncogenic PI3K/AKT pathway-induced autophagy. The overexpressed PTENP1 decoys oncomirs miR-17, miR-19b, and miR-20a, which would otherwise target PTEN, PHLPP (a negative AKT regulator), and such autophagy genes as ULK1, ATG7, and p62, indicating that PTENP1 modulates the HCC cell behavior and gene networks by miRNA regulation. Injection of the PTENP1-expressing SB-BV vector into mice bearing HCC tumors effectively mitigates the tumor growth, suppresses intratumoral cell proliferation, elicits apoptosis and autophagy, and inhibits angiogenesis [81].

A frontier study shows that an lncRNA named autophagy promoting factor (APF) can regulate autophagic cell death by targeting miR-188-3p and ATG7. The results show that miR-188-3p suppresses autophagy and myocardial infarction by targeting ATG7. Further, APF lncRNA regulates miR-188-3p and thus affects ATG7 expression, autophagic cell death, and myocardial infarction. The study reveals a novel regulating model of autophagic program, which comprises APF, miR-188-3p, and ATG7 in cardiovascular diseases [107].

High oncogenic BRAF levels have been shown to initiate autophagy, and it has also been speculated that BANCR can be involved in the regulation of autophagy, which is involved in tumor progression. BRAF-activated lncRNA (BANCR) is overexpressed in melanoma and has a potential functional role in melanoma cell migration. BANCR levels are significantly higher in the papillary thyroid carcinoma (PTC) tissues and PTC IHH-4 cells compared with the normal controls. Knockdown of BANCR in the IHH-4 cells inhibits proliferation and increases apoptosis of the cells in vitro. Further investigation of the underlying mechanisms reveals that BANCR markedly activates autophagy. Overexpression of BANCR inhibits apoptosis in the IHH-4 cells, whereas inhibition of autophagy stimulates apoptosis in the BANCR-overexpressed cells. BANCR overexpression also increases cell proliferation and the inhibition of autophagy abrogates BANCR overexpression-induced cell proliferation. In addition, the overexpression of BANCR results in an increase in the ratio of LC3-II/LC3-I, a marker for autophagy, while the knockdown of BANCR decreases the ratio of LC3-II/LC3-I. Additionally, BANCR increases PTC cell proliferation, which can activate autophagy [108].

Maternally expressed gene 3 (MEG3) is an imprinted gene that encodes an lncRNA associated with tumorigenesis. MEG3 levels are significantly reduced in bladder cancer tissues compared with normal controls, and autophagy activity is increased in bladder cancer tissues. A significant negative correlation is observed between MEG3 levels and LC3-II (autophagy marker) levels in vivo. MEG3 is further demonstrated to markedly suppress autophagy activation, whereas MEG3 knockdown activates autophagy in human bladder cancer cell lines. Downregulated expression of MEG3 inhibits cell apoptosis, whereas autophagy inhibition increases MEG3-knockdown cell apoptosis. MEG3 knockdown also increases cell proliferation. More importantly, autophagy inhibition abrogates MEG3-knockdown-induced cell proliferation. Overall data has demonstrated that downregulated MEG3 activates autophagy and increases cell proliferation in bladder cancer [109].

5.4 Other ncRNAs in Growth and Death of Cancer Cells

By sitting at the intersection of complex circuitries that integrate transcriptional, posttranscriptional, and epigenetic control, ncRNAs exert a pervasive function on cell regulation [2]. Studies have shown that tumor tissue and normal tissue exhibit distinct noncoding RNA expression profiles, and the aberrant noncoding RNAs can play an important role in cancer development [9]. However, the biological functions of most ncRNAs remain largely unknown. Recently, evidence has begun to accumulate, describing how ncRNAs are dysregulated in growth and death of cancer cells. In addition to miRNAs and lncRNAs, other ncRNA species are also gaining a greater appreciation for their role in carcinogenesis, although further investigation is needed to unveil the underlying mechanisms. To date, except the most studied miRNAs and lncRNAs, ncRNAs which have been reported involved in growth and death of cancer cells include piRNA, siRNA, circRNA, rRNA, tRNA, snRNA, and snoRNA.

rRNA, tRNA, snRNA, and snoRNA belong to the housekeeping noncoding RNA family. rRNA is the RNA component of the ribosome and essential for protein synthesis in all living organisms. It constitutes the predominant material within the ribosome. The ribosomal RNAs form two subunits, the large subunit (LSU) and small subunit (SSU). The LSU rRNA acts as a ribozyme, catalyzing peptide bond formation. The regulation of rRNA transcription is physiologically important because the rate of rRNA transcription is coupled tightly to ribosome biogenesis, which subsequently determines the capacity of cells to grow and proliferate [110]. rRNA has been reported in controlling proliferation and autophagy of cancer cells. tRNA is an adaptor molecule composed of RNA, typically 76–90 nucleotides in length, which serves as the physical link between the nucleotide sequence of nucleic acids (DNA and RNA) and the amino acid sequence of proteins by carrying an amino acid to the protein synthetic machinery of a cell as directed by a three-nucleotide sequence in a mRNA. As such, tRNAs are a necessary component of protein translation, the biological synthesis of new proteins according to the genetic code [111]. tRNA has been reported in controlling proliferation of cancer cells. snRNA is a class of small RNA molecules that are found within the splicing speckles and Cajal bodies of the cell nucleus in eukaryotic cells. The length of an average snRNA is approximately 150 nucleotides. They are transcribed by either RNA polymerase II or RNA polymerase III, and studies have shown that their primary function is in the processing of pre-messenger RNA (hnRNA) in the nucleus. They have also been shown to aid in the regulation of transcription factors (7SK RNA) or RNA polymerase II (B2 RNA) and maintaining the telomeres [112]. snRNA has been reported in controlling apoptosis of cancer cells. snoRNAs are a large class of small noncoding RNAs present in all eukaryotes sequenced thus far. They play a conserved role in ribosome biogenesis and are components of well-characterized ribonucleoprotein complexes referred to as snoRNPs. They can be divided in two main classes, the box C/D and the box H/ACA snoRNAs, which differ in terms of their characteristic motifs and structure as well as in their protein binding preferences and the chemical modifications they catalyze [113]. snoRNA has been reported in controlling proliferation and apoptosis of cancer cells.

piRNA, siRNA, and circRNA belong to the regulatory noncoding RNAs family. piRNAs are a novel class of noncoding single-strand RNAs. They are involved in germline development, silencing of selfish DNA elements, and maintaining germline DNA. The complex piRNA pathway can have regulatory functions in mammalian spermatogenesis in the timing of meiotic and postmeiotic events through transcriptional and translational repression or by supporting the search for homologous chromosomes and chromosome pairing integrity [114]. piRNA has been reported in controlling proliferation, cell cycle, and apoptosis of cancer cells. siRNAs are a class of double-stranded RNA molecules, 20–25 base pairs in length, which play many roles but are most notable in the RNAi pathway, where they interfere with the expression of specific genes with complementary nucleotide sequences. siRNA functions by causing mRNA to be broken down after transcription, resulting in no translation. It also acts in RNAi-related pathways, such as an antiviral mechanism or in shaping the chromatin structure of a genome. The complexity of these pathways is only now being elucidated [115]. siRNAs have been reported in controlling cell cycle and apoptosis of cancer cells. circRNAs are a special class of endogenous RNAs featuring stable structure and high tissue-specific expression. They are common in mammalian cells and regulate gene expression at the transcriptional or posttranscriptional level by interacting with microRNAs (miRNAs) or other molecules [116]. Unlike linear RNAs terminated with 5′ caps and 3′ tails, circular RNAs are characterized by covalently closed loop structures with neither 5′ to 3′ polarity nor polyadenylated tail. With the advent of specific biochemical and computational approaches, a large number of circRNAs have been identified in various cell lines and across different species [117]. circRNAs have been reported in controlling proliferation of cancer cells.

5.4.1 Proliferation

5.4.1.1 rRNA

Ribosome biogenesis drives cell growth and proliferation, but mechanisms that modulate this process within specific lineages remain poorly understood. A *Drosophila* RNA polymerase I (Pol I) regulatory complex is identified to be composed of underdeveloped (Udd), TAF1B, and a TAF1C-like factor. Disruption of Udd or TAF1B results in reduced ovarian germline stem cell (GSC) proliferation. Female GSCs display high levels of rRNA transcription, and Udd becomes enriched in GSCs relative to their differentiating daughters. Increasing Pol I transcription delays differentiation, whereas reducing rRNA production induces both morphological changes that accompany multicellular cyst formation and specific decreased expression of the BMP pathway component Mad. These findings demonstrate that modulating rRNA synthesis fosters changes in the cell fate, growth, and proliferation of female *Drosophila* GSCs and their daughters [118].

5.4.1.2 tRNA

Diverse and abundant small RNAs may be derived from tRNA, but the function of these molecules remains undefined. One such tRNA-derived fragment, cloned from human mature B cells and designated CU1276, is found to possess the functional characteristics of a microRNA, including a DICER1-dependent biogenesis, physical association with Argonaute proteins, and the ability to repress mRNA transcripts in a sequence-specific manner. Expression of CU1276 is abundant in normal germinal center B cells but absent in germinal center-derived lymphomas, suggesting a role in the pathogenesis of this disease. Furthermore, CU1276 represses endogenous RPA1, an essential gene involved in many aspects of DNA dynamics, and consequently, expression of this tRNA-derived microRNA in a lymphoma cell line suppresses proliferation and modulates the molecular response to DNA damage [119].

5.4.1.3 snoRNA

The snoRNA U50, mediating the methylation of C2848 in 28S rRNA, has been suggested as a potential tumor suppressor-like gene playing a role in breast and prostate cancers and B-cell lymphoma. The downregulation of U50 is observed in colon cancer cell lines as well as tumors. The relationship between U50 and proliferation is investigated in lymphocytes stimulated by phytohemagglutinin (PHA) and observed to have a strong decrease in U50 levels associated with a reduced C2848 methylation. This reduction is due to an alteration of U50 stability and to an increase of its consumption. Indeed, the blockade of ribosome biogenesis induces only an early decrease in U50 followed by a stabilization of U50 levels when ribosome biogenesis is almost completely blocked. In conclusion, the results link U50 to the cellular proliferation rate and ribosome biogenesis [120].

5.4.1.4 circRNA

Circular RNAs are a recently (re)discovered abundant RNA species with presumed function as miRNA sponges, thus becoming part of the competing endogenous RNA network. The expression of circular and linear RNAs and proliferation is analyzed in matched normal colon mucosa and tumor tissues. Interestingly, the ratio of circular to linear RNA isoforms is always lower in tumor compared to normal colon samples and even lower in colorectal cancer cell lines. Furthermore, this ratio correlates negatively with the proliferation index. The correlation of global circular RNA abundance (the circRNA index) and proliferation is validated in a noncancerous proliferative disease, idiopathic pulmonary fibrosis, ovarian cancer cells compared to cultured normal ovarian epithelial cells, and 13 normal human tissues. A global reduction of circular RNA abundance is found in colorectal cancer cell lines and cancer compared to normal tissues, and a negative correlation of global circular RNA abundance and proliferation is discovered [121].

5.4.1.5 piRNA

The hiwi gene is a human member of the piwi family, which represents the first class of genes known to be required for stem cell self-renewal in diverse organisms. The hiwi gene, located in 12q24.33, was originally isolated from a human testis cDNA library and encoded a 98.5 kDa basic protein. Like other piwi family members, the HIWI protein contains a conserved architecture with a PAZ motif in the middle and Piwi motif in the C-terminal region. Recent discoveries have shown that hiwi may participate in germ cell proliferation, and its overexpression may cause germ cell malignancy development [122]. Upregulation of Hiwi has been demonstrated to promote tumor cell proliferation in breast and cervical cancers, while its downregulation has been noted to suppress tumor cell proliferation in glioma and lung cancer [123]. HIWI expression is higher in gastric cancers than in normal mucosa or in mucosa with atrophic gastritis or intestinal metaplasia. The expression of hiwi in atrophic gastritis and intestinal metaplasia is also upregulated when compared to the normal state. Moreover, the proliferation of AGS gastric cancer cells is significantly inhibited when the endogenous hiwi is blocked with either recombinant antisense adenovirus or RNA interference. Consistent with previous studies of the piwi gene, a homolog of hiwi in *Drosophila* indicates that a certain level of somatic piwi activity is required to establish or maintain the stem cell identity, but a higher level of somatic piwi expression will not increase the number of germ stem cells. These findings reveal that endogenous HIWI is essential for the proliferation of gastric cells, but the exogenous HIWI does not significantly influence cell growth. The pattern of HIWI protein expression is similar to that of Ki67 expression, which suggests that hiwi participate in the proliferation of gastric cancer cells [17]. Hiwi is confirmed to overexpress at both the mRNA and protein level, in HCC specimens, as well as in MHCC97L and MHCC97H HCC cell lines. Hiwi downregulation mediated by shRNA reduces the proliferation of HCC cells [124]. In hematopoietic cancer (CML), Hiwi protein expression is undetectable in CML K562 cells and that lentivirus-mediated ectopic expression of Hiwi markedly suppresses the proliferation [125]. Hiwi expression suppressed by an RNA interference-based strategy has been demonstrated to inhibit tumor cell proliferation in a xenograft mouse model, which is generated by subcutaneously inoculating with lung cancer stem cell SSCloAldebr cells. Plasmids containing U6 promoter-driven shRNAs against Hiwi or control plasmids have been successfully established. In nude mice, intravenous delivery of Hiwi shRNA plasmids significantly inhibits tumor growth compared to treatment with control scrambled shRNA plasmids or the vehicle PBS. Moreover, delivery of Hiwi shRNA plasmids results in a significant suppressed expression of Hiwi and ALDH-1 in xenograft tumor samples, based on immunohistochemical analysis [126]. The effect of HiWi gene silencing on lung cancer tumor stem cell proliferation has been demonstrated using gene transfection and RNA interference in lung cancer tumor stem cells (TSCs). shRNA eukaryotic expression vectors, pGenesil-2-HiWi1, pGenesil-2-HiWi2263, and pGenesil-2-control, targeting the HiWi gene are constructed. PBS serves as the control group. The expression vector of the target HiWi gene shRNA is transfected into lung cancer TSCs with PEI as the medium. The results suggest that HiWi gene silencing decreases proliferation in lung cancer TSCs [127].

5.4.2 *Cell Cycle*

5.4.2.1 siRNA

The effects of HMGB1 expression on cell cycle of the human cervical cancer cell line HeLa is evaluated by RNA interference. Effective eukaryotic expression vectors carrying PGCsi3.0-1/HMGB1 siRNA and PGCsi3.0-3/HMGB1 siRNA have been previously constructed and screened, and then the vectors are transfected into HeLa cells. The introduction of PGCsi3.0-1/HMGB1 siRNA and PGCsi3.0-3/HMGB1 siRNA inhibits the expression of HMGB1 mRNA and protein efficiently and specifically. There is a significant difference between the siRNA groups and the control groups ($P<0.05$). Flow cytometry shows that the content of DNA in G2 phase in PGCsi3.0-1 group and PGCsi3.0-3 group is obviously more than those in PGCsi3.0-Neg group and non-transfected group, but the content in S phase is less ($P<0.01$). The progression of cell cycle is arrested from G2 to S phase. PGCsi3.0-1/HMGB1 siRNA and PGCsi3.0-3/HMGB1 siRNA can specially suppress the expression of HMGB1 gene and arrest the progression of cell cycle from G2 to S phase [128].

5.4.2.2 piRNA

piR-651, one member of the piRNA family, has been shown to be involved in carcinogenesis. The piR-651 inhibitor is transfected into gastric cancer cells to assess its influence on cell growth. Cell cycle analysis is used to reveal the cellular mechanisms of piR-651 in the genesis of gastric cancer. piR-651 expression is upregulated in gastric cancer tissues compared with paired noncancerous tissues. The levels of piR-651 are associated with TNM stage ($P = 0.032$). The expression of piR-651 in gastric, colon, lung, and breast cancer tissues is higher than that in paired noncancerous tissues. The upregulated expression of piR-651 is confirmed in several cancer cell lines including gastric, lung, mesothelium, breast, liver, and cervical cancer cell lines. The growth of gastric cancer cells is inhibited by a piR-651 inhibitor and arrested at the G2/M phase. Results indicate that a piR-651 inhibitor blocks gastric cancer cells at the G2/M phase [114]. Cell cycle analysis reveals that the suppression of hiwi results in significant G2/M arrest in gastric cancer. On the basis of the inhibition of growth of AGS gastric cancer cells by antisense hiwi and 106shRNA, their cell cycle progression is investigated in response to either overexpression or suppression of hiwi. The DNA content analysis shows the cell cycle of the cells treated with Ad-antisense hiwi, or Ad-106shRNA is arrested in G2/M phase [17]. Hiwi upregulation inhibits leukemic cell growth and induces cell cycle arrest at G0/G1 and G2/M phases in K562 cells [125]. Cells treated with betulinic acid (BA) show increased cell population in G(2)/M phase, with decreases in S phase population. The expression of Hiwi and cyclin B1 is downregulated in BA-treated AGS cells in a dose-dependent manner. G(2)/M cell cycle arrest and induction of apoptosis in AGS cells in vitro. The result suggests that BA exerts potent effect on G(2)/M cell cycle arrest possibly associated with the downregulation of Hiwi and its downstream target cyclin B1 expression [129].

5.4.3 Apoptosis

5.4.3.1 snRNA

A recent study utilizes U1 small nuclear RNA (snRNA) that binds physiologically to the 5′ splice site (5′ss) of pre-mRNA, to develop a novel vector system that permits imposed binding of antisense RNA to its target. The 5′ free end of U1snRNA is replaced with the antisense sequence against the K-ras gene to generate a hyperstable U1snRNA, whose binding stability to 5′ss of the K-ras transcript is tenfold higher than that of wild-type U1snRNA. The efficacy of such hyperstable U1snRNA is examined by transducing the expression plasmids into human pancreatic cancer cell lines. It reveals that two of the hyperstable U1snRNAs induce cell death after gene transduction and significantly reduce the number of G418-resistant colonies to less than 10% of the controls. Furthermore, hyperstable U1snRNA suppresses intraperitoneal dissemination of pancreatic cancer cells in vivo [130].

5.4.3.2 snoRNA

GAS5 has been identified as critical to the control of mammalian apoptosis and cell population growth using functional expression cloning. GAS5 transcripts are subject to complex posttranscriptional processing, and some, but not all, GAS5 transcripts sensitize mammalian cells to apoptosis inducers. In some cell lines, GAS5 expression induces growth arrest and apoptosis independently of other stimuli. GAS5 transcript levels are significantly reduced in breast cancer samples relative to adjacent unaffected normal breast epithelial tissues. The GAS5 gene has no significant protein-coding potential, but expression encodes small nucleolar RNAs (snoRNAs) in its introns [131].

5.4.3.3 piRNA

Bcl-like family includes a series of gene products that either suppress apoptosis such as Bcl-2 or promote apoptosis such as Bax. Interfering with Bcl-2 expression or function and/or activation of Bax may present therapeutic strategies against myeloid leukemias. Caspase-9 is a crucial regulator of cell apoptosis, the activation of which triggers the activation of the executioner caspase-3, leading to cell death. Promoting the activation of caspase-9 and caspase-3 could induce apoptosis in leukemia cells. Hiwi upregulation induces an obvious apoptosis in K562 cells, which is accompanied by decreased Bcl-2 expression and increased Bax, activated caspase-3 and caspase-9, and cleaved PARP. Hiwi overexpression decreases anti-apoptotic Bcl-2 expression and increases pro-apoptotic Bax, activated caspase-9 and caspase-3, and cleaved PARP in K562 cells. These results at least partly demonstrate that Hiwi can promote CML apoptosis by mediating these apoptosis-related proteins [125]. In glioma cells, silencing HIWI inhibits cell proliferation by promoting apoptosis and increases cell cycle arrest. The expression of proteins related to apoptosis and the cell cycle, including p21, cyclin D1, Bcl-2, and Bax is significantly altered [132]. In lung cancer tumor stem cells (TSCs), the effect of HiWi gene silencing on lung cancer tumor stem cell apoptosis has been demonstrated using gene transfection and RNA interference. 24 h after transfection, the apoptotic ratios in the pGenesil-2-HiWi1 and pGenesil-2-HiWi2263 groups are 26.16 ± 1.21 and $28.06 \pm 1.78\,\%$, respectively, which are higher as compared to those in the pGenesil-2-control group $2.86 \pm 0.09\,\%$ ($P < 0.01$). The results suggest that HiWi gene silencing promotes apoptosis of lung cancer TSCs [126].

5.4.4 *Necrosis*

There are no reports about ncRNA-related necrosis in cancer cells except miRNA and lncRNA.

5.4.5 Autophagy

5.4.5.1 rRNA

Several agents known to interfere with rRNA transcription and processing are first-line anticancer therapy, such as cisplatin for advanced non-small cell lung carcinomas and 5-fluorouracil for colon cancers. The list of agents inhibiting various stages of rRNA biogenesis continues to grow. Cis-noncoding rRNAs, alternative to primary rRNA, have been shown to regulate rRNA biogenesis. Bidirectional noncoding rRNAs have recently been detected carrying ribozyme-like properties. Anti-antisense oligonucleotides complementary to antisense noncoding rRNAs markedly stabilize the bidirectional transcripts and induce cell death in mouse lung cells. Autophagic activation is largely undifferentiable between the anti-antisense and other oligonucleotides and accounts for the undesired cytotoxicity in noncancer cells. Co-treatment with chloroquine, an autophagy inhibitor, reduces cytotoxicity in the noncancer cells, but retains the anti-antisense-mediated killings in cancer cells. The 86S oligonucleotide is more effective than the 84AS in the induction of apoptotic cell death in human lung cancer cells. Furthermore, the anti-antisense oligonucleotide stabilizes bidirectional noncoding rRNAs predominantly in human cancer cells and perturbs rRNA biogenesis [133].

5.5 Conclusion

Historically, both proteins and protein-coding RNAs have tended to dominate our view of the cell and the human diseases because of their abundance and the relative ease with which protein-coding genes and their gene products can be identified and studied. However, this paradigm has been undermined in recent years with the development of whole-genome and transcriptome sequencing technologies. Over the past several years, accumulated data have begun to advance the idea that ncRNAs are not just transcriptional noise or cloning artifacts but important supplements to proteins and other effectors in complex regulatory networks [13]. In this review, we present an updated vision of ncRNAs and summarize the mechanism of regulation by ncRNAs, which can control a wide range of biological functions such as cellular proliferation, cell cycle, apoptosis, necroptosis, and autophagy, providing new insight into the functional cellular roles that they may play in cancers, primarily focusing on the two most commonly studied ncRNA biotypes, lncRNAs and miRNAs.

Given the critical role miRNAs play in tumorigenesis processes and their disease-specific expression, they hold potential as therapeutic targets and novel biomarkers. Expression profiling of miRNAs has been shown to be a more accurate method of classifying cancer subtypes than using the expression profiles of protein-coding genes [25]. The differential expression of certain miRNAs in various tumors is a powerful tool to aid in the diagnosis and treatment of cancer. Different strategies based on blocking miRNA function or specific miRNA delivery to the tumor cells have already been used. Several preclinical approaches have been developed in order to block miRNAs, including anti-miRNA oligonucleotides, miRNA sponges, miRNA masks (target protectors), and small molecule inhibitors [134]. However, it is worth noting that further studies are needed to unveil the functions and mechanisms of miRNAs in the necroptosis of cancer cells. The discovery of dysregulated lncRNAs involved in cancer cellular biology represents a new layer of complexity in the molecular architecture of oncology. However, there are still many gaps in our current understanding of lncRNA function. lncRNAs can be useful as novel biomarkers for diagnosis, prognosis, and prediction of response to therapy. Better understanding of lncRNA functions will help clarify the real impact of genomic pervasive transcription on cell biology, on evolution, and, eventually, on the clinical setting for oncology. Furthermore, many more types of ncRNAs exist but their underlying mechanisms are still in infancy. For example, a growing number of reports have revealed the aberrant expression of PIWI proteins in various cancers, and it appears highly plausible that PIWI proteins are involved in tumorigenesis, in part because a considerable portion of the experimental results is derived from studies of patient samples, with limited opportunities for experimental manipulation. Further basic studies with more manipulable materials, such as cell lines and experimental animals, are urgently needed to address the possibility of PIWI as a therapeutic target. However, data for elucidating the detailed molecular role of PIWI proteins in tumorigenesis is very limited. An immediate focus is to identify PIWI protein-associating RNAs to determine whether piRNAs and/or other RNA species specifically interact with PIWI proteins in cancer cells. It is also imperative to investigate how PIWI proteins are involved in the biological functioning of cancer cells, such as transposon silencing, transcriptional or posttranscriptional regulation, DNA repair, and chromosome condensation and segregation, and to determine the roles of PIWI proteins in tumorigenesis [135].

Given the versatile, critical, and surprising regulatory functions of ncRNAs uncovered so far, forming a better understanding of the precise molecular mechanisms by which ncRNAs function in various cancers will be an exciting journey and also critical for exploring new potential strategies for early diagnosis and therapy. In the future, the field of ncRNAs research will certainly be in spotlight. Clearly, a deeper understanding of the biological effects of the various ncRNAs will need to be acquired before this possibility can be actualized into real therapeutic strategies.

References

1. Fu XD. Non-coding RNA: a new frontier in regulatory biology. Natl Sci Rev. 2014;1(2):190–204. doi:10.1093/nsr/nwu008.
2. Tordonato C, Di Fiore PP, Nicassio F. The role of non-coding RNAs in the regulation of stem cells and progenitors in the normal mammary gland and in breast tumors. Front Genet. 2015;6:72. doi:10.3389/fgene.2015.00072.

3. Hah N, Danko CG, Core L, et al. A rapid, extensive, and transient transcriptional response to estrogen signaling in breast cancer cells. Cell. 2011;145(4):622–34. doi:10.1016/j.cell.2011.03.042.
4. Cui P, Lin Q, Ding F, et al. A comparison between ribo-minus RNA-sequencing and polyA-selected RNA-sequencing. Genomics. 2010;96(5):259–65. doi:10.1016/j.ygeno.2010.07.010.
5. Yang L, Lin C, Jin C, et al. lncRNA-dependent mechanisms of androgen-receptor-regulated gene activation programs. Nature. 2013;500(7464):598–602. doi:10.1038/nature12451.
6. Kowalczykiewic D, Wrzesinski J. The role of piRNA and Piwi proteins in regulation of germ-line development. Postepy Biochem. 2011;57(3):249–56.
7. Hon GC, Hawkins RD, Caballero OL, et al. Global DNA hypomethylation coupled to repressive chromatin domain formation and gene silencing in breast cancer. Genome Res. 2012;22(2):246–58. doi:10.1101/gr.125872.111.
8. Hervouet E, Cartron PF, Jouvenot M, Delage-Mourroux R. Epigenetic regulation of estrogen signaling in breast cancer. Epigenetics. 2013;8(3):237–45. doi:10.4161/epi.23790.
9. Zhang C, Peng G. Non-coding RNAs: an emerging player in DNA damage response. Mutat Res Rev Mutat Res. 2015;763:202–11. doi:10.1016/j.mrrev.2014.11.003.
10. Li Y, Zhang Y, Li S, et al. Genome-wide DNA methylome analysis reveals epigenetically dysregulated non-coding RNAs in human breast cancer. Sci Rep. 2015;5:8790. doi:10.1038/srep08790.
11. Yan Y, Shen Z, Gao Z, et al. LncRNA specific for distant metastasis of gastric cancer is associated with TRIM16 expression and facilitates tumor cell invasion in vitro. J Gastroenterol Hepatol. 2015;30:1367–75. doi:10.1111/jgh.12976.
12. Lu T, Shao N, Ji C. Targeting microRNAs to modulate TRAIL-induced apoptosis of cancer cells. Cancer Gene Ther. 2013;20(1):33–7. doi:10.1038/cgt.2012.81.
13. Shi X, Sun M, Wu Y, et al. Post-transcriptional regulation of long non-coding RNAs in cancer. Tumour Biol. 2015;36(2):503–13. doi:10.1007/s13277-015-3106-y.
14. Bhan A, Mandal SS. Long non-coding RNAs: emerging stars in gene regulation, epigenetics and human disease. ChemMedChem. 2014;9(9):1932–56. doi:10.1002/cmdc.201300534.
15. Seth S, Johns R, Templin MV. Delivery and biodistribution of siRNA for cancer therapy: challenges and future prospects. Ther Deliv. 2012;3(2):245–61.
16. Qiao D, Zeeman AM, Deng W, Looijenga LH, Lin H. Molecular characterization of hiwi, a human member of the piwi gene family whose overexpression is correlated to seminomas. Oncogene. 2002;21(25):3988–99. doi:10.1038/sj.onc.1205505.
17. Liu X, Sun Y, Guo J, et al. Expression of hiwi gene in human gastric cancer was associated with proliferation of cancer cells. Int J Cancer. 2006;118(8):1922–9. doi:10.1002/ijc.21575.
18. Liu JJ, Shen R, Chen L, et al. Piwil2 is expressed in various stages of breast cancers and has the potential to be used as a novel biomarker. Int J Clin Exp Pathol. 2010;3(4):328–37.
19. Lim SL, Ricciardelli C, Oehler MK, et al. Overexpression of piRNA pathway genes in epithelial ovarian cancer. PLoS One. 2014;9(6), e99687. doi:10.1371/journal.pone.0099687.
20. Plosky BS. eRNAs lure NELF from paused polymerases. Mol Cell. 2014;56(1):3–4. doi:10.1016/j.molcel.2014.09.016.
21. Artandi SE, DePinho RA. Telomeres and telomerase in cancer. Carcinogenesis. 2010;31(1):9–18. doi:10.1093/carcin/bgp268.
22. Hansen TB, Kjems J, Damgaard CK. Circular RNA and miR-7 in cancer. Cancer Res. 2013;73(18):5609–12. doi:10.1158/0008-5472.CAN-13-1568.
23. Schmidt U, Keck ME, Buell DR. miRNAs and other non-coding RNAs in posttraumatic stress disorder: A systematic review of clinical and animal studies. J Psychiatr Res. 2015;65:1–8. doi:10.1016/j.jpsychires.2015.03.014.
24. Malumbres M. miRNAs and cancer: an epigenetics view. Mol Aspects Med. 2013;34(4):863–74. doi:10.1016/j.mam.2012.06.005.
25. Esquela-Kerscher A, Slack FJ. Oncomirs-microRNAs with a role in cancer. Nat Rev Cancer. 2006;6(4):259–69. doi:10.1038/nrc1840.

26. Sanchez-Mejias A, Tay Y. Competing endogenous RNA networks: tying the essential knots for cancer biology and therapeutics. J Hematol Oncol. 2015;8(1):30. doi:10.1186/s13045-015-0129-1.
27. Feitelson MA, Arzumanyan A, Kulathinal RJ, et al. Sustained proliferation in cancer: mechanisms and novel therapeutic targets. Semin Cancer Biol. 2015;35:S25–54. doi:10.1016/j.semcancer.2015.02.006.
28. Zheng W, Liu Z, Zhang W, Hu X. miR-31 functions as an oncogene in cervical cancer. Arch Gynecol Obstet. 2015;292:1083–9. doi:10.1007/s00404-015-3713-2.
29. Li R, Shi X, Ling F, et al. MiR-34a suppresses ovarian cancer proliferation and motility by targeting AXL. Tumour Biol. 2015;36:7277–83. doi:10.1007/s13277-015-3445-8.
30. Xu M, Gu M, Zhang K, et al. miR-203 inhibition of renal cancer cell proliferation, migration and invasion by targeting of FGF2. Diagn Pathol. 2015;10(1):24. doi:10.1186/s13000-015-0255-7.
31. Yang X, Chen BB, Zhang MH, Wang XR. MicroRNA-126 inhibits the proliferation of lung cancer cell line A549. Asian Pac J Trop Med. 2015;8(3):239–42. doi:10.1016/S1995-7645(14)60323-0.
32. Rebucci M, Sermeus A, Leonard E, et al. miRNA-196b inhibits cell proliferation and induces apoptosis in HepG2 cells by targeting IGF2BP1. Mol Cancer. 2015;14(1):79. doi:10.1186/s12943-015-0349-6.
33. Chen F, Hu SJ. Effect of microRNA-34a in cell cycle, differentiation, and apoptosis: a review. J Biochem Mol Toxicol. 2012;26(2):79–86. doi:10.1002/jbt.20412.
34. Schafer KA. The cell cycle: a review. Vet Pathol. 1998;35(6):461–78.
35. Malumbres M, Barbacid M. Cell cycle, CDKs and cancer: a changing paradigm. Nat Rev Cancer. 2009;9(3):153–66. doi:10.1038/nrc2602.
36. Kim HS, Lee KS, Bae HJ, et al. MicroRNA-31 functions as a tumor suppressor by regulating cell cycle and epithelial-mesenchymal transition regulatory proteins in liver cancer. Oncotarget. 2015;6(10):8089–102.
37. Lin Y, Li D, Liang Q, et al. miR-638 regulates differentiation and proliferation in leukemic cells by targeting cyclin-dependent kinase 2. J Biol Chem. 2015;290(3):1818–28. doi:10.1074/jbc.M114.599191.
38. Li LP, Wu WJ, Sun DY, et al. miR-449a and CDK6 in gastric carcinoma. Oncol Lett. 2014;8(4):1533–8. doi:10.3892/ol.2014.2370.
39. Zhao Z, Ma X, Sung D, et al. microRNA-449a functions as a tumor suppressor in neuroblastoma through inducing cell differentiation and cell cycle arrest. RNA Biol. 2015;12(5):538–54. doi:10.1080/15476286.2015.1023495.
40. Chen W, Qi J, Bao G, et al. Emerging role of microRNA-27a in human malignant glioma cell survival via targeting of prohibitin. Mol Med Rep. 2015;12(1):1515–23. doi:10.3892/mmr.2015.3475.
41. Ouyang L, Shi Z, Zhao S, et al. Programmed cell death pathways in cancer: a review of apoptosis, autophagy and programmed necrosis. Cell Prolif. 2012;45(6):487–98. doi:10.1111/j.1365-2184.2012.00845.x.
42. Shivapurkar N, Reddy J, Chaudhary PM, Gazdar AF. Apoptosis and lung cancer: a review. J Cell Biochem. 2003;88(5):885–98. doi:10.1002/jcb.10440.
43. Su Z, Yang Z, Xu Y, et al. MicroRNAs in apoptosis, autophagy and necroptosis. Oncotarget. 2015;6(11):8474–90.
44. Wang H, Li J, Chi H, et al. MicroRNA-181c targets Bcl-2 and regulates mitochondrial morphology in myocardial cells. J Cell Mol Med. 2015;19:2084–97. doi:10.1111/jcmm.12563.
45. Zhang Y, Schiff D, Park D, Abounader R. MicroRNA-608 and microRNA-34a regulate chordoma malignancy by targeting EGFR, Bcl-xL and MET. PLoS One. 2014;9(3), e91546. doi:10.1371/journal.pone.0091546.
46. Ji F, Zhang H, Wang Y, et al. MicroRNA-133a, down-regulated in osteosarcoma, suppresses proliferation and promotes apoptosis by targeting Bcl-xL and Mcl-1. Bone. 2013;56(1):220–6. doi:10.1016/j.bone.2013.05.020.

47. Lou G, Liu Y, Wu S, et al. The p53/miR-34a/SIRT1 Positive Feedback Loop in Quercetin-Induced Apoptosis. Cell Physiol Biochem. 2015;35(6):2192–202. doi:10.1159/000374024.
48. Huang G, Nishimoto K, Zhou Z, et al. miR-20a encoded by the miR-17-92 cluster increases the metastatic potential of osteosarcoma cells by regulating Fas expression. Cancer Res. 2012;72(4):908–16. doi:10.1158/0008-5472.CAN-11-1460.
49. Xie X, Hu Y, Xu L, et al. The role of miR-125b-mitochondria-caspase-3 pathway in doxorubicin resistance and therapy in human breast cancer. Tumour Biol. 2015;36:7185–94. doi:10.1007/s13277-015-3438-7.
50. Ihle MA, Trautmann M, Kuenstlinger H, et al. miRNA-221 and miRNA-222 induce apoptosis via the KIT/AKT signalling pathway in gastrointestinal stromal tumours. Mol Oncol. 2015;9:1421–33. doi:10.1016/j.molonc.2015.03.013.
51. Su Z, Yang Z, Xu Y, et al. Apoptosis, autophagy, necroptosis, and cancer metastasis. Mol Cancer. 2015;14(1):48. doi:10.1186/s12943-015-0321-5.
52. Petersen SL, Chen TT, Lawrence DA, et al. TRAF2 is a biologically important necroptosis suppressor. Cell Death Differ. 2015;22:1846–57. doi:10.1038/cdd.2015.35.
53. Liu J, van Mil A, Vrijsen K, et al. MicroRNA-155 prevents necrotic cell death in human cardiomyocyte progenitor cells via targeting RIP1. J Cell Mol Med. 2011;15(7):1474–82. doi:10.1111/j.1582-4934.2010.01104.x.
54. Wang K, Liu F, Zhou LY, et al. miR-874 regulates myocardial necrosis by targeting caspase-8. Cell Death Dis. 2013;4:e709. doi:10.1038/cddis.2013.233.
55. Rebecca VW, Amaravadi RK. Emerging strategies to effectively target autophagy in cancer. Oncogene. 2016;35:1–11. doi:10.1038/onc.2015.99.
56. Lu C, Chen J, Xu HG, et al. MIR106B and MIR93 prevent removal of bacteria from epithelial cells by disrupting ATG16L1-mediated autophagy. Gastroenterology. 2014;146(1):188–99. doi:10.1053/j.gastro.2013.09.006.
57. Zheng B, Zhu H, Gu D, Pan X, et al. MiRNA-30a-mediated autophagy inhibition sensitizes renal cell carcinoma cells to sorafenib. Biochem Biophys Res Commun. 2015;459(2):234–9. doi:10.1016/j.bbrc.2015.02.084.
58. Chang Y, Yan W, He X, et al. miR-375 inhibits autophagy and reduces viability of hepatocellular carcinoma cells under hypoxic conditions. Gastroenterology. 2012;143(1):177–87. doi:10.1053/j.gastro.2012.04.009. e178.
59. Korkmaz G, le Sage C, Tekirdag KA, et al. miR-376b controls starvation and mTOR inhibition-related autophagy by targeting ATG4C and BECN1. Autophagy. 2012;8(2):165–76. doi:10.4161/auto.8.2.18351.
60. Zhai H, Fesler A, Ju J. MicroRNA: a third dimension in autophagy. Cell Cycle. 2013;12(2):246–50. doi:10.4161/cc.23273.
61. Shen G, Li X, Jia YF, et al. Hypoxia-regulated microRNAs in human cancer. Acta Pharmacol Sin. 2013;34(3):336–41. doi:10.1038/aps.2012.195.
62. Ma Y, Yang HZ, Dong BJ, et al. Biphasic regulation of autophagy by miR-96 in prostate cancer cells under hypoxia. Oncotarget. 2014;5(19):9169–82.
63. Ao JE, Kuang LH, Zhou Y, et al. Hypoxia-inducible factor 1 regulated ARC expression mediated hypoxia induced inactivation of the intrinsic death pathway in p53 deficient human colon cancer cells. Biochem Biophys Res Commun. 2012;420(4):913–7. doi:10.1016/j.bbrc.2012.03.101.
64. Li W, Zhang X, Zhuang H, et al. MicroRNA-137 is a novel hypoxia-responsive microRNA that inhibits mitophagy via regulation of two mitophagy receptors FUNDC1 and NIX. J Biol Chem. 2014;289(15):10691–701. doi:10.1074/jbc.M113.537050.
65. Zhai H, Song B, Xu X, et al. Inhibition of autophagy and tumor growth in colon cancer by miR-502. Oncogene. 2013;32(12):1570–9. doi:10.1038/onc.2012.167.
66. Chen S, Li P, Li J, et al. MiR-144 inhibits proliferation and induces apoptosis and autophagy in lung cancer cells by targeting TIGAR. Cell Physiol Biochem. 2015;35(3):997–1007. doi:10.1159/000369755.
67. Spizzo R, Almeida MI, Colombatti A, Calin GA. Long non-coding RNAs and cancer: a new frontier of translational research? Oncogene. 2012;31(43):4577–87. doi:10.1038/onc.2011.621.

68. Prensner JR, Chinnaiyan AM. The emergence of lncRNAs in cancer biology. Can Dis. 2011;1(5):391–407. doi:10.1158/2159-8290.CD-11-0209.
69. St Laurent G, Wahlestedt C, Kapranov P. The landscape of long non-coding RNA classification. Trends Genet. 2015;31(5):239–51. doi:10.1016/j.tig.2015.03.007.
70. Mitra SA, Mitra AP, Triche TJ. A central role for long non-coding RNA in cancer. Front Genet. 2012;3:17. doi:10.3389/fgene.2012.00017.
71. Cheetham SW, Gruhl F, Mattick JS, Dinger ME. Long non-coding RNAs and the genetics of cancer. Br J Cancer. 2013;108(12):2419–25. doi:10.1038/bjc.2013.233.
72. Nie FQ, Sun M, Yang JS, et al. Long non-coding RNA ANRIL promotes non-small cell lung cancer cell proliferation and inhibits apoptosis by silencing KLF2 and P21 expression. Mol Cancer Ther. 2015;14(1):268–77. doi:10.1158/1535-7163.MCT-14-0492.
73. Yao Y, Ma J, Xue Y, et al. Knockdown of long non-coding RNA XIST exerts tumor-suppressive functions in human glioblastoma stem cells by up-regulating miR-152. Cancer Lett. 2015;359(1):75–86. doi:10.1016/j.canlet.2014.12.051.
74. Qiu JJ, Wang Y, Ding JX, et al. The long non-coding RNA HOTAIR promotes the proliferation of serous ovarian cancer cells through the regulation of cell cycle arrest and apoptosis. Exp Cell Res. 2015;333(2):238–48. doi:10.1016/j.yexcr.2015.03.005.
75. Wan J, Huang M, Zhao H, et al. A novel tetranucleotide repeat polymorphism within KCNQ1OT1 confers risk for hepatocellular carcinoma. DNA Cell Biol. 2013;32(11):628–34. doi:10.1089/dna.2013.2118.
76. Sun H, Wang G, Peng Y, et al. H19 lncRNA mediates 17beta-estradiol-induced cell proliferation in MCF-7 breast cancer cells. Oncol Rep. 2015;33:3045–52. doi:10.3892/or.2015.3899.
77. Leygue E. Steroid receptor RNA activator (SRA1): unusual bifaceted gene products with suspected relevance to breast cancer. Nucl Recept Signal. 2007;5, e006. doi:10.1621/nrs.05006.
78. Hu L, Wu Y, Tan D, et al. Up-regulation of long non-coding RNA MALAT1 contributes to proliferation and metastasis in esophageal squamous cell carcinoma. J Exp Clin Cancer Res. 2015;34(1):7. doi:10.1186/s13046-015-0123-z.
79. Wang X, Li M, Wang Z, et al. Silencing of long non-coding RNA MALAT1 by miR-101 and miR-217 inhibits proliferation, migration, and invasion of esophageal squamous cell carcinoma cells. J Biol Chem. 2015;290(7):3925–35. doi:10.1074/jbc.M114.596866.
80. Li J, Jiang K, Zhao F. Icariin regulates the proliferation and apoptosis of human ovarian cancer cells through microRNA-21 by targeting PTEN, RECK and Bcl-2. Oncol Rep. 2015;33:2829–36. doi:10.3892/or.2015.3891.
81. Chen CL, Tseng YW, Wu JC, et al. Suppression of hepatocellular carcinoma by baculovirus-mediated expression of long non-coding RNA PTENP1 and MicroRNA regulation. Biomaterials. 2015;44:71–81. doi:10.1016/j.biomaterials.2014.12.023.
82. Wang Y, He L, Du Y, et al. The long non-coding RNA lncTCF7 promotes self-renewal of human liver cancer stem cells through activation of Wnt signaling. Cell Stem Cell. 2015;16(4):413–25. doi:10.1016/j.stem.2015.03.003.
83. Jiao F, Hu H, Han T, et al. Long non-coding RNA MALAT-1 enhances stem cell-like phenotypes in pancreatic cancer cells. Int J Mol Sci. 2015;16(4):6677–93. doi:10.3390/ijms16046677.
84. Williams GH, Stoeber K. The cell cycle and cancer. J Pathol. 2012;226(2):352–64. doi:10.1002/path.3022.
85. Casimiro MC, Crosariol M, Loro E, et al. Cyclins and cell cycle control in cancer and disease. Genes Cancer. 2012;3(11–12):649–57. doi:10.1177/1947601913479022.
86. Vermeulen K, Van Bockstaele DR, Berneman ZN. The cell cycle: a review of regulation, deregulation and therapeutic targets in cancer. Cell Prolif. 2003;36(3):131–49.
87. Kitagawa M, Kitagawa K, Kotake Y, et al. Cell cycle regulation by long non-coding RNAs. Cell Mol Life Sci. 2013;70(24):4785–94. doi:10.1007/s00018-013-1423-0.
88. Leveille N, Melo CA, Rooijers K, et al. Genome-wide profiling of p53-regulated enhancer RNAs uncovers a subset of enhancers controlled by a lncRNA. Nat Commun. 2015;6:6520. doi:10.1038/ncomms7520.

89. Tripathi V, Shen Z, Chakraborty A, et al. Long non-coding RNA MALAT1 controls cell cycle progression by regulating the expression of oncogenic transcription factor B-MYB. PLoS Genet. 2013;9(3):e1003368. doi:10.1371/journal.pgen.1003368.
90. Kim T, Jeon YJ, Cui R, et al. Role of MYC-regulated long non-coding RNAs in cell cycle regulation and tumorigenesis. J Natl Cancer Inst. 2015;107(4):dju505.
91. Zhang Y, Ma M, Liu W, et al. Enhanced expression of long non-coding RNA CARLo-5 is associated with the development of gastric cancer. Int J Clin Exp Pathol. 2014;7(12):8471–9.
92. Shi Y, Liu Y, Wang J, et al. Down-regulated long non-coding RNA BANCR promotes the proliferation of colorectal cancer cells via downregulation of p21 expression. PLoS One. 2015;10(4):e0122679. doi:10.1371/journal.pone.0122679.
93. Yang Q, Xu E, Dai J, et al. A novel long non-coding RNA AK001796 acts as an oncogene and is involved in cell growth inhibition by resveratrol in lung cancer. Toxicol Appl Pharmacol. 2015;285(2):79–88. doi:10.1016/j.taap.2015.04.003.
94. Elmore S. Apoptosis: a review of programmed cell death. Toxicol Pathol. 2007;35(4):495–516. doi:10.1080/01926230701320337.
95. Sanchez Y, Segura V, Marin-Bejar O, et al. Genome-wide analysis of the human p53 transcriptional network unveils a lncRNA tumour suppressor signature. Nat Commun. 2014;5:5812. doi:10.1038/ncomms6812.
96. Yoon JH, Abdelmohsen K, Srikantan S, et al. LincRNA-p21 suppresses target mRNA translation. Mol Cell. 2012;47(4):648–55. doi:10.1016/j.molcel.2012.06.027.
97. Luo H, Sun Y, Wei G, et al. Functional characterization of long non-coding RNA Lnc_bc060912 in human lung carcinoma cells. Biochemistry. 2015;54:2895–902. doi:10.1021/acs.biochem.5b00259.
98. Lu KH, Li W, Liu XH, et al. Long non-coding RNA MEG3 inhibits NSCLC cells proliferation and induces apoptosis by affecting p53 expression. BMC Cancer. 2013;13:461. doi:10.1186/1471-2407-13-461.
99. Han L, Zhang EB, Yin DD, et al. Low expression of long non-coding RNA PANDAR predicts a poor prognosis of non-small cell lung cancer and affects cell apoptosis by regulating Bcl-2. Cell Death Dis. 2015;6:e1665. doi:10.1038/cddis.2015.30.
100. DeOcesano-Pereira C, Amaral MS, Parreira KS, et al. Long non-coding RNA INXS is a critical mediator of BCL-XS induced apoptosis. Nucleic Acids Res. 2014;42(13):8343–55. doi:10.1093/nar/gku561.
101. Pickard MR, Mourtada-Maarabouni M, Williams GT. Long non-coding RNA GAS5 regulates apoptosis in prostate cancer cell lines. Biochim Biophys Acta. 2013;1832(10):1613–23. doi:10.1016/j.bbadis.2013.05.005.
102. Hang Q, Sun R, Jiang C, Li Y. Notch 1 promotes cisplatin-resistant gastric cancer formation by up-regulating lncRNA AK022798 expression. Anticancer Drugs. 2015;26:632–40. doi:10.1097/CAD.0000000000000227.
103. Zhao H, Zhang X, Frazao JB, et al. HOX antisense lincRNA HOXA-AS2 is an apoptosis repressor in all trans retinoic acid treated NB4 promyelocytic leukemia cells. J Cell Biochem. 2013;114(10):2375–83. doi:10.1002/jcb.24586.
104. Liu Q, Sun S, Yu W, et al. Altered expression of long non-coding RNAs during genotoxic stress-induced cell death in human glioma cells. J Neurooncol. 2015;122(2):283–92. doi:10.1007/s11060-015-1718-0.
105. Kania E, Pajak B, Orzechowski A. Calcium homeostasis and ER stress in control of autophagy in cancer cells. Biomed Res Int. 2015;2015:352794. doi:10.1155/2015/352794.
106. Lorin S, Hamai A, Mehrpour M, Codogno P. Autophagy regulation and its role in cancer. Semin Cancer Biol. 2013;23(5):361–79. doi:10.1016/j.semcancer.2013.06.007.
107. Wang K, Liu CY, Zhou LY, et al. APF lncRNA regulates autophagy and myocardial infarction by targeting miR-188-3p. Nat Commun. 2015;6:6779. doi:10.1038/ncomms7779.
108. Wang Y, Guo Q, Zhao Y, et al. BRAF-activated long non-coding RNA contributes to cell proliferation and activates autophagy in papillary thyroid carcinoma. Oncol Lett. 2014;8(5):1947–52. doi:10.3892/ol.2014.2487.

109. Ying L, Huang Y, Chen H, et al. Down-regulated MEG3 activates autophagy and increases cell proliferation in bladder cancer. Mol Biosyst. 2013;9(3):407–11. doi:10.1039/c2mb25386k.
110. Hayashi Y, Kuroda T, Kishimoto H, et al. Down-regulation of rRNA transcription triggers cell differentiation. PLoS One. 2014;9(5):e98586. doi:10.1371/journal.pone.0098586.
111. Sharp SJ, Schaack J, Cooley L, et al. Structure and transcription of eukaryotic tRNA genes. CRC Crit Rev Biochem. 1985;19(2):107–44.
112. Hadjiolov AA, Venkov PV, Tsanev RG. Ribonucleic acids fractionation by density-gradient centrifugation and by agar gel electrophoresis: a comparison. Anal Biochem. 1966;17(2):263–7.
113. Dupuis-Sandoval F, Poirier M, Scott MS. The emerging landscape of small nucleolar RNAs in cell biology. Wiley Interdiscip Rev RNA. 2015;6:381–97. doi:10.1002/wrna.1284.
114. Cheng J, Guo JM, Xiao BX, et al. piRNA, the new non-coding RNA, is aberrantly expressed in human cancer cells. Clin Chim Acta. 2011;412(17–18):1621–5. doi:10.1016/j.cca.2011.05.015.
115. Agrawal N, Dasaradhi PV, Mohmmed A, et al. RNA interference: biology, mechanism, and applications. Microbiol Mol Biol Rev. 2003;67(4):657–85.
116. Li P, Chen S, Chen H, et al. Using circular RNA as a novel type of biomarker in the screening of gastric cancer. Clin Chim Acta. 2015;444:132–6. doi:10.1016/j.cca.2015.02.018.
117. Chen LL, Yang L. Regulation of circRNA biogenesis. RNA Biol. 2015;12(4):381–8. doi:10.1080/15476286.2015.1020271.
118. Zhang Q, Shalaby NA, Buszczak M. Changes in rRNA transcription influence proliferation and cell fate within a stem cell lineage. Science. 2014;343(6168):298–301. doi:10.1126/science.1246384.
119. Maute RL, Schneider C, Sumazin P, et al. tRNA-derived microRNA modulates proliferation and the DNA damage response and is down-regulated in B cell lymphoma. Proc Natl Acad Sci U S A. 2013;110(4):1404–9. doi:10.1073/pnas.1206761110.
120. Pacilli A, Ceccarelli C, Trere D, Montanaro L. SnoRNA U50 levels are regulated by cell proliferation and rRNA transcription. Int J Mol Sci. 2013;14(7):14923–35. doi:10.3390/ijms140714923.
121. Bachmayr-Heyda A, Reiner AT, Auer K, et al. Correlation of circular RNA abundance with proliferation--exemplified with colorectal and ovarian cancer, idiopathic lung fibrosis, and normal human tissues. Sci Rep. 2015;5:8057. doi:10.1038/srep08057.
122. Cox DN, Chao A, Baker J, et al. A novel class of evolutionarily conserved genes defined by piwi are essential for stem cell self-renewal. Genes Dev. 1998;12(23):3715–27.
123. Wang Y, Jiang Y, Bian C, et al. Overexpression of Hiwi inhibits the cell growth of chronic myeloid leukemia K562 cells and enhances their chemosensitivity to daunomycin. Cell Biochem Biophys. 2015;73:129–35. doi:10.1007/s12013-015-0668-7.
124. Xie Y, Yang Y, Ji D, Zhang D, et al. Hiwi down-regulation, mediated by shRNA, reduces the proliferation and migration of human hepatocellular carcinoma cells. Mol Med Rep. 2015;11(2):1455–61. doi:10.3892/mmr.2014.2847.
125. Wang Y, Jiang Y, Ma N, et al. Overexpression of Hiwi inhibits the growth and migration of chronic myeloid leukemia cells. Cell Biochem Biophys. 2015;73:117–24. doi:10.1007/s12013-015-0651-3.
126. Liang D, Dong M, Hu LJ, et al. Hiwi knockdown inhibits the growth of lung cancer in nude mice. Asian Pac J Cancer Prev. 2013;14(2):1067–72.
127. Liang D, Fang Z, Dong M, et al. Effect of RNA interference-related HiWi gene expression on the proliferation and apoptosis of lung cancer stem cells. Oncol Lett. 2012;4(1):146–50. doi:10.3892/ol.2012.677.
128. He X, Qian Y, Cai H, Wang Z. The effect of RhoC siRNA on the invasiveness and proliferation of human cervical cancer cell line SiHa cells. J Huazhong Univ Sci Technolog Med Sci. 2008;28(6):665–9. doi:10.1007/s11596-008-0611-x.
129. Yang LJ, Chen Y, Ma Q, et al. Effect of betulinic acid on the regulation of Hiwi and cyclin B1 in human gastric adenocarcinoma AGS cells. Acta Pharmacol Sin. 2010;31(1):66–72. doi:10.1038/aps.2009.177.

130. Kato K, Hitomi Y, Imamura K, Esumi H. Hyperstable U1snRNA complementary to the K-ras transcripts induces cell death in pancreatic cancer cells. Br J Cancer. 2002;87(8):898–904. doi:10.1038/sj.bjc.6600563.
131. Mourtada-Maarabouni M, Pickard MR, Hedge VL, Farzaneh F, Williams GT. GAS5, a non-protein-coding RNA, controls apoptosis and is down-regulated in breast cancer. Oncogene. 2009;28(2):195–208. doi:10.1038/onc.2008.373.
132. Wang X, Tong X, Gao H, et al. Silencing HIWI suppresses the growth, invasion and migration of glioma cells. Int J Oncol. 2014;45(6):2385–92. doi:10.3892/ijo.2014.2673.
133. Hwang CJ, Fields JR, Shiao YH. Non-coding rRNA-mediated preferential killing in cancer cells is enhanced by suppression of autophagy in non-transformed counterpart. Cell Death Dis. 2011;2:e239. doi:10.1038/cddis.2011.110.
134. Zaravinos A. The regulatory role of MicroRNAs in EMT and cancer. J Oncol. 2015;2015:865816. doi:10.1155/2015/865816.
135. Suzuki R, Honda S, Kirino Y. PIWI expression and function in cancer. Front Genet. 2012;3:204. doi:10.3389/fgene.2012.00204.

Chapter 6
Noncoding RNAs in Cancer Cell Plasticity

Jiahui Xu and Suling Liu

Abstract Accumulating evidence has shown the presence of cancer stem cells in a wide spectrum of human cancers, which have the ability to self-renew and differentiate, thus leading to tumorigenesis, proliferation, cancer dissemination, drug resistance, and tumor relapse. Cancer cell plasticity allows tumor to invade and grow at primary or distant sites. Epithelial-mesenchymal transition (EMT) is the most important mechanism of cancer cell plasticity and cancer stem cells. Substantial evidence has supported a noncoding RNA network, especially miRNA, in regulating cancer cell plasticity and cancer stem cell biology. Besides, lncRNA is also found to participate in cancer development. Understanding the mechanisms of these processes might be valuable for developing accurate targeted therapies to tackle cancer progression and cancer stem cells.

Keywords Cancer stem cells • Noncoding RNA • Cancer progression • Cell plasticity • Targeted therapy

6.1 Introduction

Cancer is a group of diseases consist of abnormally growing cells with the potential to invade and metastasize to other parts of the body. Generally, cancer grows when normal cells change as a result of accumulated mutations due to environmental factors or sometimes hereditary mutations. Mutations in normal cells lead to chromosomal instability, proliferation, and finally aggressive metastatic behavior. Owing to mutations, most of human cancers are heterogeneous diseases. There is a high degree of phenotypic and functional diversity between tumors, and even within the same tumor, divergences exist. For instance, breast tumors are diverse in their nature and responsiveness to therapies. According to gene expression molecular

J. Xu • S. Liu (✉)
Innovation Center for Cell Signaling Network and the CAS Key Laboratory of Innate Immunity and Chronic Disease, School of Life Science and Medical Center, University of Science and Technology of China, Hefei, Anhui 230027, China
e-mail: suling@ustc.edu.cn

E. Song (ed.), *The Long and Short Non-coding RNAs in Cancer Biology*, Advances in Experimental Medicine and Biology 927,
DOI 10.1007/978-981-10-1498-7_6

pattern, it could be classified into several subtypes including luminal subtype, basal subtype, HER2-overexpression subtype, and normal-like subtype. These subtypes prove to be different in their malignance and responsiveness to treatments [1, 2]. Some cancers also contain a hierarchy in which cancer stem cells (CSCs) differentiate into non-cancer stem cells (or bulk tumor cells) [3].

The cancer cell plasticity describes the ability of cancer cells to transform reversibly between distinct cell states phenotypically and genotypically, contributing to tumor growth in primary and distant sites. For example, some cancer cells, such as breast cancer cells, can transit between epithelial state and mesenchymal state. Reciprocal transition between epithelial state and mesenchymal state, which is a crucial event in embryonic development, has been confirmed to be a hallmark of cancer metastasis [4]. It is reported that, cancer cells in mesenchymal state are more competent than those in epithelial state to invade and form cancer dissemination [4, 5]. Considering the heterogeneity of cancer, Gupta et al. have found that isolated subpopulations of breast cancer cells with given phenotype will finally return to an equilibrium proportions over time, which can be explained by the Markov model, in which they suggest cell transition stochastically between states and any subpopulations of cancer cells can finally return to an equilibrium proportions over time in given conditions [6]. Of note here, cancer cells acquiring drug resistance responsive to therapy is also an important aspect of cancer cell plasticity [7]. Besides, via turning on or off some markers reversibly, cancer cells can transit between distinct states. For example, study in melanoma has revealed that dynamical expression of JRID1B, an H3K4 demethylase, endows cancer cells with tumorigenic ability. The JRID1B-positive and JRID1B-negative cells can transit to each other and the JRID1B-positive cells function in tumor maintenance [8].

The concept of CSCs refers to a subpopulation of cells within tumor possessing the ability to self-renew and differentiate into non-stem progenitor. Increasing evidence has supported the CSC hypothesis that many of human cancers are driven by the CSCs. Self-renewal of CSCs and differentiation into non-stem progeny maintains the cancer cell pool and mediates the cancer metastasis, therapy resistance, and relapse. CSCs' transition from tumorigenic state to a non-tumorigenic state is one aspect of CSC plasticity. Similar with tumor heterogeneity, there exist different states in the CSC subpopulation. Transition between these states is another important aspect of CSC plasticity [5].

Noncoding RNA (ncRNA) is a functional molecule that is not translated into a protein. Quantities of ncRNAs have been found in recent decades, including ribosomal RNA (rRNA), transfer RNA (tRNA), small nuclear RNA (snRNA), small nucleolar RNA (snoRNA), microRNA (miRNA), long noncoding RNA (lncRNA), and so on. Evidence increasingly indicates that ncRNA has a significant effect on cancer and CSC biology and may act as a potential therapeutic target. In this chapter, we will review the functions and mechanisms of ncRNA, mainly miRNA and lncRNA, in regulating cancer cell plasticity (see Fig. 6.1), tumor progression, and CSC biology.

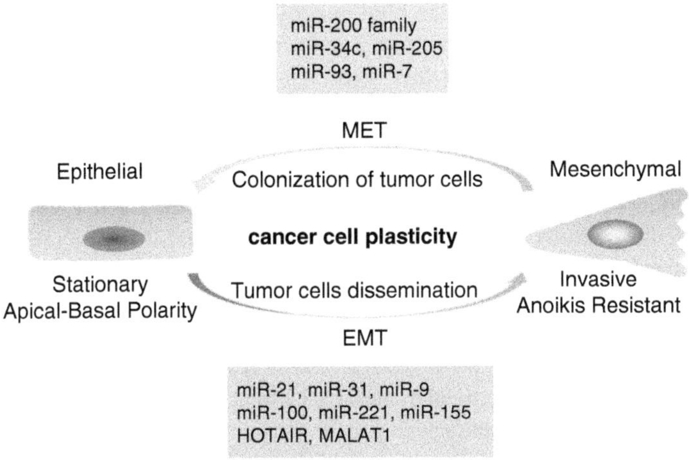

Fig. 6.1 Overview of epithelial-mesenchymal cancer cell plasticity and the involvement of important miRNA and lncRNA

6.2 Cancer Cell Plasticity, Tumor Progression, and Cancer Stem Cell Biology

Most of the cancers diagnosed are primitively derived from normal tissue cells. After a progression of changes at the cellular, genetic, and epigenetic level, the normal cells are ultimately transformed to acquired uncontrolled cell division ability and therefore tumor forming, which process we usually call carcinogenesis or tumorigenesis. Considering the similarities normal stem cells and cancer stem cells share (ability to self-renew and differentiate), there is accumulating evidence that stem cells and progenitor cells may be the targets of transformation during carcinogenesis [9]. Once transformed, cancer cells grow without control to form a mass in primary tissue and develop invasiveness. To break the tissue microenvironment, invade into the local site or even metastasize to distant organ, cancers arising from epithelial tissues need to activate a program called epithelial-mesenchymal transition (EMT) to acquire an invasive phenotype.

EMT is a complex molecular and cellular program, during which polarized epithelial cells lose their epithelial features, including cell-cell adhesion and planar and apical-basal polarity, while acquiring a mesenchymal characteristics, including enhanced motility, invasiveness, and resistance to apoptosis [10, 11]. Abundance of cellular processes and extracellular signals are engaged to initiate and regulate an EMT process, including activation of specific transcription factors, expression of specific cell surface markers, expression of specific microRNA, epithelial cell-stromal cell interaction, hypoxia, cytokines and growth factors derived from tumor environment, and so on. For instance, loss of E-cadherin, a major marker of epithelial cells, is considered essential during EMT. Transcription factors that repress E-cadherin directly or indirectly are supposed to promote EMT. For example, Snail1,

Slug, ZEB1, and SIP1 (ZEB2) can bind directly to E-cadherin promoter to repress its transcription, while Twist1 indirectly represses E-cadherin [12, 13]. Moreover, Onder et al. have disclosed that loss of E-cadherin upregulates Twist1 and ZEB1 expression, and Twist1 is a crucial downstream effector on cellular function. Therefore, they proposed a feed-forward signaling loop between Twist and E-cadherin [14]. Signals from tumor environment could also have a significant effect on EMT. According to Cannito's study, hypoxia can promote EMT via promoting Wnt/β-catenin pathway and therefore resulting Snail translocation. Furthermore, late migration and invasiveness can be sustained in a hypoxia-inducible factor-1α (HIF-1α)-dependent mechanism [15]. Reverse process of EMT is mesenchymal-epithelial transition (MET), characterized by reestablished apical-basal polarity, tight junction, and expression of cell-cell-adhesion molecules such as E-cadherin. MET is often thought to be critical in tumor growth in distant organ, which reendows tumor cells with epithelial characteristics similar to cells in primary tumor.

The evidence of CSC was first described distinctly in 1994 by Lapidot et al. as they found tumorigenic leukemic cancer stem cells and a hierarchical organization in leukemic cells [16]. Since then CSC has been gradually accepted. The CSC hypothesis raises that many human cancers, including breast cancer, colon cancer, liver cancer, glioblastoma, leukemia, pancreas cancer, melanoma, and so on, are driven by a subpopulation of cancer cells that possess stem cell properties. These cells have the ability to self-renew and differentiate into progeny without stemness, therefore driving tumor formation, maintaining the cancer cell pool, mediating metastasis, resistance to therapies, and relapse leading to therapy failure. Numerous studies have showed that, CSCs, a small subset of cancer cells within a tumor, can be identified and isolated by a distinct set of markers. For example, based on cell surface marker expression, Al-Hajj et al. have successfully distinguished the tumorigenic cells from the non-tumorigenic cells in human breast cancer and identified the CSC as $CD44^{+}CD24^{-/low}$ $lineage^{-}$ [17]. Utilizing in vitro and in vivo experimental system, Ginestier and colleagues have found another CSC marker aldehyde dehydrogenase activity 1(ALDH1). They found that, in human breast cancers, cells with high ALDH1 activity displayed tumorigenesis capable of self-renewal and recapitulated the original heterogeneity of the parental tumor [18]. Similar observation was made by Singh in brain cancer when using cell surface marker CD133. $CD133^{+}$ cell fractions are able to initiate tumor in nonobese diabetic, severe combined immunodeficient (NOD-SCID) mice brain [19]. Subsequent studies have found that CD133 highly expression cells also contain tumorigenic cells in colon cancer [20], which indicates that CSCs share something conserved between distinct cancers. What's more, combining different CSC markers identifies a more tumorigenic population. For instance, $ALDH^{+}CD133^{+}$ cells show an increased ability to initiate tumor compared with $ALDH^{+}CD133^{-}$ or $ALDH^{+}$ alone [21].

Studies have found that CSCs are regulated by both cell-intrinsic and cell-extrinsic pathways which are tightly regulated in normal cells. Accumulating evidence indicates that the core signaling pathways, including Wnt, Notch, Hedgehog, PI3K/AKT, etc., which are deregulated in cancer processes and CSCs, critically regulate survival and self-renewal of CSCs. Hedgehog signaling pathway plays a pivotal role in self-renewal and differentiation of normal stem cell and are tightly regulated by the stem cell niche. Deregulation of Hedgehog signaling pathway may play an important role in carcinogenesis, and activation of Hedgehog signaling pathway has been observed in the CSCs [22, 23]. Similarly, Notch and its downstream signaling are also critical in normal tissue stem cells or progenitor cells [24], and there is substantial evidence that abnormal Notch signaling pathway associates with cancer progressions [25, 26]. Wnt signaling pathway, a pivotal regulator of cell-fate decision, has been implicated in a variety of cancers [27–29], including ovarian cancer, breast cancer, non-small cell lung cancer, etc. Beside intracellular signaling pathway mentioned above, tumor environment also has a significant impact on tumor progression and CSCs. Cytokines derived from tumor niche, such as IL-6 and IL-8, have been observed to play a vital role in cancers [30–33].

In the recent decades, several intriguing studies have described a link between the EMT and the CSC. One study has found expression of CD44 is controlled by Wnt/β-catenin cascade [34]. Since CD44 is a marker of CSC, it might imply a role for EMT-related Wnt/β-catenin cascade in CSC maintenance. Mani et al. have observed a direct link between the EMT and the epithelial stem cell properties, and found that the induction of EMT via expression of either Twist or Snail in a non-tumorigenic state immortalizes human mammary epithelial cells (HMLEs) or that via exposure to TGF-β generates $CD44^{high}CD24^{low}$ stem cell-like cells exhibiting not only enhanced ability to form mammospheres, a property correlated with mammary stem cells, but also EMT characteristics such as loss of E-cadherin and expression of Twist, Snail, and N-cadherin. Simultaneously, stem cells isolated from normal human mammary and breast carcinomas express the EMT markers [35]. Most recently, Liu et al. has uncovered the relationship between breast cancer stem cell (BCSC) and EMT. They showed that BCSCs could exist in at least two distinct states, namely, mesenchymal-like (EMT) state and epithelial-like (EMT) state. Moreover, BCSCs in distinct state were diverse in phenotype and function. The EMT state BCSCs, expressing a set of cell surface marker $CD24^{-}CD44^{+}$, were primarily quiescent and localized to the tumor-invasive front, whereas the MET state BCSCs, characterized as $ALDH^{+}$, were proliferative and localized inside of the tumor. They proposed that the plasticity of BCSCs allowed them to undergo reversible EMT/MET transitions, which finally contributed to tumor invasiveness, metastasis, and growth at distant sites. Therefore, it's worthy to note that it may be necessary to target alternative CSC states to achieve a better curative effect [5].

6.3 MicroRNAs Regulate Cancer Cell Plasticity and Tumor Progression

MicroRNA (miRNA) is a 19–23-nucleotide-long noncoding RNA, which functions in gene silence and posttranscriptional regulation of gene expression. The target of miRNA is usually a messenger RNA (mRNA). Via base pairing with the complementary sequences, miRNA represses the translational efficiency or destabilizes the target mRNA and can act on one or more target mRNAs. miRNA has diverse functions in cell biology including cell proliferation, differentiation, and apoptosis. Deregulated miRNAs have been proved to correlate closely with cancers [36, 37]. Depending on the mRNAs they target, miRNAs can be tumor suppressive or oncogenic. As early as 10 years ago, through utilizing the bead-based flow cytometric miRNA expression-profiling method, studies have observed a general downregulation of miRNAs in cancerous tissues compared with normal tissues. Moreover, the miRNA profiles in a way imply the developmental lineage and differentiation state of cancers. Furthermore, poorly differentiated tumors can be successfully classified by miRNA profiles [36]. Researches in breast cancer also indicate a significant deregulated miRNA expression in cancer versus normal tissues [37]. Meanwhile, through miRNA expression-profiling analysis, miRNAs, such as let-7e, miR-151-5p, miR-222, miR-21, miR-155, and miR-221, have been identified to be upregulated in cancerous tissues [38–40]. All of these suggest that we can discriminate cancerous tissues from normal tissues using miRNA profiles, which prompts a potential role for miRNA in cancer diagnosis. Indeed, subsequent studies have revealed that serum miRNA signature could be a useful biomarker for tumor progression and prediction of the outcome of several cancers [41–43].

Accumulating evidence has indicated that miRNA plays a critical role in cancer formation and development [44]. The *c-Myc* oncogene, which has been proved to act as both miRNA inducer and repressor, functions in inducing multiple cancer formation. On one hand, for example, the oncogenic miR-17-92 cluster is frequently found amplified in varieties of human cancers and is regulated by the *MYC* gene in transcriptional level [45–47]. Mu et al. have disclosed that expression of endogenous miR-17-92 is indispensable for suppressing apoptosis in Myc-induced B-cell lymphomas [48]. Via directly suppressing the expression of chromatin regulatory genes Sin3b, Hbp1, Suv420h1, and Btg1 and proapoptotic gene BIM, the MYC-regulated miR-17-92 cluster sustains autonomous proliferation and survival in MYC-induced tumors and therefore maintains the neoplastic state [49]. Besides, there is a network among MYC, miR17-92, and E2F1, a transcription factor that promotes cell cycle progression, in regulation of cell cycle, in which MYC and E2F1 positively regulate each other, while MYC-induced miR-17-92 negatively regulates E2F1 [45, 50]. Another important downstream target of miR-17-92 is the tumor suppressor PTEN. On one hand, c-Myc can mediate cell proliferation and apoptosis resistance in non-small cell lung cancer (NSCLC) cells through suppressing PTEN by miR-17-92 [51]. On the other hand, which is more often, MYC represses dozens of tumor-suppressive miRNAs including Let-7, miR-23, miR-34a, and so on [52]. Let-7, for instance, cooperates with an RNA-binding protein HuR to inhibit the expression of c-Myc in an interdependent manner. HuR represses the MYC oncogene by recruiting the let-7-loaded RNA-induced silencing complex (RISC) to the 3′-untranslated region (UTR) of *c-Myc* [53]. MYC can bind to the let-7 promoter, while there are let-7 binding sites in MYC 3′UTR. Hence, there exists a negative loop between let-7 and MYC. It is reported that multiple genes regulating cell cycle, proliferation and apoptosis, are responsive to the alteration of let-7. The major targets of let-7 are *RAS* and *HMGA2* oncogenes. Johnson et al. have suggested that let-7 may function as a tumor suppressor through acting on the *RAS* oncogene, for there are multiple let-7 complementary sites in the *RAS* 3′UTR [54]. Similarly, Lee et al. have found that let-7 destabilizes HMGA2, a high-mobility group protein, via multiple target sites in 3′UTR of *HMGA2* oncogene, to repress cell proliferation [55]. Deregulation of let-7 is generally found in cancer tissues, suggesting that let-7 is poorly expressed in cancer tissues compared with normal tissues [56].

Since aberrant activation of EMT triggers malignant tumor progression, a large amount of evidence has proved an miRNA network in regulating EMT process. Much work has observed an EMT in cancer cells in response to transforming growth factor (TGF)-β. TGF-β seems to play a dominant role in EMT in advanced cancer via directly activating transcription factors ZEB, Snail, and Twist [10]. Studies have identified a number of miRNAs that possibly take part in TGF-β-induced EMT pathway, including miR-200, miR-21, miR-31, and so on. It is reported that expression level of miR-21 and miR-31 is significantly increased in response to TGF-β stimulation. MiR-21 and miR-31 synergize with TGF-β to enhance the EMT by targeting TIAM1, a guanidine exchange factor of the Rac GTPase [57].

The miR-200 family contains miR-200a, miR-200b, miR-200c, miR-141, and miR-429. There is growing evidence suggesting that miR-200 participates in tumor metastasis via regulating EMT. All five members of miR-200 family have been found to markedly decrease in cells that have undergone EMT induced by TGF-β and in invasive breast cancer cell lines, which is reported to depend on the SMAD signaling pathway [58, 59]. The major targets of miR-200 are ZEB1 and SIP1, the E-cadherin transcription repressors [58]. Through directly repressing mRNA of ZEB1 and SIP1, miR-200 maintains the expression of E-cadherin and the epithelial morphology. Meanwhile, in mesenchymal cells, ZEB1 and SIP1 act as repressors of miR-200 through binding to a conserved site at the miR-200 promoter region [60], hence forming a reciprocal negative feedback loop between miR-200 and ZEB1/SIP1. Brabletz et al. have also found that the miR-200-ZEB1 feedback loop controls the Notch signaling in cancer cells, especially in poorly differentiated invasive tumor cells [61]. The Notch signaling pathway component, such as Jagged 1, Maml2, and Maml3, is also one of the miR-200 targets. Via inhibiting miR-200, ZEB1 upregulates the Notch signaling, contributing to cancer cells properties. Similarly, Yang et al. have discovered a miR-200-dependent pathway in the Notch-induced EMT [62]. The Notch ligand Jagged 2 has been found to upregulate the expression of GATA-binding factors, which in turn suppress the miR-200, thus promoting the EMT and metastasis. Furthermore, study in lung cancer has demonstrated that ZEB1 shows altered expression level in erlotinib-sensitive cancer cells and that ectopic expression of miR-200c can alter the drug sensitivity [63], suggesting that miR-200-ZEB1 feedback loop might be a potent target for cancer therapy.

6.4 MicroRNAs Play Important Roles in Regulating Cancer Stem Cells

The tumor consists of heterogeneous cells, in which cancer stem cells (CSCs), with the ability to self-renew and differentiate, are thought to be the driving force of the tumor development, therapy resistance, and recurrence. The CSCs have been found in a wide spectrum of cancer types, among which breast cancer stem cell (BCSC) is the first described CSC in solid tumor and also the best characterized CSC so far. Numerous extracellular factors and intracellular elements have been uncovered to regulate the CSCs properties, among which microRNAs have been validated to play a key role in regulating CSC properties [64]. Through comparing miRNA profiles, several miRNA clusters, such as miR-200c-141, miR-200b-200a-429, and miR-183-96-182, have been found to differentially express between BCSCs and non-tumorigenic bulk tumor cells [65]. According to Liu et al., the BCSC subpopulation is likewise heterogeneous containing distinct groups characterized by different markers, such as $ALDH^+$ and $CD24^-CD44^+$. They have pointed out that BCSCs may exist in at least two alternative states (EMT and MET) on the basis of the CSC markers they express. The mesenchymal-like (EMT) BCSC represents the $CD24^-CD44^+$ subpopulation and is characterized as primarily quiescence and invasive marginal location. The $ALDH^+$ subpopulation is described as epithelial-like (MET) BCSCs, which are proliferative and located centrally. Moreover, they have proposed that BCSCs may transition between the EMT and MET states to achieve tumor invasion, dissemination, and growth at distant organs [5]. Interconversion of EMT state and MET state is regulated by the microRNA network in that miR-9, miR-100, miR-221, and miR-155 induce the EMT state, while miR-34c, miR-200, miR-205, and miR-93 induce the MET state [66, 67].

More recently, several intriguing studies have described the role of miRNAs in modulating the cell fate of CSC. Through evaluating expression level in different-stage breast cancer samples, miR-9 has been found to overexpress in late-stage tumors with aggressive phenotypes and associate with a $CD24^-CD44^+$ phenotype and EMT [68]. Via repressing the CSC regulatory gene SMADCA5, SMADCD1, and BMPR2, miR-100 has been found to directly regulate self-renewal and differentiation of BCSCs and reduce the $ALDH^+$ population [69]. Depending on the differentiation states, miRNA seems to have different impact on the CSCs. In the less differentiated breast cancer cells, miR-93 can induce MET accompanied by decreased expression of TGF-β and numerous stem cell regulatory genes, such as JAK1, STAT3, AKT3, SOX4, EZH1, and HMGA2, thus reducing the CSC population. However, the CSC subpopulation is increased as miR-93 expresses in a more differentiated breast cancer cells [70]. Another miRNA, let-7, has been reported as well to regulate multiple properties of BCSCs through its target gene H-RAS and HMGA2. It is interesting to point out that, via repressing H-RAS, let-7 reduces CSC self-renewal while having no effect on its differentiation. Via repressing HMGA2, let-7 enhances CSC differentiation while having no effect on its self-renewal [71]. In addition, let-7 repression seems to promote the BCSC expansion by the Wnt/β-catenin pathway through transactivation of a negative let-7 biogenesis regulator Lin28 [72]. Similar with let-7, miR-30 has also been found to regulate the self-renewal and apoptosis of BCSCs via its target Ubc9 and ITGB3, respectively [73]. More recently, a signaling axis involving Snail, miR-146a, and Numb has been identified in regulating symmetric and asymmetric cell division in colorectal cancer stem cells [74]. All these demonstrate that miRNA could be both repressor and promoter in regulating CSC properties, implying a more accurate miRNA-targeted therapy according to distinct tumor differentiation states.

Decreased expression of miR-200 family members has been proved to be important in tumor metastasis, apoptosis resistance, and drug resistance. The major targets of miR-200 are ZEB1 and SIP1. Of note here, stem cell factors, such as Sox2 and Klf4, are also candidate targets of miR-200 family members, suggesting a link between miR-200 and stem cells. Recently, molecular links between miR-200 and CSCs have drawn particular attention. Lim's study has found that expression of miR-200 gradually loses in a non-stem human mammary epithelial (HMLE) cells during its transition to a stem-like phenotype and that the restoration of its expression can promote MET and decrease the stem cell properties. Similar phenomenon has been observed in the stem cell fractions of metastatic breast cancer. Furthermore, their research has uncovered an epigenetic modification mechanism of miR-200. According to Lim, miR-200 is repressed through polycomb-group-mediated histone modifications and DNA methylation [75]. Shimono et al. has disclosed that miR-200c is downregulated in human BCSCs and its expression can inhibit the growth of breast cancer cell and breast tumor formation driven by BCSCs in vivo [65], illustrating an indispensable role of miR-200c in BCSCs. Besides, it has been reported that the polycomb repressor Bmi1, a key regulator of cancer stem cell self-renewal, is repressed by miR-200 [65, 76]. Research in melanoma has also revealed that miR-200c overexpression leads to diminished expression of Bmi1, as well as melanoma tumor growth and metastasis inhibition [77]. In addition, the miR-200-ZEB1 reciprocal negative feedback loop not only promotes tumor cells dissemination but also regulates tumor-initiating cells in pancreatic and colorectal cancer cells. Through repressing the stemness-inhibiting miRNA, such as miR-200 and miR20-203, ZEB1 connects EMT activation with stemness maintenance [76]. These results suggest that miR-200 and miR-200-ZEB1 negative loop may be critical targets for CSC-targeted therapies.

6.5 Long Noncoding RNAs (LncRNAs) in Cancer Biology and Regulating Cancer Cell Plasticity

With the advance of high-resolution microarray and genome-wide sequencing technology, a large amount of novel transcripts have been found. It is reported that about 70 % of the genome is actively transcribed [78]. Of note here, noncoding RNA has drawn particular attention. Long noncoding RNA (lncRNA) is a new class of noncoding RNAs, with length ranged from 200 bp to 100 kbp, representing a large population of the noncoding RNA. Although there are thousands of lncRNA, it is the least well understood, and the vast majority of it is functionally uncharacterized [79]. Still, recent studies have gradually discovered one of lncRNA's roles as the driver for tumor-suppressive or oncogenic function in multiple cancer types, including breast cancer, prostate cancer, pancreatic cancer, hepatocellular cancer, and so on [80, 81]. Expression of lncRNA is often altered and deregulated in tumors [82].

The HOX antisense intergenic RNA (HOTAIR), one of the best well-known lncRNAs, is a 2.2 kbp lncRNA molecule located on chromosome 12q13.13. HOTAIR is reported to be upregulated in many prevalent human cancers. There is substantial evidence that HOTAIR takes part in carcinogenesis, metastatic dissemination, and drug resistance. Expression of HOTAIR is an important prognostic marker of many cancers [83, 84]. Li et al. have discovered that high expression level of HOTAIR in laryngeal squamous cell carcinoma promotes methylation of the tumor suppressor PTEN, resulting to tumorigenesis [85]. While suppressed expression of HOTAIR could inhibit tumorigenesis and tumor proliferation [86], elevated expression level of HOTAIR was found in primary breast tumor and metastases, with its expression level in primary tumor as a powerful predictor of poor prognosis. HOTAIR increased the tumor invasiveness and metastasis in a polycomb repressive complex 2 (PRC2)-dependent manner [87]. Besides, the depletion of HOTAIR was found to associate with increased expression of E-cadherin and decreased expression of vimentin [86]. In clinical specimens of gastric cancer and colon cancer, HOTAIR inhibition was found to reverse the EMT process [88, 89]. All of these suggest that HOTAIR may act as an EMT modulator. In addition, recent research has uncovered that HOTAIR contributes to cisplatin resistance through downregulation of p21 expression [90]. Furthermore, substantial evidence indicates that HOTAIR may take part in the CSC regulation. It was found to be expressed at a much higher level in the colon CSC subpopulation ($CD133^{+}CD44^{+}$) compared with the other non-stem cancer cells. And knockdown of HOTAIR by siRNA correlated with a decreased colony forming capacity of colon and breast cancer cells [91]. These results suggest that HOTAIR may be an important regulator of cancer cell plasticity and a valuable predictor of tumor progression. HOTAIR inhibition may be a potential option for cancer prevention and CSC-targeted therapies.

Another lncRNA associated tightly with tumorigenesis is antisense noncoding RNA in the INK4 locus (ANRIL). Similar with HOTAIR, through binding to and recruiting PCR2, ANRIL represses the expression of $p15^{INK4B}$ locus, which encodes a tumor suppressor $p15^{INK4B}$ and has a pivotal role in cell cycle inhibition, senescence, and stress-induced apoptosis [92]. Besides, Nie et al. have uncovered another mechanism of ANRIL repression of p15 via silencing of KLF2 and P21 transcription [93]. These observations suggest that one of lncRNA's mechanisms in mediating tumorigenesis may be through silencing of tumor suppressor genes.

The metastasis-associated lung adenocarcinoma transcript 1 (lncRNA MALAT1) is widely expressed in normal organs, such as lung and pancreas, and upregulated in several cancer types [94–96]. Through comparing metastatic and nonmetastatic early stage non-small cell lung cancer (NSCLC) tumors, MALAT1 was first demonstrated to be significantly associated with high metastatic potential and poor patient prognosis [97]. Consequently, a number of studies began to investigate the correlation of MALAT1 and metastasis. MALAT1 was then proved to be a regulator of numerous metastasis-associated gene expressions in lung cancer [96]. Through siRNA-mediated silencing of MALAT1 in bladder cancer cells, Ying et al. have found a decreased level in EMT-associated transcription factors, such as ZEB1, SIP1, and Slug, and an increased level of E-cadherin. They further demonstrated that MALAT1 promoted EMT in a Wnt signaling pathway activation-dependent manner [98]. MALAT1 was found to function in regulating the TGF-β-induced EMT [99]. In addition, it could promote tumor growth and metastasis in osteosarcoma by activating the PI3K-AKT pathway [100]. These results indicate that MALAT1 acts as a novel EMT regulator and may be a potential therapeutic target for cancer metastasis.

Besides oncogenic lncRNA, such as HOTAIR, ANRIL, and MALAT1, there are some lncRNAs acting as tumor suppressors. For instance, lincRNA-p21, a 3.1 kbp transcript located in the proximity of the cell cycle regulator gene Cdkn1a, via physically associating with heterogeneous nuclear ribonucleoprotein K (hnRNP-K), mediates transcription repression in the p53 pathway to regulate hundreds of p53 downstream target genes and triggers apoptosis [101]. Another tumor-suppressive lncRNA is GAS5, growth arrest-specific 5, which is abundant in cells with arrested growth owing to nutrients lacking, influences the cell survival and metabolism by modulating the transcriptional activity of the glucocorticoid receptor [102]. Significant reduction of GAS5 level has been observed in breast cancer tissues relative to corresponding adjacent normal tissues [103], partially explaining cancer cell survival in nutrient-lacking environment. Collectively, these studies show that tumor-suppressive lncRNA may play critical roles in cancer biology, but the underlying mechanisms still require much more exploration.

Since both miRNA and lncRNA are critical regulators of cancer, the interaction of miRNA and lncRNA in regulating cancer properties has drawn much attention. Recently, increasing studies have described the interaction between miRNA and lncRNA. LncRNA and miRNA may act as either decoy or decay reciprocally. The well-known lncRNA HOTAIR, on one hand, was found to suppress the tumor suppressor miR-7, by modulating the expression of HoxD10, and therefore sustain the expression of C-myc, Twist, and miR-9, hence maintaining the EMT process and the CSC pool of breast cancer [104]. On the other hand, expression of HOTAIR was repressed by miR-34a via directly binding to HOTAIR mRNA sequence in prostate cancer cells [105]. Liu et al. have also found a reciprocal suppressive relation between the p53-regulated tumor suppressor loc285194 and the tumor promoter miR-211. The tumor growth inhibition mediated by loc285194 was in part due to loc285194-specific suppression of miR-211 [106]. Besides, lncRNA and miRNA might synergize with each other. Studies in embryo development have discovered that miR-675 is embedded in the lncRNA *H19*'s first exon. By controlling the release of miR-675, H19 limited the growth of placenta before birth [107].

However, whether H19 and miR-675 function in the same way in tumor, we need more exploration.

6.6 Conclusions and Perspectives

Many human tumors consist of heterogeneous components, among which cancer stem cells possess the ability to self-renew and differentiate into the non-tumorigenic progeny, therefore driving tumorigenesis, proliferation, metastatic dissemination, and drug resistance. The cancer cell plasticity enables tumor to transition between distinct morphologies and proliferate at primary or distant sites. EMT is the most important molecular mechanism of cancer cell plasticity, either in non-tumorigenic cells or in cancer stem cells. Accumulating evidence indicates that noncoding RNA plays a vital role in regulating cancer and cancer stem cell biology. Alteration in miRNA expression alone has been found to cause neoplasm [108, 109]. Herein, we have discussed that noncoding RNAs, mainly miRNA and lncRNA, may act as oncogenic or tumor suppressors in cancer formation and progression and cancer stem cell biology. There definitely exists an ncRNA network in controlling cancer cell plasticity and CSC transition through regulating EMT-associated genes and relevant signaling pathways. Interaction of miRNA and lncRNA seems to play an important role in cancer and CSC properties, prompting a potential therapeutic method by targeting both miRNAs and lncRNAs, correlative oncogenes, and signaling pathways. However, as there are large amounts of ncRNAs, many underlying mechanisms of their interactions and physiological and pathological roles are still undiscovered.

References

1. Perou CM, Sørlie T, Eisen MB, et al. Molecular portraits of human breast tumours. Nature. 2000;406(6797):747–52.
2. Sorlie T, Perou CM, Tibshirani R, et al. Gene expression patterns of breast carcinomas distinguish tumor subclasses with clinical implications. Proc Natl Acad Sci U S A. 2001;98(19):10869–74.
3. Dick JE. Breast cancer stem cells revealed. Proc Natl Acad Sci. 2003;100(7):3547–9.
4. Thiery JP, Acloque H, Huang RY, Nieto MA. Epithelial-mesenchymal transitions in development and disease. Cell. 2009;139(5):871–90.
5. Liu S, Cong Y, Wang D, et al. Breast cancer stem cells transition between epithelial and mesenchymal states reflective of their normal counterparts. Stem Cell Rep. 2014;2(1):78–91.
6. Gupta PB, Fillmore CM, Jiang G, et al. Stochastic state transitions give rise to phenotypic equilibrium in populations of cancer cells. Cell. 2011;146(4):633–44.
7. Sharma SV, Lee DY, Li B, et al. A chromatin-mediated reversible drug-tolerant state in cancer cell subpopulations. Cell. 2010;141(1):69–80.
8. Roesch A, Fukunaga-Kalabis M, Schmidt EC, et al. A temporarily distinct subpopulation of slow-cycling melanoma cells is required for continuous tumor growth. Cell. 2010;141(4):583–94.

9. Perou CM. Molecular stratification of triple-negative breast cancers. Oncologist. 2011;16 Suppl 1:61–70.
10. Kalluri R, Weinberg RA. The basics of epithelial-mesenchymal transition. J Clin Invest. 2009;119(6):1420.
11. Polyak K, Weinberg RA. Transitions between epithelial and mesenchymal states: acquisition of malignant and stem cell traits. Nat Rev Cancer. 2009;9(4):265–73.
12. Yang J, Weinberg RA. Epithelial-mesenchymal transition: at the crossroads of development and tumor metastasis. Dev Cell. 2008;14(6):818–29.
13. Peinado H, Olmeda D, Cano A. Snail, Zeb and bHLH factors in tumour progression: an alliance against the epithelial phenotype? Nat Rev Cancer. 2007;7(6):415–28.
14. Onder TT, Gupta PB, Mani SA, et al. Loss of E-cadherin promotes metastasis via multiple downstream transcriptional pathways. Cancer Res. 2008;68(10):3645–54.
15. Cannito S, Novo E, Compagnone A, et al. Redox mechanisms switch on hypoxia-dependent epithelial-mesenchymal transition in cancer cells. Carcinogenesis. 2008;29(12):2267–78.
16. Caceres-Cortes J, Mindeni M, Patersoni B, Caligiuri MA. A cell initiating human acute myeloid leukaemia after transplantation into SCID mice. Nature. 1994;367:17.
17. Al-Hajj M, Wicha MS, Benito-Hernandez A, et al. Prospective identification of tumorigenic breast cancer cells. Proc Natl Acad Sci. 2003;100(7):3983–8.
18. Ginestier C, Hur MH, Charafe-Jauffret E, et al. ALDH1 is a marker of normal and malignant human mammary stem cells and a predictor of poor clinical outcome. Cell Stem Cell. 2007;1(5):555–67.
19. Singh SK, Hawkins C, Clarke ID, et al. Identification of human brain tumour initiating cells. Nature. 2004;432(7015):396–401.
20. Ricci-Vitiani L, Lombardi DG, Pilozzi E, et al. Identification and expansion of human colon-cancer-initiating cells. Nature. 2007;445(7123):111–5.
21. Huang EH, Hynes MJ, Zhang T, et al. Aldehyde dehydrogenase 1 is a marker for normal and malignant human colonic stem cells (SC) and tracks SC overpopulation during colon tumorigenesis. Cancer Res. 2009;69(8):3382–9.
22. Liu S, Dontu G, Mantle ID, et al. Hedgehog signaling and Bmi-1 regulate self-renewal of normal and malignant human mammary stem cells. Cancer Res. 2006;66(12):6063–71.
23. di Magliano MP, Hebrok M. Hedgehog signalling in cancer formation and maintenance. Nat Rev Cancer. 2003;3(12):903–11.
24. Dontu G, Jackson KW, McNicholas E, et al. Role of Notch signaling in cell-fate determination of human mammary stem/progenitor cells. Breast Cancer Res. 2004;6(6):R605–15.
25. Radtke F, Raj K. The role of Notch in tumorigenesis: oncogene or tumour suppressor? Nat Rev Cancer. 2003;3(10):756–67.
26. Roy M, Pear WS, Aster JC. The multifaceted role of Notch in cancer. Curr Opin Genet Dev. 2007;17(1):52–9.
27. Arend RC, Londoño-Joshi AI, Straughn JM, Buchsbaum DJ. The Wnt/β-catenin pathway in ovarian cancer: a review. Gynecol Oncol. 2013;131(3):772–9.
28. Loh YN, Hedditch EL, Baker LA, et al. The Wnt signalling pathway is up-regulated in an in vitro model of acquired tamoxifen resistant breast cancer. BMC Cancer. 2013;13(1):174.
29. Bravo DT, Yang Y-L, Kuchenbecker K, et al. Frizzled-8 receptor is activated by the Wnt-2 ligand in non-small cell lung cancer. BMC Cancer. 2013;13(1):316.
30. Korkaya H, G-i K, Davis A, et al. Activation of an IL6 inflammatory loop mediates trastuzumab resistance in HER2+ breast cancer by expanding the cancer stem cell population. Mol Cell. 2012;47(4):570–84.
31. Inoue K, Slaton JW, Eve BY, et al. Interleukin 8 expression regulates tumorigenicity and metastases in androgen-independent prostate cancer. Clin Cancer Res. 2000;6(5):2104–19.
32. Ginestier C, Liu S, Diebel ME, et al. CXCR1 blockade selectively targets human breast cancer stem cells in vitro and in xenografts. J Clin Invest. 2010;120(2):485.
33. Liu S, Ginestier C, Ou SJ, et al. Breast cancer stem cells are regulated by mesenchymal stem cells through cytokine networks. Cancer Res. 2011;71(2):614–24.

34. Wielenga VJ, Smits R, Korinek V, et al. Expression of CD44 in Apc and Tcf mutant mice implies regulation by the WNT pathway. Am J Pathol. 1999;154(2):515–23.
35. Mani SA, Guo W, Liao M-J, et al. The epithelial-mesenchymal transition generates cells with properties of stem cells. Cell. 2008;133(4):704–15.
36. Lu J, Getz G, Miska EA, et al. MicroRNA expression profiles classify human cancers. Nature. 2005;435(7043):834–8.
37. Iorio MV, Ferracin M, Liu C-G, et al. MicroRNA gene expression deregulation in human breast cancer. Cancer Res. 2005;65(16):7065–70.
38. Yu J, Li A, Hong S-M, et al. MicroRNA alterations of pancreatic intraepithelial neoplasias. Clin Cancer Res. 2012;18(4):981–92.
39. Wang J, Sen S. MicroRNA functional network in pancreatic cancer: from biology to biomarkers of disease. J Biosci. 2011;36(3):481–91.
40. Yu F, Jiao Y, Zhu Y, et al. MicroRNA 34c gene down-regulation via DNA methylation promotes self-renewal and epithelial-mesenchymal transition in breast tumor-initiating cells. J Biol Chem. 2012;287(1):465–73.
41. Keller A, Leidinger P, Gislefoss R, et al. Stable serum miRNA profiles as potential tool for non-invasive lung cancer diagnosis. RNA Biol. 2011;8(3):506–16.
42. Hu Z, Chen X, Zhao Y, et al. Serum MicroRNA signatures identified in a genome-wide serum MicroRNA expression profiling predict survival of non-small-cell lung cancer. J Clin Oncol. 2010;28(10):1721–6.
43. Schultz NA, Dehlendorff C, Jensen BV, et al. MicroRNA biomarkers in whole blood for detection of pancreatic cancer. JAMA. 2014;311(4):392–404.
44. Malek NP. Another myc in the wall: MicroRNA-101 controls important functions in liver cancer formation. Hepatology. 2014;59(5):1676–7.
45. Diosdado B, Van De Wiel M, et al. MiR-17-92 cluster is associated with 13q gain and c-myc expression during colorectal adenoma to adenocarcinoma progression. Br J Cancer. 2009;101(4):707–14.
46. Xiao C, Srinivasan L, Calado DP, et al. Lymphoproliferative disease and autoimmunity in mice with increased miR-17-92 expression in lymphocytes. Nat Immunol. 2008;9(4):405–14.
47. Hayashita Y, Osada H, Tatematsu Y, et al. A polycistronic microRNA cluster, miR-17-92, is overexpressed in human lung cancers and enhances cell proliferation. Cancer Res. 2005;65(21):9628–32.
48. Mu P, Han Y-C, Betel D, et al. Genetic dissection of the miR-17~ 92 cluster of microRNAs in Myc-induced B-cell lymphomas. Genes Dev. 2009;23(24):2806–11.
49. Li Y, Choi PS, Casey SC, et al. MYC through miR-17-92 suppresses specific target genes to maintain survival, autonomous proliferation, and a neoplastic state. Cancer Cell. 2014;26(2):262–72.
50. O'Donnell KA, Wentzel EA, Zeller KI, et al. c-Myc-regulated microRNAs modulate E2F1 expression. Nature. 2005;435(7043):839–43.
51. Krysan K, Kusko R, Grogan T, et al. PGE2-driven expression of c-Myc and oncomiR-17-92 contributes to apoptosis resistance in NSCLC. Mol Cancer Res. 2014;12(5):765–74.
52. Dang CV. MYC on the path to cancer. Cell. 2012;149(1):22–35.
53. Kim HH, Kuwano Y, Srikantan S, et al. HuR recruits let-7/RISC to repress c-Myc expression. Genes Dev. 2009;23(15):1743–8.
54. Johnson SM, Grosshans H, Shingara J, et al. RAS is regulated by the let-7 microRNA family. Cell. 2005;120(5):635–47.
55. Lee YS, Dutta A. The tumor suppressor microRNA let-7 represses the HMGA2 oncogene. Genes Dev. 2007;21(9):1025–30.
56. Takamizawa J, Konishi H, Yanagisawa K, et al. Reduced expression of the let-7 microRNAs in human lung cancers in association with shortened postoperative survival. Cancer Res. 2004;64(11):3753–6.
57. Cottonham CL, Kaneko S, Xu L. miR-21 and miR-31 converge on TIAM1 to regulate migration and invasion of colon carcinoma cells. J Biol Chem. 2010;285(46):35293–302.

58. Gregory PA, Bert AG, Paterson EL, et al. The miR-200 family and miR-205 regulate epithelial to mesenchymal transition by targeting ZEB1 and SIP1. Nat Cell Biol. 2008;10(5):593–601.
59. Xiong M, Jiang L, Zhou Y, et al. The miR-200 family regulates TGF-β1-induced renal tubular epithelial to mesenchymal transition through Smad pathway by targeting ZEB1 and ZEB2 expression. AM J Physiol-Ren Physiol. 2012;302(3):F369–79.
60. Bracken CP, Gregory PA, Kolesnikoff N, et al. A double-negative feedback loop between ZEB1-SIP1 and the microRNA-200 family regulates epithelial-mesenchymal transition. Cancer Res. 2008;68(19):7846–54.
61. Brabletz S, Bajdak K, Meidhof S, et al. The ZEB1/miR-200 feedback loop controls Notch signalling in cancer cells. EMBO J. 2011;30(4):770–82.
62. Yang Y, Ahn Y-H, Gibbons DL, et al. The Notch ligand Jagged2 promotes lung adenocarcinoma metastasis through a miR-200-dependent pathway in mice. J Clin Invest. 2011;121(4):1373.
63. Bryant J, Britson J, Balko J, et al. A microRNA gene expression signature predicts response to erlotinib in epithelial cancer cell lines and targets EMT. Br J Cancer. 2012;106(1):148–56.
64. Liu C, Kelnar K, Liu B, et al. The microRNA miR-34a inhibits prostate cancer stem cells and metastasis by directly repressing CD44. Nat Med. 2011;17(2):211–5.
65. Shimono Y, Zabala M, Cho RW, et al. Down-regulation of miRNA-200c links breast cancer stem cells with normal stem cells. Cell. 2009;138(3):592–603.
66. Yu S, Liu Y, Wang J, Guo Z, Zhang Q, Yu F, et al. Circulating microRNA profiles as potential biomarkers for diagnosis of papillary thyroid carcinoma. J Clin Endocrinol Metab. 2012;97(6):2084–92.
67. Liu S, Clouthier SG, Wicha MS. Role of microRNAs in the regulation of breast cancer stem cells. J Mammary Gland Biol Neoplasia. 2012;17(1):15–21.
68. Gwak JM, Kim HJ, Kim EJ, et al. MicroRNA-9 is associated with epithelial-mesenchymal transition, breast cancer stem cell phenotype, and tumor progression in breast cancer. Breast Cancer Res Treat. 2014;147(1):39–49.
69. Deng L, Shang L, Bai S, et al. MicroRNA100 inhibits self-renewal of breast cancer stem-like cells and breast tumor development. Cancer Res. 2014;74(22):6648–60.
70. Liu S, Patel SH, Ginestier C, et al. MicroRNA93 regulates proliferation and differentiation of normal and malignant breast stem cells. PLoS Genet. 2012;8(6):e1002751.
71. Yu F, Yao H, Zhu P, et al. let-7 regulates self renewal and tumorigenicity of breast cancer cells. Cell. 2007;131(6):1109–23.
72. Cai W-Y, Wei T-Z, Luo Q-C, et al. The Wnt-β-catenin pathway represses let-7 microRNA expression through transactivation of Lin28 to augment breast cancer stem cell expansion. J Cell Sci. 2013;126(13):2877–89.
73. Yu F, Deng H, Yao H, et al. Mir-30 reduction maintains self-renewal and inhibits apoptosis in breast tumor-initiating cells. Oncogene. 2010;29(29):4194–204.
74. Lerner RG, Petritsch C. A microRNA-operated switch of asymmetric-to-symmetric cancer stem cell divisions. Nat Cell Biol. 2014;16(3):212–4.
75. Lim Y-Y, Wright JA, Attema JL, et al. Epigenetic modulation of the miR-200 family is associated with transition to a breast cancer stem-cell-like state. J Cell Sci. 2013;126(10):2256–66.
76. Wellner U, Schubert J, Burk UC, et al. The EMT-activator ZEB1 promotes tumorigenicity by repressing stemness-inhibiting microRNAs. Nat Cell Biol. 2009;11(12):1487–95.
77. Liu S, Tetzlaff MT, Cui R, Xu X. miR-200c inhibits melanoma progression and drug resistance through down-regulation of BMI-1. Am J Pathol. 2012;181(5):1823–35.
78. Djebali S, Davis CA, Merkel A, et al. Landscape of transcription in human cells. Nature. 2012;489(7414):101–8.
79. Guttman M, Amit I, Garber M, et al. Chromatin signature reveals over a thousand highly conserved large non-coding RNAs in mammals. Nature. 2009;458(7235):223–7.

80. Cheetham S, Gruhl F, Mattick J, Dinger M. Long non-coding RNAs and the genetics of cancer. Br J Cancer. 2013;108(12):2419–25.
81. Prensner JR, Chinnaiyan AM. The emergence of lncRNAs in cancer biology. Cancer Discov. 2011;1(5):391–407.
82. Calin GA, C-g L, Ferracin M, et al. Ultraconserved regions encoding ncRNAs are altered in human leukemias and carcinomas. Cancer Cell. 2007;12(3):215–29.
83. Nie Y, Liu X, Qu S, et al. Long non-coding RNA HOTAIR is an independent prognostic marker for nasopharyngeal carcinoma progression and survival. Cancer Sci. 2013;104(4):458–64.
84. Lv X-B, Lian G-Y, Wang H-R, et al. Long non-coding RNA HOTAIR is a prognostic marker for esophageal squamous cell carcinoma progression and survival. PLoS One. 2013;8(5):e63516. doi:10.1371/journal.pone.0063516.
85. Li D, Feng J, Wu T, et al. Long intergenic non-coding RNA HOTAIR is overexpressed and regulates PTEN methylation in laryngeal squamous cell carcinoma. Am J Pathol. 2013;182(1):64–70.
86. Wu Y, Liu J, Zheng Y, et al. Suppressed expression of long non-coding RNA HOTAIR inhibits proliferation and tumourigenicity of renal carcinoma cells. Tumor Biol. 2014;35(12):11887–94.
87. Gupta RA, Shah N, Wang KC, et al. Long non-coding RNA HOTAIR reprograms chromatin state to promote cancer metastasis. Nature. 2010;464(7291):1071–6.
88. Wu Z-H, Wang X-L, Tang H-M, et al. Long non-coding RNA HOTAIR is a powerful predictor of metastasis and poor prognosis and is associated with epithelial-mesenchymal transition in colon cancer. Oncol Rep. 2014;32(1):395–402.
89. Xu Z-Y, Yu Q-M, Du Y-A, et al. Knockdown of long non-coding RNA HOTAIR suppresses tumor invasion and reverses epithelial-mesenchymal transition in gastric cancer. Int J Biol Sci. 2013;9(6):587.
90. Liu Z, Sun M, Lu K, et al. The long non-coding RNA HOTAIR contributes to cisplatin resistance of human lung adenocarcinoma cells via downregulation of p21 (WAF1/CIP1) expression. PLoS One. 2013;8(10):e77293.
91. Alves CP, Fonseca AS, Muys BR, et al. The lincRNA Hotair is required for epithelial-to-mesenchymal transition and stemness maintenance of cancer cells lines. Stem Cells. 2013;31(12):2827–32.
92. Kotake Y, Nakagawa T, Kitagawa K, et al. Long non-coding RNA ANRIL is required for the PRC2 recruitment to and silencing of p15INK4B tumor suppressor gene. Oncogene. 2011;30(16):1956–62.
93. F-q N, Sun M, Yang J-s, et al. Long non-coding RNA ANRIL promotes non-small cell lung cancer cell proliferation and inhibits apoptosis by silencing KLF2 and P21 expression. Mol Cancer Ther. 2015;14(1):268–77.
94. Schmidt LH, Spieker T, Koschmieder S, et al. The long non-coding MALAT-1 RNA indicates a poor prognosis in non-small cell lung cancer and induces migration and tumor growth. J Thorac Oncol. 2011;6(12):1984–92.
95. Jiao F, Hu H, Yuan C, et al. Elevated expression level of long non-coding RNA MALAT-1 facilitates cell growth, migration and invasion in pancreatic cancer. Oncol Rep. 2014;32(6):2485–92.
96. Gutschner T, Hämmerle M, Eißmann M, et al. The non-coding RNA MALAT1 is a critical regulator of the metastasis phenotype of lung cancer cells. Cancer Res. 2013;73(3):1180–9.
97. Ji P, Diederichs S, Wang W, et al. MALAT-1, a novel non-coding RNA, and thymosin β4 predict metastasis and survival in early-stage non-small cell lung cancer. Oncogene. 2003;22(39):8031–41.
98. Ying L, Chen Q, Wang Y, et al. Up-regulated MALAT-1 contributes to bladder cancer cell migration by inducing epithelial-to-mesenchymal transition. Mol Biosyst. 2012;8(9):2289–94.
99. Fan Y, Shen B, Tan M, et al. TGF-β-induced up-regulation of malat1 promotes bladder cancer metastasis by associating with suz12. Clin Cancer Res. 2014;20(6):1531–41.

100. Dong Y, Liang G, Yuan B, et al. MALAT1 promotes the proliferation and metastasis of osteosarcoma cells by activating the PI3K/Akt pathway. Tumor Biol. 2014;36(3):1477–86.
101. Huarte M, Guttman M, Feldser D, et al. A large intergenic non-coding RNA induced by p53 mediates global gene repression in the p53 response. Cell. 2010;142(3):409–19.
102. Kino T, Hurt DE, Ichijo T, et al. Non-coding RNA Gas5 is a growth arrest and starvation-associated repressor of the glucocorticoid receptor. Sci Signal. 2010;3(107):ra8.
103. Mourtada-Maarabouni M, Pickard M, Hedge V, et al. GAS5, a non-protein-coding RNA, controls apoptosis and is down-regulated in breast cancer. Oncogene. 2009;28(2):195–208.
104. Zhang H, Cai K, Wang J, et al. MiR-7, inhibited indirectly by LincRNA HOTAIR, directly inhibits SETDB1 and reverses the EMT of breast cancer stem cells by down-regulating the STAT3 pathway. Stem Cells. 2014;32(11):2858–68.
105. Chiyomaru T, Yamamura S, Fukuhara S, et al. Genistein inhibits prostate cancer cell growth by targeting miR-34a and oncogenic HOTAIR. PLoS One. 2013;8(8):e70372.
106. Liu Q, Huang J, Zhou N, et al. LncRNA loc285194 is a p53-regulated tumor suppressor. Nucleic Acids Res. 2013;41(9):4976–87.
107. Keniry A, Oxley D, Monnier P, et al. The H19 lincRNA is a developmental reservoir of miR-675 that suppresses growth and Igf1r. Nat Cell Biol. 2012;14(7):659–65.
108. Medina PP, Nolde M, Slack FJ. OncomiR addiction in an in vivo model of microRNA-21-induced pre-B-cell lymphoma. Nature. 2010;467(7311):86–90.
109. Klein U, Lia M, Crespo M, et al. The DLEU2/miR-15a/16-1 cluster controls B cell proliferation and its deletion leads to chronic lymphocytic leukemia. Cancer Cell. 2010;17(1):28–40.

Chapter 7
Noncoding RNAs in Regulation of Cancer Metabolic Reprogramming

Dongdong Yang, Linchong Sun, Zhaoyong Li, and Ping Gao

Abstract Since the description of the Warburg effect 90 years ago, metabolic reprogramming has been gradually recognized as a major hallmark of cancer cells. Mounting evidence now indicates that cancer is a kind of metabolic disease, quite distinct from conventional perception. While metabolic alterations in cancer cells have been extensively observed in glucose, lipid, and amino acid metabolisms, its underlying regulatory mechanisms are still poorly understood. Noncoding RNA, also known as the "dark matter in life," functions through various mechanisms at RNA level regulating different biological pathways. The last two decades have witnessed the booming of noncoding RNA study on microRNA (miRNA), long noncoding RNA (lncRNA), circular RNA (circRNA), PIWI-interacting RNA (piRNA), etc. In this chapter, we will discuss the regulatory roles of noncoding RNAs on cancer metabolism.

Keywords Noncoding RNA • MicroRNA • Long noncoding RNA • Glucose • Lipid • Amino acid • Cancer metabolism

Abbreviations

G6P	Glucose-6-phosphate
3PG	3-Phosphoglucerate
PEP	Phosphoenolpyruvate
Pyr	Pyruvate
Lac	Lactate
Ac-CoA	Acetyl-CoA
Cit	Citrate

D. Yang • L. Sun • Z. Li • P. Gao (✉)
Hefei National Laboratory for Physical Sciences at Microscale, CAS Key Laboratory of Innate Immunity and Chronic Disease, Innovation Center for Cell Signaling Network, and School of Life Science, University of Science and Technology of China, Hefei 230027, China
e-mail: pgao2@ustc.edu.cn

E. Song (ed.), *The Long and Short Non-coding RNAs in Cancer Biology*, Advances in Experimental Medicine and Biology 927,
DOI 10.1007/978-981-10-1498-7_7

α-KG	α-Ketoglutarate
Glu	Glutamine
Pro	Proline
BCAAs	Branched-chain amino acids
HMG-CoA	3-Hydroxy-3-methylglutaryl-coenzyme A
MVA	Mevalonic acid
IPP	Isopentenyl pyrophosphate
SE	Squalene epoxide
Chol	Cholesterol
7-DHC	7-Dehydrocholesterol
IGF1-R	Insulin-like growth factor 1 receptor
GLUTs	Glucose transporters
G6PC	Glucose-6-phosphatase
HK2	Hexokinase 2
G6PD	Glucose-6-phosphate dehydrogenase
PKM2	Pyruvate kinase M2
LDHA	Lactate dehydrogenase A
PDK1	Pyruvate dehydrogenase kinase 1
PDH	Pyruvate dehydrogenase
ACLY	ATP citrate lyase
FASN	Fatty acid synthase
HMGCS1	3-Hydroxy-3-methylglutaryl-CoA synthase 1
HMGCR	3-Hydroxy-3-methylglutaryl-CoA reductase
DHCR7	7-Dehydrocholesterol reductase
CPT1A	Carnitine palmitoyltransferase 1A
GLS	Glutaminase
POX	Proline dehydrogenase
DBT	Dihydrolipoyl branched-chain acyltransferase
SQLE	Squalene epoxidase

7.1 Introduction

7.1.1 Metabolic Reprogramming in Cancer

Metabolism is generally considered as one of the most fundamental features of biological life. It refers to a series of enzyme-catalyzed reactions within the cells, allowing substance and energy exchange between the living organism and the external environment. Metabolism is usually divided into two categories, catabolism (dissimilation) and anabolism (assimilation), representing harvesting energy through cellular respiration and using energy to construct cellular components, respectively. Metabolic pattern under pathological process such as tumorigenesis is quite different from that under physiological status, a phenomenon called metabolic reprogramming. The principle of metabolic reprogramming aims at balancing energy demand and expenditure and doing the best to facilitate the synthesis of micromolecules to promote cancer cell proliferation.

In the 1920s, Otto Warburg discovered that, compared to their normal counterparts, tumor cells prefer to consume more glucose and use glycolytic pathway to produce more lactate, even under oxygen-rich conditions [1, 2]. This phenomenon is named Warburg effect or aerobic glycolysis. Nevertheless, sporadic attention has been drawn to the field of cancer metabolism for many years until about two decades ago, when evidence has been accumulated to demonstrate that most if not all cancers possess the character of Warburg effect which, in fact, is now widely regarded as one of the emerging hallmarks of cancer cells [3]. Metabolic reprogramming allows cancer cells to employ distinct metabolic strategy to meet the demands of proliferation or survival under different scenarios. Because of rapid cell growth and abnormal intratumoral blood vessels [4, 5], blood flow within solid tumors is both spatial and temporal heterogeneity, resulting in a high degree of metabolic heterogeneity [6]. Regional metabolic patterns vary largely according to the particularity of the microenvironment, especially the differential supply of nutrients and oxygen. Under nutrient-replete conditions, cancer cells show a propensity for synthesizing macromolecules such as lipids, proteins, and nucleic acids in order to sustain the rapid proliferation phenotype, while under nutrient-limiting conditions, tumor cells tend to perform catabolism: many alternative energy sources, such as acetate [7–10], lactate [11], and branched-chain amino acid [12], have been identified to provide energy as well as carbon skeleton materials in extreme conditions.

Alterations in cancer cell metabolism not only happen in glucose metabolism but also in lipid and amino acid metabolism [13, 14]. Adaptive changes in tumor cell metabolism include a switch from oxidative phosphorylation to glycolysis and a shift from glucose to glutamine as the major carbon source for fatty acid synthesis, especially under specific microenvironment such as hypoxia [15]. While nutrients are limiting, increased protein scavenging, lipid scavenging, and fatty acid oxidation could be observed [6]. As a general consequence, changes of metabolic pathways in cancer cells would eventually lead to the activation of oncogenes, inactivation of tumor suppressor genes, and their comprehensive integrations. For instances, LKB1/AMPK [16], c-Myc [17–20], HIF-1 [20–22], p53 [20, 23], and other signaling pathways are extensively involved in these processes. However, we are still far from a complete understanding of how metabolic reprogramming is regulated in cancer cells.

7.1.2 Noncoding RNAs in Cancer

RNAs can be categorized into two groups: protein-coding RNAs which can be translated into functional proteins (i.e., messenger RNA or mRNA) and noncoding RNAs which function only at RNA levels [24, 25]. Generally speaking, noncoding RNAs include transfer RNA (tRNA), ribosomal RNA (rRNA), microRNA (miRNA)

[26], small nuclear RNA (snRNA), small nucleolar RNA (snoRNA), antisense RNA (atRNA), heterogeneous nuclear RNA (hnRNA), small interfering RNA (siRNA) [27, 28], guide RNA (gRNA), enhancer RNA (eRNA), signal recognition particle (SNP) RNA, transfer-messenger RNA (tmRNA), PIWI-interacting RNA (piRNA) [29–31], etc. In terms of their size, noncoding RNAs can be divided into small noncoding RNAs [32] about 20–200 nucleotides such as miRNAs, piRNAs, siRNAs, etc., and long noncoding RNAs (lncRNAs) [33, 34] which are over 200 nucleotides. Only 1–2 % of the human genome is responsible for protein encoding [35], and the rest of the noncoding area was previously thought to be useless, i.e., "junk DNA" whose transcript products are noncoding RNAs. Although known as the "dark matter in life," noncoding RNAs have now been validated to function in various biological processes by mounting evidence, including regulation of mRNA stability, protein translation and transport, RNA processing and modification, chromatin structure adjustment, and so on [36–38].

The genetic regulatory roles of noncoding RNAs were first discovered about two decades ago when two microRNAs Lin-4 and Let-7 were reported in 1993 [39, 40]. More and more evidence indicates that noncoding RNAs, either miRNAs or lncRNAs, alter in the initiation and progression of human cancer [41] and may also act as tumor suppressors or oncogenes [42]. As we all know, each miRNA or lncRNA can have hundreds of cancer-related target genes, indicating that they can participate in various pathophysiological processes of cancer cells. Recent studies have identified miR-34 is regulated by p53 after DNA damage and expression of miR-34 leads to cell cycle arrest [43, 44]. Another study has shown that twist-induced miR-10b [45] promotes cell migration and invasion in mouse and human breast cancer cells [21]. MiR-21 was the first miRNA being coined as an oncomiR due to its universal upregulation and pro-growth effect in cancer [46]. In accordance, miR-21 overexpression in vivo resulted in elevated tumor outgrowth [47]. MiR-155 has also been found to be overexpressed in many cancer types including hematopoietic cancers, lung, breast, and colon cancer [48]. Mouse models lacking miR-155 show impaired function of B and T lymphocytes and dendritic cells [49]. HOTAIR, an lncRNA discovered by Howard Chang's group, recruits polycomb repressive complex 2 (PRC2) to transcriptionally corepress the expression of the homeobox gene D cluster (HOXD), thus reprogramming chromatin state to promote cancer metastasis [50]. In a word, noncoding RNAs are extensively involved in various processes of tumorigenesis.

7.1.3 The Interplay Between Noncoding RNAs and Cancer Metabolism

It becomes well known during the last decade that noncoding RNAs act as key regulators of cancer metabolism [51–53]. Cancer cells display increased metabolic flexibility compared to non-transformed cells, such as taking up nutrients avidly and regulating major metabolic pathways to support fast growth and proliferation [54]. MiRNAs, owing to their small sizes, can target specifically certain metabolic or regulatory genes to exert precise modulation at posttranscription levels. For instance, the regulatory roles of miR-19a/miR-19b [55] in glucose metabolism, miR-33a/miR-33b [56, 57] in lipid metabolism, and miR-21 [58] in amino acid metabolism have greatly enriched our view on microRNA-modulated cancer metabolism. Long noncoding RNAs can have multiple ways of gene regulation such as functioning at transcriptional level or posttranscriptional level or through epigenetic status modulation, because of their structure flexibility to interact with different protein factors [53, 59–61]. The finding that lincRNA-p21 regulates Warburg effect directly links long noncoding RNAs to cancer metabolic regulation [62]. Here in this chapter, we will discuss in detail the important roles noncoding RNAs play in the regulation of cancer metabolism.

7.2 MicroRNAs Regulate Cancer Metabolic Reprogramming

Among noncoding RNAs, miRNAs are the best characterized. They are a family of non-protein-coding, single-stranded RNAs which can negatively regulate target gene expression. MiRNA-encoding genes at intergenic, intronic, or polycistronic genomic loci are transcribed into long primary miRNAs (pri-miRNAs) by RNA polymerase II or, in some cases, by RNA polymerase III [63–65]. The pri-miRNA transcript forms a stem-loop structure that is recognized and processed by the microprocessor, Drosha and DGCR8 RNase III complex [66]. The generated precursor miRNA (pre-miRNA) is then ready to be exported to the cytoplasm with the help of nuclear export receptor exportin 5 (XPO5), which is a RanGTP-dependent transporter protein [67, 68]. In the cytoplasm, pre-miRNAs are cleaved near the terminal loop by the RNase III enzyme Dicer, subsequently releasing the 20–24 nucleotides mature miRNA duplexes [69, 70]. MicroRNAs play their roles through partial or complete complementary pairing with sequences usually located in the 3′ untranslated region (UTR) of target mRNAs in two ways: inhibiting protein translation or facilitating mRNA degradation [71, 72]. MiRNAs can manage the expression of genes involved in various biologic processes, including cell cycle, development, signal transduction, metabolism, and metastasis, as well as cell proliferation, differentiation, and apoptosis.

7.2.1 *MiRNAs and Glucose Metabolism*

As a major carbon source, glucose is pivotal to cancer cell survival. Hence, the processes of its uptake, transportation, consumption, and synthesis are tightly regulated to sustain glucose homeostasis during cancer cell metabolic reprogramming [73–77]. Herein, we will discuss the effect of microRNAs exerting on glucose metabolism [78], including glucose source and expenditure. As for the source of glucose, three primary processes should be considered: (1) insulin receptor signaling-mediated glucose homeostasis, (2) glucose transporter-mediated uptake of nutrient from blood circulation, and (3) transformation of glucose from nonsugar substances via gluconeogenesis. The consumption of glucose mainly involves aerobic glycolysis, pentose phosphate pathway, mitochondrial respiration, or tricarboxylic acid cycle, all of which will be specified respectively in the following paragraphs.

7.2.1.1 Source of Glucose

First of all, as a vital hormone playing a fundamental role on balancing circulating glucose level, insulin ensures cellular glucose homeostasis sustaining through insulin signaling pathway [79]. Dysregulation of this signaling pathway will lead to lack of energy fuel in the cell. Wang et al. have reported that in glioma cells miR-7, a tumor-suppressive miRNA, directly targets the 3'-UTR of insulin-like growth factor 1 receptor (IGF1-R) gene to inhibit cellular growth and glycolysis [80]. Clinical investigation of individuals shows that miR-26a regulates insulin sensitivity and metabolism of glucose and lipids. Overweight individuals will have decreased expression of liver miR-26a compared with lean individuals. Moreover, global or liver-specific overexpression of miR-26a improves insulin sensitivity, decreases hepatic glucose production in mice with high-fat diet, and vice versa [81]. Another study demonstrates that obesity-induced upregulation of miR-143, which targets oxysterol-binding protein-related protein 8 (ORP8), inhibits insulin-stimulated AKT activation and impairs glucose homeostasis [82]. MiR-143 has also been reported to downregulate hexokinase 2 (HK2) and inhibit glycolysis and cancer progression [83].

Glucose transportation into cancer cells is mediated by various glucose transporters (GLUTs). Thirteen members of the mammalian glucose transporter family have been identified, belonging to the solute carrier 2A (SLC2A) family [78, 84]. Glut1 and Glut3 with high affinities for glucose are overexpressed in transformed cells rather than Glut2 and Glut5 which have low affinities for glucose [84]. Glut4, another glucose transporter, is reported to be closely related to insulin resistance. Those GLUTs mentioned above are extensively regulated by miRNAs in cancer cells. In renal cell carcinoma, miR-1291 inhibits Glut1 and displays a tumor-suppressive effect [85]. Another study in renal cell carcinoma has demonstrated that decreased expression of miR-199a, miR-138, miR-150, and miR-532-5p is correlated with Glut1 upregulation, while increased expression of miR-130b, miR-19a, miR-19b, and miR-301a is associated with downregulation of Glut1 [55]. MiR-133 reduces the protein level of KLF15 and its downstream effector Glut4 [86]. MiR-223, which is consistently upregulated in insulin-resistant cardiomyocyte, increases glucose uptake via inducing Glut4 protein expression. In addition, Glut4 can also be regulated and inhibited by miR-93 [87]. When cancer cells are cultured under nutrient starvation conditions, especially when glucose is short of supply, gluconeogenesis will be upregulated to generate glucose. It has been reported that miR-33a and miR-33b inhibit G6PC as well as PCK1 through directly targeting the two key enzymes of hepatic gluconeogenesis [88].

7.2.1.2 Consumption of Glucose

Cancer cells avidly uptake glucose for catabolism to meet the energy demands, meanwhile providing synthetic micromolecules [73–77]. Glycolysis is upregulated in most cancer cell types, being a striking feature of cancer and providing a possibility by positron emission computed tomography (PET) [89, 90]. It can also provide cancer cells with macromolecules as well as reducing power by strengthening the pentose phosphate pathway to meet the need of rapid proliferation and reduce reactive oxygen species (ROS) production from oxidative phosphorylation (OXPHOS) [91, 92]. As Warburg has previously suggested, mitochondrial defect is not a common case in most cancer cells [92]. Noncoding RNAs, especially microRNAs, contribute to the switch from mitochondrial respiration to glycolysis by directly or indirectly targeting the key metabolic enzymatic nodes in these important pathways [93]. MiR-143, suppressed by mTOR activation, reduces glucose metabolism and inhibits cancer cell proliferation by targeting HK2 [94, 95]. Pyruvate kinase M2 (PKM2), a rate-limiting glycolytic enzyme, is also regulated by microRNAs. MiR-122 reprograms glucose metabolism in pre-metastatic niche to promote breast cancer metastasis by directly targeting and inhibiting PKM2 [96]. Let-7a-c-Myc-hnRNPA1 feedback loop has also been reported to be indirectly involved in PKM2-promoting glycolysis and cell growth in gliomas [97]. Although pyruvate is produced during glycolysis process, its fate is still undetermined: being converted to lactate by LDHA, entering TCA cycle for mitochondrial respiration, or being a source for alanine synthesis. One of the strategies that cancer cells use to achieve glycolytic switch is through controlling PDHX and PDK. It has been reported that miR-26a targets PDHX in colorectal cancer cells [98] and that miR-375 regulates cell survival by targeting PDK1 in gastric carcinomas [99]. Interestingly, another powerful microRNA, miR-34a, targets many glycolytic enzymes including HK1, HK2, LDHA and GPI, as well as PDK1 [100, 101].

7.2.1.3 Glucose-Related Signaling Pathways

Apart from those mentioned above, signal transduction pathways related to glucose metabolism are also largely involved. The tumor-suppressive miR-218 targets the mTOR component Rictor and inhibits subsequently AKT phosphorylation in oral cancer, resulting in glucose metabolism impairment [102]. LKB1/AMPK signaling is well known for its powerful energy sensing ability of sensitively and rapidly responding to AMP/ATP ratio. MiR-451 regulates LKB1/AMPK signaling and allows an adaptation under metabolic stress conditions in glioma cells [103]. A recent study has demonstrated that upregulation of Lin-28A/Lin-28B and downregulation of Let-7 facilitate aerobic glycolysis in various cancer cell lines and PDK1 expression is critical for Lin-28A-/Lin-28B-mediated cancer proliferation both in vitro and in vivo [104]. Moreover, the regulation of aerobic glycolysis via Lin-28/Let-7/PDK1 axis is in a hypoxia- or HIF-1-independent manner [104].

7.2.2 MiRNAs and Lipid Metabolism

Lipids are water-insoluble molecules mainly consisting of triglycerides and lipoids such as phospholipids and sterols [105, 106]. Among all the important roles lipids serve at cellular and organismal levels, there are three primary ones: comprising energy storage, signal transduction, and structural components of cell membranes [107, 108]. Fatty acids are basic building blocks to synthesize triglycerides mainly for energy storage. Phospholipids and sterols, together with glycolipids, represent three major classes of bio-membranous lipids. Lipids can also have important roles in signaling, functioning as second messengers and as hormones [109]. Specific alterations in cancer cell lipid metabolism include de novo lipid synthesis rather than directly uptaking exogenous sources, storing lipid rather than oxidation for usage, and forming cholesterol ester from free cholesterols to promote tumor cell migration. To achieve the regulation on lipid synthesis and homeostasis pathways, cancer cells need to integrate various transcription factors and different oncogenic signaling pathways which, to some degree, are regulated by noncoding RNAs as well as protein factors. Herein, we focus on microRNAs that regulate the metabolism of fatty acids and cholesterol, two vital classes of lipids, and discuss the transcription factors and signaling pathways involved.

7.2.2.1 Fatty Acid Metabolism

Compared with glucose metabolism, lipid metabolism enjoyed less attention during the previous years. Nevertheless, the importance of lipid metabolism reprogramming in cancer cells has been increasingly recognized with extensive studies in fatty acid metabolism fields [110]. Fatty acid uptake, synthesis, storage, metabolism, and homeostasis are of vital importance for they are the basic building blocks required to produce phospholipids (PLs), phosphatidylcholine (PC), phosphatidylethanolamine (PE), sterols, sphingolipids, and lyso-PLs [111]. The source of fatty acid synthesis is either from uptaking from the environment or from de novo synthesis, the latter of which is preferred by cancer cells. Referring to fatty acid metabolism, four steps should be taken into consideration: (1) fatty acid synthesis, ATP citrate lyase (ACLY), acetyl-CoA carboxylase (ACC), fatty acid synthase (FASN), and acyl-CoA synthetase (ACS) are involved; (2) fatty acid degradation, carnitine palmitoyltransferase 1A (CPT1A) is involved; (3) fatty acid storage, glycerol-3-phosphate acyltransferase (GPAT), 1-acylglycerol-3-phosphate-O-acyltransferase (AGPAT), phosphatidic acid phosphatase (PAP), and diacylglycerol acyltransferase

(DGAT) are involved; and (4) fatty acid release from storage, adipose triacylglyceride lipase (ATGL), hormone-sensitive lipase (HSL), and monoacylglycerol lipase (MAGL) are involved. Those enzymes mentioned above can be the major direct targets of microRNAs. MiR-33, an extensively studied sterol regulatory element-binding transcription factor 2 (SREBF2)-embedded microRNA, inhibits the translation process of several transcripts encoding enzymes in fatty acid β-oxidation such as CPT1A, CROT, and HADHB, thus disturbing fatty acid degradation [112]. MiR-33-mediated inhibition of SIRT6 expression also results in increased chromatin acetylation and derepression of sterol regulatory element-binding protein (SREBP)-dependent fatty acid biosynthesis genes, thus increasing lipogenesis [113, 114]. The hypoxia-inducible microRNA cluster miR-199~214 has been reported to target myocardial PPARδ and impair mitochondrial fatty acid oxidation process [115]. Sterol regulatory element-binding proteins (SREBPs) are important transcription factors that regulate fatty acid and cholesterol metabolism [116–119], and PPARα is the major transcriptional regulator of fatty acid oxidation for its activity that induces oxidation in the mitochondrion [120–122]. The influences of miRNAs exerting on fatty acid metabolism are largely dependent on the regulation of these two factors which we will discuss more in the following.

7.2.2.2 Cholesterol Metabolism

Cholesterol is required for the cancer cell growth and survival for being a basic component of cell membranes and an obligate precursor molecule of various steroid hormones, bile acids [123], and vitamin D [124]. Its dietary absorption, synthesis, release, transportation, and storage are all tightly regulated [125]. Free cholesterol and triglyceride are secreted from the liver to other parts of the body in the form of VLDL with the help of the chaperone protein called microsomal triglyceride transfer protein (MTP) [126]. Reverse cholesterol transport (RCT) is a complicated process relying on the interaction between apolipoprotein A1 (ApoA1) and ATP-binding cassette transporter A1 (ABCA1) to form a nascent HDL particle. ATP-binding cassette transporter G1 (ABCG1) further effluxes cholesterol to nascent HDL, eventually forming the mature HDL particle [127, 128]. HDL then returns to the liver and is absorbed by scavenger receptor B1 (SRB1) for recycling [127]. SREBP2 is a transcription factor that responds to low cellular cholesterol levels, increasing cellular cholesterol levels by promoting uptake and synthesis while reducing cholesterol efflux. In contrast, the liver X receptor (LXR) family of transcription factors are stimulated under cholesterol excess conditions, decreasing cellular cholesterol levels by upregulating cholesterol efflux and reducing uptake and synthesis [126]. Almost every process discussed above can be targeted by versatile microRNAs.

MiR-122 is the most abundant microRNA in the liver, comprising approximately 70 % of total miRNAs [79]. Inhibition of miR-122 by antisense oligonucleotides in mice results in increased hepatic fatty acid oxidation and decreased cholesterol synthesis. Meanwhile, total plasma cholesterol is reduced by 25–35 %, which is reflected by the changing levels of both the LDL and HDL fractions. MiR-122 inhibition can also indirectly cause a decrease on cholesterol synthesis enzymes including 3-hydroxy-3-methylglutaryl-CoA synthase 1 (HMGCS1), 3-hydroxy-3-methylglutaryl-CoA reductase (HMGCR), 7-dehydrocholesterol reductase (DHCR7), and squalene epoxidase (SQLE) [129, 130]. The SREBP transcription factors act coordinately with their intronic miRNAs miR-33a and miR-33b to regulate fatty acids, triglyceride, and cholesterol homeostasis. Several recent studies have reported the discovery of the miRNA-host gene circuit, demonstrating that miR-33a cooperates with the SREBP2 to upregulate intracellular cholesterol levels [131, 132]. MiR-33a and miR-33b are also found to have a crucial role in the post-transcriptional repression of ABCA1, which promotes the efflux of free cholesterol from within the cell to ApoA1 and is essential for the formation of HDL [133]. Other microRNAs such as miR-758 and miR-106b also act as vital roles in regulating cholesterol homeostasis [134, 135].

7.2.2.3 SREBPs, LXR, PPARα, and Signal Transduction Pathways

SREBPs, with the basic helix-loop-helix leucine zipper motif, activated under cholesterol limit conditions, are the major regulators of fatty acid and cholesterol homeostasis [125]. SREBP2 enhances cholesterol uptake and synthesis by activating LDL receptor (LDLR), scavenger receptor B1 (SRB1), 3-hydroxy-3-methylglutaryl-CoA reductase (HMG-CoA-R), 3-hydroxy-3-methylglutaryl-CoA synthase 1 (HMG-CoA-S1), and methylsterol monoxygenase (SC4MOL) and decreases cholesterol efflux via suppressing ABCA1, ABCG1, ATP-binding cassette transporter G5/G8 (ABCG5/ABCG8), and cholesterol 7-alpha-monooxygenase also known as cytochrome P450 7A1 (CYP7A1). Conversely, LXR activation increases cholesterol efflux and catabolism by stimulating ABCA1, ABCG1, ABCG5/ABCG8, and CYP7A1 accordingly while decreasing cholesterol uptake and synthesis by inhibiting LDLR, SRB1, HMG-CoA-R, and HMG-CoA-S1. AKT and mTORC1 [136–139] are two major upstream regulators of those transcription factors mentioned above in fatty acid metabolism and cholesterol metabolism. Activated PI3K/AKT signaling increases overall fatty acid synthesis by upregulating SREBPs and its downstream enzymes FASN and SCD-1 [140–143]. AKT inhibits the expression of CPT1A, one of the principal mitochondrial membrane long-chain fatty acid (LCFA) importers, therefore impairing β-oxidation [144, 145]. PTEN, the endogenous tumor suppressor for PI3K/AKT, is downregulated by monounsaturated fatty acid (MUFA)-induced miR-21, thus increasing AKT-dependent FASN expression [146, 147]. AMPK also inhibits SREBPs indirectly via LXR or via direct phosphorylation [148]. Therefore, downregulation of AMPKα1 by miR-33a and miR-33b may relieve AMPK inhibition of both SREBPs and their target genes to coordinately boost intracellular levels of cholesterol, fatty acids, and other lipids [149, 150].

7.2.3 MiRNAs and Amino Acid Metabolism

7.2.3.1 Glutamine and Proline

Amino acids have been dichotomized into essential and nonessential. Several nonessential amino acids become quite critical for cancer cells to respond to microenvironmental stress. Glutamine (Gln), the most abundant free nonessential amino acid in human blood, complements glucose to meet the demands in a metabolic platform that maximizes anabolism in growing tumor cells. Glutaminase (GLS) is the first enzyme that converts glutamine to glutamate, which is further converted to α-ketoglutarate for metabolism in the TCA cycle. In 2009, Gao and colleagues found that mitochondrial GLS protein was induced in c-Myc high-expressing human P493 B lymphocytes. Further analysis of the regulation mechanism has indicated that c-Myc transcriptionally represses miR-23a and miR-23b, releasing the expression of their target protein GLS. Thus, glutaminolysis pathway is enhanced, and the product glutamate enters into the TCA cycle for the production of ATP and the synthesis of glutathione [19, 151]. Additionally, NF-κB p65/Rel A subunit also inhibits miR-23a expression via directly binding to its promoter in human leukemic Jurkat cells, leading to increased basal GLS protein expression and reinforced glutamine metabolism [152].

Although less well-recognized, proline is another nonessential amino acid interconvertible with glutamate and arginine. The steps between proline and glutamate are as follows: glutamate → glutamic-γ-semialdehyde (GSA) → Δ1-pyrroline-5-carboxylic acid (P5C) → proline. Proline oxidase, also called proline dehydrogenase (POX/PRODH), is a mitochondrial inner-membrane enzyme involved in the degradation of proline and catalyzes the first step in proline catabolism. Liu et al. have found that, in c-Myc-inducible human P493 B lymphocytes and PC3 human prostate cancer cells, POX suppression is necessary for c-Myc-mediated cancer cell proliferation and survival and c-Myc suppresses POX at the transcriptional level indirectly through upregulating miR-23b*. Importantly, c-Myc markedly increases the biosynthesis of proline from glutamine. The metabolic link between noncoding RNAs and glutamine as well as proline modulated by c-Myc emphasizes the importance of these connections [153–155].

7.2.3.2 One-Carbon Metabolism

Compared to the profound understanding of glycolysis and glutaminolysis in cancer cells, we are just beginning to appreciate the critical impact of one-carbon metabolism, which has been long considered as "housekeeping" process during cancer progression. One-carbon metabolism includes folate and methionine cycles and serine synthesis pathway from different cellular nutritional statuses and impacts the balance of redox status; the biosynthesis of lipids, nucleotides, and proteins; and so on [156–158].

Methionine adenosyltransferase (MAT) is an indispensable enzyme responsible for the biosynthesis of S-adenosyl methionine (SAM), a principal biological methyl donor in all mammalian cells [159]. In mammals, MAT1A and MAT2A, two different genes encoding for two homologous MAT catalytic subunits α1 and α2, are liver markers at different developmental stages. MAT1A is predominantly expressed in normal liver and often silenced in HCC, accompanied by opposite variation of MAT2A, known as the MAT1A/MAT2A switch [160–163]. In hepatocellular carcinoma Hep3B and HepG2 cells, overexpression of miR-664, miR-485-3p, and miR-495 negatively regulates MAT1A expression at mRNA level. Conversely, decreased cell growth ability and increased cellular apoptosis are induced when these miRNAs are knocked down. Higher nuclear MAT1A expression robustly reinforces global CpG methylation and Lin-28B promoter methylation, thus reducing Lin-28B expression and resulting in lower probability of tumorigenesis, invasion, and metastasis [164]. MAT2B, the gene encoding a MAT2A regulatory subunit, can result in decreased SAM levels and provide a growth advantage to hepatoma cells. Berberine, an isoquinoline alkaloid isolated from various medicinal herbs such as *Coptis chinensis*, induces miR-21-3p expression which upregulates intracellular SAM contents in HepG2 cells through directly targeting MAT2A and MAT2B 3'-UTRs [165]. During 2-acetylaminofluorene (2-AAF)-induced rat hepatocarcinogenesis, the inhibition of MAT1A and 5,10-methylenetetrahydrofolate reductase (MTHFR) gene expression is mediated by miR-29 and miR-22, respectively. Downregulation of MAT1A and MTHFR leads to an increase in histone H3 lysine 27 trimethylation and a decrease in histone H3 lysine 18 acetylation at the promoter/first exon of the gene, accompanied with alteration of one-carbon metabolism, such as SAM, SAH, and SAM/SAH ratio [166]. Using computational miRNA target prediction methods and Monte Carlo-based statistical analyses, Stone and colleagues have identified two candidate miRNA "master regulators" (miR-22 and miR-125) and one candidate pair of "master coregulators" (miR-344-5p/miR-484 and miR-488) that may influence the expression of a significant number of genes (such as MTHFD2, MTHFR, SHMT2, DNMT3A/DNMT3B, and so on) involved in one-carbon metabolism [167]. Recently, it is found that mosquito-specific miR-1174 targets serine hydroxymethyltransferase (SHMT2), a mitochondrial enzyme taken part in serine synthesis pathway, and complete miR-1174 elimination will lead to serious defects in sugar absorption [168]. However, whether the regulatory mechanism also exists in tumor cells remains unknown.

7.2.3.3 Branched-Chain Amino Acids

The branched-chain amino acids (BCAAs), leucine, isoleucine, and valine, are essential amino acids because they cannot be synthesized de novo. These three kinds of essential amino acids make up approximately 1/3 of skeletal muscle in the human body and constitute about 40 % of the preformed amino acids required by mammals [169]. Exercise promotes BCAA catabolism [170], and BCAAs play vital roles in synapse function, insulin secretion, and protein turnover [171, 172].

In mammals, branched-chain α-ketoacid dehydrogenase (BCKD) catalyzes the irreversible reaction in BCAA catabolism and thus commits to a decrease in the concentration of these amino acids within cancer cell. MiR-29b, one of the members of human miR-29 family members, targets mRNA of dihydrolipoyl branched-chain acyltransferase (DBT), which is a vital component of BCKD. Thus, miR-29b maintains BCAA homeostasis which is crucial given that these amino acids account for up to 20 % of the amino acids found in most proteins [173].

7.3 Long Noncoding RNAs Regulate Cancer Metabolism Reprogramming

In 2002, Okazaki and colleagues identified a novel transcriptional class, later referred to as long noncoding RNAs, when analyzing the mouse transcriptome based on functional annotation of 60,770 full-length cDNA libraries [174]. Long noncoding RNAs (lncRNAs) are abundant, poorly conserved, non-polyadenylated non-protein-coding transcripts arbitrarily defined as being longer than 200 nucleotides [175]. For the better part of the past decade, particular attention has focused on the booming of lncRNA studies owing to the advancement of genomic functional research. LncRNAs have been extensively studied in various physiological and pathological processes such as X-chromosome inactivation and gastrointestinal cancers [176]. Because of the size and specifically transcription by RNA polymerase III, lncRNAs play different roles in transcriptional, posttranscriptional, and epigenetic gene regulation. Referring to transcription regulation, lncRNAs can act as coregulators, modify transcription factor activities, or associate with coregulators to regulate specific gene transcription. Meanwhile, lncRNAs may interact with the component of basal transcription machinery to regulate global transcriptional pattern. Mechanism of posttranscriptional regulation involves modulating splicing and siRNA-directed transcript degradation and translation process. Besides, epigenetic states will also have influences on transcription pattern and cancer biology, such as imprinting, Xist and X-chromosome inactivation, and telomeric regulation, which can also be affected by lncRNAs [14, 177–179].

A lot of lncRNAs are indicated in cancer progression. For example, enhanced expression of lncRNA MALAT1 promotes cell proliferation and migration in pancreatic cancer [180]. Another lncRNA, PVT1, promotes tumorigenesis in non-small cell lung cancer and induces multidrug resistance in gastric cancer cells [181]. In addition, c-Myc-activated lncRNA CCAT1 promotes proliferation and invasion in colon cancer cells [182].

Moreover, the study of lncRNAs and cancer metabolism is a combination and extension of two supremely hot-studied fields. There have been many indications of lncRNA playing a role in regulating cancer metabolism before the first direct interactions of lncRNA with metabolic enzymes was found. For instances, a large quantity of lncRNAs act as miRNA sponges to impair the function of miRNAs, which relates lncRNAs to the paragraphs we have discussed above. Another example is that the H19 lncRNA, which is implicated in development and growth control and is associated with cancer, can act as a molecular sponge by inhibiting Let-7 [183]. H19 expression is remarkably elevated in a variety of human cancers. Recently, H19 RNA has also been reported to be upregulated in hypoxic stress and to demonstrate oncogenic properties, which can be prevented by wild-type p53 and promoted by HIF-1α overexpression [184]. However, H19 has been found to be significantly decreased in muscle cells of human with type 2 diabetes, which leads to increased bioavailability of Let-7 and subsequent diminishment of Let-7 targets and results in impaired insulin signaling and decreased glucose uptake [185]. In addition, H19 can function as a precursor of miR-675, which represses cell migration in normal prostate epithelial cells by targeting TGFBI, an extracellular matrix protein enhancing cancer metastasis [186]. Another lncRNA, UCA1, has been shown to promote glycolysis in bladder cancer cells through mTOR-STAT3/microRNA-143-HK2 cascade, revealing a link between lncRNA and the dysregulated glucose metabolism in cancer [187].

A remarkable work on lncRNA regulating cancer metabolism published recently is that hypoxia-induced lincRNA-p21 regulates Warburg effect via reciprocal mechanism. Yang et al. have showed that lincRNA-p21 is hypoxia-responsive and critical to enhanced glycolysis under hypoxic conditions via binding to HIF-1α and VHL and disrupting VHL-HIF-1α interaction. Thus, lincRNA-p21 attenuates VHL-mediated HIF-1α ubiquitination and leads to HIF-1α stabilization and accumulation. The positive feedback loop between lincRNA-p21 and HIF-1α can promote tumor growth which is validated in vitro and in vivo [62]. Another novel lncRNA, colorectal neoplasia differentially expressed (CRNDE), which is activated early in colorectal cancer, responds to insulin/IGF signaling and involves in metabolism regulation. CRNDE potentially interacts with chromatin-modifying complexes, thus affecting epigenetic status. In colorectal cancer cells, insulin treatment and insulin-like growth factors (IGFs) suppress CRNDE nuclear transcripts, which can be reversed by PI3K/AKT/mTOR or Raf/MAPK pathway inhibitors. SiRNA-mediated knockdown of CRNDE affects the expression of genes involved in insulin/IGF signaling pathway, including glucose and lipid metabolism, which suggests that CRNDE expression can facilitate the switch to aerobic glycolysis in cancer cells [188]. Other lncRNAs, such as liver-specific lncRNA HULC, can be transcriptionally stabilized by the IGF2 mRNA-binding protein 1 (IGF2BP1). HULC is reported to modulate aberrant lipid metabolism via targeting miR-9-mediated RXRA signaling in hepatoma cells [189].

Central signaling pathways involving in cancer metabolic reprogramming are also regulated by lncRNAs, providing good hints for investigating the roles lncRNAs play in cancer cell metabolism. Liu et al. have identified that lncRNA NKILA interacts with NF-κB/IκB complex, repressing NF-κB signaling and cancer-associated inflammation. Low expression of NKILA is also found to be associated with breast cancer metastasis and poor clinical prognosis [190]. Growth arrest-specific 5 (GAS5) lncRNA, an androgen receptor repressor, promotes apoptosis of prostate cancer cells and declines in castrate-resistant prostate cancer cells. Inhibition of mTOR will enhance GAS5 expression, and GAS5 itself is required for mTOR inhibitor action, suggesting the reciprocal regulation of GAS5 lncRNA and mTOR inhibitor [191]. Prostate cancer gene expression marker 1 (PCGEM1) is an androgen-induced lncRNA overexpressed within prostate cancer and shows its tumorigenic potential due to its ability of androgen receptor activation. Hung et al. have reported that PCGEM1 takes part in cancer metabolism regulation and promotes glucose uptake through activating c-Myc. Glucose then can be shunted to pentose phosphate pathway to facilitate biosynthesis of macromolecules and reduced force for maintaining redox balance [192]. Interestingly, studies also suggest that lncRNAs involve in p53 regulatory network, i.e., MALAT1, MEG3, Wrap53, PANDA, RoR [184, 193, 194], etc. RoR, an extracellular lncRNA, is also reported to modulate hypoxia signaling pathways [195]. All those interactions between lncRNAs and important players suggest the potential roles of lncRNAs in regulating tumor metabolism.

7.4 Other Noncoding RNAs in Cancer Metabolism Reprogramming

7.4.1 PIWI-interacting RNAs

PIWI-interacting RNA (piRNA), one of the largest classes of small ncRNAs, forms RNA-protein complexes via association with PIWI proteins. Compared to miRNAs, piRNAs have distinct characteristics of decreased sequence conservation and increased complexity. PiRNA-protein complexes have been involved in both post-transcriptional gene silencing and epigenetic status in germ line cells, particularly in spermatogenesis [29–31]. Surprisingly, Huang and colleagues show that zuc/MitoPLD activity is required for piRNA-mediated silencing of transposable elements both in fly and mouse germ lines [196]. As we all know that mitochondria is a central component of cell metabolism, the interaction between mitochondrial signaling and piRNA pathway suggests the potential roles piRNAs may play in tumor metabolism.

7.4.2 Small Nucleolar RNAs

Small nucleolar RNAs (snoRNAs), another class of small RNA molecules, have a basic function of guiding chemical modifications on other RNAs, such as transfer RNAs, ribosomal RNAs, and small nuclear RNAs. There are two major categories of snoRNAs: C/D box snoRNAs associated with methylation and H/ACA box snoRNAs associated with pseudouridylation [197–200]. Through genetic screening, Jinn et al. have identified snoRNA U17 as a regulator of cellular cholesterol homeostasis by altering intracellular cholesterol trafficking via targeting HUMMR mRNA and modulating ER-mitochondria contacts. The newly identified function of U17 other than ribosomal processing directly links snoRNAs to tumor metabolism and suggests the important role of U17 snoRNA-HUMMR pathway in tumorigenesis [201].

7.4.3 Circular RNAs

The word “circular RNA (circRNA)” was first proposed by Sanger et al. in 1976, suggesting that viroids are single-stranded, thermally stable, covalently closed circular RNA molecules which are pathogenic to several higher plants [202]. Circular RNAs can be formed by direct ligation of 5’ and 3’ ends of linear RNAs or by “back-splicing” a splicing form of a downstream 5’ splice site joining to an upstream 3’ splice site [203, 204]. It has been identified that competing endogenous RNA and circular RNA act as important regulators of miRNA activity. Hansen and colleagues have identified that ciRS-7 acts as a specific miR-7 sponge, inhibiting the expression of several oncogenes regulated by miR-7 and suggesting the important roles ciRS-7/miR-7 axis plays in cancer-related pathways and for clinical usage [205].

7.5 Conclusions and Future Perspectives

As illustrated in the figure (see Fig. 7.1), noncoding RNAs are extensively involved in the nutrient metabolism of malignancy. Cancer is a kind of metabolic disease. On one hand, metabolic reprogramming has been recognized as distinct characteristics of cancer cells, providing basis for cancer monitoring, diagnosis, and antitumor therapies. On the other hand, it is of vital importance for us to study the differential metabolic profiles of cancerous and normal cells globally and to find out pivotal metabolic enzymatic nodes that contribute to tumorigenesis. Current studies on tumor metabolism have focused on finding the “Achilles’” heel of cancer and greatly enriched our understanding of tumor nature. Meanwhile, many issues in this field remain to be explored. For example, what are the substantive characteristics of aberrant metabolism of tumor cells? What causes the abnormal metabolic pattern and what’s the mechanism? What consequences will metabolic reprogramming bring to cancer cells? How can we target the characteristic metabolic changes to cure the disease? Besides the regular hotspots in noncoding RNA research fields, the interplay between noncoding RNAs and cancer metabolism has attracted much attention. Noncoding RNAs, owing to its small size and specificity on gene regulation in cancer biology, provide an attractive antitumor therapeutic approach through targeting cancer metabolic reprogramming. Hopefully in the near future, interplay between noncoding RNAs and cancer metabolism will depict a new landscape of cancerous disease and lead to promising clinical therapies.

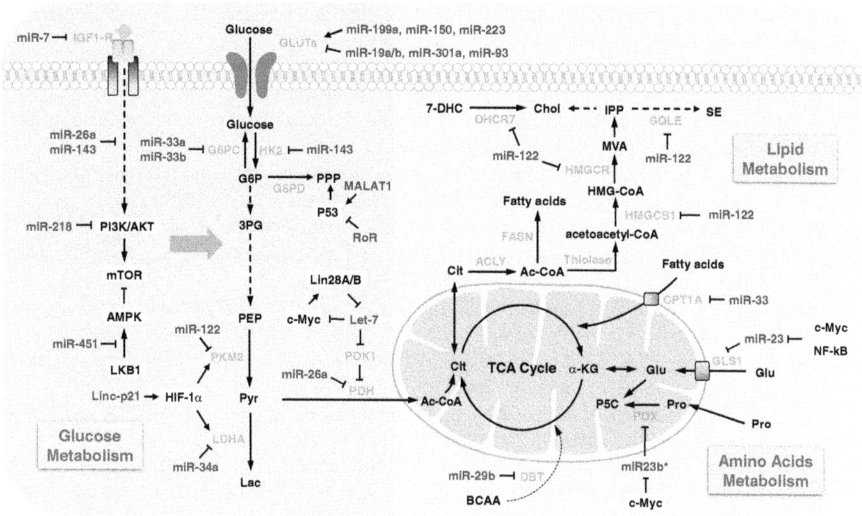

Fig. 7.1 Noncoding RNAs are extensively involved in the nutrient metabolism of cancer cells. Glucose metabolism, which intimately interacts with cancer signaling such as HIF-1α, c-Myc, and PI3K/AKT, includes glycolytic pathway, pentose phosphate pathway (*PPP*), and mitochondrial respiration pathway. Fatty acid synthesis and cholesterol synthesis are depicted as lipid metabolism. Glutamine and BCAA metabolism are shown as examples for amino acid metabolism. Oncogenes and tumor suppressors involved in the pathways are illustrated in *black* with white background, while microRNAs are marked in *red*, and the major metabolic enzymes are marked in *green*

Acknowledgments This work is supported in part by the National Basic Key Research Program of China (2014CB910600), National Science Foundation of China (31372148, 81530076, 31571472), Chinese Academy of Sciences (XDA01010404), and Research Fund for the Doctoral Program of Higher Education of China (2133402110014).

References

1. Warburg O. On the origin of cancer cells. Science. 1956;123:309–14.
2. Warburg O, Wind F, Negelein E. The metabolism of tumors in the body. J Gen Physiol. 1927;8:519–30.
3. Hanahan D, Weinberg RA. Hallmarks of cancer: the next generation. Cell. 2011;144:646–74.
4. Fukumura D, Jain RK. Tumor microenvironment abnormalities: causes, consequences, and strategies to normalize. J Cell Biochem. 2007;101:937–49.
5. Zhang HF, Qian DZ, Tan YS, et al. Digoxin and other cardiac glycosides inhibit HIF-1 alpha synthesis and block tumor growth. Proc Natl Acad Sci U S A. 2008;105:19579–86.
6. Boroughs LK, DeBerardinis RJ. Metabolic pathways promoting cancer cell survival and growth. Nat Cell Biol. 2015;17:351–9.
7. Comerford SA, Huang Z, Du X, et al. Acetate dependence of tumors. Cell. 2014;159:1591–602.
8. Lyssiotis CA, Cantley LC. Acetate fuels the cancer engine. Cell. 2014;159:1492–4.

9. Mashimo T, Pichumani K, Vemireddy V, et al. Acetate is a bioenergetic substrate for human glioblastoma and brain metastases. Cell. 2014;159:1603–14.
10. Schug ZT, Peck B, Jones DT, et al. Acetyl-CoA synthetase 2 promotes acetate utilization and maintains cancer cell growth under metabolic stress. Cancer Cell. 2015;27:57–71.
11. Feron O. Pyruvate into lactate and back: from the Warburg effect to symbiotic energy fuel exchange in cancer cells. Radiother Oncol. 2009;92:329–33.
12. Tonjes M, Barbus S, Park YJ, et al. BCAT1 promotes cell proliferation through amino acid catabolism in gliomas carrying wild-type IDH1. Nat Med. 2013;19:901–8.
13. Hsu PP, Sabatini DM. Cancer cell metabolism: Warburg and beyond. Cell. 2008;134:703–7.
14. Li Z, Zhang H. Reprogramming of glucose, fatty acid and amino acid metabolism for cancer progression. Cell Mol Life Sci. 2015;73(2):377–92.
15. Semenza GL. HIF-1 mediates metabolic responses to intratumoral hypoxia and oncogenic mutations. J Clin Invest. 2013;123:3664–71.
16. Hardie DG, Alessi DR. LKB1 and AMPK and the cancer-metabolism link – ten years after. BMC Biol. 2013. doi:10.1186/1741-7007-11-36.
17. Dang CV. Therapeutic targeting of Myc-reprogrammed cancer cell metabolism. Cold Spring Harb Symp Quant Biol. 2011;76:369–74.
18. Dang CV. MYC on the path to cancer. Cell. 2012;149:22–35.
19. Dang CV, Le A, Gao P. MYC-induced cancer cell energy metabolism and therapeutic opportunities. Clin Cancer Res. 2009;15:6479–83.
20. Yeung SJ, Pan J, Lee MH. Roles of p53, Myc and HIF-1 in regulating glycolysis – the seventh hallmark of cancer. Cell Mol Life Sci. 2008;65:3981–99.
21. Semenza GL. Regulation of cancer cell metabolism by hypoxia-inducible factor 1. Semin Cancer Biol. 2009;19:12–6.
22. Semenza GL. HIF-1: upstream and downstream of cancer metabolism. Curr Opin Genet Dev. 2010;20:51–6.
23. Kruiswijk F, Labuschagne CF, Vousden KH. P53 in survival, death and metabolic health: a lifeguard with a licence to kill. Nat Rev Mol Cell Biol. 2015;16:393–405.
24. Mattick JS. Challenging the dogma: the hidden layer of non-protein-coding RNAs in complex organisms. Bioessays. 2003;25:930–9.
25. Mattick JS. The hidden genetic program of complex organisms. Sci Am. 2004;291:60–7.
26. Heinrichs A. MicroRNAs get a boost. Nat Rev Mol Cell Biol. 2009;10:302.
27. Baumann K. Gene expression: RNAi as a global transcriptional activator. Nat Rev Mol Cell Biol. 2014;15:298.
28. Dillon CP, Sandy P, Nencioni A, et al. RNAI as an experimental and therapeutic tool to study and regulate physiological and disease processes. Annu Rev Physiol. 2005;67:147–73.
29. Ross RJ, Weiner MM, Lin HF. PIWI proteins and PIWI-interacting RNAs in the soma. Nature. 2014;505:353–9.
30. Sato K, Siomi MC. Piwi-interacting RNAs: biological functions and biogenesis. Essays Biochem. 2013;54:39–52.
31. Weick EM, Miska EA. PiRNAs: from biogenesis to function. Development. 2014;141:3458–71.
32. Kim VN, Han J, Siomi MC. Biogenesis of small RNAs in animals. Nat Rev Mol Cell Biol. 2009;10:126–39.
33. Batista PJ, Chang HY. Long non-coding RNAs: cellular address codes in development and disease. Cell. 2013;152:1298–307.
34. Bonasio R, Shiekhattar R. Regulation of transcription by long non-coding RNAs. Annu Rev Genet. 2014;48:433–55.
35. Liz J, Esteller M. LncRNAs and microRNAs with a role in cancer development. Biochim Biophys Acta. 2016;1859(1):169–76. doi:10.1016/j.
36. Fu XD. Non-coding RNA: a new frontier in regulatory biology. Natl Sci Rev. 2014;1:190–204.
37. Kugel JF, Goodrich JA. Non-coding RNAs: key regulators of mammalian transcription. Trends Biochem Sci. 2012;37:144–51.

38. Mondal T, Kanduri C. Maintenance of epigenetic information: a non-coding RNA perspective. Chromosom Res. 2013;21:615–25.
39. Bartel DP. MicroRNAs: genomics, biogenesis, mechanism, and function. Cell. 2004;116:281–97.
40. Lee RC, Feinbaum RL, Ambros V. The C. elegans heterochronic gene lin-4 encodes small RNAs with antisense complementarity to lin-14. Cell. 1993;75:843–54.
41. Calin GA, Croce CM. MicroRNA signatures in human cancers. Nat Rev Cancer. 2006;6:857–66.
42. Esquela-Kerscher A, Slack FJ. Oncomirs – microRNAs with a role in cancer. Nat Rev Cancer. 2006;6:259–69.
43. Chang TC, Wentzel EA, Kent OA, et al. Transactivation of miR-34a by p53 broadly influences gene expression and promotes apoptosis. Mol Cell. 2007;26:745–52.
44. He L, He X, Lim LP, et al. A microRNA component of the p53 tumour suppressor network. Nature. 2007;447:1130–4.
45. Ma L, Teruya-Feldstein J, Weinberg RA. Tumour invasion and metastasis initiated by microRNA-10b in breast cancer. Nature. 2007;449:682–8.
46. Volinia S, Calin GA, Liu CG, et al. A microRNA expression signature of human solid tumors defines cancer gene targets. Proc Natl Acad Sci U S A. 2006;103:2257–61.
47. Hatley ME, Patrick DM, Garcia MR, et al. Modulation of K-Ras-dependent lung tumorigenesis by MicroRNA-21. Cancer Cell. 2010;18:282–93.
48. Faraoni I, Antonetti FR, Cardone J, et al. miR-155 gene: a typical multifunctional microRNA. Biochim Biophys Acta. 2009;1792:497–505.
49. Rodriguez A, Vigorito E, Clare S, et al. Requirement of bic/microRNA-155 for normal immune function. Science. 2007;316:608–11.
50. Gupta RA, Shah N, Wang KC, et al. Long non-coding RNA HOTAIR reprograms chromatin state to promote cancer metastasis. Nature. 2010;464:1071–6.
51. Chan B, Manley J, Lee J, et al. The emerging roles of microRNAs in cancer metabolism. Cancer Lett. 2015;356:301–8.
52. Pulito C, Donzelli S, Muti P, et al. MicroRNAs and cancer metabolism reprogramming: the paradigm of metformin. Ann Transl Med. 2014;2:58.
53. Zhao XY, Lin JD. Long non-coding RNAs: a new regulatory code in metabolic control. Trends Biochem Sci. 2015;40:586–96.
54. Chen B, Li H, Zeng X, et al. Roles of microRNA on cancer cell metabolism. J Transl Med. 2012;10:228.
55. Chow TF, Mankaruos M, Scorilas A, et al. The miR-17-92 cluster is over expressed in and has an oncogenic effect on renal cell carcinoma. J Urol. 2010;183:743–51.
56. Lefebvre P, Staels B. SREBF2-embedded mir33 links the nuclear bile acid receptor FXR to cholesterol and lipoprotein metabolism. Arterioscler Thromb Vasc Biol. 2015;35:748–9.
57. Ono K, Horie T, Nishino T, et al. MicroRNA-33a/b in lipid metabolism - novel "thrifty" models. Circ J. 2015;79:278–84.
58. Lo TF, Tsai WC, Chen ST. MicroRNA-21-3p, a berberine-induced miRNA, directly down-regulates human methionine adenosyltransferases 2A and 2B and inhibits hepatoma cell growth. PLoS ONE. 2013;8:e75628.
59. Haemmerle M, Gutschner T. Long non-coding RNAs in cancer and development: where do we go from here? Int J Mol Sci. 2015;16:1395–405.
60. Karapetyan AR, Buiting C, Kuiper RA, et al. Regulatory roles for long ncRNA and mRNA. Cancers. 2013;5:462–90.
61. Prensner JR, Chinnaiyan AM. The emergence of lncRNAs in cancer biology. Cancer Discov. 2011;1:391–407.
62. Yang F, Zhang H, Mei Y, et al. Reciprocal regulation of HIF-1alpha and lincRNA-p21 modulates the Warburg effect. Mol Cell. 2014;53:88–100.
63. Borchert GM, Lanier W, Davidson BL. RNA polymerase III transcribes human microRNAs. Nat Struct Mol Biol. 2006;13:1097–101.

64. Lee Y, Jeon K, Lee JT, et al. MicroRNA maturation: stepwise processing and subcellular localization. EMBO J. 2002;21:4663–70.
65. Lee Y, Kim M, Han J, et al. MicroRNA genes are transcribed by RNA polymerase II. EMBO J. 2004;23:4051–60.
66. Yang JS, Lai EC. Alternative miRNA biogenesis pathways and the interpretation of core miRNA pathway mutants. Mol Cell. 2011;43:892–903.
67. Bohnsack MT, Czaplinski K, Gorlich D. Exportin 5 is a RanGTP-dependent dsRNA-binding protein that mediates nuclear export of pre-miRNAs. RNA. 2004;10:185–91.
68. Yi R, Qin Y, Macara IG, et al. Exportin-5 mediates the nuclear export of pre-microRNAs and short hairpin RNAs. Genes Dev. 2003;17:3011–6.
69. Hutvagner G, McLachlan J, Pasquinelli AE, et al. A cellular function for the RNA-interference enzyme Dicer in the maturation of the let-7 small temporal RNA. Science. 2001;293:834–8.
70. Ketting RF, Fischer SEJ, Bernstein E, et al. Dicer functions in RNA interference and in synthesis of small RNA involved in developmental timing in C-elegans. Genes Dev. 2001;15:2654–9.
71. Cannell IG, Kong YW, Bushell M. How do microRNAs regulate gene expression? Biochem Soc Trans. 2008;36:1224–31.
72. Lim LP, Lau NC, Garrett-Engele P, et al. Microarray analysis shows that some microRNAs down-regulate large numbers of target mRNAs. Nature. 2005;433:769–73.
73. Bensinger SJ, Christofk HR. New aspects of the Warburg effect in cancer cell biology. Semin Cell Dev Biol. 2012;23:352–61.
74. Cairns RA, Harris IS, Mak TW. Regulation of cancer cell metabolism. Nat Rev Cancer. 2011;11:85–95.
75. Elf SE, Chen J. Targeting glucose metabolism in patients with cancer. Cancer. 2014;120:774–80.
76. Upadhyay M, Samal J, Kandpal M, et al. The Warburg effect: insights from the past decade. Pharmacol Ther. 2013;137:318–30.
77. Wu W, Zhao S. Metabolic changes in cancer: beyond the Warburg effect. Acta Biochim Biophys Sin. 2013;45:18–26.
78. Yu C, Xue J, Zhu W, et al. Warburg meets non-coding RNAs: the emerging role of ncRNA in regulating the glucose metabolism of cancer cells. Tumour Biol. 2015;36:81–94.
79. Sun L, He X, Cao Y, et al. MicroRNAs and energy metabolism in cancer cells. In: Sadegh B, editor. MicroRNAs: key regulators of oncogenesis. Switzerland: Springer; 2014. p. 83–95.
80. Wang B, Sun F, Dong N, et al. MicroRNA-7 directly targets insulin-like growth factor 1 receptor to inhibit cellular growth and glucose metabolism in gliomas. Diagn Pathol. 2014;9:211. doi:10.1186/s13000-014-0211-y.
81. Fu XH, Dong BN, Tian Y, et al. MicroRNA-26a regulates insulin sensitivity and metabolism of glucose and lipids. J Clin Investig. 2015;125:2497–509.
82. Jordan SD, Kruger M, Willmes DM, et al. Obesity-induced overexpression of miRNA-143 inhibits insulin-stimulated AKT activation and impairs glucose metabolism. Nat Cell Biol. 2011;13:434–46.
83. Fang R, Xiao T, Fang Z, et al. MicroRNA-143 (miR-143) regulates cancer glycolysis via targeting hexokinase 2 gene. J Biol Chem. 2012;287:23227–35.
84. Macheda ML, Rogers S, Best JD. Molecular and cellular regulation of glucose transporter (GLUT) proteins in cancer. J Cell Physiol. 2005;202:654–62.
85. Burant CF, Bell GI. Mammalian facilitative glucose transporters: evidence for similar substrate recognition sites in functionally monomeric proteins. Biochemistry. 1992;31:10414–20.
86. Horie T, Ono K, Nishi H, et al. MicroRNA-133 regulates the expression of GLUT4 by targeting KLF15 and is involved in metabolic control in cardiac myocytes. Biochem Bioph Res Commun. 2009;389:315–20.
87. Joost HG, Bell GI, Best JD, et al. Nomenclature of the GLUT/SLC2A family of sugar/polyol transport facilitators. Am J Physiol Endocrinol Metab. 2002;282:E974–6.

88. Ramirez CM, Goedeke L, Rotllan N, et al. MicroRNA 33 regulates glucose metabolism. Mol Cell Biol. 2013;33:2891–902.
89. Busk M, Horsman MR, Jakobsen S, et al. Cellular uptake of PET tracers of glucose metabolism and hypoxia and their linkage. Eur J Nucl Med Mol Imaging. 2008;35:2294–303.
90. Kelloff GJ, Hoffman JM, Johnson B, et al. Progress and promise of FDG-PET imaging for cancer patient management and oncologic drug development. Clin Cancer Res. 2005;11:2785–808.
91. Kuehne A, Emmert H, Soehle J, et al. Acute activation of oxidative pentose phosphate pathway as first-line response to oxidative stress in human skin cells. Mol Cell. 2015;59:359–71.
92. Ward PS, Thompson CB. Metabolic reprogramming: a cancer hallmark even warburg did not anticipate. Cancer Cell. 2012;21:297–308.
93. Gao P. MicroRNAs and cancer metabolism. In: Cho WCS, editor. MicroRNAs in cancer translational research. Switzerland: Springer; 2011. p. 485–97.
94. Gregersen LH, Jacobsen A, Frankel LB, et al. MicroRNA-143 down-regulates Hexokinase 2 in colon cancer cells. BMC Cancer. 2012;12:232.
95. Peschiaroli A, Giacobbe A, Formosa A, et al. MiR-143 regulates hexokinase 2 expression in cancer cells. Oncogene. 2013;32:797–802.
96. Fong MY, Zhou WY, Liu L, et al. Breast-cancer-secreted miR-122 reprograms glucose metabolism in premetastatic niche to promote metastasis. Nat Cell Biol. 2015;17:183–94.
97. Luan WK, Wang YY, Chen XC, et al. PKM2 promotes glucose metabolism and cell growth in gliomas through a mechanism involving a let-7a/c-Myc/hnRNPA1 feedback loop. Oncotarget. 2015;6:13006–18.
98. Chen B, Liu YL, Jin XW, et al. MicroRNA-26a regulates glucose metabolism by direct targeting PDHX in colorectal cancer cells. BMC Cancer. 2014;14:443. doi:10.1186/1471-2407-14-443.
99. Tsukamoto Y, Nakada C, Noguchi T, et al. MicroRNA-375 is down-regulated in gastric carcinomas and regulates cell survival by targeting PDK1 and 14-3-3 zeta. Cancer Res. 2010;70:2339–49.
100. Kaller M, Liffers ST, Oeljeklaus S, et al. Genome-wide characterization of miR-34a induced changes in protein and mRNA expression by a combined pulsed SILAC and microarray analysis. Mol Cell Proteomics. 2011;10(8):M111.010462. doi:10.1074/mcp.M111.010462.
101. Kim HR, Roe JS, Lee JE, et al. P53 regulates glucose metabolism by miR-34a. Biochem Biophys Res Commun. 2013;437:225–31.
102. Uesugi A, Kozaki K, Tsuruta T, et al. The tumor suppressive microRNA miR-218 targets the mTOR component Rictor and inhibits AKT phosphorylation in oral cancer. Cancer Res. 2011;71:5765–78.
103. Godlewski J, Nowicki MO, Bronisz A, et al. MicroRNA-451 regulates LKB1/AMPK signaling and allows adaptation to metabolic stress in glioma cells. Mol Cell. 2010;37:620–32.
104. Ma X, Li C, Sun L, et al. Lin28/let-7 axis regulates aerobic glycolysis and cancer progression via PDK1. Nat Commun. 2014;5:5212.
105. Medes G, Thomas A, Weinhouse S. Nutritional factors in fatty acid synthesis by tissue slices in vitro. J Biol Chem. 1952;197:181–91.
106. Santos CR, Schulze A. Lipid metabolism in cancer. FEBS J. 2012;279:2610–23.
107. Ameer F, Scandiuzzi L, Hasnain S, et al. De novo lipogenesis in health and disease. Metabolism. 2014;63:895–902.
108. Currie E, Schulze A, Zechner R, et al. Cellular fatty acid metabolism and cancer. Cell Metab. 2013;18:153–61.
109. Huang CF, Freter C. Lipid metabolism, apoptosis and cancer therapy. Int J Mol Sci. 2015;16:924–49.
110. Fernandez-Hernando C, Suarez Y, Rayner KJ, et al. MicroRNAs in lipid metabolism. Curr Opin Lipidol. 2011;22:86–92.
111. Wakil SJ, Abu-Elheiga LA. Fatty acid metabolism: target for metabolic syndrome. J Lipid Res. 2009;50:S138–43.

112. Davalos A, Goedeke L, Smibert P, et al. MiR-33a/b contribute to the regulation of fatty acid metabolism and insulin signaling. Proc Natl Acad Sci U S A. 2011;108:9232–7.
113. Rottiers V, Najafi-Shoushtari SH, Kristo F, et al. MicroRNAs in metabolism and metabolic diseases. Cold Spring Harb Symp Quant Biol. 2011;76:225–33.
114. Tao R, Xiong X, DePinho RA, et al. Hepatic SREBP-2 and cholesterol biosynthesis are regulated by FoxO3 and Sirt6. J Lipid Res. 2013;54:2745–53.
115. El Azzouzi H, Leptidis S, Dirkx E, et al. The hypoxia-inducible microRNA cluster miR-199a approximately 214 targets myocardial PPARdelta and impairs mitochondrial fatty acid oxidation. Cell Metab. 2013;18:341–54.
116. Jeon TI, Osborne TF. SREBPs: metabolic integrators in physiology and metabolism. Trends Endocrinol Metab. 2012;23:65–72.
117. Shao W, Espenshade PJ. Expanding roles for SREBP in metabolism. Cell Metab. 2012;16:414–9.
118. Shimano H. SREBPs: physiology and pathophysiology of the SREBP family. FEBS J. 2009;276:616–21.
119. Soyal SM, Nofziger C, Dossena S, et al. Targeting SREBPs for treatment of the metabolic syndrome. Trends Pharmacol Sci. 2015;36:406–16.
120. Contreras AV, Torres N, Tovar AR. PPAR-alpha as a key nutritional and environmental sensor for metabolic adaptation. Adv Nutr. 2013;4:439–52.
121. Gao J, Yuan S, Jin J, et al. PPARalpha regulates tumor progression, foe or friend? Eur J Pharmacol. 2015;765:560–4.
122. Li TG, Chiang JYL. Regulation of bile acid and cholesterol metabolism by PPARs. Dig PPAR Res. 2009. doi:10.1155/2009/501739.
123. Li TG, Chiang JYL. Bile acid signaling in metabolic disease and drug therapy. Pharmacol Rev. 2014;66:948–83.
124. Koszowska AU, Nowak J, Dittfeld A, et al. Obesity, adipose tissue function and the role of vitamin D. Cen Eur J Immunol. 2014;39:260–4.
125. Lefebvre P, Staels B. SREBF2-embedded mir33 links the nuclear bile acid receptor FXR to cholesterol and lipoprotein metabolism. Arterioscl Throm Vasc Biol. 2015;35:748–9.
126. DiMarco DM, Fernandez ML. The regulation of reverse cholesterol transport and cellular cholesterol homeostasis by microRNAs. Biology. 2015;4:494–511.
127. Annema W, Tietge UJ. Regulation of reverse cholesterol transport – a comprehensive appraisal of available animal studies. Nutr Metab. 2012;9:25.
128. Hill SA, McQueen MJ. Reverse cholesterol transport-a review of the process and its clinical implications. Clin Biochem. 1997;30:517–25.
129. Krutzfeldt J, Rajewsky N, Braich R, et al. Silencing of microRNAs in vivo with 'antagomirs'. Nature. 2005;438:685–9.
130. Rotllan N, Fernandez-Hernando C. MicroRNA regulation of cholesterol metabolism. Cholesterol. 2012;2012:847849.
131. Najafi-Shoushtari SH, Kristo F, Li Y, et al. MicroRNA-33 and the SREBP host genes cooperate to control cholesterol homeostasis. Science. 2010;328:1566–9.
132. Rayner KJ, Suarez Y, Davalos A, et al. MiR-33 contributes to the regulation of cholesterol homeostasis. Science. 2010;328:1570–3.
133. Niesor EJ, Schwartz GG, Perez A, et al. Statin-induced decrease in ATP-binding cassette transporter A1 expression via microRNA33 induction may counteract cholesterol efflux to high-density lipoprotein. Cardiovasc Drugs Ther. 2015;29:7–14.
134. Kim J, Yoon H, Ramirez CM, et al. MiR-106b impairs cholesterol efflux and increases Abeta levels by repressing ABCA1 expression. Exp Neurol. 2012;235:476–83.
135. Ramirez CM, Davalos A, Goedeke L, et al. MicroRNA-758 regulates cholesterol efflux through posttranscriptional repression of ATP-binding cassette transporter A1. Arterioscl Throm Vasc Biol. 2011;31:2707–14.
136. Han CC, Wei SH, He F, et al. The regulation of lipid deposition by insulin in goose liver cells is mediated by the PI3K-AKT-mTOR signaling pathway. PLoS ONE. 2015;10(5):e0098759. doi:10.1371/journal.pone.0098759.

137. Hsieh YC, Chen YM, Li CY, et al. To complete its replication cycle, a shrimp virus changes the population of long chain fatty acids during infection via the PI3K-Akt-mTOR-HIF1alpha pathway. Dev Comp Immunol. 2015;53:85–95.
138. Li J, Huang Q, Long X, et al. CD147 reprograms fatty acid metabolism in hepatocellular carcinoma cells through Akt/mTOR/SREBP1c and P38/PPARalpha pathways. J Hepatol. 2015;63(6):1378–89.
139. Song MS, Salmena L, Pandolfi PP. The functions and regulation of the PTEN tumour suppressor. Nat Rev Mol Cell Biol. 2012;13:283–96.
140. Dobrzyn P, Jazurek M, Dobrzyn A. Stearoyl-CoA desaturase and insulin signaling-what is the molecular switch? Biochim Biophys Acta. 2010;1797:1189–94.
141. Krycer JR, Sharpe LJ, Luu W, et al. The Akt-SREBP nexus: cell signaling meets lipid metabolism. Trends Endocrinol Metab. 2010;21:268–76.
142. Porstmann T, Santos CR, Griffiths B, et al. SREBP activity is regulated by mTORC1 and contributes to Akt-dependent cell growth. Cell Metab. 2008;8:224–36.
143. Teran-Garcia M, Adamson AW, Yu G, et al. Polyunsaturated fatty acid suppression of fatty acid synthase (FASN): evidence for dietary modulation of NF-Y binding to the Fasn promoter by SREBP-1c. Biochem J. 2007;402:591–600.
144. Deberardinis RJ, Lum JJ, Thompson CB, et al. Phosphatidylinositol 3-kinase-dependent modulation of carnitine palmitoyltransferase 1A expression regulates lipid metabolism during hematopoietic cell growth. J Biol Chem. 2006;281:37372–80.
145. Schlaepfer IR, Rider L, Rodrigues LU, et al. Lipid catabolism via CPT1 as a therapeutic target for prostate cancer. Mol Cancer Ther. 2014;13:2361–71.
146. Gori M, Arciello M, Balsano C. MicroRNAs in nonalcoholic fatty liver disease: novel biomarkers and prognostic tools during the transition from steatosis to hepatocarcinoma. Bio Med Res Int. 2014;2014:741465.
147. Vinciguerra M, Carrozzino F, Peyrou M, et al. Unsaturated fatty acids promote hepatoma proliferation and progression through down-regulation of the tumor suppressor PTEN. J Hepatol. 2009;50:1132–41.
148. Lee J, Hong SW, Park SE, et al. AMP-activated protein kinase suppresses the expression of LXR/SREBP-1 signaling-induced ANGPTL8 in HepG2 cells. Mol Cell Endocrinol. 2015;414:148–55.
149. Fernandez-Hernando C, Ramirez CM, Goedeke L, et al. MicroRNAs in metabolic disease. Arterioscler Thromb Vasc Biol. 2013;33:178–85.
150. Zheng Y, Jiang S, Zhang Y, et al. Detection of miR-33 expression and the verification of its target genes in the fatty liver of geese. Int J Mol Sci. 2015;16:12737–52.
151. Gao P, Tchernyshyov I, Chang TC, et al. C-Myc suppression of miR-23a/b enhances mitochondrial glutaminase expression and glutamine metabolism. Nature. 2009;458:762–5.
152. Rathore MG, Saumet A, Rossi JF, et al. The NF-kappa B member p65 controls glutamine metabolism through miR-23a. Int J Biochem Cell Biol. 2012;44:1448–56.
153. Liu W, Le A, Hancock C, et al. Reprogramming of proline and glutamine metabolism contributes to the proliferative and metabolic responses regulated by oncogenic transcription factor c-MYC. Proc Natl Acad Sci U S A. 2012;109:8983–8.
154. Liu YM, Borchert GL, Donald SP, et al. Proline oxidase functions as a mitochondrial tumor suppressor in human cancers. Cancer Res. 2009;69:6414–22.
155. Phang JM, Liu W, Zabirnyk O. Proline metabolism and microenvironmental stress. Annu Rev Nutr. 2010;30:441–63.
156. Koning T, Fuchs F, Klomp L. Serine, glycine, and threonine. In: Lajtha A, editor. Handbook of neurochemistry and molecular neurobiology-amino acids and peptides in the nervous system. 3rd ed. Berlin: Springer; 2007. p. 23–45.
157. Locasale JW. Serine, glycine and one-carbon units: cancer metabolism in full circle. Nat Rev Cancer. 2013;13:572–83.
158. Tedeschi PM, Markert EK, Gounder M, et al. Contribution of serine, folate and glycine metabolism to the ATP, NADPH and purine requirements of cancer cells. Cell Death Dis. 2013;4, e877.

159. Lu SC, Mato JM. S-Adenosylmethionine in cell growth, apoptosis and liver cancer. J Gastroenterol Hepatol. 2008;23(Supplement s1):S73–7.
160. Avila MA, Berasain C, Torres L, et al. Reduced mRNA abundance of the main enzymes involved in methionine metabolism in human liver cirrhosis and hepatocellular carcinoma. J Hepatol. 2000;33:907–14.
161. Cai J, Sun WM, Hwang JJ, et al. Changes in S-adenosylmethionine synthetase in human liver cancer: molecular characterization and significance. Hepatology. 1996;24:1090–7.
162. Frau M, Tomasi ML, Simile MM, et al. Role of transcriptional and posttranscriptional regulation of methionine adenosyltransferases in liver cancer progression. Hepatology. 2012;56:165–75.
163. Wang W, Peng JX, Yang JQ, et al. Identification of gene expression profiling in hepatocellular carcinoma using cDNA microarrays. Dig Dis Sci. 2009;54:2729–35.
164. Yang HP, Cho ME, Li TWH, et al. MicroRNAs regulate methionine adenosyltransferase 1A expression in hepatocellular carcinoma. J Clin Investig. 2013;123:285–98.
165. Lo TF, Tsai WC, Chen ST. MicroRNA-21-3p, a berberine-induced miRNA, directly downregulates human methionine adenosyltransferases 2A and 2B and inhibits hepatoma cell growth. PLoS ONE. 2013;8(9):883–8.
166. Koturbash I, Melnyk S, James SJ, et al. Role of epigenetic and miR-22 and miR-29b alterations in the down-regulation of Mat1a and Mthfr genes in early preneoplastic livers in rats induced by 2-acetylaminofluorene. Mol Carcinog. 2013;52:318–27.
167. Stone N, Pangilinan F, Molloy AM, et al. Bioinformatic and genetic association analysis of microRNA target sites in one-carbon metabolism genes. PLoS ONE. 2011;6:e21851.
168. Liu S, Lucas KJ, Roy S, et al. Mosquito-specific microRNA-1174 targets serine hydroxymethyltransferase to control key functions in the gut. Proc Natl Acad Sci U S A. 2014;111:14460–5.
169. Shimomura Y, Murakami T, Nagasaki M, et al. Regulation of branched-chain amino acid metabolism and pharmacological effects of branched-chain amino acids. Hepatol Res. 2004;30S:3–8.
170. Shimomura Y, Murakami T, Nakai N, et al. Exercise promotes BCAA catabolism: effects of BCAA supplementation on skeletal muscle during exercise. J Nutr. 2004;134:1583S–7.
171. Kimball SR. Regulation of global and specific mRNA translation by amino acids. J Nutr. 2002;132:883–6.
172. Yudkoff M. Brain metabolism of branched-chain amino acids. Glia. 1997;21:92–8.
173. Mersey BD, Jin P, Danner DJ. Human microRNA (miR29b) expression controls the amount of branched chain alpha-ketoacid dehydrogenase complex in a cell. Hum Mol Genet. 2005;14:3371–7.
174. Okazaki Y, Furuno M, Kasukawa T, et al. Analysis of the mouse transcriptome based on functional annotation of 60,770 full-length cDNAs. Nature. 2002;420:563–73.
175. Geisler S, Coller J. RNA in unexpected places: long non-coding RNA functions in diverse cellular contexts. Nat Rev Mol Cell Biol. 2013;14:699–712.
176. Roth A, Diederichs S. Rap and chirp about X inactivation. Nature. 2015;521:170–1.
177. Fatica A, Bozzoni I. Long non-coding RNAs: new players in cell differentiation and development. Nat Rev Genet. 2014;15:7–21.
178. Mercer TR, Dinger ME, Mattick JS. Long non-coding RNAs: insights into functions. Nat Rev Genet. 2009;10:155–9.
179. Wilusz JE. Long non-coding RNAs: re-writing dogmas of RNA processing and stability. Biochim Biophys Acta. 2015;1859(1):128–38.
180. Wu XS, Wang XA, Wu WG, et al. MALAT1 promotes the proliferation and metastasis of gallbladder cancer cells by activating the ERK/MAPK pathway. Cancer Biol Ther. 2014;15:806–14.
181. Zhang XW, Bu P, Liu L, et al. Overexpression of long non-coding RNA PVT1 in gastric cancer cells promotes the development of multidrug resistance. Biochem Biophys Res Commun. 2015;462:227–32.

182. He XL, Tan XM, Wang X, et al. C-Myc-activated long non-coding RNA CCAT1 promotes colon cancer cell proliferation and invasion. Tumor Biol. 2014;35:12181–8.
183. Kallen AN, Zhou XB, Xu J, et al. The imprinted H19 lncRNA antagonizes let-7 microRNAs. Mol Cell. 2013;52:101–12.
184. Matouk IJ, Mezan S, Mizrahi A, et al. The oncofetal H19 RNA connection: hypoxia, p53 and cancer. Biochem Biophys Acta (BBA) Mol Basis Dis. 2010;1803:443–51.
185. Gao Y, Wu FJ, Zhou JC, et al. The H19/let-7 double-negative feedback loop contributes to glucose metabolism in muscle cells. Nucleic Acids Res. 2014;42:13799–811.
186. Zhu MJ, Chen Q, Liu X, et al. lncRNA H19/miR-675 axis represses prostate cancer metastasis by targeting TGFBI. FEBS J. 2014;281:3766–75.
187. Li ZK, Li X, Wu SZ, et al. Long non-coding RNA UCA1 promotes glycolysis by upregulating hexokinase 2 through the mTOR-STAT3/microRNA143 pathway. Cancer Sci. 2014;105:951–5.
188. Ellis BC, Graham LD, Molloy PL. CRNDE, a long non-coding RNA responsive to insulin/IGF signaling, regulates genes involved in central metabolism. Biochem Biophys Acta (BBA) Mol Basis Dis. 2014;1843:372–86.
189. Hammerle M, Gutschner T, Uckelmann H, et al. Posttranscriptional destabilization of the liver-specific long non-coding RNA HULC by the IGF2 mRNA-binding protein 1 (IGF2BP1). Hepatology. 2013;58:1703–12.
190. Liu BD, Sun LJ, Liu Q, et al. A cytoplasmic NF-kappa B interacting long non-coding RNA blocks I kappa B phosphorylation and suppresses breast cancer metastasis. Cancer Cell. 2015;27:370–81.
191. Yacqub-Usman K, Pickard MR, Williams GT. Reciprocal regulation of GAS5 lncRNA levels and mTOR inhibitor action in prostate cancer cells. Prostate. 2015;75:693–705.
192. Hung CL, Wang LY, Yu YL, et al. A long non-coding RNA connects c-Myc to tumor metabolism. Proc Natl Acad Sci U S A. 2014;111:18697–702.
193. Liu Q, Huang JG, Zhou NJ, et al. LncRNA loc285194 is a p53-regulated tumor suppressor. Nucleic Acids Res. 2013;41:4976–87.
194. Zhang A, Xu M, Mo YY. Role of the lncRNA-p53 regulatory network in cancer. J Mol Cell Biol. 2014;6:181–91.
195. Takahashi K, Yan IK, Haga H, et al. Modulation of hypoxia-signaling pathways by extracellular linc-RoR. J Cell Sci. 2014;127:1585–94.
196. Aravin AA, Chan DC. PiRNAs meet mitochondria. Dev Cell. 2011;20:287–8.
197. Mannoor K, Liao JP, Jiang F. Small nucleolar RNAs in cancer. Biochim Biophys Acta. 2012;1826:121–8.
198. Stepanov GA, Filippova JA, Komissarov AB, et al. Regulatory role of small nucleolar RNAs in human diseases. Dig Biomed Res Int. 2015. doi:10.1155/2015/206849.
199. Watkins NJ, Bohnsack MT. The box C/D and H/ACA snoRNPs: key players in the modification, processing and the dynamic folding of ribosomal RNA. Wiley Interdiscip Rev RNA. 2012;3:397–414.
200. Yu YT, Meier UT. RNA-guided isomerization of uridine to pseudouridine-pseudouridylation. RNA Biol. 2014;11:1483–94.
201. Jinn S, Brandis KA, Ren A, et al. SnoRNA U17 regulates cellular cholesterol trafficking. Cell Metab. 2015;21:855–67.
202. Sanger HL, Klotz G, Riesner D, et al. Viroids are single-stranded covalently closed circular RNA molecules existing as highly base-paired rod-like structures. Proc Natl Acad Sci U S A. 1976;73:3852–6.
203. Lasda E, Parker R. Circular RNAs: diversity of form and function. RNA. 2014;20:1829–42.
204. Vicens Q, Westhof E. Biogenesis of circular RNAs. Cell. 2014;159:13–4.
205. Hansen TB, Kjems J, Damgaard CK. Circular RNA and miR-7 in cancer. Cancer Res. 2013;73:5609–12.

Chapter 8
Noncoding RNAs in Tumor Angiogenesis

Azam Khorshidi, Preet Dhaliwal, and Burton B. Yang

Abstract Solid tumors require angiogenesis to grow beyond 2 mm in size. In most cases, tumor cells undergo angiogenic switch and secrete substances that are required for generation of new capillary sprouting from existing blood vessels. Tumor angiogenesis is driven by a complex interplay between pro-angiogenic (VEGF/VEGFR, PDGF/PDGFR) and anti-angiogenic factors (TSP-1/TSP-2) within the tumor microenvironment. In addition, control of tissue remodeling and degradation by matrix metalloproteinases (MMPs) and tissue inhibitor of metalloproteinases (TIMPs) contribute to tumor angiogenesis. Furthermore, tumor suppressors or oncogenes that control cellular motility and maintain or promote hypoxia (HIFs and MYC) are also actively playing roles in tumor angiogenesis. Noncoding RNAs (ncRNAs), including microRNAs, are a novel class of regulatory molecules that control the gene expression in a posttranscriptional manner. MicroRNAs regulate important physiological processes, such as proliferation, apoptosis, and differentiation, as well as pathological conditions including oncogenesis. Accumulating evidence suggests that microRNAs directly modulate the process of angiogenesis by targeting important angiogenic factors and signaling molecules. Understanding the molecular mechanism behind the regulation of angiogenesis by microRNAs is important due to their therapeutic potential which may lead to improving outcome for cancer patients. Besides, ncRNAs with a regulatory role in angiogenesis, such as long noncoding RNAs (lncRNAs), have been identified in the genome. However, the mechanisms of the vast majority of lncRNAs are currently unknown. For the few lncRNAs characterized at the functional level, accumulating evidence shows that they play important roles in malignant diseases. The function and mechanism in angiogenesis will be described in this chapter.

A. Khorshidi • P. Dhaliwal • B.B. Yang (✉)
Sunnybrook Research Institute, Sunnybrook Health Sciences Centre,
2075 Bayview Avenue, M4N 3M5 Toronto, Canada

Department of Laboratory Medicine and Pathobiology, University of Toronto,
Toronto, Canada
e-mail: byang@sri.utoronto.ca

E. Song (ed.), *The Long and Short Non-coding RNAs in Cancer Biology*,
Advances in Experimental Medicine and Biology 927,
DOI 10.1007/978-981-10-1498-7_8

Keywords Angiogenesis • Noncoding RNA • Tumor

8.1 Introduction

Angiogenesis occurs through a highly organized series of events by which new blood vessels form through the growth of existing blood vessels. During angiogenesis, quiescent endothelial cells, which cover the luminal side of all blood vessels, are activated in response to environmental triggers and start to proliferate, migrate, and organize themselves in tubular structures. The term angiogenesis can be distinguished from vasculogenesis, which refers to de novo production of endothelial cells from endothelial precursor cells (angioblast) during embryonic development. Vasculogenesis is typically followed by classical angiogenesis during prenatal development which leads to the growth and remodeling of the primitive vascular network into a complex network. In the adult body, the vascular endothelium acquires essentially a quiescent, non-angiogenic state and serves mainly as a nonthrombogenic surface to conduct nutritive blood flow to organs. These cells, however, retain considerable growth potential and are responsive to pro- and anti-angiogenic factors, which is essential for vascular remodeling during physiological conditions, like wound healing, inflammation, and pregnancy [1]. The same angiogenic pathways have also been adapted during pathological conditions such as disease as diverse as cancer, macular degeneration, psoriasis, diabetic retinopathy, and inflammatory disorders like arthritis and atherosclerosis [2]. Therefore, the control of reversing dormant endothelial cells and restoring their proliferative state must be regulated by a complex milieu of stimulatory and inhibitory signals and requires a number of molecular and cellular events to be temporarily and spatially orchestrated [3].

8.2 Tumor Angiogenesis

During tumorigenesis, malignant cells acquire multiple characteristics that provide a growth advantage over normal cells. Tumor angiogenesis is one of the hallmarks of cancer that drive tumor growth beyond a diffusion limit size and enhance metastasis [4]. In response to hypoxia, tumor cells tilt the balance toward stimulatory angiogenic factors, like vascular endothelial growth factor (VEGF) and angiopoietin, to facilitate neovascularization [5–7]. This process is called "angiogenic switch" and leads to activation of endothelial cells in nearby vessels. Subsequently, degradation of the extracellular matrix in activated endothelial cells by different proteolytic enzymes results in migration of endothelial cells toward chemotactic clues that come from the tumor tissue and the formation of vessel-like structures [8]. These newly formed vessels are premature and fragile. Subsequent inhibition of endothelial cell growth and the recruitment of pericytes and smooth muscle cells to form capillary tubes lead to maturation of vessels. Even though these vessels are disorganized and irregular in structure, they can still provide the growing tumor mass with required nutrients and metabolites [9].

Despite the wealth of data on pathological angiogenesis, it is still not clear what molecular mechanisms govern angiogenic switch in tumor angiogenesis. Recently, new opportunities for better understanding of tumor biology have come with the discovery of noncoding RNAs as a novel class of gene regulatory molecules [10]. These noncoding RNAs exert their gene regulatory function at many different levels, including posttranscriptional and posttranslational level. Here, we will review the functions and mechanisms of noncoding RNAs, mainly microRNAs and long noncoding RNAs, in angiogenesis and vasculature remodeling in cancer, as well as their significance in cancer development.

8.3 Importance of MicroRNAs in Regulation of Tumor Angiogenesis

MicroRNAs (miRNAs) are a class of small noncoding RNAs (~22 nucleotides) which play an important role in all biological pathways in multicellular organisms including mammals [11, 12]. MiRNAs regulate gene expression by binding to a target messenger RNA (mRNA), leading to either degradation or translational repression [13]. MiRNAs are generated by the act of two RNAse III endonucleases, Dicer and Drosha, in a two-step processing pathway [14]. The first evidence showing the role of microRNAs in the regulation of vascular development and angiogenesis came from studies on Dicer-deficient homozygous mice. Knockout mice die between 12.5 and 14.5 days of gestation due to lack of angiogenesis [15]. In addition, hypomorphic Dicer1 allele (Dicer d/d) mouse models are found to be infertile due to corpus luteum insufficiency resulted from impaired vascular formation in the ovary [16]. Similarly, Dicer mutants of zebra fish show disrupted blood circulations [17]. These data have been confirmed in vitro using short interfering RNA (siRNA) against Dicer in endothelial cells. Genetic silencing of Dicer in endothelial cells leads to downregulation of several key regulators of angiogenic phenotype, including reduced endothelial cell migration, capillary sprouting, and tube formation in vitro and in vivo. Experiments with cultured endothelial cells have revealed the important role of Dicer in several angiogenic pathways, including EC migration, proliferation, and capillary tube formation [18–20].

In contrast to Dicer, knockdown of Drosha does not lead to any major problem in angiogenesis in vivo. Even though knockdown of Drosha in cultured endothelial cells with siRNA results in significant reduction in tube formation and capillary sprouting, no significant blockade of angiogenesis has been observed in vivo [19]. This difference might be due to the presence of an alternative Drosha-independent

Table 8.1 MiRNAs with known functions in angiogenesis

MiRNAs	Targets	Molecular mechanism	Reference
MiR-17 family	TSP-1, CTGF, TIMP-1, ITGB-8, LATS2, β-TRCP2, STAT3, HIF-1α	Endothelial cell proliferation/ migration, pro-angiogenic molecules	[27–29, 36–40]
MiR-378	Sufu, Fus-1	VEGF, Ang-1, and Ang-2	[41]
MiR-98	–	Anti-angiogenic mechanisms	[48, 49]
MiR-126	SPRED1, PIK3R2	VEGF level, endothelial cell proliferation	[50, 52, 53, 55, 56]
MiR-221/ miR-222	c-kit, cyclin G1, p27, p57	Endothelial cell proliferation/ migration	[16, 18, 19, 58]
MiR-15b/ miR-16	Bcl-2	VEGF level, cell proliferation/survival	[59]
MiR-130a	HOXA5, GAX	Endothelial cell proliferation/ migration, tube formation	[60]
MiR-210	Ephrin-A3	Endothelial cell proliferation/ migration	[64]
MiR-296	HGS	VEGFR2 and PDGFR-β	[73]

miRNA processing pathway that compensates for the lack of Drosha [21–23] or the involvement of Dicer in other cellular pathways including the regulation of heterochromatin formation [24, 25].

8.3.1 Important miRNAs in Angiogenesis

Expression of miRNAs is strictly regulated in a tissue- and organ- specific manner. Three studies have been performed in an attempt to identify miRNAs involved in control of endothelial cell functions. Eight miRNAs, including let-7b, miR-16, miR-21, miR-23a, miR-29, miR-100, and miR-221, and miR-222, are shown to be highly expressed in human umbilical cord endothelial cells by all three sets of data. Meanwhile, only two out of the three studies find that let-7a, let-7d, miR-20, miR-99a, miR-126, miR-181a, and miR-320 are highly expressed in endothelial cells [18, 19, 26]. Only a few highly expressed miRNAs in endothelial cells have been functionally characterized in angiogenesis (see Table 8.1), which we will discuss in this section.

8.3.1.1 MicroRNA-17 Gene Family

The miR-17~92 cluster, also named oncomiR-1, is the first identified tumor-promoting miRNA [27]. This cluster consists of seven miRNAs, including miR-17, miR-5p/miR-3p, miR-18a, miR-19a, miR-20a, miR-19b-1, and miR-92a-1 that originate from the intronic region of c13orf25 on chromosome 13. Two paralogs of miR-17 ~ 92, miR-106a ~ 363 and miR-106b ~ 25, also exist in mammals. It has been shown that a higher degree of tumor vascularization is observed in vivo after overexpression of miR-17 ~ 92 cluster in ras-positive tumor cells, which provides the first clue to the importance of the member of this family in tumor angiogenesis [28], while inhibition of miR-17 ~ 92 in vitro represses EC sprouting and tube formation in Matrigel [29]. The angiogenic role of miR-17 confers this miRNA to play an important role in tumorigenesis [30–34]. One mechanism by which this cluster controls neovascularization is through modulating the production of angiogenic factors. For example, miR-18 and miR-19 preferentially suppress the expression of connective tissue growth factor (CTGF) and thrombospondin-1 (TSP-1), respectively, both of which inhibit angiogenesis. Meanwhile, knockdown of miR-17 ~ 92 cluster can partially restore the expression of TSP-1 and CTGF. Another member of this family, miR-17, targets tissue inhibitor of metalloproteinase 3 (TIMP-3) to modulate migration and proliferation of endothelial cells [35]. In fact, lack of angiogenesis in the corpus luteum in the hypomorphic Dicer allele (Dicer d/d) mice is attributed to the lack of miR-17-5p and let-7b [16].

Another microRNA in the microRNA-17 gene family with pro-angiogenic activity is miR-93 which belongs to the miR-106b ~ 25 cluster. The miRNA-106b ~ 25 cluster is composed of the highly conserved miRNA-106b, miRNA-93, and miRNA-25. Different studies have shown different molecular pathways through which miR-93 modulates angiogenesis. It has been demonstrated that overexpression of miR-93 in U87 glioblastoma cell line increases proliferation, migration, and tube formation of cocultured endothelial cells in vitro and enhances angiogenesis in vivo by modulating integrin signaling pathway through downregulation of integrin beta 8 (ITGB-8) [36]. Similarly, miR-93 increases tumor angiogenesis and metastasis in MT-1 breast carcinoma cell line by targeting large tumor suppressor, homology 2 (LATS2), which is involved in Hippo tumor suppressor pathway [37]. Hazarika et al. have reported enhanced proliferation and tube formation in ECs following miR-93 overexpression, which is caused by the downregulation of P21 and E2F1, the regulators of cell cycle pathway [38]. In addition, overexpression of miR-93 in non-small cell lung cancer H1299 cell line favors tube formation in endothelial cells in coculture studies. It is suggested that miR-93 can modulate angiogenesis in lung cancer cells by targeting beta-transducin repeat containing protein 2 (β-TRCP2) in ubiquitin proteasome pathway [39].

MiR-20b, a member of the miR-106a ~ 363 cluster, appears to have anti-angiogenic activity by modulating signals within tumor microenvironment. Mir-20b targets signal transducer and activator of transcription 3 (STAT3) and hypoxia-inducible factor-1 alpha (HIF-1α), leading to reduction in vascular endothelial growth factor A (VEGF-A) expression [40].

8.3.1.2 MiR-378

Mir-378 is an oncogene that enhances tumor growth, survival, and angiogenesis through targeting tumor suppressor genes, suppressor of fused (Sufu), and tumor suppressor candidate 2 (TUSC2), Sox2, fibronectin, and Nodal [41–45]. Sufu is a negative regulator of Sonic hedgehog (Shh) signaling, whose level of expression is inversely related to miR-378 expression in many cell lines tested. It has been known that Shh induces large blood vessel formation by promoting the expression of angiogenic cytokines including VEGF and angiopoietin-1 (Ang-1) and angiopoietin-2 (Ang-2) [46]. Thus, miR-378 enhances tumor angiogenesis by targeting Sufu and TUSC2 which are repressors of angiogenic cytokine production. Injection of miR-378 overexpressing cancer cells to nude mice results in much larger tumors with more blood vessels compared to control cells [41]. This data is consistent with a report that miR-378 enhances VEGF expression by competing with miR-125a for the same seed region in the 3'-UTR of the VEGF gene [41, 47]. Conversely, when cells are transfected with a construct expressing an antisense sequence against miR-378, the function of miR-378 in cell survival and angiogenesis will be reversed [41].

8.3.1.3 MiR-98

MiR-98 belongs to the let-7 family and has been shown to have anti-angiogenic function. Overexpression of miR-98 in highly invasive breast carcinoma cell lines inhibits tumor angiogenesis and invasion in vitro and in vivo by targeting active receptor-like kinase 4 (ALK4) and matrix metalloproteinase-11 (MMP11). Repressed ALK4 and MMP11 expression affect endothelial cell activity and prevent them from proliferation, spreading, and tube formation [48]. These results are consistent with a study that shows ectopic expression of miR-98 inhibits B16-F1 cell migration as well as in vivo metastasis and tumor angiogenesis by reducing interleukin-6 (IL-6) level [49].

8.3.1.4 MiR-126

MiR-126 is one of the best studied microRNAs in angiogenesis which is highly and exclusively expressed in endothelial cells (ECs) [50]. In both mouse and zebra fish, miR-126 is enriched in organs with high density of vascular component, like the heart and lung. It is encoded by intron 7 of the epidermal growth factor-like domain 7 (Egfl7), which encodes an EC-specific secreted peptide as an inhibitor and chemoattractant of smooth muscle cells [51]. MiR-126 regulates many aspects of EC biology, including, migration, sprouting, cytoskeleton organization, and capillary network stability. Even though reduction of miR-126 in zebra fish does not affect vascular patterning, it compromises the integrity of blood vessels as is shown by increased hemorrhage and vessel collapse [52]. Similarly, disruption of miR-126 in mice causes leaky vessels and hemorrhage leading to 50 % embryonic lethality [53]. Of the mutant embryos that survive birth, impaired angiogenesis is displayed during both physiological and pathologic process, indicating the unique role of miR-126 in neoangiogenesis of adult tissues in response to injury [53]. In line with these data, mice treated with high dose of antagomir against miR-126 have shown significant impaired angiogenic responses [54]. Different groups have proposed different mechanisms underlying the angiogenic activity of miR-126. For instance, Fish et al. have identified and validated three targets for miR-126 with respect to endothelial biology, including Sprouty-related EVH domain-containing protein (SPRED1), PI3 kinase regulatory subunit 2 (PIK3R2), and vascular cell adhesion molecule 1 (VCAM1). The first two mRNAs are negative regulators of VEGF signaling, and the latter gene helps recruit leukocyte to the vessel walls. VCAM1 and SPRED1 have been validated by Wang's research team through their microarray analysis [50], while PIK3R2 has been described as a target for miR-126 by another team [53, 55]. However, the role of miR-126 in tumor angiogenesis is somewhat controversial. It has been observed that downregulation of miR-126 inversely correlates with an increased microvessel density (MVD) and vascular endothelial growth factor A (VEGF-A) expression in gastric cancer tissues [56]. Similarly, miR-126 has been reported to be downregulated in oral cancer which induces angiogenesis and lymphangiogenesis by restoration of VEGF-A level [55]. Nevertheless, others have found that miR-126 significantly enhances lung tumor angiogenesis, including increased EC proliferation and migration, by targeting VEGF-A (restoration of miR-126 downregulates VEGF and inhibits the growth of lung cancer cell lines).

8.3.1.5 MiR-221 and MiR-222

MiR-221 and miR-222 are located in close proximity on chromosome X11.3 which has been detected in endothelial cell by many miRNA profiling studies [18]. However, their expressions are not restricted to the endothelium. These two microRNAs belong to the same family and have common targets. Prediction algorithm suggests that miR-221 and miR-222 target c-kit mRNA in ECs. C-kit is a tyrosine kinase receptor for stem cell factor (SCF), which is a growth factor shown to be involved in angiogenesis by promoting survival, proliferation, migration, and tube formation in human umbilical vein cells (HUVECs). Interestingly, overexpression of miR-221 and miR-222 in HUVECs decreases cell proliferation, migration, and wound healing in response to SCF. In addition, high glucose treatment of HUVECs reduces c-kit expression by inducing the expression of miR-221, which impairs cell migratory response to SCF. These finding suggests that miR-221 can be an important regulator of diabetes-associated vascular dysfunction. It has also been shown that miR-221 and miR-222 overexpression in Dicer knockdown ECs restores the elevated level of endothelial nitric oxide synthase (eNOS), which is essential for endothelial cell function and vascular integrity. Since the 3'-UTR of eNOS has no target site for miR-221 and miR-222, it is proposed that the regulation of eNOS protein level by these miRNAs is likely to be indirect. All in all, miR-221 and miR-222 appear to function as anti-angiogenic factors in endothelial cells [57, 58].

Even though the overexpression of miR-221 and miR-222 has been shown to inhibit proliferation of endothelial cells, it promotes proliferation in cancer cells by targeting p27, a member of the cyclin-dependent kinase inhibitor, indicating cell type-specific function of these miRNAs.

8.3.1.6 MiR-15b and MiR-16

MiR-15b and miR-16 are located in the same cluster on chromosome 3. Although the role of these two miRNAs has not been investigated in endothelial cells, they might be involved in angiogenesis. It has been shown that hypoxia represses the expression of miR-15b and miR-16 in CNE cells from human nasopharyngeal carcinoma cell line [47]. Moreover, transfection of cells with miR-15b and miR-16 results in reduced VEGF protein expression. Therefore, hypoxia-induced downregulation of miR-15b and miR-16 contributes to VEGF expression which is a fundamental regulator of normal and abnormal angiogenesis. Meanwhile, miR-15b and miR-16 can target antiapoptotic protein, Bcl-2, to induce apoptosis in leukemic cells. Thus, overexpression of miR-15b and miR-16 can be a fascinating therapeutic approach to target tumor cell death and block VEGF-mediated angiogenesis [59].

8.3.1.7 MiR-130a

MiR-130a is one of the microRNAs frequently detected in ECs but has limited available data on its function. Upon exposure of ECs to serum, the level of miR-130 is found to rapidly increase [60]. MiR-130 has been shown to target two anti-angiogenic proteins, growth arrest-specific homeobox (GAX) and homeobox A5 (HOXA5). GAX is an important regulator of EC phenotype in response to pro- or anti-angiogenic factors and is expressed both in ECs and smooth muscle cells. Therefore, miR-130a is a pro-angiogenic microRNA, whose overexpression can antagonize the inhibitory effect of GAX on EC proliferation, migration, and tube formation and the inhibitory effect of HOXA5 on tube formation [60].

8.3.1.8 MiR-210

Hypoxia occurs during pathological condition, where cancer triggers an adaptive response to low oxygen by upregulation of genes that are essential for new blood vessel formation. It has been shown that the expression of miR-210 is induced under low-oxygen environment which drives the angiogenic response in endothelial cells. It is believed that miR-210 stimulates migration, proliferation, and the formation of capillary-like structure in ECs, whereas downregulation of miR-210 blocks cell migration and tube formation in response to VEGF [61]. In fact, upregulation of miR-210 is an essential element in response to hypoxia in ECs, affecting migration, survival, and differentiation. MiR-210 regulates angiogenesis mainly by targeting hypoxia-induced factor-1 alpha (HIF-1α) and ephrin-A3 (Eph-A3). Ephrin molecules have been known for their essential roles in vasculature and lymphatic vessel remodeling as well as EC, pericyte, and smooth muscle cell function [62]. It has been shown that HIF-1α induces the expression of miR-210 which leads to downregulation of Eph-A3 [63]. Repression of Eph-A3 is necessary and sufficient to induce tubelike structure and chemotactic migration of ECs in response to VEGF [64], while expression of Eph-A3 allele that is not targeted by miR-210 blocks the pro-angiogenic effect of miR-210 in ECs.

In addition to Eph-A3, miR-210 can target protein-tyrosine phosphatase 1B (PTP1B) [65, 66] which is a negative regulator of VEGF signaling. PTP1B can dephosphorylate VEGF receptor 2 (VEGFR2) in endothelial cells. Downregulation of PTP1B by miR-210 allows for activation of VEGF signaling under hypoxic condition [67].

8.3.1.9 Let-7 Family

Let-7 and its family members are highly conserved microRNAs across species which have been found highly expressed in HUVECs [18, 19]. The role of the let-7 family in angiogenesis was first revealed by the observation that let-7a, let-7b, let-7c, let-7f, and let-7g were reduced by more than 30 % after Drosha and Dicer knockdown [16, 18, 19]. This inhibition of let-7 family members leads to significant sprout formation in vitro [68]. Many angiogenesis-related genes are predicted to be the targets of let-7 family members, including thrombospondin-1 [16, 18], thrombospondin-2 [69], TIMP-1 [16], Nrp-2 and c-Met [26], TEK/Tie-2, KDR/VEGFR2, and Tie-1 [18]. Onc study has showed that let-7 can be involved in hypoxia-inducible factor-1α (HIF-1α)/let-7/argonaute 1(AGO1)/VEGF signal pathway in hypoxia-induced angiogenesis. HIF-1α, as a transcription factor, upregulates let-7 expression which in turn decreases the expression of AGO1. This will lead to a desupression of VEGF translation and an increase in angiogenesis [70].

8.3.1.10 MiR-296

MiR-296, also called angiomiR, is one of the important regulators of the angiogenic process [71, 72]. The knockdown and the overexpression of miR-296 inhibit and promote morphologic characteristics associated with angiogenesis of human ECs, respectively. Possible role for miR-296 in tumor angiogenesis is supported by the experiments showing reduced angiogenesis in tumor xenograft after inhibition of miR-296 using antagomirs. MiR-296 functions as a pro-angiogenic factor by inducing the expression of VEGF receptor (VEGFR2) and platelet-derived growth factor receptor (PDGFR) in angiogenic blood vessels. It also targets the hepatocyte growth factor-regulated tyrosine kinase substrate (HGS) which is involved in the degradative sorting of PDGFR, EGFR, and VEGFR. An expression analysis also shows that when HUVECs are cocultured with U87 glioma cells, the expression of miR-296 is upregulated. Moreover, miR-296 upregulation has been detected in ECs isolated from human brain tumors compared to ECs isolated from normal brain. Consistently, the expression of HGS is downregulated, while VEGFR2 and PDGFR are upregulated in these glioma blood vessel samples [73]. Altogether, these findings support a pro-angiogenic role for miR-296 in tumors.

8.3.2 Other miRNAs Related to Angiogenesis

It has been shown that miR-9 can regulate tumor angiogenesis by targeting VEGF-A. In breast cancer cells, MYC and MYCN transcription factors induce the expression of miR-9, which targets E-cadherin and hence increases cell motility and invasiveness. Downregulation of E-cadherin will activate β-catenin signaling, which in turn upregulates VEGF-A expression and increases in tumor angiogenesis [74]. The miR-143-145 cluster is highly expressed in smooth muscle cells (SMCs). Not only can this cluster regulate the vascular homeostasis but also play a role in neighboring endothelial cells. It has been shown that miR-143/miR-145 improves the angiogenic and vessel stabilization properties of ECs by regulating angiotensin-converting enzyme (ACE) and tropomyosin 4 (Tpm4) [75]. Another microRNA, miR-132, is highly expressed in human tumors and hemangiomas, which promote angiogenesis in endothelial cells by suppressing p120RasGAP, a molecular brake for RAS [71, 76]. MiR-29 is another functionally characterized microRNA with the role in regulating cell cycle and angiogenic phenotype of endothelial cells. This microRNA is upregulated in response to hypoxic stimuli in HUVECs. It has been shown that miR-29 promotes the proliferation and tube formation of HUVECs by targeting HBP1, a suppressor transcription factor [77]. Another study shows that miR-29 is regulated by TGF-β/Smad4 signaling in human and mice endothelial cells. Overexpression of miR-29 by TGF-β leads to downregulation of the phosphatase and tensin homolog (PTEN) in endothelial cells, which is a target of miR-29, and activates the AKT pathway, which will eventually lead to enhancement of angiogenesis [78].

8.3.3 Tumor-Specific MicroRNAs

Another area which has not been thoroughly looked into is tumor-specific expression of microRNAs [79]. This particular area can uncover the mechanisms and functions of different microRNAs, especially those that behave differently in different cell types. Below we will discuss some microRNAs that have been shown to have specific and differential expressions in certain cancer cell types.

8.3.3.1 In Colorectal Cancer Cells

Analysis of microRNAs in colorectal cancer has showed a significant decrease of three specific microRNAs, which include miR-145 [80, 81], miR-22 [82], and miR-126 [80, 82, 83]. All of these miRNAs share the same target protein p70S6K1 kinase, which activates HIF-1α and VEGF downstream, both strong promoters of angiogenesis [80]. The decrease in these microRNAs allows for an increase in the target mRNAs, which are sp70S6K1, HIF-1α, and VEGF, and increases angiogenesis in colorectal cancer cells.

8.3.3.2 In Glioblastoma Cells

The levels of miR-218 are significantly decreased in necrotic mesenchymal glioblastoma cells [80, 84]. MicroRNA-218 targets several mRNAs that are involved in the receptor tyrosine kinase (RTK) pathway. Thus, a decrease in miR-218 causes an increase in target RTK pathway-associated mRNA, which then leads to an increase in downstream targets, mainly HIF-2α. HIF-2α is responsible for promoting cell survival and tumor angiogenesis [80, 84].

8.3.3.3 In Human Gastric Cells

In human gastric cells, there is a significant upregulation of miR-382 and miR-18a [80, 85, 86]. Both of these microRNAs act to inhibit tumor angiogenesis, albeit through different mechanisms. MiR-382 has sequences matching to the 3'-UTR of PTEN mRNA, and this similarity leads to the inhibition of tumor angiogenesis [85]. On the other side, miR-18a binds to and inactivates targets in the mTOR signaling pathway [86]. Both of these miRNAs work to inhibit angiogenesis and stunt tube formation.

8.3.3.4 In Prostatic Cancer Cells

In prostatic cancer cells, it has been reported that there is a significant upregulation of miR-21, which targets PTEN mRNA. The increase in miR-21 leads to an increase in the AKT and ERK1/ERK2 signaling pathways, which increases the levels of HIF-1α and VEGF expression downstream. This cascade leads to tumor progression and angiogenesis [87].

8.3.4 Conclusion

More than 700 miRNAs have been identified in the human genome so far. However, the functions of a few specific miRNAs have been validated in regulating the functions of endothelial cells and angiogenesis. Moreover, miRNAs that have so far been studied are those highly expressed in endothelial cells. However, miRNAs that are expressed in smaller amounts under physiological conditions might have equally important functions in maintenance of the physiological state of endothelial cells. A single miRNA can target multiple mRNAs, whereas a single gene may be regulated by multiple miRNAs. Understanding the complex interaction between miRNAs and their targets will be an important area of investigation for the future, since it will lead us toward the development of miRNA drugs designed against specific molecular targets for clinical application.

8.4 Role of Long Noncoding RNAs in Regulation of Tumor Angiogenesis

Long noncoding RNAs (lncRNAs) are a class of noncoding RNAs that are over 200 nucleotides in size [88]. They are generally found in both the nucleus and the cytoplasm and have an array of effects on the cells. Studies have shown that lncRNA are involved in a plethora of cellular events, which include chromatin remodeling, protein scaffolding, translational control, splicing regulation, and microRNA sponges. However, recent studies have indicated that lncRNAs also have a noticeable impact on tumor progression through angiogenesis – a hallmark of cancer. LncRNAs can be found as natural antisense transcripts (NATs) that regulate their sense proteins, or they can be found between protein-coding genes [89, 90]. Evidence has shown that the downregulation of different lncRNAs leads to an abnormal gene expression that will promote tumor progression in various types of cancers [91]. However, the most convincing data shows that lncRNA interacts with critical angiogenesis regulators such as the VEGF pathway [92–97]. They also interact with other angiogenesis regulators such as phosphoglycerate kinase 1 (PGK1) [93]. Below, we will look further into some lncRNAs that have been associated with tumor angiogenesis.

8.4.1 Long Noncoding RNA MALAT1

Metastasis-associated lung adenocarcinoma transcript 1 (MALAT1) is an lncRNA that has been associated with tumor angiogenesis [88]. The function of this lncRNA is to sustain endothelial cell proliferation. It is interesting to know that it is one of the few lncRNAs that are relatively well conserved between mice and humans [90].

MALAT1 is found in high amounts in the nucleus of the cell whose expression has been shown to increase under hypoxic conditions (in vitro). It has also been shown that MALAT1 is upregulated in many human tumors and promotes tumor cell invasion and metastasis [89, 91–95].

Recent studies have examined the effect of knocking down MALAT1 in cancer cell lines. It is found that decreased MALAT1 leads to a decrease in cell number with increased apoptosis, while MALAT1-deficient cells show increased sprouting and migration of endothelial cells. However, these sprouts do not have complete extensions, which show an ability defect of the cells to form new vascular networks [88, 89]. Additionally, when these cells have been treated with VEGF, no improvements in the outgrowth of the sprouts are observed [89]. Such indicates that MALAT1-deficient cells have a decreased ability to react to VEGF and undergo angiogenesis. Furthermore, MALAT1 knockout models have been tested in the developing mouse retina, where MALAT1-deficient cells show decreased vascular proliferation that leads to a reduced vascular network compared to wild-type mice retinas. Further in vivo experiments have shown decreased neovascularization and blood flow recovery in MALAT1-deficient mice [89]. In general, the loss of MALAT1 decreases the proliferative potential of a cell and increases its migratory behavior, which in turn decreases the cell's ability to undergo angiogenesis. The mechanism of how MALAT1 works is through the deregulation of cell cycle-related factors. This lncRNA lowers the expression of endothelial cyclins CCNA2 and CCNB1/CCNB2, which are important factors in S-phase of the cell cycle, while the inhibitory factors of S-phase including the kinase p21 and kinase inhibitor p27Kip1 are upregulated [89].

Studies with pancreatic cancer cells have also found another mechanism for the function of MALAT1, which is inducing angiogenesis. Increased levels of MALAT1 promote cells to undergo epithelial to mesenchymal transitions (EMT), which may cause cancer cells to obtain stem cell-like properties. Since MALAT1 is overexpressed in many tumors, it has been shown to increase the number of cancer stem cells (CSCs), which very closely interact with angiogenesis [95]. Other findings show that upregulated MALAT1 causes an increased endothelial tube formation, an increased cell migration, and an increased amount of VEGF. These all lead to an enhanced amount of vascularization occurring. Another increased factor is the amount of CD31, which is another important indicator of tumor angiogenesis [49]. It has been hypothesized that the CSCs express many pro-angiogenic factors such as VEGF, which lead to the increased amount of angiogenesis in tumor cells. An increase in angiogenesis allows for further growth of CSCs, thus forming its positive feedback loop. There is also evidence that sex-determining region Y-box 2 (SOX2) may play a role in this mechanism because in MALAT1 knockdown studies, there is a large decrease in SOX2, which is important for cells to maintain their stemness. So by extension, if cells lose their stemness due to the loss of SOX2, we will also see a decrease in the amount of angiogenesis because of the decrease in cancer stem cells releasing pro-angiogenic factors [95].

Overall, decreasing MALAT1 in endothelial cells disrupts mechanisms linked to endothelial cell cycle progression and increased migratory behavior, which in turn negatively impacts the ability for these cells to form new vasculature. Moreover, there is a decrease in the number of CSCs. However, more studies are needed to clarify the function of MALAT1. Inhibition of MALAT1 is proposed as a treatment to prevent tumor growth and metastasis due to its pro-angiogenic properties.

8.4.2 Long Noncoding RNA MVIH

Hepatocellular carcinoma (HCC) is the fifth most common solid cancer in the world and the most common form of liver cancer. Unfortunately, it only has a 50 % survival rate after 5 years in for groups aging from 17 to 69 [91–93]. It is clearly a very deadly cancer mainly for its rapid growth caused in part by very active angiogenesis, which unavoidably leads to metastasis.

The cause of the increased angiogenesis has been linked to the long noncoding RNA associated with microvascular invasion in HCC (lncRNA MVIH). This lncRNA is located within the intron of the RPS24 gene, which is a ribosomal protein [90, 91, 94]. Using tissue samples from patients with HCC, it has been determined that lncRNA MVIH is overexpressed in HCC tumor cells, compared to non-tumor cells. Using RNA pull-down methods, it has been found that lncRNA MVIH is associated with phosphoglycerate kinase 1 (PGK1) [91]. PGK1 is an enzyme encoded by the PGK1 gene, which can be secreted by tumor cells. However, PGK1 acts to suppress angiogenesis. Therefore, it is imperative for a tumor cell to inhibit this anti-angiogenic factor in order to grow. LncRNA MVIH overexpressed by tumor cells will bind to PGK1, affectively reducing its function. Without PGK1's presence to prevent angiogenesis, tumor cells in HCC will gain increased microvessel density, leading to rapid growth of tumors [91, 93, 98]. This downward spiral continues because the increased angiogenesis leads to increased microvascular invasion or metastasis – in particular intrahepatic metastasis [91]. This makes HCC very deadly.

In all, the lncRNA MVIH plays a crucial role in the tumorigenesis of hepatocellular carcinomas. These tumor cells upregulate lncRNA MVIH in order to increase angiogenesis and to eventually achieve metastasis. With this knowledge, the focus can be shifted to create a novel medicine that down regulates lncRNA MVIH, which may be able to reduce angiogenesis, thus decreasing cancer growth and metastasis.

8.4.3 The Long Noncoding RNA HOXD-AS1

LncRNA HOXD-AS1 is encoded in the HOXD gene cluster and can be found equally in the nucleus and cytoplasm. It has recently been shown that lncRNA HOXD-AS1 is a marker of neuroblastoma (NB) progression. Many lncRNAs have been studied to determine their differential expression in aggressive NB vs. noncancerous tissues, when treated with retinoic acid (RA). Of the many noncoding

RNAs tested, lncRNA HOXD-AS1 is the only one upregulated substantially. RA is the first-line drug used to battle NB and works as a differentiating agent that typically arrests the growth of NB cells, making them more vulnerable to chemotherapeutic drugs [92, 93, 98]. This lncRNA is located between the HOXD1 and the HOXD3 genes, but it is antisense to both of these (hence being called the AS1). It is also highly conserved within hominids, but not so much with other primates [92].

Studies have indicated that the expression level of lncRNA HOXD-AS1 increases with progressing stage/aggressiveness of neuroblastoma, thus possibly playing a factor in its increased tumorigenesis [92]. The function of HOXD-AS1 has been assessed through knockdown experiments via siRNA. Many genes have been observed and found differential expressions involved with inflammation and angiogenesis. To be specific, the increased expressions of many cytokines, such as, CX311, CCL20, TNF, and GD15, have been found to be important for extracellular matrix communication. There is also a change in the expression of matrix remodeling genes LOX and ADAMTS3 and key regulators of angiogenesis and lymphangiogenesis ANG and PROX1 [92]. This shows that lncRNA HOXD-AS1 affects in some way angiogenesis in tumor cells, adding to the aggressiveness of the cancer. There is a significant increase in the JAK/STAT pathway, which is related to inflammation and angiogenesis. However, the PI3K/AKT pathway is found to be the main regulator of expression of HOXD-AS1. Furthermore, the expression of HOXD-AS1 is correlated to the expression of HOXD1 and HOXD2 genes, which implies common regulatory mechanisms. This is also another indicator of malignancy because the aberrant expression of HOX genes in tumor cells has been linked to malignancy [92].

Overall, the lncRNA HOXD-AS1 has many implications in the tumorigenicity of neuroblastoma. Although the mechanisms are not completely clear, it is evident that this lncRNA affects many regulators of angiogenesis. Thus, the increased amount of lncRNA HOXD-AS1 in increasing aggressiveness of tumors can be caused due to the increased expression of angiogenic factors, leading increased growth and metastasis of cancers. More studies need to be done in order to uncover the exact mechanisms of action of this lncRNA, which may enlighten its use as a target of a therapeutic drug. For now, lncRNA HOXD-AS1 remains a reliable biomarker of neuroblastoma.

8.4.4 The Long Noncoding RNA HIF-1A-AS2

Hypoxia-inducible factor-1 alpha subunit antisense RNA 2 (lncRNA HIF-1A-AS1) is an lncRNA that is involved in tumor angiogenesis. LncRNA HIF-1A-AS2 is upregulated in non-papillary clear cell renal carcinomas and is a marker of poor prognosis in breast cancer [4, 94–97].

This lncRNA has been shown to be involved in angiogenic regulatory pathways, which may be the reason for its impact on tumor progression. LncRNA HIF-1A-AS2 has been shown to negatively regulate hypoxia-inducible factor-1 alpha (HIF-1α), which is a critical regulator of angiogenesis [4]. HIF-1α is increased in response to hypoxia and activates many genes that increase the amount of angiogenesis, which is vital for growing tumors to obtain sufficient nutrients and oxygen, while also excreting waste. However, in some cancers, the level of lncRNA HIF-1A-AS2 also increases and works by binding and causing the degradation of HIF-1α mRNA [4]. The lncRNA acts in a negative feedback manner to decrease the amount of angiogenesis. This is a prime example of a NAT regulating the expression of its sense protein.

In general, lncRNA HIF-1A-AS2 is involved in angiogenic pathways. However, the exact mechanisms and its true purpose have not yet been uncovered. It is overexpressed in some cancers, yet seems to act to counter angiogenesis by downregulating HIF-1α. This could be a possible reaction to the increase in angiogenesis caused by HIF-1α. More studies are needed to look into this particular lncRNA.

8.4.5 The Long Noncoding RNA MEG3

Maternally expressed gene 3 (MEG3) is an lncRNA that is expressed in many cells and tissues. MEG3 expression is lost in many different tumors, whether it is through gene deletion or hypermethylation of the promoter or other regions of the gene [4, 96]. Studies have shown that the re-expression of MEG3 in tumors causes inhibition of tumor cell proliferation through the accumulation of p53 and downstream activation of p53 genes. P53 acts as a transcription factor for many tumor suppressor genes. Therefore, when MEG3 function is lost, cells also lose the function of p53, leading to aggressive cancers. The lncRNA MEG3 may function as a novel tumor suppressor since its downregulation and/or deletion is largely associated with aggressive cancers [4].

Furthermore, the loss of MEG3 coincides with an increase in the expression of pro-angiogenic genes, which may be a main cause in the increased aggressiveness of these tumors. Studies using the mouse ortholog Meg3 have shown that several genes affecting angiogenesis are upregulated when Meg3 is knocked out. It has also been observed that VEGF-A and its receptor VEGFR1, which are bona fide primary regulators of angiogenesis, are significantly increased. Thus, when Meg3 is lost, angiogenesis increases dramatically [4, 96]. Since new blood vessels are vital for tumor growth, the inactivation of MEG3 is one way by which tumors can continue to develop.

In addition to VEGF pathway genes, there is also an increase in genes encoding for adherens junctions, which are critical for endothelial cell-to-cell interactions and interactions with the cell matrix. These allow for stable vessel formation [4, 97]. In addition, there is an increase in hemophilic cell adhesion, GTPase activator activity, and actin cytoskeleton organization and biogenesis, all of which relate to an increased vessel formation and angiogenesis. Moreover, there is an increase in Notch signaling, which also aids in vessel stability [4, 97].

All in all, the lncRNA MEG3 is a tumor suppressor, as its presence greatly decreases tumor cell proliferation and its downregulation is a fundamental step in tumor growth. MEG3 is heavily linked to angiogenesis, and in its absence, angiogenesis occurs undeterred. Consequently, in normal cells, it may be that MEG3 acts to suppress aberrant angiogenesis from occurring. There are many possibilities in using MEG3 in therapeutic settings to suppress tumor growth. More studies are needed to determine how MEG3 is exactly downregulated or deleted and how it can be used to battle tumor growth.

8.4.6 Other Long Noncoding RNAs

Besides the lncRNAs previously mentioned, there are a few other lncRNAs that may show some promise in uncovering more information about lncRNA association with angiogenesis.

One such lncRNA is sONE or NOS3AS, which happens to be another natural antisense transcript (NAT), much like lncRNA HIF-1A-AS2. This lncRNA regulates the expression of endothelial nitric oxide synthase (eNOS), under normal oxygen conditions and hypoxic conditions. Not much else is known about this molecule, and it has not been completely determined whether it acts as RNA or a protein, despite a possible protein product has been discovered [52]. However, there is potential link to angiogenesis due to its effect on endothelial cells involved in blood vessels. More studies are needed to determine whether this molecule acts as RNA or if it in fact codes for a functional protein and then to test its function on angiogenesis. Another NAT lncRNA is Tie-1-AS, which is the antisense transcript for tyrosine kinase containing immunoglobulin and epidermal growth factor homology domain 1 (tie-1). This lncRNA is highly conserved in humans, mice, and zebra fish [52]. It acts by binding to tie-1 mRNA transcripts and decreasing its levels. The decrease in tie-1 causes defects in endothelial cell junctions, which lead to poor vessel formation and angiogenesis [4, 99, 100]. This lncRNA is a very good candidate for further studies involving tumor angiogenesis. One large class of noncoding RNAs is pseudogenes. Pseudogenes can play important roles in angiogenesis and tumorigenesis [106, 107]. In some mRNAs, there are long fragments of 3'-untranslated regions (3'-UTRs), which may interact with miRNAs and function similarly to the long noncoding RNAs [101–105, 108].

8.5 Interactions Between MicroRNAs and Long Noncoding RNAs

Up to this point, we have discussed how different miRNAs and lncRNAs affect tumor angiogenesis. However, the interplay between noncoding RNAs and angiogenesis are more complicated when we realize the large amount of interactions that exist between different noncoding RNAs. These interactions add another layer of complexity, where the activity of a specific miRNA can alter the effect of an lncRNA and vice versa. This allows for certain noncoding RNA to affect tumor angiogenesis indirectly. There are many ways in which miRNA and lncRNA can affect each other. For instance, (1) miRNA can trigger the decay of lncRNA, (2) lncRNA can act as miRNA sponges, (3) miRNA and lncRNA can compete for the same mRNA, and (4) some lncRNA can generate miRNA. Regardless of the mechanism used, it is clear that there are significant interactions between these noncoding RNAs, thus offering a greater potential for control of tumor angiogenesis. Although this area is relatively new, there are some interactions discovered that are relevant to tumor angiogenesis.

8.5.1 MicroRNA Interactions with LncRNA MALAT1

Long noncoding RNA MALAT1 has been previously discussed and determined to be a pro-angiogenic factor. An increase of lncRNA MALAT1 causes a significant increase in the number of CSCs which induce pro-angiogenic effects, whereas MALAT1 deficiency leads to reduced levels of angiogenesis. Studies have shown that MALAT1 may also function as an miRNA sponge for miR-200c and miR-145. By decreasing the effects of the miRNAs (miR-200c and miR-145), their target gene Sox2 is upregulated, leading to an increase in stem cell-like properties. This is another mechanism for MALAT1 to elicit pro-angiogenic effects through miRNA interactions. Furthermore, other studies have shown that the overexpression of miR-9 decreases the levels of MALAT1, through which miRNA triggers lncRNA decay. More specifically, this occurs through miR-9 binding to MALAT1 and targeting it for AGO2-mediated degradation, which has been demonstrated in the Hodgkin lymphoma cell line L428 and glioblastoma cell line U87MG [87, 109]. Moreover, miR-9 also affects angiogenesis through other mechanisms, such as targeting VEGF-A, which has been discussed earlier. This example shows clear cross talk between two noncoding RNAs that have both been implicated in affecting tumor angiogenesis.

8.5.2 MicroRNA Let-7 Interactions with LncRNA

MicroRNAs in the let-7 family have been shown to be important in regulating angiogenesis with their anti-angiogenic effects. Let-7 miRNA targets many angiogenic genes, and their inhibition may promote tumor progression and angiogenesis. In addition, let-7 miRNA also interacts with many different lncRNAs which may affect the overall activity of let-7, thus affecting its function to regulate tumor angiogenesis.

Moreover, lncRNA-p21 inhibits translation, unlike most known lncRNAs. It is activated by the tumor suppressor protein p53. It has been shown to be negatively regulated in human cervical cancer cells by HuR, AGO2, and microRNA let-7b. Let-7b overexpression causes lncRNA-p21 degradation [87, 109, 110], and lncRNA HOTAIR is also impacted by let-7 miRNA in a similar fashion to lncRNA-p21 [87, 109–111].

Furthermore, long noncoding Nras functional RNA (ncNRFR) is an lncRNA that promotes tumorigenesis. Studies have shown that this lncRNA has a 22 nt sequence that perfectly matches the sequence of miRNA let-7a. Moreover, this sequence only differs from other members of the let-7 family by a few nucleotides, which includes let-7b to -7 g, let-7i, and miR-98. Studies have showed that increasing lncRNA ncNRFR leads to a decrease in let-7 miRNA function, which consequently increases let-7 target mRNAs [112]. Overall, this interaction leads to tumor promotion through suppression of let-7 microRNA, which is important in reducing tumor angiogenesis.

8.5.3 MicroRNA Interactions with LncRNA-RoR

The long noncoding RNA regulator of reprogramming (lncRNA-RoR) is found to be in high concentrations in embryonic stem cells. It is interesting to note that many miRNAs that are involved in angiogenesis interact with this particular lncRNA by binding and decreasing its function. MicroRNA-145 is known to interact with lncRNA-RoR, and previous studies have shown that this miRNA is involved in the stabilization of vessels in endothelial cells [113, 114]. Furthermore, miR-99 and miR-181 have also been shown to be important in angiogenesis and are both implicated in interacting with lncRNA-RoR [114].

The above examples show the complexity and many possible interactions that can occur between different noncoding RNAs. The involvement of microRNA and long noncoding RNA in tumor angiogenesis becomes much more complicated when they readily interact with each other. This means that some miRNAs that are very important in tumor angiogenesis may be influenced significantly by lncRNA and vice versa. These interactions need to be delved into for a discovery of new mechanisms and potential novel ways to combat tumor progression.

8.6 Conclusion

So far, researches into lncRNAs and their association to tumor angiogenesis have been promising. LncRNAs such as MALAT1, MVIH, HIF-A-AS2, and MEG3 have all shown large involvement in tumor progression through affecting angiogenesis. There are many other lncRNAs that have been indicated in affecting tumor progression, and more researches are justified to uncover novel lncRNAs and the mechanisms of their action. The study of lncRNA in tumor progression and specifically angiogenesis is still relatively new, which gives a potential for a wealth of new information. One area that can be further explored is the interactions between lncRNAs and microRNAs. LncRNAs generally function as microRNA sponges, and there are many microRNAs found affecting tumor angiogenesis [115–117]. Thus, there may be lncRNAs that affect angiogenesis by extension through microRNAs. Given that lncRNA interactions are complex and cover many biological pathways, it offers great potential for further therapeutic discoveries.

Acknowledgments This work was supported by a Discovery Grant from the Natural Sciences and Engineering Research Council of Canada (NSERC; 227937–2012) to BBY, who is the recipient of a Career Investigator Award (CI 7418) from the Heart and Stroke Foundation of Ontario.

References

1. Carmeliet P. Angiogenesis in life, disease and medicine. Nature. 2005;438(7070):932–6.
2. Carmeliet P, Peter C. Angiogenesis in health and disease. Nat Med. 2003;9(6):653–60.
3. Yancopoulos GD, et al. Vascular-specific growth factors and blood vessel formation. Nature. 2000;407(6801):242–8.
4. Hanahan D, Douglas H, Weinberg RA. The hallmarks of cancer. Cell. 2000;100(1):57–70.
5. Folkman J, Judah F. Angiogenesis in cancer, vascular, rheumatoid and other disease. Nat Med. 1995;1(1):27–30.
6. Zheng PS, Wen J, Ang LC, et al. Versican/PG-M G3 domain promotes tumor growth and angiogenesis. FASEB J. 2004;18(6):754–6.
7. Hanahan D, Folkman J. Patterns and emerging mechanisms of the angiogenic switch during tumorigenesis. Cell. 1996;86(3):353–64.
8. Carmeliet P. Mechanisms of angiogenesis and arteriogenesis. Nat Med. 2000;6(4):389–95.
9. Folkman J, Judah F. Role of angiogenesis in tumor growth and metastasis. Semin Oncol. 2002;29(6Q):15–8.
10. Dalmay T, Edwards DR. MicroRNAs and the hallmarks of cancer. Oncogene. 2006;25(46):6170–5.
11. Garzon R, Calin GA, Croce CM. MicroRNAs in cancer. Annu Rev Med. 2009;60(1):167–79.
12. Lee YS, Anindya D. MicroRNAs in cancer. Annu Rev Pathol. 2009;4(1):199–227.
13. Bartel DP. MicroRNAs: genomics, biogenesis, mechanism, and function. Cell. 2004;116(2):281–97.
14. Graves P, Zeng Y. Biogenesis of mammalian microRNAs: a global view. Genomics Proteomics Bioinformatics. 2012;10(5):239–45.

15. Yang WJ, Yang DD, Songqing N, et al. Dicer is required for embryonic angiogenesis during mouse development. J Biol Chem. 2005;280(10):9330–5.
16. Otsuka M, Zheng M, Hayashi M, et al. Impaired microRNA processing causes corpus luteum insufficiency and infertility in mice. J Clin Invest. 2008;118(5):1944–54.
17. Giraldez AJ, Cinalli RM, Glasner ME, et al. MicroRNAs regulate brain morphogenesis in zebrafish. Science. 2005;308(5723):833–8.
18. Suárez Y, Fernández-Hernando C, Pober JS, et al. Dicer dependent microRNAs regulate gene expression and functions in human endothelial cells. Circ Res. 2007;100(8):1164–73.
19. Kuehbacher A, Urbich C, Zeiher AM, et al. Role of Dicer and Drosha for endothelial microRNA expression and angiogenesis. Circ Res. 2007;101(1):59–68.
20. Shilo S, Roy S, Khanna S, et al. Evidence for the involvement of miRNA in redox regulated angiogenic response of human microvascular endothelial cells. Arterioscler Thromb Vasc Biol. 2008;28(3):471–7.
21. Graham RJ, Jan CH, Bartel DP, et al. Intronic microRNA precursors that bypass Drosha processing. Nature. 2007;448(7149):83–6.
22. Barik S, Sailen B, Titus B. Intronic microRNA: creation, evolution and regulation. In: Gusev Y, editor. MicroRNA profiling in cancer: a bioinformatics perspective. Singapore: Pan Stanford; 2009. p. 117–31.
23. Okamura K, Hagen JW, Hong D, et al. The mirtron pathway generates microRNA-class regulatory RNAs in Drosophila. Cell. 2007;130(1):89–100.
24. Pushpavalli SN, Bag I, Pal-Bhadra M, et al. Drosophila Argonaute-1 is critical for transcriptional cosuppression and heterochromatin formation. Chromosome Res. 2012;20(3):333–51.
25. Fukagawa T, Nogami M, Yoshikawa M, et al. Dicer is essential for formation of the heterochromatin structure in vertebrate cells. Nat Cell Biol. 2004;6(8):784–91.
26. Poliseno L, Tuccoli A, Mariani L, et al. MicroRNAs modulate the angiogenic properties of HUVECs. Blood. 2006;108(9):3068–71.
27. He L, Thomson JM, Hemann MT, et al. A microRNA polycistron as a potential human oncogene. Nature. 2005;435(7043):828–33.
28. Dews M, Homayouni A, Yu D, et al. Augmentation of tumor angiogenesis by a Myc-activated microRNA cluster. Nat Genet. 2006;38(9):1060–5.
29. Suárez Y, Fernandez-Hernando C, Yu J, et al. Dicer-dependent endothelial microRNAs are necessary for postnatal angiogenesis. Proc Natl Acad Sci U S A. 2008;105(37):14082–7.
30. Khorshidi A, Dhaliwal P, Yang BB. Anti-tumor activity of miR-17 in melanoma. Cell Cycle. 2015;14(16):2549–50.
31. Li H, Gupta S, Du WW, et al. MicroRNA-17 inhibits tumor growth by stimulating T-cell mediated host immune response. Oncoscience. 2014;1(7):531–9.
32. Fang L, Li H, Wang L, et al. MicroRNA-17-5p promotes chemotherapeutic drug resistance and tumour metastasis of colorectal cancer by repressing PTEN expression. Oncotarget. 2014;5(10):2974–87.
33. Li H, Yang BB. MicroRNA-regulated stress response in cancer and its clinical implications. Cell Cycle. 2013;12(13):1983–4.
34. Shan SW, Fang L, Shatseva T, et al. Mature miR-17-5p and passenger miR-17-3p induce hepatocellular carcinoma by targeting PTEN, GalNT7 and vimentin in different signal pathways. J Cell Sci. 2013;126(Pt 6):1517–30.
35. Yang X, Du WW, Li H, et al. Both mature miR-17-5p and passenger strand miR-17-3p target TIMP3 and induce prostate tumor growth and invasion. Nucleic Acids Res. 2013;41(21):9688–704.
36. Fang L, Deng Z, Shatseva T, et al. MicroRNA miR-93 promotes tumor growth and angiogenesis by targeting integrin-β8. Oncogene. 2011;30(7):806–21.
37. Fang L, Du WW, Yang W, et al. MiR-93 enhances angiogenesis and metastasis by targeting LATS2. Cell Cycle. 2012;11(23):4352–65.

38. Hazarika S, Farber CR, Dokun AO, et al. MicroRNA-93 controls perfusion recovery after hindlimb ischemia by modulating expression of multiple genes in the cell cycle pathway. Circulation. 2013;127(17):1818–28.
39. Savita U, Karunagaran D. MicroRNA-106b-25 cluster targets β-TRCP2, increases the expression of Snail and enhances cell migration and invasion in H1299 (non small cell lung cancer) cells. Biochem Biophys Res Commun. 2013;434(4):841–7.
40. Cascio S, D'Andrea A, Ferla R, et al. MiR-20b modulates VEGF expression by targeting HIF-1 alpha and STAT3 in MCF-7 breast cancer cells. J Cell Physiol. 2010;224(1):242–9.
41. Lee DY, Deng Z, Wang CH, et al. MicroRNA-378 promotes cell survival, tumor growth, and angiogenesis by targeting SuFu and Fus-1 expression. Proc Natl Acad Sci U S A. 2007;104(51):20350–5.
42. Deng Z, Yang X, Fang L, et al. Misprocessing and functional arrest of microRNAs by miR-Pirate: roles of miR-378 and miR-17. Biochem J. 2013;450(2):375–86.
43. Deng Z, Du WW, Fang L, et al. The intermediate filament vimentin mediates microRNA miR-378 function in cellular self-renewal by regulating the expression of the Sox2 transcription factor. J Biol Chem. 2013;288:319–31.
44. Liu F, Lv Q, Du WW, et al. Specificity of miR-378a-5p targeting rodent fibronectin. Biochim Biophys Acta. 2013;1833(12):3272–85.
45. Luo L, Ye G, Nadeem L, et al. MicroRNA-378a-5p promotes trophoblast cell survival, migration and invasion by targeting Nodal. J Cell Sci. 2012;125(Pt 13):3124–32.
46. Pola R, Ling LE, Silver M, et al. The morphogen Sonic hedgehog is an indirect angiogenic agent upregulating two families of angiogenic growth factors. Nat Med. 2001;7(6):706–11.
47. Hua Z, Lv Q, Ye W, et al. MiRNA-directed regulation of VEGF and other angiogenic factors under hypoxia. PLoS ONE. 2006;1, e116.
48. Siragam V, Rutnam ZJ, Yang W, et al. MicroRNA miR-98 inhibits tumor angiogenesis and invasion by targeting activin receptor-like kinase-4 and matrix metalloproteinase-11. Oncotarget. 2012;3(11):1370–85.
49. Li F, Li XJ, Qiao L, et al. MiR-98 suppresses melanoma metastasis through a negative feedback loop with its target gene IL-6. Exp Mol Med. 2014;46:e116.
50. Wang S, Aurora AB, Johnson BA, et al. The endothelial-specific microRNA miR-126 governs vascular integrity and angiogenesis. Dev Cell. 2008;15(2):261–71.
51. Soncin F, Mattot V, Lionneton F, et al. VE-statin, an endothelial repressor of smooth muscle cell migration. EMBO J. 2003;22(21):5700–11.
52. Fish JE, Santoro MM, Morton SU, et al. MiR-126 regulates angiogenic signaling and vascular integrity. Dev Cell. 2008;15(2):272–84.
53. Kuhnert F, Mancuso MR, Hampton J, et al. Attribution of vascular phenotypes of the murine Egfl7 locus to the microRNA miR-126. Development. 2008;135(24):3989–93.
54. Van Solingen C, Seghers L, Bijkerk R, et al. Antagomir-mediated silencing of endothelial cell specific microRNA-126 impairs ischemia-induced angiogenesis. J Cell Mol Med. 2009;13(8A):1577–85.
55. Sasahira T, Kurihara M, Bhawal UK, et al. Downregulation of miR-126 induces angiogenesis and lymphangiogenesis by activation of VEGF-A in oral cancer. Br J Cancer. 2012;107(4):700–6.
56. Chen H, Li L, Wang S, et al. Reduced miR-126 expression facilitates angiogenesis of gastric cancer through its regulation on VEGF-A. Oncotarget. 2014;5(23):11873–85.
57. Li Y, Song YH, Li F, et al. MicroRNA-221 regulates high glucose-induced endothelial dysfunction. Biochem Biophys Res Commun. 2009;381(1):81–3.
58. Minami Y, Satoh M, Maesawa C, et al. Effect of atorvastatin on microRNA 221 / 222 expression in endothelial progenitor cells obtained from patients with coronary artery disease. Eur J Clin Investig. 2009;39(5):359–67.
59. Guo CJ, Pan Q, Li DG, et al. MiR-15b and miR-16 are implicated in activation of the rat hepatic stellate cell: an essential role for apoptosis. J Hepatol. 2009;50(4):766–78.

60. Chen Y, Gorski DH. Regulation of angiogenesis through a microRNA (miR-130a) that downregulates antiangiogenic homeobox genes GAX and HOXA5. Blood. 2008;111(3):1217–26.
61. He J, Wu J, Xu NH. MiR-210 disturbs mitotic progression through regulating a group of mitosis-related genes. Nucleic Acids Res. 2013;41(1):498–508.
62. Kuijper S, Turner CJ, Adams RH. Regulation of angiogenesis by Eph–ephrin interactions. Trends Cardiovasc Med. 2007;17(5):145–51.
63. Ivan M, Harris AL, Martelli F, et al. Hypoxia response and microRNAs: no longer two separate worlds. J Cell Mol Med. 2008;12(5A):1426–31.
64. Fasanaro P, D'alessandra Y, Stefanoet VD, et al. MicroRNA-210 modulates endothelial cell response to hypoxia and inhibits the receptor tyrosine kinase ligand ephrin-A3. J Biol Chem. 2008;283(23):15878–83.
65. Fasanaro P, Greco S, Lorenzi M, et al. An integrated approach for experimental target identification of hypoxia-induced miR-210. J Biol Chem. 2009;284(50):35134–43.
66. Hu S, Huang M, Li Z, et al. MicroRNA-210 as a novel therapy for treatment of ischemic heart disease. Circulation. 2010;122(11 Suppl):S124–31.
67. Nakamura Y, Patrushev N, Inomata N, et al. Role of protein tyrosine phosphatase 1B in vascular endothelial growth factor signaling and cell-cell adhesions in endothelial cells. Circ Res. 2008;102(10):1182–91.
68. Yang X, Rutnam ZJ, Jiao CW, et al. An anti-let-7 sponge decoys and decays endogenous let-7 functions. Cell Cycle. 2012;11(16):3097–108.
69. Bae ON, Wang JM, Baek SH, et al. Oxidative stress-mediated thrombospondin-2 upregulation impairs bone marrow-derived angiogenic cell function in diabetes mellitus. Arterioscler Thromb Vasc Biol. 2013;33(8):1920–7.
70. Chen Z, Lai TC, Jan YH, et al. Hypoxia-responsive miRNAs target argonaute 1 to promote angiogenesis. J Clin Invest. 2013;123(3):1057–67.
71. Anand S, Cheresh DA. MicroRNA-mediated regulation of the angiogenic switch. Curr Opin Hematol. 2011;18(3):171–6.
72. Wang S, Olson EN. AngiomiRs-key regulators of angiogenesis. Curr Opin Genet Dev. 2009;19(3):205–11.
73. Würdinger T, Tannous BA, Saydam O. MiR-296 regulates growth factor receptor overexpression in angiogenic endothelial cells. Cancer Cell. 2008;14(5):382–93.
74. Ma L, Young J, Prabhala H, et al. MiR-9, a MYC/MYCN-activated microRNA, regulates E-cadherin and cancer metastasis. Nat Cell Biol. 2010;12(3):247 56.
75. Parmacek MS. MicroRNA-modulated targeting of vascular smooth muscle cells. J Clin Invest. 2009;119(9):2526–8.
76. Anand S, Majeti BK, Acevedo LM. MicroRNA-132–mediated loss of p120RasGAP activates the endothelium to facilitate pathological angiogenesis. Nat Med. 2010;16(8):909–14.
77. Yang Z, Wu L, Zhu X, et al. MiR-29a modulates the angiogenic properties of human endothelial cells. Biochem Biophys Res Commun. 2013;434(1):143–9.
78. Wang J, Wang Y, Wang Y. Transforming growth factor β-regulated microRNA-29a promotes angiogenesis through targeting the phosphatase and tensin homolog in endothelium. J Biol Chem. 2013;288(15):10418–26.
79. Rutnam ZJ, Wight TN, Yang BB. miRNAs regulate expression and function of extracellular matrix molecules. Matrix Biol. 2013;32(2):74–85.
80. Wang W, Zhang E, Lin C. MicroRNAs in tumor angiogenesis. Life Sci. 2015;136:28–35.
81. Xu Q, Liu LZ, Qian X, et al. MiR-145 directly targets p70S6K1 in cancer cells to inhibit tumor growth and angiogenesis. Nucleic Acids Res. 2012;40(2):761–74.
82. Watanabe HS. Horizons in cancer research. Sunrise: Nova Science Publisher; 2014.
83. Zhang Y, Wang XY, Xu BH, et al. Epigenetic silencing of miR-126 contributes to tumor invasion and angiogenesis in colorectal cancer. Oncol Rep. 2013;30(4):1976–84.
84. Mathew LK, Skuli N, Mucaj V, et al. MiR-218 opposes a critical RTK-HIF pathway in mesenchymal glioblastoma. Proc Natl Acad Sci U S A. 2014;111(1):291–6.

85. Seok JK, Skuli N, Mucaj V, et al. MicroRNA-382 induced by HIF-1α is an angiogenic miR targeting the tumor suppressor phosphatase and tensin homolog. Nucleic Acids Res. 2014;42(12):8062–72.
86. Zheng Y, Li S, Ding Y, et al. The role of miR-18a in gastric cancer angiogenesis. Hepato-Gastroenterology. 2013;60(127):1809–13.
87. Liu LZ, Li C, Chen Q, et al. MiR-21 induced angiogenesis through AKT and ERK activation and HIF-1α expression. PLoS ONE. 2011;6(4), e19139.
88. Thum T, Fiedler J. LINCing MALAT1 and angiogenesis. Circ Res. 2014;114(9):1366–8.
89. Michalik KM, You X, Manavski Y, et al. Long non-coding RNA MALAT1 regulates endothelial cell function and vessel growth. Circ Res. 2014;114(9):1389–97.
90. Yang G, Lu X, Yuan L. LncRNA: a link between RNA and cancer. Biochim Biophys Acta. 2014;1839(11):1097–109.
91. Yuan SX, Yang F, Yang Y, et al. Long non-coding RNA associated with microvascular invasion in hepatocellular carcinoma promotes angiogenesis and serves as a predictor for hepatocellular carcinoma patients' poor recurrence-free survival after hepatectomy. Hepatology. 2012;56(6):2231–41.
92. Yarmishyn AA, Batagov AO, Tan JZ, et al. HOXD-AS1 is a novel lncRNA encoded in HOXD cluster and a marker of neuroblastoma progression revealed via integrative analysis of non-coding transcriptome. BMC Genomics. 2014;15 Suppl 9:S7.
93. Huang JL, Zheng L, Hu YW, et al. Characteristics of long non-coding RNA and its relation to hepatocellular carcinoma. Carcinogenesis. 2014;35(3):507–14.
94. Kunej T, Obsteter J, Pogacar Z, et al. The decalog of long non-coding RNA involvement in cancer diagnosis and monitoring. Crit Rev Clin Lab Sci. 2014;51(6):344–57.
95. Jiao F, Hu H, Ha T, et al. Long non-coding RNA MALAT-1 enhances stem cell-like phenotypes in pancreatic cancer cells. Int J Mol Sci. 2015;16(4):6677–93.
96. Zhou Y, Zhang X, Klibanski A. MEG3 non-coding RNA: a tumor suppressor. J Mol Endocrinol. 2012;48(3):R45–53.
97. Gordon FE, Nutt CL, Cheunsuchon P, et al. Increased expression of angiogenic genes in the brains of mouse Meg3Null embryos. Endocrinology. 2010;151(6):2443–52.
98. Im JH, Muschel RJ. New evidence of lncRNA role in tumor progression and metastasis. Hepatobiliary Surg Nutr. 2012;1(1):55–6.
99. Iaconetti C, Gareri C, Polimeni A, et al. Non-coding RNAs: the "Dark Matter" of cardiovascular pathophysiology. Int J Mol Sci. 2013;4(10):19987–20018.
100. Derrien T, Guigó R, Johnson R, et al. The long non-coding RNAs: a new (P)layer in the "Dark Matter". Front Genet. 2012;2:107. doi:10.3389/fgene.2011.00107. eCollection 2011.
101. Rutnam ZJ, Yang BB. The non-coding 3' UTR of CD44 induces metastasis by regulating extracellular matrix functions. J Cell Sci. 2012;125(Pt 8):2075–85.
102. Fang L, Du WW, Yang X, et al. Versican 3'-untranslated region (3'-UTR) functions as a ceRNA in inducing the development of hepatocellular carcinoma by regulating miRNA activity. FASEB J. 2013;27(3):907–19.
103. Lee SC, Fang L, Wanget CH, et al. A non-coding transcript of nephronectin promotes osteoblast differentiation by modulating microRNA functions. FEBS Lett. 2011;585(16):2610–6.
104. Jeyapalan Z, Deng ZQ, Shatseva T, et al. Expression of CD44 3'-untranslated region regulates endogenous microRNA functions in tumorigenesis and angiogenesis. Nucleic Acids Res. 2011;39(8):3026–41.
105. Kahai S, Lee SC, Lee DY, et al. MicroRNA miR-378 regulates nephronectin expression modulating osteoblast differentiation by targeting GalNT-7. PLoS ONE. 2009;4(10):e7535.
106. Rutnam ZJ, Du WW, Yang W, et al. The pseudogene TUSC2P promotes TUSC2 function by binding multiple microRNAs. Nat Commun. 2014;5:2914.
107. Poliseno L, Salmena L, Zhang JW, et al. A coding-independent function of gene and pseudogene mRNAs regulates tumour biology. Nature. 2010;465(7301):1033–8.

108. Lee DY, Shatseva T, Jeyapalan Z, et al. A 3'-untranslated region (3'UTR) induces organ adhesion by regulating miR-199a* functions. PLoS ONE. 2009;4(2):e4527.
109. Leucci E, Patella F, Waageet J, et al. MicroRNA-9 targets the long non-coding RNA MALAT1 for degradation in the nucleus. Sci Rep. 2013;3:2535. doi:10.1038/srep02535.
110. Yoon JH, Abdelmohsen K, Srikantan S, et al. LincRNA-p21 suppresses target mRNA translation. Mol Cell. 2012;47(4):648–55.
111. Yoon JH, Abdelmohsen K, Kim J, et al. Scaffold function of long non-coding RNA HOTAIR in protein ubiquitination. Nat Commun. 2013;4:2939.
112. Franklin JL, Rankinb CR, Levyet S, et al. Malignant transformation of colonic epithelial cells by a colon-derived long non-coding RNA. Biochem Biophys Res Commun. 2013;440(1):99–104.
113. Franco-Zorrilla JM, Valli A, Todesco M, et al. Target mimicry provides a new mechanism for regulation of microRNA activity. Nat Genet. 2007;39(8):1033–7.
114. Wang Y, Xu Z, Jiang J, et al. Endogenous miRNA sponge lincRNA-RoR regulates Oct4, Nanog, and Sox2 in human embryonic stem cell self-renewal. Dev Cell. 2013;25(1):69–80.
115. Yoon JH, Abdelmohsen K, Gorospe M. Functional interactions among microRNAs and long non-coding RNAs. Semin Cell Dev Biol. 2014;34:9–14.
116. Thum T, Condorelli G. Long non-coding RNAs and microRNAs in cardiovascular pathophysiology. Circ Res. 2015;116(4):751–62.
117. Paraskevopoulou MD, Georgakilas G, Kostoulas N, et al. DIANA-LncBase: experimentally verified and computationally predicted microRNA targets on long non-coding RNAs. Nucleic Acids Res. 2013;41(Database issue):D239–45.

Chapter 9
Noncoding RNAs in Cancer Immunology

Qian Li and Qiang Liu

Abstract Cancer immunology is the study of interaction between cancer cells and immune system by the application of immunology principle and theory. With the recent approval of several new drugs targeting immune checkpoints in cancer, cancer immunology has become a very attractive field of research and is thought to be the new hope to conquer cancer. This chapter introduces the aberrant expression and function of noncoding RNAs, mainly microRNAs and long noncoding RNAs, in tumor-infiltrating immune cells, and their significance in tumor immunity. It also illustrates how noncoding RNAs are shuttled between tumor cells and immune cells in tumor microenvironments via exosomes or other microvesicles to modulate tumor immunity.

Keywords Noncoding RNAs • Cancer immunology • Long noncoding RNAs • MicroRNAs

9.1 Introduction

9.1.1 History of Cancer Immunology

The history of cancer immunology went back to 1909 when Paul Ehrlich indicated that human natural immune system probably could distinguish tumor cells from normal cells and eliminated them. In 1943, it was shown that syngeneic mice immunized against tumors in the same inbred strain could reject a subsequent tumor challenge [1]. This was confirmed by many later studies that demonstrated the importance of cellular immunology as a mediator of allograft rejection as well as protection against the transfer of mouse tumors. At the end of 1950s, Burnet and Thomas

---

Q. Li • Q. Liu (✉)
Sun Yat-sen Memorial Hospital, Sun Yat-sen University,
107, Yanjiang West Road, Guangzhou 510120, China
e-mail: victorlq@hotmail.com

E. Song (ed.), *The Long and Short Non-coding RNAs in Cancer Biology*,
Advances in Experimental Medicine and Biology 927,
DOI 10.1007/978-981-10-1498-7_9

discovered "oncological immunosurveillance" that the immune system could pick up the malignancy and destroy it to suppress tumor development [2].

In 1970s, cancer immunology experienced a period of fast development when natural killer cells (NK cells) were recognized. A new concept—"immune evasion"—that the tumor would "actively use" or "edit" the pathway of immune system to avoid being hunted down by host immunity appeared. Since 1995, several preclinical studies successfully showed that dendritic cells (DCs), when appropriately activated and induced to present tumor-derived peptides, could effectively elicit tumor-specific T-cell response [3–5]. A number of following clinical trials in different cancer types also demonstrated the induction of antitumor immune responses with clinical responses in some cases [6–9].

With decades of hard work on cancer immunology, astonishing results were recently obtained in the clinical trials using immune checkpoint inhibitors and chimeric antigen receptor T-cell therapy. Cancer immunotherapy was selected as the top "Breakthrough of the Year" by American Association of Science in 2013 [10]. These exciting results inspired tremendous research interest, and cancer immunotherapy was selected by American Society of Clinical Oncology as the "Advance of the Year" in 2016.

9.1.2 Current Understanding of Cancer Immunology

During the past few decades, extensive research has revealed several key aspects of cancer immunology: Immunosurveillance of tumors definitely exists. "Successful" tumors are often immunoselected so that they can evade immune elimination. Both the adaptive and innate immune systems are involved in tumor recognition and clearance. Tumor elimination involves the same recognition mechanisms that are used to combat pathogens [11].

It is now known that both innate and adaptive immune cells, effector molecules, and pathways can act as an extrinsic or intrinsic tumor suppressor. More importantly, their deregulation can promote tumor progression through tumor immunoediting. Table 9.1 shows the immune organs, cells, and related immune molecules that are important in normal immune system as well as cancer immunology.

It is generally believed that the battle between immune system and cancer is a dynamic process. The immune system plays a dual role in the complicated interaction between tumors and the host, which prompts a refinement of cancer immunosurveillance hypothesis into "cancer immunoediting." During this process, the immune system actively modifies the immunogenic phenotype of tumors as they develop [12]. Cancer immunoediting is composed of three main phases.

Table 9.1 Immune organs, cells, and molecules in cancer immunology

Immune organ		Immune cells	Immune molecules	
Center	Periphery		Membrane molecules	Secretory molecules
Thymus	Spleen	Stem cells	TCR	Immunoglobulin
Bone marrow	Lymphonodus	Lymphocyte	BCR	Cytokines
	Mucosa-associated lymphoid tissue	Mononuclear phagocyte	CDs	Complement system
	Skin immune system	Antigen-presenting cell	MHC	
		Others (mast cells, granulocyte, erythrocyte, thrombocyte)	Adherence factor	
			Others	

9.1.2.1 Elimination Phase

This is a phase that immune cells such as cytotoxic T lymphocytes and NK cells, tumor-specific antigens, and cytokines of both innate and adaptive immunity work in concert to eliminate cancer cells at early stage.

9.1.2.2 Equilibrium Phase

Emerging evidence suggests that heterogeneous cancer cells can have some variants to avoid being arrested by immune system. When cancer cells obtain a non-immunogenic phenotype, they can escape from the elimination phase and enter into the equilibrium phase that hosts immunity, after which tumor cells come to a dynamic balance. This phase is the longest of the three processes in cancer immunoediting and may occur over a period of many years [13]. The equilibrium phase explains why tumor cells can exist for a long time in the host body, remaining dormant and subclinical apparent.

9.1.2.3 Escape Phase

During this period of Darwinian selection, new cancer cell variants that acquire new mutations to further increase the growth and immunoresistance will emerge. In this phase, cancer cells continue to grow and expand in an uncontrolled manner and eventually lead to clinically apparent tumors.

9.1.3 Noncoding RNA

After the Human Genome and ENCODE projects were finished, it came into surprise that only 2 % or less of human genome could be transcribed into mRNAs and later translated into proteins. The rest of the genome initially thought to be "junk DNA" was later discovered to have various functional and regulatory roles.

Among the noncoding RNAs, microRNAs (miRNAs) are one of the best-studied groups. miRNA is a category of small noncoding and single-stranded RNAs with the length of about 22 nts. miRNAs mainly negatively regulate protein expression by interacting with the 3′-untranslated region of mRNA to inhibit its translation or induce mRNA degradation.

Long noncoding RNAs (lncRNAs) are a group of noncoding RNAs with the length of more than 200 nts. Accumulating evidence suggests that lncRNAs play important roles in many aspects of cell biology. It is estimated that there are at least 20,000 human lncRNA transcripts. Although mechanistically complicated, lncRNAs mainly regulate gene expression or signal transduction by directing interaction with transcriptional factors, signal modulators, etc.

In this chapter, we will mainly discuss microRNA and lncRNA in cancer immunology.

9.2 MicroRNAs and Cancer Immunology

9.2.1 MicroRNAs and Macrophage

Macrophage is one of the most abundant immune cells in tumor microenvironment and plays an important role in regulating tumor initiation, proliferation, invasion, and metastasis [14]. miRNAs are critical in regulating monocyte differentiation or maturation.

9.2.1.1 miR-17-5p, miR-20a, and miR-106a

Fontana's group has reported that the levels of microRNA-17-5p, microRNA-20a, and microRNA-106a are different between monocyte and macrophage. Transfection with miRNA-17-5p, miRNA-20a, and miRNA-106a suppresses AML1 protein expression, which then results in M-CSF receptor (M-CSFR) downregulation, increased blast proliferation, and inhibition of monocytic differentiation and maturation [15]. This suggests that monocytopoiesis is controlled by a circuitry involving sequentially miRNA-17-5p, miRNA-20a, and miRNA-106a, AML1, and M-CSFR.

9.2.1.2 miR-155

Costinean et al. have reported that overexpression of miR-155 in hematopoietic cells induces malignancy [16]. It has also been shown that miR-155 is overexpressed in multiple types of cancer including breast cancer, lung cancer, thyroid cancer, and pancreatic cancer [17, 18]. Transcriptome analysis has identified a large number of miR-155-regulated genes, including cytokines, chemokines, and transcription factors, suggesting that miR-155 may play an important role in regulating cancer immunology.

The targets of miR-155 include TLR2, TLR3, TLR4, or TLR9, which are increased in macrophage when simulated by bacteria and viruses [19, 20]. Androulidaki A and Baltimore D have demonstrated that miR-155's negative regulation of inflammatory is directly through suppression of cytokine signaling 1 and phosphatidylinositol-3,4,5-trisphosphate 5-phosphatase 1 (SOCS1 and SHIP1) [21]. Ceppi et al. have demonstrated that miR-155 simulates TRAF6/IRAK1 in NF-κB pathway. Additionally, TAB2, an important signal transducer, is directly controlled by miR-155 in TLR/IL-1 signaling pathways in DC. miR-155 is highly expressed during macrophage maturation, indicating that it may play a critical role during macrophage differentiation [22].

9.2.1.3 Other miRNAs

Berman and his colleagues have used two different techniques of microRNA array to identify 199 miRs expressed in human's primary macrophages, including let-7a, miR-16, miR-23a, miR-30b, miR-103, miR-146a, miR-212, and miR-378. Nevertheless, the function and mechanism of these miRNAs remain unclear [23].

It has been reported that miR-147 acts as a negative regulator to prevent inflammatory responses via binding with TLR2, TLR3, or TLR4 in Myd88 and TRIF-dependent ways [24]. Ectopic expression of miR-125b in macrophages increases their responsiveness to IFN-γ and leads to a more effective killing of EL4 tumor [25]. miR-21 has also been shown to regulate the expression of PDCD4, a proinflammatory tumor suppressor. And it can also bind to TLR7 and TLR8 to trigger immune response, leading to tumor growth and metastasis [26]. More recently, Zhu D has demonstrated that miR-17, miR-20a, and miR-106a regulate SIRPα synthesis and SIRPα-mediated macrophage inflammatory responses in a redundant fashion [27].

9.2.2 MicroRNA and NK Cells

NK cells kill cancer cells by using killer-activating receptors (KARs) expressed on the cell surface to interact with tumor-derived ligands such as MICA and MICB. Therefore, even when cancer cells loss their major histocompatibility complex (MHC), NK cells can still cause their death.

Bezman et al. have used the mice with conditional abatement of Dicer and DiGeorge syndrome critical region 8 (Dgcr8) to dissect the roles of miRNAs in regulating the activation, survival, and function in NK cells. While the Dgcr8 or Dicer was deleted in NK cells, the apoptosis of NK cells in periphery increased, suggesting that miRNAs may control the survival of NK cells. Potential mechanisms include mitotic defects due to centromere dysfunctions, defects in heterochromatin maintenance, and aberrant overexpression of proapoptotic proteins [28]. Yamanaka and his colleagues have recently found that inhibition of miR-21 in the human NK cell line leads to increased apoptosis associated with upregulation of proapoptotic miR-21 targets such as PTEN, PDCD4, and Bim, indicating that miR-21 may be a critical miR in controlling of NK cell survival [29].

As is already known, low expression of CD27, active proliferation, and high level of CD11b are markers of mature NK cells. In Bezman's study, they have found that more immature $CD27^{high}$ $CD11b^{low}$ NK cells accumulate in mice with the deletion of Dgcr8 or Dicer, indicating that miRNAs also impact the maturation of NK cells. miR150 has also been shown to be dynamically regulated during NK cell maturation [28, 30].

NKG2D (natural killer group 2, member D) is thought to play an important role in mediating the activation of anticancer immune response in NK cells. Nevertheless, there is no significant effect on the level of NKG2D in mice with deletion of Dgcr8 or Dicer. Meanwhile, Bezman's study has indicated that miRNAs regulate the expression of a NKG2D ligand, Rae-1. It is possible that a specific miRNA, although currently unknown, targets Rae-1 mRNA [28].

Heinemann et al. have identified another ligand of NKG2D, ULBP2, as a strong prognostic marker in human malignant melanoma. They showed that the tumor-suppressive microRNAs (miRNAs) miR-34a and miR-34c controlled ULBP2 expression by directly targeting the 3′-untranslated region of ULBP2 mRNA. The level of miR-34a was inversely correlated with the expression of ULBP2 surface molecules [31].

NK cells have longer life than naïve, effector, and memory $CD8^+$ T cells [32], but their miRNA profiles are similar. miR-142-3p, miR-142-5p, miR-150, miR-16, miR-23a, miR-15b, miR-29a, miR-29b, miR-30b, and miR-26a are highly expressed in both naïve NK cells and $CD8^+$ T cells. miR-21, miR-221, and miR-222 are expressed in both NK and effector $CD8^+$ T cells, whereas miR-146a is found in both NK cells and memory $CD8^+$ T cells. Similar to NK cells, the frequency and number of naïve $CD8^+$ T cells are preferentially reduced in Dicer- and Dgcr8-deficient mice. Thus, there may be common miRNAs and similar regulatory mechanisms among NK cells and naïve, effector, and memory $CD8^+$ T cells [33, 34].

miR-29 has three isolated forms (a, b, c) that are highly expressed in normal tissues but are downregulated in a broad range of solid tumors, including neuroblastoma, sarcomas, and brain tumors. B7-H3 (CD276), a co-stimulatory molecule that plays a potent role in immune responses, is a type I transmembrane glycoprotein aberrantly expressed in numerous types of cancer and associated with poor prognosis. Interestingly, B7-H3 can be transcribed in both normal tissues and tumors, but B7-H3 protein can only be expressed in tumor tissues. Cheung and his colleagues have showed that B7-H3 protein expression is inversely correlated with miR-29. Using luciferase reporter assay, miR-29a was shown to directly target B7-H3 3′-untranslated region. Knock-in and knockdown of miR-29a led to downregulation and upregulation of B7-H3 in both NK cells and T cells [35].

9.2.3 *MicroRNA and T Cells*

T cells mediate "cellular" immunity by directly interacting with cancer cells. When specific tumor antigens are presented, activated CD4+ and CD8+ T cells traffic to the tumor site, release cytotoxic granules, lyse cancer cells via Fas–FasL interaction, and recruit secondary effectors.

9.2.3.1 miR-146a and miR-146b

It has been reported that miR-146a is expressed in mouse peripheral $CD4^+$ T cells, particularly in $CD4^+CD25^+$ T_{reg} cells. High expression of miR-146a has also been found in mouse $CD8^+$ T cells, as well as non-T_{reg} $CD4^+$ T cells ($CD4^+CD25^-$). Sheppard and his colleagues have showed that miR-146a is upregulated in naïve CD8+ T cells after IL-2 or IL-15 treatment even without TCRs, but decreases after being exposed to IL-7 [36]. Additionally, in T-cell acute lymphoblastic leukemia, ectopic expression of miR-146a results in significant upregulation of PU.1, c-Fos, CCAAT/enhancer-binding protein alpha (C/EBPα), and GATA3 and slight upregulation of Foxp3 and Runx1. There is also a significant, moderate downregulation in the expressions of Notch1, LIM-domain only (Lmo2), Son of Sevenless 1 (SOS1), Ikaros, and signal transducer and activator of transcription 3 (STAT3). These results indicate that miR-146a is associated with regulation of Th1 response [37].

Yang et al. have demonstrated that T cells ($CD4^+$ T cells and $CD8^+$ T cells) of mice lacking miR-146a expression are more sensitive and active in inflammatory responses. In this study, TCR-driven NF-κB activation upregulated the expression of miR-146a, which in turn downregulated NF-κB activity to function as a negative feedback regulator. This effect was at least partly through repressing the NF-κB signaling transducers TRAF6 and IRAK1 [38]. Lu and his colleagues have identified that deficiency of miR-146a in Treg cells leads to upregulation of signal transducer and activator of transcription 1 (STAT1), downregulation of SOCS1, and breakdown of IFN-γ-dependent immunity [39]. Liu et al. have identified several miRNAs specifically upregulated in premalignant PTEN-deficient thymocytes, including miR-146a and miR-146b. Overexpression of miR-146a and miR-146b delayed induction of c-myc oncogene, a key driver of transformation. Additionally, miR-146b-induced c-myc suppression in mature CD4 T cells impaired TCR-mediated proliferation. Together, these results suggest that miR-146a and miR-146b may inhibit progression to a malignancy [40].

9.2.3.2 miR-155

miR-155 has been shown to be important in the differentiation of T cells, especially CD4+ T cells. Rodriguez et al. have indicated that deletion of miR-155 in naïve T cells results in polarized differentiation preferentially into Th2 cells, with substantial production of Th2 cytokines including IL-4, IL-5, and IL-10. The target of miR-155 was shown to be c-Maf, a transcription factor to promote transcription of IL-4 [41]. Banerjee and his colleagues have found that miR-155 is induced upon T-cell activation and, it promotes Th1 differentiation when overexpressed in activated $CD4^+$ T cells. Antagonism of miR-155 resulted in induction of IFN-γRα, which was identified as another miR-155 target in T cells, suggesting that miR-155 contributed to Th1 differentiation in $CD4^+$ T cells by inhibiting IFN-γ signaling. miR-155 not only regulated effector T cell functions but also involved in the development of Treg cells [42]. Lu et al. have found that Foxp3, a transcription factor, is also a target of miR-155. In the absence of miR-155($Foxp3^+$) in mice, the number of Tregs significantly reduced, but the suppressor activity of Tregs remained unchanged, indicating that miR-155 was critical to Treg development with a mechanism different from miR-146a [43].

Ji et al. have showed that knockdown of miR-155 increases the expression of SOCS1 and suppresses STAT3 expression to inhibit cancer cell growth, migration, and invasion in laryngeal squamous cell carcinoma (LSCC). Meanwhile, overexpression of miR-155 promoted cell growth, migration, and invasion [44].

9.2.3.3 miR-181a

T-cell sensitivity to antigen is tightly regulated during maturation to ensure proper development of immunity and tolerance. In 2007, Li et al. showed that increasing miR-181a expression in mature T cells enhanced their sensitivity to peptide antigens, while inhibiting miR-181a expression in the immature T cells decreased their sensitivity and impaired both positive and negative selection. These effects were partly achieved by downregulating multiple phosphatases, which led to an increase of steady-state phosphorylated intermediates and a decrease of TCR signaling threshold [45, 46].

Additionally, Mori et al. have showed that the expression of miR-181a is associated with poor prognosis in colorectal cancer patients [47].

9.2.3.4 miR-17-92

miR-17-92 is shown to be expressed in many types of human tumors including lung cancer, breast cancer, and lymphoma, mainly due to gene amplification and Myc-mediated transcriptional upregulation. Xiao et al. have demonstrated that B-cell-specific miR-17-92 transgenic mice develop lymphomas with high penetrance and Myc-driven lymphomagenesis stringently requires two intact alleles of miR-17-92.

Further analysis has shown that miR-17-92 drives lymphomagenesis by suppressing the expression of multiple negative regulators of the PI3K and NF-κB pathways and by inhibiting the mitochondrial apoptosis pathway [48].

Sasaki et al. performed miRNA microarray analysis and found that the levels of miR-17-92 is significantly overexpressed in Th1 than that in Th2. The inhibition of IL-4 pathway by neutralizing antibody or inhibitors of STAT6 has restored the expression of miR-17-92 in Th2 cells. These results suggest that the type-2-skewing tumor microenvironment induces the downregulation of miR-17-92 expression in T cells, thus decreasing the persistence of tumor-specific T cells and tumor control. Sasaki also suggested that genetic engineering of T cells to express miR-17-92 might represent a promising approach for cancer immunotherapy. miR-17-92 regulated the maturation of CD8+ T cells and inhibited the TGF-β-dependent differentiation of Treg cells. Moreover, it was required for differentiation of Th17 [49].

The transcription of miR-17-92 cluster was regulated by E2F, STAT3, c-Myc, and sonic hedgehog pathway. The target of miR-17-92 contained E2F, p21, anti-angiogenic thrombospondin-1, connective tissue growth factor, and phosphatase and tensin homolog (PTEN). These suggest that miR-17-92 plays a key role in controlling apoptosis and proliferation, consistent with the findings in miR-17-92 transgenic mice that have stronger capacity for survival and proliferation.

Another target of miR-17-92 is HIF-1α, identified in lung cancer by Taguchi [50]. Unlike other cells relying on HIF-1α to survive, T cells only use HIF-1a as a negative regulator in inflammatory. Knockdown of HIF-1α increased the secretion of IFN-γ in T cells. Moreover, miR-17-92 expression was decreased in T cells from tumor-bearing mice. Collectively, miR-17-92 may participate in regulating tumor immune evasion.

9.2.3.5 Other miRNAs

miR-326 has been found to target Ets, a transcription factor, which is a negative regulator of Th17 development. Increased expression of miR-326 results in the promotion of Th17 maturation and enhanced IL-17 secretion [51].

Meggetto et al. have recently found that miR-150, epigenetically silenced by STAT3/DNMT1, is involved in the development of hemopathies and downregulated in T-lymphomas [52].

9.2.4 MicroRNA and B cells

B cells kill cancer cells mainly by releasing antibodies specifically targeting cancer cells.

9.2.4.1 miR-181b

The first miRNA studied in B cells is miR-181b, which is overexpressed during the development of B cells. PCDC10, GATA6, HIF-1α, and AID have been demonstrated to be the targets of miR-181b. Konski has recently found that miR-181b inhibits breast cancer metastasis by directly downregulating the proinflammatory cytokines CXCL1 and CXCL2 [53, 54].

9.2.4.2 miR-150

miR-150, a mediator during the development of B cells, was found to be lowly expressed in immature B cells, but highly expressed in mature ones. C-Myc [55], FOXP1 [55], and KFL2 [56] are found to be the targets of miR-150. Zhou et al. have reported that premature expression of miR-150 in B cell blocks the translation of pro-B cells to pre-B cells [57].

9.2.4.3 miR-155

Similarly important to its role in T cells, miR-155 was found to play a crucial role in B cell. Overexpression of miR-155 was shown to lead to B preleukemia. Further clinical observation indicates that sustained high expression of miR-155 leads to myeloproliferative disorder. Recent studies have indicated that miR-155 is associated with hematopoietic lineage differentiation. Moreover, knockdown of miR-155 in B cells results in reduced germinal center responses and less production of IgG1 antibodies. Several studies have found that the targets of miR-155 in B cells are quite similar to those in T cell, including Pu.1, SHIP1, and AID (activation-induced cytidine deaminase). Dorsett et al. have established transgenic mice models using AID with mutated miR-155-binding site. They have demonstrated that AID is required for immunoglobulin gene diversification in B lymphocytes, and it also promotes chromosomal translocations, suggesting that miR-155 can act as a tumor suppressor by reducing potentially oncogenic translocations generated by AID [58]. Wang et al. have recently demonstrated that knockdown of miR-155 disrupts the B-cell-activating factor (BAFF)-R-related signaling pathway and reduces the translocation of nuclear factor (NF)-κB into the nucleus [59].

In 2013, Neilsen et al. assembled a comprehensive list with 140 genes and regulatory proteins targeted by miR-155 in myelopoiesis and leukemogenesis (AICDA, ETS1, JARID2, SPI1, etc.), inflammation (BACH1, FADD, IKBKE, INPP5D, MYD88, RIPK1, SPI1, SOCS, etc.), and known tumor suppressors (C/EBPβ, IL17RB, PCCD4, TCF12, ZNF652, etc.) [60].

In addition to AP-1 family members, JunB and FosB have been identified to be recruited to the *MIR155HG* promoter region to regulate miR-155 expression following B-cell receptor activation [61].

9.3 lncRNA and Cancer Immunology

lncRNAs have rapidly become a hot topic of research in recent 3–5 years. Not only are the quantities of lncRNAs much higher than those of miRNAs, but the mechanism and functional diversity of lncRNAs are also much more complicated than those of miRNAs. The technical difficulties do not hinder the enthusiasm about lncRNAs for they have been shown to be very important in many aspects of biology including immunology. The lncRNAs discussed below are mostly discovered and play important roles in models of immunology. It is very likely that they are also very important in certain circumstances of cancer immunology.

9.3.1 lincRNA-cox2

The induction of inflammatory gene expression is key to the antimicrobial defenses, which is controlled by a collaboration involving activation of transcription factors, transcriptional co-regulators, and chromatin-modifying factors. Recently Fitzgerald et al. have identified a long noncoding RNA (lncRNA), lincRNA-COX2, which acts as a key regulator of this inflammatory response. They have also demonstrated that lincRNA-cox2 mediate both the activation and repression of distinct classes of immune genes.

lincRNA-Cox2 is named because of its adjacent location to Cox2 (prostaglandin–endoperoxide synthase 2, PTGS2) gene. The inductions of Cox2 and lincRNA-Cox2 are both dependent on MyD88, which is a key adaptor protein in the TLR signaling. However, lincRNA-Cox2 does not regulate the transcription of cox2 gene because silencing lincRNA-Cox2 cannot change the expression of its neighboring gene Cox2.

A number of chemokines (Ccl5, Cx3cl1), chemokine receptors (Ccrl), and interferon-stimulated genes (ISGs) (Irf7, Oas1a, Oas1l, Oas2, Ifi204, and Isg15) are upregulated when lincRNA-Cox2 is silenced in unstimulated cells. And when the same cells are stimulated by Pam_3CSK_4, lincRNA-Cox2 induces 713 gene expressions including Tlr1, Il6, *and* Il23a.

While the detailed mechanism remains unclear, lincRNA-Cox2 localized in both the cytoplasm and nucleus interacts with heterogeneous nuclear ribonucleoproteins (hnRNPs) A/B and A2/B1 to regulate the transcription of the related immune genes [62].

9.3.2 PACER

In 2014, Krawczyk and Emerson found that an lncRNA, named P50-associated cyclooxygenase-2 (COX-2) extragenic RNA (PACER), activated the expression of COX-2 gene. COX2 protein is overexpressed in many types of cancer and is also found to play an important role in the differentiation of monocytes into macrophages.

Using primary human mammary epithelial cells and monocyte/macrophage cell lines, Krawczyk and Emerson showed that the chromatin boundary/insulator factor CCCTC-binding factor (CTCF) established an open chromatin domain and induced the expression of PACER within the upstream promoter region of COX-2. Upon induction of COX-2 expression, PACER is associated with p50, a repressive subunit of NF-κB, and occluded it from the COX-2 promoter, potentially facilitating its interaction with activation-competent NF-κB p65/p50 dimers. This enabled the recruitment of p300 histone acetyltransferase, a domain-wide increase in histone acetylation and assembly of RNA polymerase II initiation complexes. These findings unveil an unexpected mechanism of gene control by lncRNA-mediated repressor occlusion and identify the COX-2-lncRNA, PACER, as a new potential target for COX-2 modulation in inflammation and cancer [63].

Recently it has been shown that Cox2 is expressed in tumor-associated macrophages (TAMs) that play an important role in tumor microenvironment. Both in vivo and in vitro experiments demonstrated that PACER acted as a positive regulator to promote the expression of Cox2, in both epithelial cells and monocyte-derived macrophages.

9.3.3 NKILA

In 2015, Song and his colleagues identified an NF-κB-interacting lncRNA (NKILA), which was upregulated by NF-κB, bound to NF-κB/IκB, and directly masked phosphorylation motifs of IκB, thereby inhibiting IKK-induced IκB phosphorylation and NF-κB activation. Interestingly, unlike DNA that was dissociated from NF-κB by IκB, NKILA interacted with NF-κB/IκB to form a stable complex.

NKILA was found to be mainly expressed in the cytoplasm of cancer cells and upregulated by the inflammatory signals in tumor microenvironment. As a negative feedback regulator of NF-κB, it turned out to be essential to prevent the over-activation of NF-κB pathway in inflammation-stimulated breast epithelial cells. Furthermore, low NKILA expression was found to associate with breast cancer metastasis and poor patient prognosis. This study uncovers an important mechanism of lncRNAs by directly interacting with functional domains of signaling proteins to modulate the activity of signaling pathways. NKILA has become a new class of NF-κB modulators to suppress cancer metastasis and inflammation [64].

9.3.4 lnc-DC

In 2014, Cao and his colleagues identified lnc-DC, which was exclusively highly expressed in human conventional dendritic cells (DCs), including almost every subsets such as $Lin^{-}MHC\text{-}II^{+}CD11c^{+}$ cDCs and so on from the blood and skin. RNA-seq experiment has showed that lnc-DC does not exist in stem cells, progenitor cells, monocytes, or T/B cells. Thus, lnc-DC is a specific and stable marker of DCs.

When lnc-DC was knock downed, the expressions of many immune genes changed. For example, CD40, CD80, CD86, and HLA-DR, which were important factors for T cells, were downregulated. Meanwhile, CD14, a marker of monocytes, increased. Additionally, DCs' ability to uptake antigen got lost. The proliferation of $CD4^{+}T$ cells and production of immune cytokines significantly reduced. These results indicate that lnc-DC may play a role in regulating the differentiation of DCs.

Indeed lnc-DC is essential in regulating the differentiation of DCs from human monocytes in vitro and from mouse bone marrow cells in vivo. Knockdown of lnc-DCs inhibits the capacity of DCs to stimulate the activation of T cells. lnc-DC mediates these effects by activating the transcription factor STAT3 (signal transducer and activator of transcription 3). lnc-DC, mainly expressed in the cytoplasm of DCs, directly binds to STAT3 and promotes STAT3 phosphorylation on tyrosine-705 by preventing STAT3 binding to and being dephosphorylated by SHP1. This study has identified a lncRNA that regulates the differentiation of DCs and also broadened the known mechanisms of lncRNA action [65].

9.3.5 THRIL

To understand whether lncRNAs play a role in regulating cell defense mechanisms and host–pathogen interactions, Li et al. have designed a custom human lincRNA microarray to detect genome-wide changes of the expression of lincRNAs in a classical model of innate immune cell activation. THRIL (TNFα and hnRNPL-related immunoregulatory lincRNA) is one of the lncRNAs differentially expressed in the macrophage-like cells that are differentiated from human THP1 monocyte cell line and stimulated with a synthetic lipopeptide ligand of TLR2.

THRIL was shown to be expressed in many human tissues and required for the induction of TNFα expression. It could bind specifically to heterogeneous nuclear ribonucleoprotein L (hnRNPL) and form a functional THRIL–hnRNPL complex that regulated the transcription of the TNFα gene by binding to its promoter.

Interestingly, while silencing THRIL decreases TNFα expression, overexpression of THRIL first increases TNFα expression and later downregulates not only TNFα but also itself. The high secretion of TNFα results in THRIL downregulation, suggesting a feedback loop between THRIL and TNFα.

Transcriptome analysis has showed that THRIL is required for the expression of many immune-responsive genes, including other cytokines such as IL-8, CXCL-10, CCL1, and CSF1, and transcriptional and posttranscriptional regulators of TNFα expression. Knockdown of THRIL leads to deregulation of these genes during innate activation of THP1 macrophages.

More importantly, THRIL expression is correlated with the severity of symptoms in patients with Kawasaki disease, an acute inflammatory disease in children. This study provides evidence that lincRNAs and their binding proteins can regulate

TNFα expression and may play important roles in the innate immune response and inflammatory diseases in humans.

9.3.6 *lnc-IL7R*

lnc-IL7R is another lncRNA identified through lncRNA microarray technique. lnc-IL7R, which overlaps with the 3′-untranslated region (3′-UTR) of the human interleukin-7 receptor α-subunit gene (IL7R gene), is significantly upregulated in THP1 macrophages treated with LPS or Pam.

lnc-IL7R is capable of diminishing the LPS-induced inflammatory response because knockdown of lnc-IL7R increases the expression of LPS-induced E-selectin, VCAM-1, IL-6, and IL-8. However, although lnc-IL7R overlaps with 3′-UTR of IL7R gene, overexpression of lnc-IL7R has no effect on IL7R expression or LPS-induced expression of proinflammatory cytokines, such as TNFα, IL-8, and IL-6 in THP1 cells. Interestingly, knockdown of both lnc-IL7R and IL7R, instead of IL7R, significantly enhances LPS-induced expression of the proinflammatory mediators such as E-selectin, VCAM-1, IL-6, and IL-8, suggesting that lnc-IL7R functions independently of IL7R in regulating the inflammatory response to LPS.

Cui et al. have found that lnc-IL7R knockdown diminishes the trimethylation of histone H3 at lysine 27 (H3K27me3), a hallmark of silent transcription, at the proximal promoters of the inflammatory mediators. These data suggest that lnc-IL7R contributes another layer of complexity in regulation of the inflammatory response [66].

9.3.7 *Lethe*

Pseudogenes are genes that have lost their protein-coding abilities as a result of mutations accumulated during evolution. They were previously referred to as "dead genes" because they were thought to have lost their function completely, even lacking the ability to encode RNA. Nevertheless, recent studies have showed that pseudogenes are in fact transcribed into long noncoding RNAs, and these are now a new focus of research.

Rapicavoli et al. have reported that specific and distinct pseudogene-derived long RNAs are made when cells are exposed to different kinds of infections. Lethe (named after the "river of forgetfulness" in Greek mythology) is a pseudogene lncRNA selectively induced in mouse embryonic fibroblast (MEF) cells by proinflammatory cytokines via NF-κB or glucocorticoid receptor agonist. It binds to NF-κB and prevents it from interacting with DNA, thereby reducing the production of many inflammatory factors and functioning as a negative feedback signaling to NF-κB.

Lethe is the first pseudogene that has been shown to have an active role in regulating signaling pathways involved in inflammation. It is possible that other pseudogenes may also have active roles in regulating distinct signaling pathways. This suggests that many novel functions for pseudogenes and long noncoding RNAs are yet to be found [67].

9.3.8 IFNG-AS1

In 2010, when lncRNA research was at its beginning phase, Collins et al. demonstrated an ~1 kb-long element located at the upstream of IFN-γ gene that had no protein-coding ability and bound to Runx3 to recruit RNA polymerase II to IFN-γ gene to regulate its expression. Two years later, Collins found more noncoding DNA regions around IFN-γ gene and named them conserved noncoding sequences (CNSs). He showed that these CNSs were associated with IFN-γ gene expression. CNS-30, CNS-4, and CNS+20 exerted their function in different stages of Th1 differentiation, with CNS-4 necessary for the induction of IFN-γ in effector Th1 cells. Moreover, CNS-16, CNS-4, and CNS+20 were each partially needed for IFN-γ induction in NK cells, while CNS-16 was a repressor in regulating IFN-γ.

Then Collier et al. identified the first lncRNA and named it TMEVPG1 (Theiler's murine encephalomyelitis virus persistence candidate gene 1) because it was initially identified as a candidate gene for the control of Theiler's virus persistence. It had also been named as NeST (nettoie Salmonella pas Theiler's) before. Recently, it has often been called IFNG-AS1 because it is indicated to have a new role in the differentiation of Th1 related to IFN-γ.

According to the findings by Collier and his colleagues, the expressions of both IFNG-AS1 and IFN-γ were high in Th1 but low in Th2. Silencing IFN-γ had no affect to IFNG-AS1, but knockdown of IFNG-AS1 led to a markedly decrease of IFN-γ. Also, overexpression of IFNG-AS1 upregulates the transcription of IFN-γ, indicating IFNG-AS1 regulated the expression of IFN-γ to facilitate the differentiation of Th1 cells [68].

It has also been demonstrated that IFNG-AS1 functions in $CD8^+$ T cells via the interaction with WDR subunit to increase the H3K4me at IFN-γ locus [69].

9.3.9 lincR-Ccr2-5'AS

To further understand the role of lncRNAs in the differentiation of T cells, Hu et al. used 42 T-cell samples to analyze the changes of lncRNA during the differentiation from early T-cell progenitors to terminally differentiated T helper cells. The analysis shows highly dynamic and cell-specific expression patterns for lincRNAs during T-cell differentiation.

While STAT4 and T-bet activated Th1-preferred lincRNAs, GATA3 and STAT6 regulated Th2-preferred lincRNAs. Many lincRNAs highly induced in Th cells were located next to protein-coding genes critical to T-cell function, indicating a possible coevolution of lincRNA and protein-coding genes for the control of immune function.

lincR-Ccr2-5′AS, a lncRNA located between Ccr3 and Ccr2 genes and transcribed in the opposite direction of the Ccr2 gene, was expressed specifically in Th2 cells. Moreover, the lncRNA was associated with genes in the chemokine-mediated signaling pathway, suggesting a co-regulation of their expression. Knockdown of lincR-Ccr2-5′AS in Th2 cells did not change the IL-4 level, but markedly decreased the expression of Ccr1, Ccr2, Ccr3, and Ccr5, which were located in the vicinity of the lincR-Ccr2-5′AS gene. Moreover, knockdown of lincR-Ccr2-5′AS significantly decreases the number of Th2 cells migrated to the lung. These results indicate that lincR-Ccr2-5′AS can modulate the expression of several chemokine receptors and so contributes to Th2 cell migration [70].

9.3.10 GATA3-AS1

GATA3 is an important transcription factor that regulates Th2 differentiation. Zhang et al. have reported that a lncRNA located at the antisense strand of transcriptional start site of GATA3, named GATA3-AS1, is specifically expressed in primary Th2 cells but not in other immune cells. They have found that GATA3-AS1 belongs to the category of divergent lncRNAs and is transcribed in the opposite direction from GATA3 in both mouse and human genomes. Also, both GATA3-AS1 and GATA3 are co-expressed in mouse and human Th2 cells, although the function of GATA3-AS1 remains unknown. This systematic genome-wide analysis of human primary CD4+ T-cell subset helps to identify novel CD4+ T lineage-specific genes. lincRNA GATA3-AS1 might serve as a specific indicator of Th2 response and Th2-associated diseases and might be involved in Th2 cell differentiation [71].

9.3.11 Th2-LCR

In 2003, Hwang et al. reported that the expression of Th2 cytokines was coordinately regulated by the Th2 locus control region (Th2-LCR) located in the 3′-region of the RAD50 gene [72]. In 2015, Spurlock and his colleagues used the whole genome sequencing (RNA-seq) to identify lncRNAs differentially expressed in T cells differentiated under Th1-, Th2-, or Th17-polarizing conditions. Interestingly, the majority of lineage-specific lncRNA genes were co-expressed with nearby lineage-specific protein-coding genes. These lncRNAs were predominantly intragenic with co-expressed protein-coding genes and transcribed in both sense and antisense directions with almost equal frequencies. Genes encoding Th lineage-specific mRNAs were not randomly distributed across the genome but highly enriched in genomic regions containing genes encoding Th lineage-specific lncRNAs.

This study has identified a cluster of antisense lncRNAs specifically expressed in Th2 lineage that regulate the expression of Th2 cytokines, IL-4, IL-5, and IL-13. This lncRNA cluster in human overlapped with the RAD50 locus and thus was contiguous with the previously described Th2-LCR in mouse. Given the same genomic location with the mouse Th2-LCR, this lncRNA cluster is referred to as Th2-LCR lncRNA.

Th2-LCR lncRNA is consisted with four spliced transcripts. Targeting individual alternatively spliced Th2-LCR lncRNAs with siRNAs resulted in partial inhibition of IL-4, IL-5, or IL-13 cytokine expression, but siRNA-targeted knockdown of all Th2-LCR lncRNA alternatively spliced isoforms caused a greater degree of inhibition, indicating that alternatively spliced lncRNAs functioned cooperatively to drive Th2 cytokine expression.

Further, knockdown of the Th2-LCR lncRNA results in a significant decrease in H3K4Me3 at the promoter, intron, and a distal conserved region of IL-4 and the promoter and a distal conserved element of IL-3. This suggests that Th2-LCR lncRNA plays a role in facilitating the formation of H3K4Me3 marks at multiple genomic positions at IL-4 and IL-13 loci [73].

9.3.12 *linc-MAF-4*

In 2015, Ranzani et al. investigated the expression of lincRNAs in 13 subsets of T and B lymphocytes by RNA-seq analysis and de novo transcriptome reconstruction. They found that the expression of a lincRNA specific to the TH1 subset of helper T cells, named linc-MAF-4, was inversely correlated with the expression of MAF, a TH2-associated transcription factor.

Further, knockdown of linc-MAF-4 in active naive T cells upregulates the expression of MAF and skews T-cell differentiation toward the Th2 phenotype. In addition, linc-MAF-4 has been shown to bind with MAF's promoter and act as a scaffold to recruit two repressors, EZH2 and LSD1, to the promoter. This study suggests that linc-MAF-4 downregulates the expression of MAF by recruiting EZH2 and LSD1 to modulate the activity of enzymatic EZH2 on MAF's promoter [74].

9.3.13 *NRON*

NRON is one of the first identified lncRNAs that play an important role in biology. In 2005, Willingham et al. identified a lncRNA functioning as a repressor of NFATs (nuclear factor of activated T cells) and named it NRON (noncoding RNA repressor of NFAT).

NFAT is a Ca^{2+}-regulated transcription factor which is normally heavily phosphorylated and stays in the cytoplasm of resting cells. When cells are stimulated by a rise in intracellular Ca2+, NFAT proteins are dephosphorylated by the Ca2+/calmodulin-dependent phosphatase calcineurin and then translocate to the nucleus to activate the expression of target genes.

In 2011, Sharma et al. found that phosphorylated NFAT1 is present in a large cytoplasmic RNA–protein scaffold complex that contains NRON, a scaffold protein IQGAP, and three NFAT kinases. Knockdown of both NRON and IQGAP1 increases NFAT dephosphorylation and nuclear import exclusively after stimulation, without changing the rate of NFAT rephosphorylation and nuclear export. Both NRON-depleted T cells and T cells from IQGAP1-deficient mice showed increased expression of NFAT-dependent cytokines. This study indicates that a complex of lincRNA and protein forms a scaffold to bring a transcription factor to the proximity of its regulatory kinases to regulate its activation [75].

9.3.14 *Fas-AS1*

In 2014, Sehgal et al. reported that the alternative splicing of Fas in lymphomas was tightly regulated by a long noncoding RNA corresponding to an antisense transcript of Fas and named it Fas-AS1.

Soluble Fas (sFas) receptor inhibits apoptosis by sequestering Fas ligand, and its serum level is associated with poor prognosis of non-Hodgkin's lymphomas. Sehgal et al. found that the levels of Fas-AS1 correlated inversely with the production of sFas, and Fas-AS1 binding to the RBM5 inhibited the expression of sFas. EZH2, often mutated or overexpressed in lymphomas, hyper-methylated the Fas-AS1 promoter and repressed its expression. Relieved repression of Fas-AS1 promoter or ectopic expression of Fas-AS1 decreased the expression of sFas. Treatment with Bruton's tyrosine kinase inhibitor or EZH2 knockdown decreased the expression of EZH2, RBM5, and sFas, thereby enhancing Fas-mediated apoptosis. This study is the first report showing a lncRNA regulates the function of Fas, which provides a rationale for the use of EZH2 inhibitors or ibrutinib in combination with chemotherapeutic agents that recruit Fas for effective cell killing [76].

9.4 Outlook

Cancer immunology is thought to be the new hope to defeat cancer. Noncoding RNAs, especially miRNAs and lncRNAs, emerge as important regulators in many aspects of biology including cancer immunology. However, only a small portion of lncRNAs has been studied so far, and new models of lncRNA function continue to surprise us. It is possible that other types of ncRNAs, such as circular RNA and snoRNA, may also be essential in regulating cancer immune response. In addition to the functions of ncRNAs, the prognostic and therapeutic values of these ncRNAs in cancer immunology are waiting to be explored.

References

1. Gross L. Intradermal immunization of C3H mice against a sarcoma that originated in an animal of the same line. Cancer Res. 1942;3(5):326–33.
2. Burnet M. Cancer—a biological approach. I. The processes of control. Br Med J. 1957;1(5022):779–86.
3. Mayordomo JI, Zorina T, Storkus WJ, et al. Bone marrow-derived dendritic cells pulsed with synthetic tumour peptides elicit protective and therapeutic antitumour immunity. Nat Med. 1995;1(12):1297–302.
4. Flamand V, Sornasse T, Thielemans K, et al. Murine dendritic cells pulsed in vitro with tumor antigen induce tumor resistance in vivo. Eur J Immunol. 1994;24(3):605–10. doi:10.1002/eji.1830240317.
5. Celluzzi CM, Mayordomo JI, Storkus WJ, et al. Peptide-pulsed dendritic cells induce antigen-specific CTL-mediated protective tumor immunity. J Exp Med. 1996;183(1):283–7.
6. Brossart P, Wirths S, Brugger W, Kanz L. Dendritic cells in cancer vaccines. Exp Hematol. 2001;29(11):1247–55.
7. Steinman RM, Dhodapkar M. Active immunization against cancer with dendritic cells: the near future. Int J Cancer. 2001;94(4):459–73.
8. Byrne SN, Halliday GM. Dendritic cells: making progress with tumour regression? Immunol Cell Biol. 2002;80(6):520–30. doi:10.1046/j.1440-1711.2002.01122.x.
9. Parmiani G, Castelli C, Dalerba P, et al. Cancer immunotherapy with peptide-based vaccines: what have we achieved? where are we going? J Natl Cancer Inst. 2002;94(11):805–18.
10. Couzin-Frankel J. Breakthrough of the year 2013. Cancer immunotherapy. Science. 2013;342(6165):1432–3. doi:10.1126/science.342.6165.1432.
11. Parish CR. Cancer immunotherapy: the past, the present and the future. Immunol Cell Biol. 2003;81(2):106–13. doi:10.1046/j.0818-9641.2003.01151.x.
12. Dunn GP, Old LJ, Schreiber RD. The three Es of cancer immunoediting. Annu Rev Immunol. 2004;22:329–60. doi:10.1146/annurev.immunol.22.012703.104803.
13. Kim R, Emi M, Tanabe K. Cancer immunoediting from immune surveillance to immune escape. Immunology. 2007;121(1):1–14. doi:10.1111/j.1365-2567.2007.02587.x.
14. Su S, Liu Q, Chen J, et al. A positive feedback loop between mesenchymal-like cancer cells and macrophages is essential to breast cancer metastasis. Cancer Cell. 2014;25(5):605–20. doi:10.1016/j.ccr.2014.03.021.
15. Fontana L, Pelosi E, Greco P, et al. MicroRNAs 17-5p–20a–106a control monocytopoiesis through AML1 targeting and M-CSF receptor up-regulation. Nat Cell Biol. 2007;9(7):775–87. doi:10.1038/ncb1613.
16. Costinean S, Zanesi N, Pekarsky Y, et al. Pre-B cell proliferation and lymphoblastic leukemia/high-grade lymphoma in E(mu)-miR155 transgenic mice. Proc Natl Acad Sci U S A. 2006;103(18):7024–9. doi:10.1073/pnas.0602266103.
17. Calin GA, Croce CM. MicroRNA signatures in human cancers. Nat Rev Cancer. 2006;6(11):857–66. doi:10.1038/nrc1997.
18. Tili E, Croce CM, Michaille JJ. MiR-155: on the crosstalk between inflammation and cancer. Int Rev Immunol. 2009;28(5):264–84. doi:10.1080/08830180903093796.
19. O'Connell RM, Chaudhuri AA, Rao DS, Baltimore D. Inositol phosphatase SHIP1 is a primary target of miR-155. Proc Natl Acad Sci U S A. 2009;106(17):7113–18. doi:10.1073/pnas.0902636106.

20. O'Connell RM, Taganov KD, Boldin MP, et al. MicroRNA-155 is induced during the macrophage inflammatory response. Proc Natl Acad Sci U S A. 2007;104(5):1604–9. doi:10.1073/pnas.0610731104.
21. Androulidaki A, Iliopoulos D, Arranz A, et al. The kinase Akt1 controls macrophage response to lipopolysaccharide by regulating microRNAs. Immunity. 2009;31(2):220–31. doi:10.1016/j.immuni.2009.06.024.
22. Ceppi M, Pereira PM, Dunand-Sauthier I, et al. MicroRNA-155 modulates the interleukin-1 signaling pathway in activated human monocyte-derived dendritic cells. Proc Natl Acad Sci U S A. 2009;106(8):2735–40. doi:10.1073/pnas.0811073106.
23. Luers AJ, Loudig OD, Berman JW. MicroRNAs are expressed and processed by human primary macrophages. Cell Immunol. 2010;263(1):1–8. doi:10.1016/j.cellimm.2010.03.011.
24. Liu G, Friggeri A, Yang Y, et al. MiR-147, a microRNA that is induced upon Toll-like receptor stimulation, regulates murine macrophage inflammatory responses. Proc Natl Acad Sci U S A. 2009;106(37):15819–24. doi:10.1073/pnas.0901216106.
25. Tili E, Michaille JJ, Cimino A, et al. Modulation of miR-155 and miR-125b levels following lipopolysaccharide/TNF-alpha stimulation and their possible roles in regulating the response to endotoxin shock. J Immunol. 2007;179(8):5082–9.
26. Sheedy FJ, Palsson-McDermott E, Hennessy EJ, et al. Negative regulation of TLR4 via targeting of the proinflammatory tumor suppressor PDCD4 by the microRNA miR-21. Nat Immunol. 2009;11(2):141–7. doi:10.1038/ni.1828.
27. Zhu D, Pan C, Li L, et al. MicroRNA-17/20a/106a modulate macrophage inflammatory responses through targeting signal-regulatory protein α. J Allergy Clin Immunol. 2013;132(2):426–36. doi:10.1016/j.jaci.2013.02.005.
28. Bezman NA, Cedars E, Steiner DF, et al. Distinct requirements of MicroRNAs in NK cell activation, survival, and function. J Immunol. 2010;185(7):3835–46. doi:10.4049/jimmunol.1000980.
29. Yamanaka Y, Tagawa H, Takahashi N, et al. Aberrant overexpression of microRNAs activate AKT signaling via down-regulation of tumor suppressors in natural killer-cell lymphoma/leukemia. Blood. 2009;114(15):3265–75. doi:10.1182/blood-2009-06-222794.
30. Bezman NA, Chakraborty T, Bender T, Lanier LL. MiR-150 regulates the development of NK and iNKT cells. J Exp Med. 2011;208(13):2717–31. doi:10.1084/jem.20111386.
31. Heinemann A, Zhao F, Pechlivanis S, et al. Tumor suppressive microRNAs miR-34a/c control cancer cell expression of ULBP2, a stress-induced ligand of the natural killer cell receptor NKG2D. Cancer Res. 2012;72(2):460–71. doi:10.1158/0008-5472.CAN-11-1977.
32. Sun JC, Beilke JN, Lanier LL. Adaptive immune features of natural killer cells. Nature. 2009;457(7229):557–61. doi:10.1038/nature07665.
33. Wu H, Neilson JR, Kumar P, et al. miRNA profiling of naive, effector and memory CD8 T cells. Plos One. 2007;2(10):e1020. doi:10.1371/journal.pone.0001020.
34. Landgraf P, Rusu M, Sheridan R, et al. A mammalian microRNA expression atlas based on small RNA library sequencing. Cell. 2007;129(7):1401–14. doi:10.1016/j.cell.2007.04.040.
35. Xu H, Cheung IY, Guo HF, Cheung NK. MicroRNA miR-29 modulates expression of immunoinhibitory molecule B7-H3: Potential implications for immune based therapy of human solid tumors. Cancer Res. 2009;69(15):6275–81. doi:10.1158/0008-5472.CAN-08-4517.
36. Sheppard HM, Verdon D, Brooks AE, et al. MicroRNA regulation in human CD8+ T cell subsets--cytokine exposure alone drives miR-146a expression. J Transl Med. 2014;12:292. doi:10.1186/s12967-014-0292-0.
37. Saki N, Abroun S, Soleimani M, et al. The roles of miR-146a in the differentiation of Jurkat T-lymphoblasts. Hematology. 2014;19(3):141–7. doi:10.1179/1607845413Y.0000000105.
38. Yang L, Boldin MP, Yu Y, et al. MiR-146a controls the resolution of T cell responses in mice. J Exp Med. 2012;209(9):1655–70. doi:10.1084/jem.20112218.
39. Lu LF, Boldin MP, Chaudhry A, et al. Function of miR-146a in controlling Treg cell-mediated regulation of Th1 responses. Cell. 2010;142(6):914–29. doi:10.1016/j.cell.2010.08.012.

40. Burger ML, Xue L, Sun Y, et al. Premalignant PTEN-deficient thymocytes activate microRNAs miR-146a and miR-146b as a cellular defense against malignant transformation. Blood. 2014;123(26):4089–100. doi:10.1182/blood-2013-11-539411.
41. Rodriguez A, Vigorito E, Clare S, et al. Requirement of bic/microRNA-155 for normal immune function. Science. 2007;316(5824):608–11. doi:10.1126/science.1139253.
42. Banerjee A, Schambach F, DeJong CS, et al. Micro-RNA-155 inhibits IFN-gamma signaling in CD4+ T cells. Eur J Immunol. 2010;40(1):225–31. doi:10.1002/eji.200939381.
43. Lu LF, Thai TH, Calado DP, et al. Foxp3-dependent microRNA155 confers competitive fitness to regulatory T cells by targeting SOCS1 protein. Immunity. 2008;30(1):80–91. doi:10.1016/j.immuni.2008.11.010.
44. Zhao XD, Zhang W, Liang HJ, Ji WY. Overexpression of miR -155 promotes proliferation and invasion of human laryngeal squamous cell carcinoma via targeting SOCS1 and STAT3. Plos One. 2013;8(2), e56395. doi:10.1371/journal.pone.0056395.
45. Li QJ, Chau J, Ebert PJ, et al. MiR-181a is an intrinsic modulator of T cell sensitivity and selection. Cell. 2007;129(1):147–61. doi:10.1016/j.cell.2007.03.008.
46. Li G, Yu M, Lee WW, et al. Decline in miR-181a expression with age impairs T cell receptor sensitivity by increasing DUSP6 activity. Nat Med. 2012;18(10):1518–24. doi:10.1038/nm.2963.
47. Nishimura J, Handa R, Yamamoto H, et al. MicroRNA-181a is associated with poor prognosis of colorectal cancer. Oncol Rep. 2012;28(6):2221–6. doi:10.3892/or.2012.2059.
48. Xiao C, Srinivasan L, Calado DP, et al. Lymphoproliferative disease and autoimmunity in mice with increased miR-17-92 expression in lymphocytes. Nat Immunol. 2008;9(4):405–14. doi:10.1038/ni1575.
49. Sasaki K, Kohanbash G, Hoji A, et al. MiR-17-92 expression in differentiated T cells – implications for cancer immunotherapy. J Transl Med. 2010;8:17. doi:10.1186/1479-5876-8-17.
50. Taguchi A, Yanagisawa K, Tanaka M, et al. Identification of hypoxia-inducible factor-1 alpha as a novel target for miR-17-92 microRNA cluster. Cancer Res. 2008;68(14):5540–5. doi:10.1158/0008-5472.CAN-07-6460.
51. Du C, Liu C, Kang J, et al. MicroRNA miR-326 regulates TH-17 differentiation and is associated with the pathogenesis of multiple sclerosis. Nat Immunol. 2009;10(12):1252–9. doi:10.1038/ni.1798.
52. Hoareau-Aveilla C, Valentin T, Daugrois C, et al. Reversal of microRNA-150 silencing disadvantages crizotinib-resistant NPM-ALK(+) cell growth. J Clin Invest. 2015;125(9):3505–18. doi:10.1172/JCI78488.
53. de Yebenes VG, Belver L, Pisano DG, et al. MiR-181b negatively regulates activation-induced cytidine deaminase in B cells. J Exp Med. 2008;205(10):2199–206. doi:10.1084/jem.20080579.
54. Xu X, Ge S, Jia R, et al. Hypoxia-induced miR-181b enhances angiogenesis of retinoblastoma cells by targeting PDCD10 and GATA6. Oncol Rep. 2015;33(6):2789–96. doi:10.3892/or.2015.3900.
55. Mraz M, Chen L, Rassenti LZ, et al. MiR-150 influences B-cell receptor signaling in chronic lymphocytic leukemia by regulating expression of GAB1 and FOXP1. Blood. 2014;124(1):84–95. doi:10.1182/blood-2013-09-527234.
56. Manoharan P, Basford JE, Pilcher-Roberts R, et al. Reduced levels of microRNAs miR-124a and miR-150 are associated with increased proinflammatory mediator expression in Kruppel-like factor 2 (KLF2)-deficient macrophages. J Biol Chem. 2014;289(45):31638–46. doi:10.1074/jbc.M114.579763.
57. Zhou B, Wang S, Mayr C, et al. MiR-150, a microRNA expressed in mature B and T cells, blocks early B cell development when expressed prematurely. Proc Natl Acad Sci U S A. 2007;104(17):7080–5. doi:10.1073/pnas.0702409104.
58. Dorsett Y, McBride KM, Jankovic M, et al. MicroRNA-155 suppresses activation-induced cytidine deaminase-mediated Myc-Igh translocation. Immunity. 2008;28(5):630–8. doi:10.1016/j.immuni.2008.04.002.

59. Wang YZ, Tian FF, Yan M, et al. Delivery of an miR155 inhibitor by anti-CD20 single-chain antibody into B cells reduces the acetylcholine receptor-specific autoantibodies and ameliorates experimental autoimmune myasthenia gravis. Clin Exp Immunol. 2014;176(2):207–21. doi:10.1111/cei.12265.
60. Neilsen PM, Noll JE, Mattiske S, et al. Mutant p53 drives invasion in breast tumors through up-regulation of miR-155. Oncogene. 2013;32(24):2992–3000. doi:10.1038/onc.2012.305.
61. Yin Q, Wang X, McBride J, et al. B-cell receptor activation induces BIC/miR-155 expression through a conserved AP-1 element. J Biol Chem. 2008;283(5):2654–62. doi:10.1074/jbc.M708218200.
62. Carpenter S, Aiello D, Atianand MK, et al. A long non-coding RNA mediates both activation and repression of immune response genes. Science. 2013;341(6147):789–92. doi:10.1126/science.1240925.
63. Krawczyk M, Emerson BM. P50-associated COX-2 extragenic RNA (PACER) activates COX-2 gene expression by occluding repressive NF-kappaB complexes. Elife. 2014;3:e1776.
64. Liu B, Sun L, Liu Q, et al. A cytoplasmic NF-kappaB interacting long non-coding RNA blocks IkappaB phosphorylation and suppresses breast cancer metastasis. Cancer Cell. 2015;27(3):370–81. doi:10.1016/j.ccell.2015.02.004.
65. Wang P, Xue Y, Han Y, et al. The STAT3-binding long non-coding RNA lnc-DC controls human dendritic cell differentiation. Science. 2014;344(6181):310–13. doi:10.1126/science.1251456.
66. Cui H, Xie N, Tan Z, et al. The human long non-coding RNA lnc-IL7R regulates the inflammatory response. Eur J Immunol. 2014;44(7):2085–95. doi:10.1002/eji.201344126.
67. Rapicavoli NA, Qu K, Zhang J, et al. A mammalian pseudogene lncRNA at the interface of inflammation and anti-inflammatory therapeutics. Elife. 2013;2:e762. doi:10.7554/eLife.00762.
68. Collier SP, Henderson MA, Tossberg JT, Aune TM. Regulation of the th1 genomic locus from ifng through tmevpg1 by t-bet. J Immunol. 2014;193(8):3959–65. doi:10.4049/jimmunol.1401099.
69. Gomez JA, Wapinski OL, Yang YW, et al. The NeST long ncRNA controls microbial susceptibility and epigenetic activation of the interferon-γ locus. Cell. 2013;152(4):743–54. doi:10.1016/j.cell.2013.01.015.
70. Hu G, Tang Q, Sharma S, et al. Expression and regulation of intergenic long non-coding RNAs during T cell development and differentiation. Nat Immunol. 2013;14(11):1190–8. doi:10.1038/ni.2712.
71. Zhang H, Nestor CE, Zhao S, et al. Profiling of human CD4+ T-cell subsets identifies the TH2-specific non-coding RNA GATA3-AS1. J Allergy Clin Immunol. 2013;132(4):1005–8. doi:10.1016/j.jaci.2013.05.033.
72. Hwang SS, Kim K, Lee W, Lee GR. Aberrant expression of IFN-gamma in Th2 cells from Th2 LCR-deficient mice. Biochem Biophys Res Commun. 2012;424(3):512–18. doi:10.1016/j.bbrc.2012.06.146.
73. Spurlock CF, Tossberg JT, Guo Y, et al. Expression and functions of long non-coding RNAs during human T helper cell differentiation. Nat Commun. 2015;6:6932. doi:10.1038/ncomms7932.
74. Ranzani V, Rossetti G, Panzeri I, et al. The long intergenic non-coding RNA landscape of human lymphocytes highlights the regulation of T cell differentiation by linc-MAF-4. Nat Immunol. 2015;16(3):318–25. doi:10.1038/ni.3093.
75. Sharma S, Findlay GM, Bandukwala HS, et al. Dephosphorylation of the nuclear factor of activated T cells (NFAT) transcription factor is regulated by an RNA-protein scaffold complex. Proc Natl Acad Sci U S A. 2011;108(28):11381–6. doi:10.1073/pnas.1019711108.
76. Sehgal L, Mathur R, Braun FK, et al. FAS-antisense 1 lncRNA and production of soluble versus membrane Fas in B-cell lymphoma. Leukemia. 2014;28(12):2376–87. doi:10.1038/leu.2014.126.

Chapter 10
Noncoding RNAs in Therapeutic Resistance of Cancer

Lintao Jia and Angang Yang

Abstract Despite the encouraging advances made to date in cancer therapy, the benefits to patients are frequently offset by the development of resistance to therapeutics. Given their involvement in regulating multiple aspects of gene expression and cell signaling that dictates the behaviors of malignant cells, it is not surprising that noncoding RNAs (ncRNAs) play pivotal roles in the resistance of cancers to clinically available therapeutics. Aberrant expression of these ncRNAs, attributed to inherent defects or stress-responsive variations, mediates cellular signaling that compensates for unfavorable molecular events elicited by the therapeutics, thereby preventing the pharmaceuticals from exerting their desired effects on their cellular targets; alternatively, ncRNAs may regulate cancer therapeutic sensitivity by affecting drug accessibility to neoplastic cells and in vivo drug metabolism. In addition, dysregulation of ncRNA expression in cancer stromal cells can impair the responsiveness of neoplastic cells to appropriate therapies. In this chapter, we will describe ncRNA-related mechanisms underlying cancer resistance to routine therapeutics, hopefully providing rationales for the development of drug-sensitizing strategies targeted against or based on these ncRNAs.

Keywords Drug resistance • Noncoding RNA • Chemotherapy • Radiotherapy • Targeted therapy • Tumor microenvironment • Drug accessibility • ABC transporter

L. Jia (✉)
Department of Biochemistry and Molecular Biology, Fourth Military Medical University, Xi'an 710032, China
e-mail: jialth@fmmu.edu.cn

A. Yang
Department of Immunology, Fourth Military Medical University, Xi'an 710032, China
e-mail: agyang@fmmu.edu.cn

E. Song (ed.), *The Long and Short Non-coding RNAs in Cancer Biology*, Advances in Experimental Medicine and Biology 927,
DOI 10.1007/978-981-10-1498-7_10

10.1 Introduction

In the history of modern medicine, the wish to cure human malignancies has provided primitive and persistent impetus for the mechanistic studies on cancers. Thanks to the novel insights achieved in these studies, the clinical practice of cancer therapy has gained substantial improvement in the past decades, which is evidenced by the sustained decrease in cancer-related mortality worldwide albeit the ascending incidence of many types of malignancies in recent years [1]. However, the frequently occurring cancer resistance to the regular and innovative treatment leads to compromised efficacy of these therapeutics. As a large category of gene transcripts which fulfill a role without being translated into proteins, noncoding RNAs (ncRNAs) coordinate with proteins to regulate almost every detail of the intracellular signaling machinery. Consistently, accumulating studies have disclosed the indispensable roles of ncRNAs, particularly microRNAs (miRNAs) and long noncoding RNAs (lncRNAs), in regulation of cancer therapeutic resistance [2].

10.2 Conventional and Targeted Therapy of Cancer

Because of the lack of effective therapies, cancers occurring in diverse tissue types remain among the most life-threatening human disorders. However, chemotherapy and radiotherapy, in addition to surgery, have in many cases significantly improved the progression-free survival (PFS) of cancer patients [3, 4]. Moreover, a cohort of molecular targeted pharmaceuticals, e.g., monoclonal antibodies and small-molecule inhibitors, has been approved as first-line treatments for malignancies and has improved the outcomes of various cancers when applied alone or in combination with conventional therapies [5]. Cancer biotherapy, which is characterized by the delivery of a therapeutic gene or immunoregulatory protein or by the transfer of modified cells, is moving from a vision to clinical reality, thus providing additional options for personalized therapy of cancers [3].

10.2.1 Chemo- and Radiotherapy

Chemotherapy is a drug therapy that kills cancer cells or stops them from multiplying [6]. Most chemotherapeutic drugs are cytotoxic, in that they restrain cell division (mitosis), and are therefore more effective against fast-dividing cells such as cancer cells. These drugs block cell proliferation via various mechanisms involving DNA damage or inhibition of cellular machinery (exemplified by rearrangement of the cytoskeleton), usually culminating in a form of programmed cell death known as apoptosis [7].

Chemotherapeutic reagents currently used in clinical practice can be classified as follows:

1. Alkylating agents, which bind covalently to DNA and cross-link DNA strands via their alkyl groups, causing DNA strand breaks and apoptosis. These agents work in a cell cycle-independent manner. This category of agents includes nitrogen mustards, nitrosoureas, tetrazines, aziridines, cisplatins, and their derivatives.
2. Antimetabolites, further subcategorized into antifolates, fluoropyrimidines, deoxynucleoside analogs, or thiopurines, include nucleobase or nucleoside analogs that impair DNA synthesis by competitively binding to polymerases and/or cause DNA damage upon incorporation into growing DNA strands. Antimetabolites selectively inhibit carcinoma cells in S phase of the cell cycle.
3. Anti-microtubule agents, e.g., vinca alkaloids, taxanes, and podophyllotoxin, are plant-derived or semisynthetic chemicals that block cell division by interfering with the function of cytoskeletal proteins, in particular, the assembly and disassembly of microtubules. They also inhibit angiogenesis in solid tumors.
4. Topoisomerase inhibitors, such as irinotecan, topotecan, etoposide, doxorubicin, mitoxantrone, teniposide, novobiocin, merbarone, and aclarubicin, can affect the DNA binding or catalytic activity of two enzymes, topoisomerases I and II, which are critical for DNA unwinding required for replication.
5. Cytotoxic antibiotics, such as anthracyclines (doxorubicin, daunorubicin, etc.), actinomycin, bleomycin, plicamycin, and mitomycin, are a large group of drugs with various mechanisms of action. The combination of these cytotoxic agents leads to numerous chemotherapeutic regimens that can be used against cancers of different types or different clinical stages [7].

Radiation therapy (radiotherapy) is commonly applied to cancerous tissue either alone or as part of adjuvant therapy [4]. To avoid injury to normal tissue, shaped radiation beams from several angles are aimed such that they intersect at the tumor (or the draining lymph nodes, if the tumor cells may have spread). Ionizing radiation triggers cell death by damaging DNA via release of two types of energy, photons and charged particles. Photon therapy kills cells mainly through production of free radicals, which cause severe and irreparable DNA damage, including double-stranded DNA breaks. Failure to repair damaged DNA leads to apoptotic cell death. Cancer cells reproduce more rapidly at the expense of a diminished ability to repair chromosomal abnormalities, allowing sublethal damage to accumulate. Consequently, the cells die or divide more slowly. By contrast, charged particles, e.g., protons and ions of boron, carbon, and neon, cause DNA damage in malignant cells through direct energy transfer independent of tumor oxygen supply and can be more tightly focused on the tumor due to their relatively large mass. Thanks to improved tumor targeting and attenuated cytotoxicity against healthy tissues, radiotherapy is used in clinical treatment of a growing list of malignancies [8].

10.2.2 Molecular Targeted Therapy

The past decade has witnessed the emerging and rapid expansion of molecular targeted therapy of various malignancies. In these approaches, therapeutic agents selectively bind and functionally inhibit dominant oncogenic proteins [5]. To date, two classes of oncoproteins have been targeted for therapeutic purposes:

1. Cell surface or matrix proteins such as growth factors, receptors, or leukocyte differentiation antigens, which are accessible to antibodies. Examples of monoclonal antibodies approved for cancer therapy (and their respective targets) include trastuzumab/Herceptin (erbB2/HER2), cetuximab (EGFR), rituximab (CD20), and bevacizumab/Avastin (VEGF) [9].
2. Cancer-promoting enzymes such as the protein tyrosine kinases (PTKs), which are key components of cellular signaling pathways that drive uncontrolled proliferation and apoptosis resistance. The activity of these PTKs can be selectively inhibited by a cohort of small-molecule compounds. Tyrosine kinase inhibitors (TKIs) already in the oncologist's armamentarium include gefitinib, erlotinib, and Gleevec, which target EGFR; lapatinib, which targets both EGFR and HER2; and sorafenib, which inhibits the Raf/MEK/ERK pathway; and several multitarget inhibitors [5]. Whereas TKIs exert their anticancer effects solely by inhibiting kinase activity (and consequently attenuating downstream signaling), therapeutic antibodies may play inhibitory roles on tumors both by ablating the function of the target protein and by eliciting tumoricidal immune responses such as antibody-dependent cell-mediated cytotoxicity (ADCC). The latter depends strongly on the cancer microenvironment [5, 9].

The concept of molecular targeting in cancer therapy is also reminiscent of the long-standing clinical use of antihormone therapy against specific types of cancers. Because estrogens and androgens play pivotal roles in the occurrence of some categories of breast, ovarian, and prostate cancers, compounds that competitively bind steroid hormone receptors and block hormone/receptor interactions have been used for prevention and treatment of these malignancies [10, 11]. Of note are the selective estrogen receptor modulators (SERMs), e.g., tamoxifen, which are prescribed for and are effective against estrogen receptor (ER)-positive invasive breast cancers. Similarly, steroidal antiandrogens can counteract the carcinogenic effect of androgens by targeting the androgen receptor (AR) and have consequently been used for clinical treatment of androgen-dependent prostate cancers [10, 11].

10.2.3 Gene, Cell, and Immune Therapy

Cancers are characterized by genetic alterations leading to aberrant expression of oncogene or tumor suppressors, as well as by failures of immunosurveillance and eradication of transformed cells [12]. Suppression of cancers can be achieved via delivery of tumor-suppressive or cytotoxic protein-coding genes. Alternatively, autologous or allogenic cells can be propagated or modified in vitro, endowed with tumor-inhibitory capacities, and transferred to individual patients for adoptive therapy. The anticancer proteins, genes, and cells used for these purposes include those that can elicit immunological responses or attenuate immune tolerance to cancers. Although very few technical breakthroughs have been obtained to date in regard to clinical treatment, these novel therapeutics hold great promise for improving patient outcomes when applied as adjuvants or in personalized therapy [13].

10.3 General Mechanisms of Resistance to Cancer Therapeutics

Cancer resistance to clinical treatment arises from the failure of therapeutics to inhibit the malignant phenotypes of neoplastic cells. In principle, cancer cells become unresponsive to a drug for one or more of the following reasons:

1. Inaccessibility of cells or molecular targets to the drug. Although most small-molecule anticancer pharmaceuticals can easily diffuse into dividing neoplastic cells where they impair the structure or biosynthesis of macromolecules and thereby trigger programmed cell death, cells also express machinery for outward transport of drug molecules. Of note in this regard are the members of ATP-binding cassette transporter (ABC transporter) superfamily of transmembrane proteins, which utilize the energy of ATP to transport a variety of substrates, including exogenous chemicals, across the membrane and out of the cell [14].
2. Inability of drugs to cause lethal or suppressive molecular events. This phenomenon can be attributed to detoxification of the drug by the cell, reduced production of cytotoxic mediators such as reactive oxygen species (ROS), and dysfunction in the machinery involved in DNA repair or cell death. In such cases, a specific population of cells within a heterogeneous population can continue to survive and proliferate despite the presence of an anticancer drug [14].
3. Alternative signaling that compensates for the impairment caused by the drug. Although carcinogenesis is driven mainly by key genetic variation(s), cancer cells may harbor multiple abnormalities in gene expression and intracellular signaling. More importantly, their unstable genomes may give rise to new variations that facilitate the maintenance of malignant phenotypes. Therefore, although a tumoricidal drug can successfully target a single signal pathway that drives cell survival or proliferation, activation of alternative or branched pathways may suffice to activate common downstream signaling events that deregulate oncogene and tumor-suppressor effectors [14].

The critical roles of ncRNAs in mediating these signaling events and conferring resistance to routine and molecular targeted therapies are now being characterized. In cancer cells, these ncRNAs directly regulate the intracellular signaling that

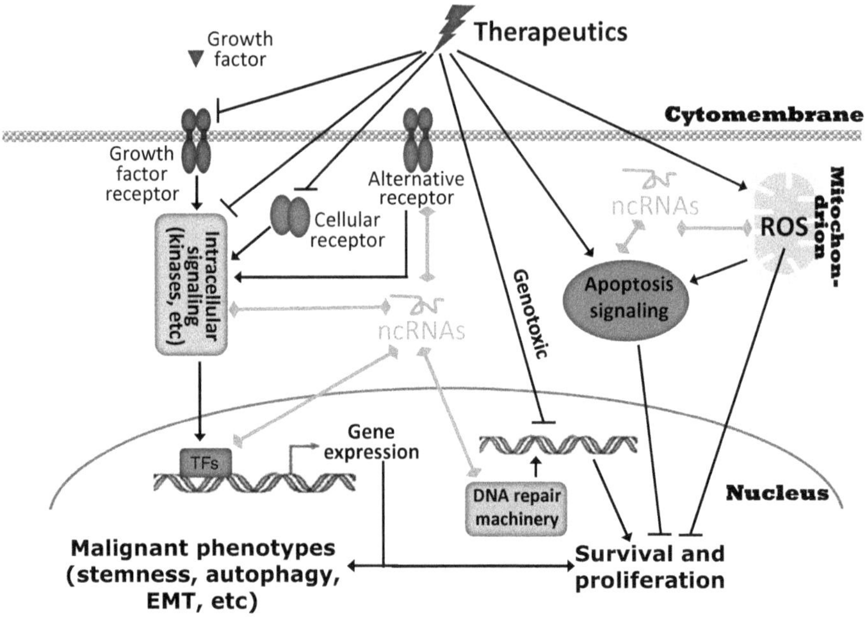

Fig. 10.1 ncRNAs regulate intracellular signal pathways that counteract the cytotoxicity of anti-cancer therapeutics. Therapeutics impede cancer cell survival, proliferation, and other malignant phenotypes by suppressing intracellular signaling that leads to the expression of pro-survival and pro-proliferative genes, as well as by damaging DNA or the cytoskeleton, producing ROS, or impairing metabolism. Ultimately, these events trigger apoptotic cell death. Therapeutic-resistant cancer cells circumvent these detrimental events via constitutive activation of downstream or alternative receptor-mediated pro-survival and pro-proliferative signaling or through blockade of apoptotic signaling. All of these processes are potentially regulated by ncRNAs, including miRNAs and lncRNAs

determines cell responsiveness to various therapeutics. Alternatively, ncRNAs may regulate the behaviors of stromal or immune cells in the tumor microenvironment, thereby affecting drug sensitivity in a non-cell-autonomous manner [14–16].

10.4 Counteracting Roles of Therapy-Evoked and ncRNA-Related Signaling Events in Cancer Cells

Although therapeutics trigger macromolecule damage and stresses that are detrimental to the survival of cancer cells, drug-refractory cell populations develop mechanisms to circumvent growth inhibition or death induced by cytotoxic drugs. Frequently, these signaling processes are regulated by ncRNAs [14] (see Fig. 10.1).

10.4.1 Canonical Intracellular Pathways for Cell Survival and Division

Cancer arises from the aberrant activation of cellular signaling pathways that promote survival and proliferation. Upon stimulation by environmental factors and coupled in many cases to intracellular messengers, these pathways initiate a cascade of kinase activation, thereby inducing activation and nuclear translocation of transcription factors or the assembly of complexes of transacting factors. Ultimately, these events culminate in the expression of genes responsible for cell survival, cell cycle entry, migration, and other behaviors [17]. Although routine therapeutic approaches such as chemotherapy, radiation, and molecular targeted pharmaceuticals elicit different upstream events, they may converge on blockade of the same pathways to inhibit cancer progression. Accordingly, sustained activation of these pathways may underlie cancer resistance to clinical therapeutics [14].

The phosphatidylinositol-3 kinase (PI3K)/Akt pathway, in which PI3K phosphorylates inositol ring 3′-OH groups in inositol phospholipids to generate the second messenger phosphatidylinositol-3,4,5-trisphosphate (PIP_3) and activate Akt, is among the most frequently activated pro-survival and pro-mitotic pathways [18]. Arcaroli et al. have found that a mutation in the PI3K catalytic subunit PIK3CA reduces its affinity to miR-520a and increases the sensitivity of colorectal cancers to Src inhibitors, suggesting that crosstalk between the Src and PI3K pathways contributes to regulation of malignant behaviors of such tumors [19]. The tumor-suppressor PTEN negatively regulates the PI3K/Akt pathway by dephosphorylating PIP_3. Numerous studies have demonstrated the involvement of PTEN deregulation in therapeutic resistance of cancers. In particular, Meng et al. have screened for miRNAs that regulate the chemosensitivity of cholangiocarcinomas. They found that miR-21 and miR-200b increased sensitivity to gemcitabine and that PTEN was a direct target of miR-21 [20]. In non-small cell lung cancers (NSCLCs) and hepatocellular carcinomas (HCCs), miR-221 and miR-222, both of which are induced by Met activation of c-Jun, can target PTEN, thereby activating Akt signaling and imparting resistance to apoptosis triggered by tumor necrosis factor-related apoptosis-inducing ligand (TRAIL) or Apo 2 ligand (Apo2L) [21]. In addition, miR-214 promotes cell survival and cisplatin resistance by targeting PTEN in ovarian cancer [22]. In hepatocellular carcinomas (HCCs), miR-216a/217 targets PTEN and Smad7 to reinforce the PI3K/Akt and TGF-β signaling, respectively, thus maintaining the malignant and stem-like phenotypes of HCC cells even under TKI treatment [23]. Although widely recognized as a tumor suppressor, miR-200c activates Akt and induces chemoresistance by targeting PPP2R1B, a subunit of protein phosphatase 2A, in esophageal cancers [23]. In prostate and breast cancers, miR-95 confers resistance to radiotherapy by targeting the sphingolipid phosphatase SGPP1, an antagonist of sphingosine-1-phosphate signaling downstream of the canonical PI3K-Akt pathway [24]. The mammalian target of rapamycin (mTOR) acts downstream of PI3K/Akt to maintain the key malignant behaviors of cancer cells. MiR-199a-39, which targets mTOR and c-Met, is downregulated in various malignancies including HCC, contributing to resistance of malignant cells to chemotherapeutics such as doxorubicin [25].

The Ras/mitogen-activated protein kinase (MAPK) pathway is another fundamental pathway required for cell growth and transformation. Ras is a small GTPase that responds to upstream signaling and elicits the cascade of Raf/MEK/MAPK kinase activation, and mutation or constitutive activation of Ras has been implicated in the development of various malignancies and the acquisition of cancer resistance to different therapeutics. Weidhaas et al. have highlighted the role of the let-7 family in improving the radiosensitivity of cancer cells by targeting Ras and other oncogenes [26]. In lung carcinomas, miR-27a modulates chemosensitivity by targeting the Raf kinase inhibitory protein (RKIP) [27, 28].

Other documented intracellular signal pathways also play important roles in potentiating cell growth and conferring therapeutic resistance to cancers. In this regard, let-7 can directly repress the interleukin-6 (IL-6)-activated JAK/STAT prosurvival pathway, and its expression correlates with a relatively optimistic prognosis for esophageal squamous cell carcinoma patients receiving cisplatin treatment [29]. Phosphodiesterase 8A (PDE8A) and UV radiation resistance-associated gene (UVRAG), which negatively regulate cAMP/PKA and Notch signaling, respectively, are targets of miR-33a in glioblastoma [30]. In addition, miR-155-3p is involved in Toll-like receptor (TLR)-mediated resistance to the anti-chronic lymphocytic leukemia (CLL) drug, fludarabine [31]. Thus, ncRNAs play critical roles in regulating therapeutic susceptibility of carcinomas by fine-tuning the potency and cross talk of canonical intracellular pathways.

10.4.2 Oncogenic Ligands and Receptors

The signals that drive survival and proliferation of cells originate from the extracellular matrix. The engagement of ligands with their receptors transfers environmental signals into the cell, where intracellular pathways are activated to support cell growth. When these signals are exaggerated or uncontrollable, they lead to malignant transformation. Growth factors, ontogenesis-related ligands, steroid hormones, and their specific receptors are representative initiators of oncogenic signaling [17].

10.4.2.1 Growth Factors/Receptors

Numerous growth factors and receptors drive oncogenic signaling and malignant transformation of cells, making them potential targets for cancer therapeutics. Human epidermal growth factor receptors (HER) are well-characterized biomarkers of various cancers. These proteins form heterologous dimers in response to binding of growth factors and subsequently phosphorylate downstream substrate proteins to activate classical signal pathways [32]. HER1/EGFR-targeted TKIs are most commonly used for treatment of lung cancers, whereas the monoclonal antibody cetuximab is approved for clinical treatment of colorectal cancers and squamous cell carcinoma of the head and neck (SCCHN) [32]. Garofalo et al. have determined the mechanisms underlying TKI resistance of EGFR-positive lung cancers and identified a cohort of downstream miRNAs that repress the master regulators of cell survival and division [33]. Rai et al. have observed that overexpression of miR-7 in TKI-resistant lung cancers increases drug sensitivity in cancers harboring an EGFR mutation (T790M) by targeting EGFR, insulin receptor substrate-1 (IRS-1), and Raf-1 [34]. An analysis of the miRNA transcriptome and global network structure in colorectal carcinoma suggests that downregulation of the K-Ras-targeting miRNAs let-7b and let-7e and upregulation of miR-17* are candidate molecular markers for cetuximab resistance [35].

The humanized HER2/erbB2 antibody, trastuzumab (Herceptin), is a pioneer antitumor antibody that expedites revolutionary progress in treatment of breast cancers and, more recently, advanced gastric cancers. Nevertheless, the majority of patients with HER2-positive cancers exhibit resistance to primary trastuzumab treatment or develop acquired resistance upon repeated administration. Recent studies have revealed that both cancer cell-autonomous mechanisms, e.g., inaccessibility or decreased affinity of HER2 for the antibody or activation of alternative growth factor pathways, and modifications of the tumor microenvironment that suppress antibody-elicited immunological responses may underlie resistance to trastuzumab [36]. To decipher the role of miRNAs in mediating trastuzumab resistance of breast cancers, our laboratory screened for miRNAs differentially expressed in trastuzumab-refractory and trastuzumab-sensitive neoplastic cells. We found that miR-200c downregulation decreased trastuzumab responsiveness by alleviating suppression of transforming growth factor-β (TGF-β) signaling, whereas downregulation of miR-375 and consequently depression of its target gene, insulin-like growth factor 1 receptor (IGF1R), maintained cell growth in the context of blocked HER2 signaling [37, 38]. These studies demonstrate that miRNAs play a regulatory role in cancer resistance to molecular targeted drugs by modulating drug-targeted or alternative growth factor pathways.

Other growth factors involved in carcinogenesis include platelet-derived growth factors (PDGFs), hepatic growth factor (HGF), IGF1R, and (very rarely) bone morphogenetic proteins (BMPs). For instance, the active A receptor type 1 (ACVR1), a key receptor in BMP signaling, is targeted by miR-148 in hepatocytes. Meanwhile, downregulation of miR-148 defines a cancer stem cell-like, aggressive, and therapy-resistant subtype of hepatocellular carcinoma via the miR-148a–ACVR1–BMP–Wnt regulatory circuit [39]. Thus, failure to abolish driving or alternative growth factor signaling is a common mechanism of drug resistance regulated by ncRNAs.

10.4.2.2 Ontogenesis-Related Ligands/Receptors

Aberrant signaling through canonical pathways involved in embryonic development, e.g., the Wnt, Notch, and Hedgehog pathways, can drive the transformation of various types of cells. Meanwhile, reactivation of these pathways may underlie resistance to clinical cancer treatments [40]. Wnt signaling is activated by the binding of a Wnt-protein ligand to a Frizzled family receptor, which transfers the biological signal to the Dishevelled protein inside the cell. The canonical Wnt pathway triggers accumulation and nuclear translocation of β-catenin, coactivating TCF/LEF family of transcription factors to switch on gene expression [41]. The miRNA-mediated regulation of the Wnt pathway is involved in therapeutic resistance in a wide range of malignancies. In colorectal cancers, asymmetric cell division (ACD) and stem cell homeostasis are disrupted, thereby facilitating carcinogenesis, via a regulatory loop involving miR-146a. The transcriptional factor Snail upregulates miR-146a through the β-catenin-TCF4 complex, whereas miR-146a targets Numb to stabilize β-catenin, maintaining Wnt activity and driving symmetrical cell division. This mechanism is critically involved in the resistance of colorectal cancer to molecular targeted drugs [42]. In pancreatic ductal adenocarcinoma cells, Smad4 deficiency ablates TGF-β-triggered expression of miR-494, which in turn upregulates FoxM1, an miR-494 target, and facilitates nuclear translocation of β-catenin, leading to oncogenesis and resistance to gemcitabine chemotherapy [43].

The Notch signaling pathway is a fundamental signaling system used by neighboring cells to communicate with each other. Notch receptors are single-pass transmembrane proteins whose ligands include members of the Delta-like (DLL1, DLL3, DLL4) and Jagged (JAG1, JAG2) families. Ligand binding causes cleavage of Notch and release of the Notch intracellular domain (NICD), which undergoes nuclear translocation and associates with the CSL (CBF1/Su[H]/Lag-1) transcription factor complex, resulting in activation of the canonical Notch target genes. Notch signaling is involved in carcinogenesis and cancer drug resistance, although it plays disparate roles in various malignancies [41]. Park et al. have found that miR-34a levels are reduced in p53-deficient breast cancers, contributing to resistance to conventional chemotherapy by upregulating the miR-34a target Notch1 [44].

The Hedgehog (Hh) signaling pathway is one of the key regulators of animal development and cell lineage commitment. In the absence of Hh ligands, the cell surface transmembrane protein Patched (PTCH) suppresses the activity and expression of the receptor Smoothened (SMO). PTCH engagement by Hh (e.g., Sonic Hedgehog [SHH], the best-studied ligand) leads to the dissociation and activation of SMO, which in turn activates the GLI transcription factors to initiate downstream gene expression. The Hh pathway has been implicated in the development of various cancers, including basal cell carcinoma and medulloblastoma [41]. Recent studies have revealed that Hh signaling, which is regulated by miRNAs, is also involved in resistance to routine cancer treatment. For example, miR-9 contributes to temozolomide resistance by targeting PTCH in glioblastoma [45]. Drugs that specifically target Hedgehog signaling are being developed for treatment of these malignancies. Thus, the classical ontogenesis-related pathways, which are fine-tuned by miRNAs, also contribute to carcinogenesis and the occurrence of drug resistance.

10.4.2.3 Steroid Hormone and Receptors

Depending on the homology relationships of their specific receptors, steroid hormones are classified as glucocorticoids, mineralocorticoids, androgens, estrogens, or progestogens [46]. By binding to and prompting the nuclear translocation of a class of intracellular receptors, they transcriptionally activate a cohort of genes that participate in cell metabolism, inflammation, immunity, and development of sexual characteristics. The exaggerated signaling by overexpression of ERs and ARs plays an important role in the development of mammary and genital carcinomas. Hence, antihormone therapeutics using estrogen antagonists such as SERMs and antiandrogens like flutamide and bicalutamide have emerged as first-line treatments for breast cancer and prostate cancer, respectively [46]. However, neoplastic cells have evolved intricate signaling mechanisms to circumvent the cytotoxic effect of these antagonists, leading to acquisition of resistance to antihormone therapeutics [47].

The involvement of ncRNAs in cancer resistance to tamoxifen, the most-prescribed SERM, has been intensively investigated. Consistent with the reported suppression of ER expression by hyperactivation of MAPKs in breast cancer, Miller et al. have identified an MAPK-regulated miRNA signature that associates significantly with reduced ER expression and poor response to tamoxifen, suggesting that miRNAs can be targeted to reverse resistance to hormone therapy [48]. Maillot has determined miRNA profiles that are regulated by estrogen signaling or altered by antiestrogen therapy in breast cancers, highlighting the role of individual miRNAs in conferring antiestrogen resistance on breast cancers [49]. ER-α can be directly targeted and inhibited by miR-221/miR-222 in breast cancers, compromising the therapeutic efficacy of tamoxifen and enabling ER-α-independent growth of tamoxifen-resistant cancer cells [50]. Other miRNAs play regulatory roles in tamoxifen responsiveness of breast cancers by affecting alternate molecular machineries that govern cell cycle entry, cell survival, and metastasis [49]. Aberrant expression of a set of miRNAs and the lncRNA BCAR4 predicts poor response to tamoxifen, whose effectiveness in breast cancer relies on expression of HER2 [49, 51]. As a direct target of ER-mediated transcriptional repression, the lncRNA HOTAIR is upregulated by tamoxifen and compensatorily increases the level of ER protein, ultimately resulting in resistance of breast cancer to chemotherapy [52]. The alternative approach to blocking ER signaling is the use of inhibitors of aromatase, a rate-limiting enzyme in the conversion of androgens such as testosterone and androstenedione into estrogens. However, breast cancer resistance to aromatase inhibitors (e.g., letrozole) arises concurrently with overexpression of miR-128a and miR-181a or downregulation of miR-125b and let-7c. Letrozole treatment also increases expression of let-7f, which downregulates aromatase, thereby desensitizing breast cancer cells to subsequent letrozole treatment [53]. In terms of cancer resistance to antiandrogen therapy, miR-616 induces androgen-independent growth of prostate cancer cells by suppressing expression of tissue factor pathway inhibitor 2 (TFPI-2), thereby contributing to drug resistance of prostate cancers [54]. In addition, two lncRNAs, PRNCR1 (also known as PCAT8) and PCGEM1, can bind and cooperate with ARs to transcriptionally activate target genes independently of ligand engagement, resulting in prostate cancer resistance to castration or antiandrogen therapy [55]. As a miRNA that mediates androgen-dependent growth of prostate cancer cells, miR-21 is also sufficient to induce castration resistance of prostate cancers [56]. Taken together, these observations show that ncRNAs play diverse roles in conferring or counteracting resistance to antihormone therapy of cancers by modulating sex steroid pathways or coordinated signaling involved in cancer progression.

10.4.3 Key Transcriptional Factors

Oncogenic and differentiation-determining transcriptional factors may promote cancer progression and drug resistance following activation by upstream signals or acquisition of constitutive activity upon mutation [57]. The oncoprotein c-Myc is overexpressed in various malignancies and is correlated with poor outcomes of routine clinical therapies. In non-Hodgkin B-cell lymphoma, stromal adhesion promotes cell survival and imparts resistance to cytotoxic drugs like mitoxantrone via an amplification loop in which c-Myc induces epigenetic silencing of miR-548m and subsequently increases the expression of the miR-548m targets c-Myc and HDAC6 [58]. Numerous other transcriptional factors that expedite drug resistance are also regulated by miRNAs. For instance, glioma cells acquire chemoresistance as a result of inhibitor of differentiation 4 (ID4) depression of miR-9-mediated suppression of Sox2 [59]. Downregulation of transcriptional factors that drive differentiation also underlies cancer resistance to clinical therapeutics, as exemplified by forkhead box O3a (FOXO3a), which is targeted and silenced by miR-153, thus attenuating platinum-induced apoptosis of colorectal cancers [60]. Therefore, transcription factors, which can both be regulated by miRNAs and dictate the expression of specific miRNAs, may play distinct roles in therapeutic resistance, depending on the repertoires of their transcriptional targets.

10.4.4 Cell Cycle Progression

Cell proliferation requires continuous entry into and progression of the cell cycle, which is divided into different phases with checkpoints controlled by numerous factors [61]. Although anticancer therapeutics may trigger cell cycle arrest through intracellular signaling, refractory subsets of malignant cells can develop miRNA-mediated regulatory mechanisms that facilitate cell cycle progression. Pouliot et al. have found that miR-155 and miR-15 improve the sensitivity of epidermoid carcinoma cells to cisplatin by targeting and repressing the cell cycle kinases WEE1 and CHK1 [62]. Salerno et al. have found in a mouse model of chronic lymphocytic leukemia (CLL) that exogenous miR-15a and miR-16-1, which target cyclin D1, improve the responses of cells to nutlin, a mouse double minute 2 (MDM2) antagonist, and genistein, a tyrosine kinase inhibitor [63]. MiR-122 sensitizes HCC to doxorubicin by modulating cyclin G1, thereby influencing p53 protein stability and transcriptional activity [64]. Thus, ncRNAs may contribute to the etiology of cancer drug resistance by governing cell cycle progression in the context of various clinical treatments.

10.4.5 Apoptotic Machinery

Both ontogenesis of the organism and maintenance of tissue homeostasis involve the removal of senescent or aberrant cells through programmed cell death [65]. In contrast to necrosis, which occurs under stressful conditions like tissue injury, apoptosis represents the most common pattern of physiological cell death. Inadequate apoptosis underlies carcinogenesis in multiple tissues, and desensitization of cells to chemotherapy- or radiotherapy-triggered apoptosis accounts for therapeutic resistance of a variety of clinical cancers [65]. The apoptotic machinery consists of two major pathways:

1. In the extrinsic pathway, extracellular ligands such as Fas ligand, tumor necrosis factor-α (TNF-α), and TRAIL bind to and trigger the oligomerization and activation of death receptors such as Fas, TNFR, and death receptor 4 (DR4). Signaling from these receptors leads in turn to the sequential processing and activation of initiator and effector caspases and ultimately to widespread degradation proteins and the collapse of the entire cell.
2. The intrinsic pathway, which senses intracellular stress signals like DNA damage, causes permeabilization of the mitochondria and release of cytochrome c into the cytoplasm, thereby initiating activation of the caspase cascade via caspase-9. Bcl-2 family members fine-tune apoptotic signaling via pore formation on the mitochondrial membrane and reciprocal interactions to determine the fate of individual cells. In addition, negative regulators of apoptosis such as the inhibitors of apoptosis (IAP) and FLICE-inhibitory protein (FLIP) impede caspase activation in the context of various apoptotic signals [65]. The aforementioned apoptosis executioners and regulators, which play critical roles in determining responses to cytotoxic therapeutics, can be targeted by ncRNAs in various types of malignancies. In particular, the sensitivity of osteosarcoma cells to FasL is regulated by miR-20a, which targets the death receptor Fas [66]. In cholangiocarcinoma, miR-25 is upregulated by Hedgehog signaling, which desensitizes neoplastic cells to TRAIL-induced apoptosis by targeting DR4 [67]. In addition, miR-21 silencing also exerts synergistic cytotoxicity with TRAIL in gliomas [68]. Conversely, miR-212 increases TRAIL sensitivity in non-small cell lung cancer by targeting the antiapoptotic protein PED/PEA-15 [69].

The cancer response to cytotoxic therapeutics also involves intrinsic apoptotic signaling, which is likewise regulated by ncRNAs. A natural product, oridonin, increases the sensitivity of leukemia to chemotherapy by downregulating miR-17 and miR-20a and thus restoring expression of their common target, the S variant of

BIM, resulting in promotion of mitochondrial apoptotic signaling [70]. STAT3 signaling maintains the expression of miR-17 and suppresses its target BIM, thereby conferring MEK inhibitor resistance on lung cancers, suggesting the cooperative antitumor potential of STAT3 and MEK inhibitors [71]. Signaling from chemokine receptor CXCR4 in acute myeloid leukemia (AML) cells downregulates let-7a through the transcription factor Yin Yang 1, resulting in chemoresistance due to increased expression of let-7a targets such as Bcl-xL [72]. Lam et al. have identified miRNA modulators of colorectal cancer responsiveness to the Bcl-2 inhibitor ABT-263 (navitoclax) and found that a majority of these miRNAs sensitize neoplastic cells by downregulating the pro-survival Bcl-2 family member Mcl-1 [73]. Hepatitis C virus increases HCC sensitivity to sorafenib via miR-193b targeting of Mcl-1, thereby promoting apoptosis of HCC cells [74]. These findings suggest that modulation of apoptotic signaling by ncRNAs causes altered responses to clinical therapeutics.

10.4.6 Genotoxic Stress

Although cells have evolved machinery for comprehensive genome surveillance and DNA repair, deficiencies in these machineries (or, alternatively, severe DNA injury) may prevent restoration of genomic homeostasis [75]. From the standpoint of the tumor cell, DNA abnormalities are a double-edged sword. On one hand, genomic DNA instability and mutation are the key drivers of carcinogenesis: activation of oncogenes or dysfunction of tumor suppressors elicits uncontrolled mitosis and apoptosis resistance, explaining the intimate relationship between DNA repair defects and tumorigenesis. On the other hand, irreparable DNA damage triggers cell death to maintain the purity of the genetic material, providing the rationale for radiation therapy and the large proportion of chemotherapeutic drugs that kill cells by extensively damaging the DNA [75]. The types of DNA damage include undesired modification or mismatch of bases, single-strand damage, and double-strand break (DSB). Sensors of DNA damage establish checkpoints prior to the initiation of DNA repair. Once activated by damaged DNA, these checkpoints halt the cell cycle and give the cell time to repair the damage. Checkpoint activation is controlled by two master kinases, ataxia telangiectasia mutated (ATM) and ataxia telangiectasia and Rad3 related (ATR), which phosphorylate downstream targets in a signal transduction cascade and eventually induce cell cycle arrest [76]. In addition, checkpoint mediator proteins including BRCA1, MDC1, and 53BP1 are required for transmission of the checkpoint activation signal to downstream proteins. miRNAs are involved in these processes. MiR-205 inhibits DNA damage repair by targeting zinc finger E-box-binding homeobox 1 (ZEB1) and the ubiquitin-conjugating enzyme Ubc13, thus acting as a tumor radiosensitizer by targeting the DNA repair machinery [77]. Notably, however, radiotherapy downregulates miR-205 through ATM and ZEB1 in breast cancer. In NSCLCs, miR-181a and miR-630 promote and reduce cisplatin-triggered cell death, respectively, in the former case via regulation of the intrinsic apoptotic pathway and in the latter case via miR-630-mediated blockade of early manifestations of the DNA damage response such as activation of ATM [78].

As a type of severe DNA damage, DSBs can be repaired via three mechanisms: nonhomologous end joining (NHEJ), microhomology-mediated end joining (MMEJ), and homologous recombination (HR) [76]. Poly ADP ribose polymerase 1 (PARP1) plays crucial roles in DNA repair by preventing the development of DSBs from single-strand breaks and by participating in the MMEJ repair of DSBs. Because MMEJ is an error-prone repair pathway, PARP1 overexpression has been detected in various malignancies and therapy-resistant cancer cells. The lncRNA PCAT1 sensitizes prostate cancers to genotoxic drugs, e.g., inhibitors of PARP1, by posttranscriptionally repressing the DSB repair protein BRCA2 [79]. Although miR-223 supports the aggressive phenotype of esophageal adenocarcinomas, it also improves the response of malignant cells to genotoxic drugs by directly targeting PARP1 [80]. RAD51 is critically involved in HR of DNA during DSB repair. DNA repair in malignant cells is also attenuated by miR-96, which targets RAD51 and the trans-lesion synthesis DNA polymerase REV1, increasing the sensitivity of cancers to the interstrand cross-linking drug cisplatin and PARP1 inhibitors [81].

The tumor-suppressor p53 responds to diverse cellular stresses to regulate expression of target genes, thereby inducing cell cycle arrest, apoptosis, senescence, or metabolic changes. Most importantly, p53 serves as a guardian of the genome by coupling DNA damage to the cellular DNA repair machinery or to apoptotic cell death when repair fails [82]. In p53-deficient cancers, the functional balance and cross talk between p73, which mediates chemosensitivity, and p63, which promotes cell survival, proliferation, and cell survival, are crucial for cancer progression. This phenomenon is at least partially mediated by miRNAs, such as miR-193a-5p, which targets p73, and is itself regulated by both p63 and p73. Chemotherapy causes p63/p73-dependent induction of this miRNA, thereby inducing chemoresistance due to miRNA-mediated feedback inhibition of p73 [83]. In liver tumor-initiating cells, miR-130b maintains cell growth, self-renewal, and chemotherapy resistance by targeting tumor protein 53-induced nuclear protein 1 (TP53INP1) [84]. Cisplatin-induced apoptosis of testicular cancer cells is counteracted by cytoplasmic $p21^{WAF1/CIP1}$, a p53 target that accumulates due to reduced Oct4 transactivation of miR-106b and miR-17-5p [85]. Collectively, ncRNAs are strongly implicated in the regulation of susceptibility to cancer therapeutics that elicit genotoxic stress.

10.4.7 Oxidative Stress

Cells produce ROS during the course of normal metabolism and eliminate them via various mechanisms [86]. Oxidative stress arises from a dynamic imbalance between the systemic manifestation of ROS and a biological system's ability to detoxify these reactive intermediates [86]. Although severe oxidative stress is cytotoxic, oxidative stress underlies carcinogenesis, and the insusceptibility of carcinoma cells to oxidative stress leads to drug resistance [87]. The physiological

ROS-scavenging systems include intracellular antioxidants such as glutathione and a variety of antioxidant enzymes such as superoxide dismutase (SOD). Meanwhile, the transcription factor nuclear factor erythroid 2-related factor 2 (Nrf2) is a master regulator of the body's antioxidant response: Nrf2 is activated by different sensors of redox status and constitutively degraded by the key regulator Kelch-like ECH-associated protein 1 (KEAP1). Activated Nrf2 binds to the antioxidant response element (ARE) and switches on expression of a wide range of detoxification or stress-response genes [87]. These machineries, which are implicated in cell detoxification and therapeutic resistance, are also targeted by ncRNAs. Drayton et al. have found that miR-27a improves the responsiveness of bladder cancers to cisplatin by targeting the cystine/glutamate exchanger SLC7A11, thereby disrupting glutathione biosynthesis [88]. In addition, histone deacetylase inhibition can overcome lung cancer resistance to polyamines by upregulating miR-200c, which in turn upregulates Nrf2-mediated transcription of the polyamine catabolic enzyme spermidine/spermine N(1)-acetyltransferase (SSAT) by directly targeting KEAP1 [89].

10.4.8 Malignant Phenotypes: Cancer Stem Cells (CSCs), Autophagy, and Epithelial–Mesenchymal Transition (EMT)

According to the CSC theory, cancers are initiated and maintained by a small subset of stem-like or cancer-initiating cells that are capable of self-renewal and differentiation into other populations of the tumor mass [41]. CSCs are also the primary cause of distal metastasis and therapeutic resistance. The properties and behaviors of this specific subset of cancer cells are regulated by miRNAs, a topic that is beyond the scope of this chapter [41]. In addition, expression of the lncRNA X-inactive specific transcript (XIST) is a biomarker that predicts the response of breast cancer to HDAC inhibitors, although the underlying mechanisms remain to be elucidated [90].

Cancer cells undergo the EMT to acquire the migratory and invasive properties required for metastasis. In addition, the EMT plays a vital role in acquisition of resistance to anticancer therapeutics [91]. As a master regulator of the EMT, TGF-β signaling plays essential roles in regulating malignant phenotypes, e.g., drug resistance, of various cancers. MiR-200 family members, especially miR-200c, are downregulated in various cancers that are refractory to chemotherapy or treatment with TKIs or monoclonal antibodies [92, 93]. This phenomenon is intimately related to miR-200c regulation of TGF-β signaling and the EMT via targeting of the transcription factors ZEB1 and ZNF217 and mesenchymal genes such as FN1, NTRK2, and QKI [94–97]. MiR-200 also inhibits EGFR-independent cell growth by targeting MIG6, thereby conferring resistance to EGFR-targeted therapeutics [92, 94]. In lung adenocarcinomas, the miR-134/487b/655 cluster regulates TGF-β-induced EMT and drug resistance to gefitinib by targeting MAGI2, a scaffold protein required for PTEN stabilization [98]. MiR-34a sensitizes head and neck cancers to

EGFR TKIs by targeting the protein tyrosine kinase receptor Axl and repressing the EMT [99]. MiR-30c antagonizes breast cancer chemoresistance by targeting the EMT-related cytoskeleton proteins actin-binding protein twinfilin 1 (TWF1) and vimentin. In addition to mediating the EMT, TWF1 also desensitizes cancer cells to chemotherapy by promoting IL-11 production [100].

Autophagy is an intracellular process of macromolecule and organelle recycling or turnover. Targeted cytoplasmic constituents are isolated within a double-membraned vesicle known as an autophagosome, which subsequently fuses with a lysosome, where the cargo is degraded [101]. Autophagy enables cells to survive stress from the external environment, such as nutrient deprivation, and also allows them to withstand internal stresses like accumulation of damaged organelles and invasion by pathogens. Moreover, autophagy can cause programmed cell death, depending on the cell type and the context of intracellular signaling [101]. Autophagy maintains cellular homeostasis and prevents early transformation of cells by eliminating superfluous or damaged proteins, enhancing host defense against pathogens and circumventing precancerous chronic tissue damage; by contrast, after the onset of cancer, autophagy facilitates tumor progression, including the development of therapeutic resistance [101]. In this regard, miR-23b sensitizes pancreatic cancers to radiotherapy by targeting ATG12 and blocking radiation-initiated cell-protective autophagy [102]. However, it remains unclear to what extent the various ncRNAs responsible for the regulation of autophagy are involved in therapeutic resistance of cancers.

10.5 Drug Accessibility Regulated by ncRNA in Cancer Treatment

The cytotoxicity of anticancer drugs can be diminished by limiting the access of pharmaceutical molecules to malignant cells. This can occur when cancer cells develop mechanisms to pump out drugs via transporter proteins on the cell membrane or when cells manage to shield target proteins from drug engagement. Both paradigms are regulated by ncRNAs [14] (see Fig. 10.2). The detailed miRNA-mediated regulation of ABC transporters has been reviewed elsewhere [103].

10.5.1 Drug Export

Eukaryotes express a class of transporter proteins on the cell membrane that pump out xenobiotics, toxins, and drugs from inside the cell. The efflux of cytotoxic drugs decreases intracellular drug concentrations and represents a common mechanism by which neoplastic cells acquire resistance to anticancer drugs [104].

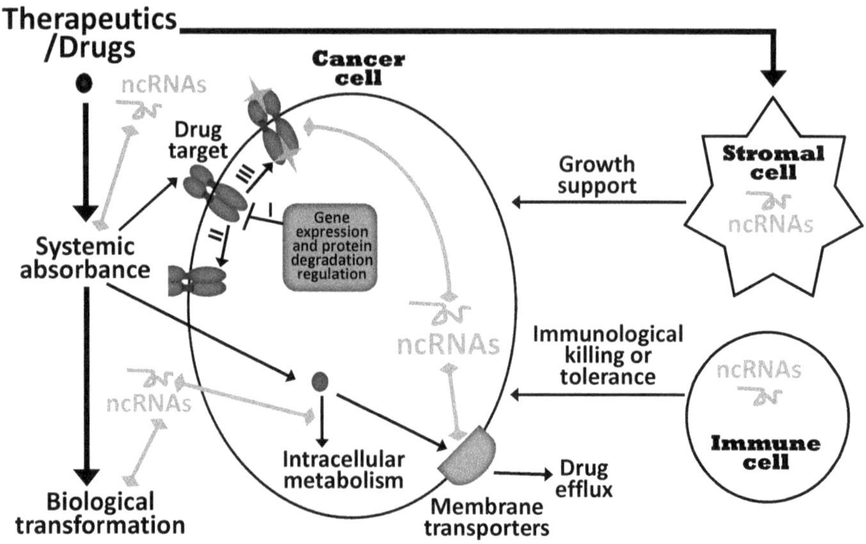

①Intracellular mechanisms that downregulate drug-targeted protein; ② Mutations that lower target protein affinity to the drug; ③Molecules that shield drug binding sites on the target protein.

Fig. 10.2 ncRNAs regulate therapeutic accessibility and pharmacokinetics of drugs, as well as the cancer microenvironment. Cancer cells develop autonomous resistance to therapeutics or anticancer drugs by reducing therapeutic/drug accessibility. In addition, the systemic absorbance and biological transformation of anticancer drugs, as well as the microenvironment, which may be educated by cancer cells or modulated by the therapeutics, influence therapeutic outcomes. A wide range of genes involved in these machineries are candidate targets of ncRNAs, which include (I) intracellular mechanisms that downregulate drug-targeted proteins, (II) mutations that lower target protein affinity to the drug, and (III) molecules that shield drug-binding sites on the target protein

10.5.1.1 ATP-Binding Cassette (ABC) Transporters

ABC transporters, a group of active transporter proteins ubiquitously expressed by mammalian cells, hydrolyze ATP to ADP and use the energy to drive the efflux of intracellular substrates against a concentration gradient. The 48 members of the ABC transporter family identified to date have been divided into seven subfamilies: ABCA through ABCG. ABC transporter proteins are composed of two nucleotide-binding domains (NBDs) and two transmembrane domains (TMDs) [105]. The classification is based on the sequence of the NBDs, also known as ABC domains, which are mainly involved in hydrolyzing ATP, binding physiological and xenobiotic substrates, and extruding them out of the cell. The majority of ABC transporters are full or complete transporters, although some (e.g., the ABCG subfamily) are half transporters that contain only one NBD and TMD per protein [105]. Two NBDs are required for normal transporter activity, consistent with the observation that ABCB1 (P-gp or MDR1) hydrolyzes two ATPs in a stepwise process during drug trafficking. The hydrolysis of the first ATP structurally modifies the TMDs by flipping the inner leaf to the outer side of the cell membrane, resulting in efflux of the drug from the cell. The hydrolysis of the second ATP restores the structure of the transporter to its original high-affinity state [105].

ABC transporters are responsible for outward transportation of xenobiotics and numerous agents including amino acids, cholesterol and its derivatives, carbohydrates, vitamins, peptides, lipids, certain important proteins, hydrophobic drugs, and antibiotics [105]. Given their capability to potentiate efflux of anticancer agents, ABC transporters play a pivotal role in conferring resistance to chemotherapeutic and molecular targeted drugs on neoplastic cells. However, depending on their individual structures, different members of the ABC transporters are involved in the efflux of different tumoricidal drugs. ABC drug transporters increase the efflux of their substrates (e.g., anticancer agents), thereby reducing the intracellular concentration of drugs and resulting in an MDR phenotype [105]. Meanwhile, the expression of ABC transporters is regulated in neoplastic cells through multiple mechanisms, including posttranscriptional silencing by ncRNAs. Borel et al. have identified 13 miRNAs that regulate the ABC transporter family in HCCs. Deregulation of these miRNAs contributed to significant upregulation of drug efflux pumps and MDR of HCCs [106]. Jaiswal et al. have found that multidrug resistance (MDR) can be transferred intercellularly by delivery of the transcripts and regulatory miRNAs of drug efflux proteins, including ABC transporters, via microparticles derived from membrane budding, thereby "retemplating" the transcriptional landscape of recipient cells from MDR donor cells to drug-sensitive recipient cells [107].

ABCB1

ABCB1/P-glycoprotein (P-gp/MDR1) is a 160–170 kDa protein encoded by the *MDR1* gene. As an apical membrane transporter localized in cells of the kidney, placenta, liver, adrenal glands, intestine, and blood-brain barrier, ABCB1 transports xenobiotics and cellular toxicants not only out of the cell but also into the urine and bile, thereby facilitating their excretion from the body. ABCB1 overexpression confers resistance to a variety of anticancer compounds like vinblastine (VLB), vincristine (VCR), paclitaxel (PTX), and colchicine (COL). ABCB1 also imparts TKI resistance to carcinoma cells [108]. Kovalchuk et al. have found that miR-451 antagonizes chemoresistance of the breast cancer cell line, MCF-7, by directly targeting ABCB1 [109]. The H19 mRNA, which is encoded by the imprinted *H19* gene and is thought to function as an RNA component of the ribonucleoprotein, is expressed at significantly higher levels in breast, lung, or hepatocellular cancer cells refractory to chemotherapeutic drugs like doxorubicin. H19 is implicated in ABCB1 expression through the control of promoter methylation [110, 111].

ABCCs

The ABCC/multidrug resistance protein (MRP) family can be further subdivided into three groups: long ABCCs such as ABCC1 (MRP1), ABCC2 (MRP2), ABCC3 (MRP3), ABCC6 (MRP6), and ABCC10 (MRP7); short ABCCs such as ABCC4 (MRP4), ABCC5 (MRP5), ABCC11 (MRP8), and ABCC12 (MRP9); and ABCC7 to ABCC9, which are components of ion channels rather than transporters. These ABCCs are critical mediators of drug resistance arising in various types of carcinomas [108]. In particular, ABCC1 overexpression correlates with doxorubicin resistance of leukemia and lung cancer, whereas ABCC10 expression confers resistance to various anticancer drugs including docetaxel, PTX, VCR, VLB, cytarabine, gemcitabine, 2′,3′-dideoxycytidine, 9-(2-phosphonyl-methoxyethyl) adenine (PMEA), and epothilone B. Both ABCC1 and ABCC2 increase the efflux of TKIs such as imatinib and sorafenib, whereas imatinib exposure causes further upregulation of ABCC1, thus conferring TKI resistance on various malignancies [108]. All of these ABCCs have been verified as targets of miRNAs [103].

ABCG2

ABCG2 is also known as breast cancer resistance protein (BCRP), mitoxantrone resistance protein (MXR), or ABC transporter expressed in placenta (ABCP) [108]. As a half transporter with one TMD and one NBD, it must homodimerize or oligomerize with other transporters to exhibit transporter activity and mediate MDR. ABCG2 is expressed in the placenta, small intestines, colon, liver, and blood vessels, where it protects cells or tissues against toxins and xenobiotics. ABCG2 also transports organic anion conjugates, nucleoside analogs, organic dyes, TKIs, anthracyclines, and topoisomerase I inhibitors and is responsible for cancer resistance to mitoxantrone (MX) and doxorubicin (DX). In addition, mutations of ABCG2 may result in significant conformational changes and alter the drug-binding and efflux capacity of the transporter [108]. By demonstrating that two miRNAs, miR-519c and miR-520h, target ABCG2, To et al. have demonstrated that the acquisition of MX resistance in various cancers can be attributed to the shortening of the ABCG2 3′ UTR, resulting in loss of miRNA binding sites or sequestering of the miRNA by highly expressed ABCG2 mRNA [112].

10.5.1.2 Nucleoside Transporter (NT) Proteins

NTs are integral membrane proteins involved in the salvage of natural nucleobases and nucleosides involved in nucleic acid synthesis [113]. They belong to solute carrier families 28 and 29 (SLC28 and SLC29), which encode human concentrative NTs (hCNTs) and equilibrative NT proteins (hENTs), respectively. Localized on the apical membrane of polarized epithelia, these NTs are required for uptake of numerous nucleoside and nucleobase analogs currently used for treatment of cancers and viral infections and are therefore determinants of drug action. hCNTs prompt the influx of nucleoside drugs coupled to the influx of sodium ions [113]. Different members of the SLC28 gene family exhibit preferences for pyrimidine or purine nucleosides and their derivatives as substrates, as exemplified by hCNT1, a high-affinity pyrimidine nucleoside transporter involved in intracellular delivery of chemotherapeutics such as gemcitabine. Similarly, hENT family members are responsible for transport of natural nucleosides and nucleoside-derived drugs. SLC22, which encodes human organic cation transporters (hOCTs) and organic anion transporters (hOATs), plays a predominant role in the uptake of nucleoside-derived drugs with specific structural variations, e.g., lack of 3′-OH [113]. Among the growing number of ncRNAs known to modulate the expression of NTs, several miRNAs including miR-122, miR-214, miR-339-3p, and miR-650 target hCNT1/SLC28A1, suggesting that these ncRNAs are involved in acquisition of chemoresistance by pancreatic cancers [114].

10.5.2 Blockade of Drug–Target Interactions

The therapeutic efficacy of anticancer drugs relies on efficient drug–target interactions. Consequently, cancer cells have developed various mechanisms to suppress drug binding to target proteins [115, 116]. For instance, a well-documented mutation (T790M) in the kinase domain of EGFR dramatically decreases the receptor's affinity for TKIs, thereby imparting resistance to these drugs [115]. Acquisition of resistance to trastuzumab occurs in a subset of HER2-positive breast cancer cells expressing mucin 1 or mucin 4. In addition to promoting cell invasion and enhancing HER2–HER3 signaling, these *O*-glycosylated transmembrane proteins interfere with trastuzumab targeting by masking the antibody-binding epitope of HER2 [116]. These situations can be ameliorated by miRNA-mediated suppression of the mucin proteins [117, 118]. Meanwhile, although miRNAs targeting HER family oncogenes can impair the onset of malignancies, they can also facilitate growth factor-independent cancer progression and resistance of advanced tumors to therapies targeting these cancer drivers [119]. In addition, Boni et al. have found that miR-192 and miR-215 directly repress thymidylate synthase (TYMS), thereby imparting resistance to TYMS-targeted chemotherapeutic agents such as 5-fluorouracil (5-FU) in gastrointestinal cancers [120].

10.6 Drug Pharmacokinetics Controlled by ncRNAs

The tumor-inhibitory potency of a chemical drug is determined by drug pharmacokinetics and metabolism, which together control the time the drug is retained in tumor tissue [121]. It is worth noting that the aforementioned ABC transporters and nucleoside transporter proteins play vital roles in regulating the pharmacokinetics of tumoricidal drugs. Moreover, these transporters are not expressed exclusively by malignant cells, but are ubiquitously present in the intestine, kidney, liver, and blood-brain barrier, which determine the absorption, in vivo distribution, and renal or hepatic processing of drugs [108, 121] (see Fig. 10.2).

Upon exerting a cytotoxic role in the desired tissue, a drug may undergo biotransformation prior to excretion. Consequently, the key enzymes responsible for inactivation of anticancer compounds dictate the half-life and persistence of drugs [121]. In particular, miR-27a and miR-27b sensitize malignant cells to 5-FU by targeting and repressing dihydropyrimidine dehydrogenase (DPD), a key uracil catabolic enzyme responsible for conversion of 5-FU to the inactive metabolite 5-dihydrofluorouracil [122]. Persson et al. have found that RNAs in the vault particle, a conserved organelle, are implicated in multidrug resistance of malignant cells. One of these so-called small vault RNAs (svRNAs), svRNAb, negatively regulates the expression of CYP3A4, which encodes a cytochrome P450 enzyme crucially involved in the metabolism of many chemotherapeutic compounds and almost 60 % of all marketed drugs [123].

10.7 Drug-Refractory Cancer Microenvironment Modulated by Noncoding RNAs

Numerous cutting-edge studies highlight the role of the microenvironment on the development, progression, and therapeutic responsiveness of cancers [124]. In theory, the tumor-suppressive efficacy of therapeutics represents the combined outcome of direct cytotoxicity to neoplastic cells and the modulation of the tumor microenvironment by the drug [125] (see Fig. 10.2). The microenvironment includes the extracellular matrix (ECM), stromal fibroblasts, immune cells, and blood vessels supplying solid tumors, all of which affect cancer progression via direct cell–cell contact or secretion of diverse factors [124, 125]. Cells in the microenvironment are extremely important for the tumor-inhibitory action of monoclonal antibodies, which in addition to their cancer cell-autonomous mechanisms elicit antitumor immunity [125]. This is exemplified by miR-27a/miR-27b, which efficiently induces the transformation of normal fibroblasts into cancer-associated fibroblasts (CAF), as evidenced by induction of α-smooth muscle actin (α-SMA) expression and TGF-β production, thereby conferring cisplatin resistance of esophageal cancers [126]. In addition, attenuated miR-142-3p suppression of the ectonucleoside triphosphate diphosphohydrolase CD39 leads to a reduction of ATP levels in regulatory T (Treg) cells relative to those in conventional T cells, explaining the vulnerability of Tregs to low-dose cyclophosphamide. This observation has implications for overcoming immune tolerance to carcinomas receiving chemotherapy [127].

10.8 Complicated Solo Performance: Combined Versus Unknown Targets

Regulation of gene expression by ncRNAs is characterized by the ability of individual ncRNAs, e.g., miRNAs, to simultaneously target multiple mRNAs. Conversely, a given transcript can be concurrently inhibited by several miRNAs [128]. In this regard, a growing number of miRNAs have been determined to target various genes that synergistically regulate sensitivity to therapeutics. For example, miR-128 downregulation accounts for drug resistance of breast cancer-initiating cells, because it directly targets both the stem cell transcriptional factor Bmi-1 and the ABC transporter ABCC5 [129]. Giovannetti et al. have demonstrated that miR-21-mediated gemcitabine resistance in pancreatic ductal adenocarcinoma (PDAC) can be attributed to the modulation of apoptosis, Akt phosphorylation, and expression of genes involved in invasive behavior [130]. Using a genome-wide screening approach, Ziliak et al. have identified an SNP (rs1649942) that significantly affects platinum sensitivity. They attributed this effect to changes in the miRNA profile and specifically to altered expression of miR-193b, which targeted a set of platinum-associated genes including CRIM1, IFIT2, OAS1, KCNMA1, and GRAMD1B [131]. MiR-301 mediates various malignant phenotypes of breast cancers, including tamoxifen resistance, through multiple targets including FOXF2, BBC3, PTEN, and COL2A1 [132]. Alternatively, a single ncRNA involved in therapeutic resistance may regulate several pathways by targeting a multifunctional gene. In this regard, Eto et al. have found that miR-223 is highly expressed in trastuzumab-resistant gastric cancers. MiR-223 directly targeted F-box and WD repeat domain-containing 7 (FBXW7), the substrate recognition component of an evolutionarily conserved SCF (complex of SKP1, CUL1, and F-box protein)-type ubiquitin ligase complex, thereby attenuating FBXW7-dependent degradation of oncoproteins including cyclin E, c-Myc, Notch, c-Jun, mTOR, and Mcl-1 [133]. These studies suggest that ncRNAs may play a more important role than protein-coding genes in determining the therapeutic responsiveness of cancers due to their ability to target multiple functional genes.

The extensive roles of ncRNAs in the therapeutic resistance of cancers are far from completely elucidated. In addition to the numerous undefined targets of miRNAs that demarcate therapy-refractory cell subsets, many lncRNAs are believed to determine therapeutic responses via mechanisms yet to be characterized [14, 134]. For example, expression of the inactive XIST, a spliced noncoding polyadenylated RNA and the only transcript expressed exclusively from the inactive X chromosome, correlates with high sensitivity to Taxol in ovarian cancers. However, its mode of action remains poorly understood [135]. Future breakthroughs in deciphering the characteristics of ncRNAs characters will provide novel functional annotations for these RNA species in the context of therapeutic resistance of cancers.

10.9 Strategies for Overcoming ncRNA-Mediated Therapeutic Resistance

The critical involvement of ncRNAs in regulating the therapeutic resistance of various cancers warrants the development of strategies based on or targeting ncRNAs in order to reverse refractory phenotypes of carcinomas [15, 16]. First, ncRNAs can be directly manipulated to improve the sensitivity of cancers to specific therapeutics. ncRNAs or their antisense inhibitors (in particular, miRNAs and antagomirs) can be synthesized and introduced into cultured cells for therapeutic purposes or delivered in vivo through nonviral carriers such as liposomes or positively charged agents that encapsulate the RNAs in nanoparticles. Moreover, cancer-targeted delivery of small RNAs can be achieved via generation of an RNA delivery system using antibodies or ligands that recognize tumor-specific antigens or receptors [136]. ncRNAs such as miRNAs and their inhibitors can also be expressed from eukaryotic expression cassettes and then expressed ectopically in malignant cells via viral or nonviral delivery of the cassettes [136]. Second, ncRNA-regulated pathways can be targeted, providing important guidance for selection and optimization of combined medication or therapy [137]. Finally, ncRNAs can be used as biomarkers for drug responsiveness and for the relapse or prognosis of cancers after treatment targeting the drug-resistant cell populations [15, 16]. These strategies will be beneficial to the development of adjuvant therapy and will potentially increase the efficacy of routine cancer treatment.

10.10 Future Perspectives

In light of the immense diversity of anticancer therapeutics per se and the paradigms by which they eliminate malignant cells, cancer cells need to evolve widely varied mechanisms to survive cytotoxic attacks. Recent studies have underscored the critical involvement of ncRNAs in regulating the therapeutic susceptibilities of different malignancies. Nevertheless, the full regulatory network underlying therapeutic resistance of cancers (e.g., the ways in which therapeutics exert selective pressure for or even fuel the development of the molecular machineries of therapeutic resistance, presumably via ncRNAs), the hierarchy of regulators (including multiple ncRNAs) involved in drug resistance, and the roles of ncRNAs in mediating cross talk between various drug resistance pathways remain to be fully understood. Except for the regulation of drug transport or metabolism, the roles most ncRNAs play in therapeutic resistance are shared by those they conduct in regulating other malignant phenotypes of carcinomas. Therefore, future investigations should seek to demarcate these roles of ncRNAs for each type of malignancy. In addition, in the context of personalized medicine, it is desirable to determine the individual variations and underlying genetic discrepancies that govern the importance of particular ncRNAs in determining the therapeutic responses of different patient populations. Finally, in contrast to the substantial participation and definitive role of miRNAs in regulating therapeutic sensitivity by posttranscriptionally silencing target genes, the contribution of most lncRNAs to drug resistance of cancers remains elusive. Moreover, the few lncRNAs so far shown to regulate the therapeutic response of cancers represent an incomplete repertoire of functional patterns. Despite the challenges scientists have encountered in this area, future studies will help to illustrate the roles of ncRNAs as key nodes of the regulatory network and precisely define the landscape of molecules or signaling events involved in cancer therapeutic responses, ultimately yielding beneficial outcomes by facilitating the development of ncRNA-based interventions against therapeutic resistance of cancers.

References

1. Harris RE. Global epidemiology of cancer. Burlington: Jones and Bartlett Publishers; 2015.
2. Fabbri M. Non-coding RNAs and cancer. New York: Springer; 2014.
3. Ramakrishnan R, Gabrilovich DI. Novel mechanism of synergistic effects of conventional chemotherapy and immune therapy of cancer. Cancer Immunol Immunother CII. 2013;62:405–10.
4. Schaue D, McBride WH. Opportunities and challenges of radiotherapy for treating cancer. Nat Rev Clin Oncol. 2015;12:527–40.
5. Dong B, Zhu YM. Molecular-targeted therapy for cancer. Chin J Cancer. 2010;29:340–5.
6. Pinedo HM, Giaccone G. Chemotherapy. Lancet. 1997;349 Suppl 2:SII7–9.
7. Perry MC. The chemotherapy source book. Philadelphia: Lippincott Williams & Wilkins; 2007.
8. Chapman JD, Nahum AE. Radiotherapy treatment planning: linear-quadratic radiobiology. Boca Raton: CRC Press; 2015.
9. Sliwkowski MX, Mellman I. Antibody therapeutics in cancer. Science. 2013;341:1192–8.
10. Bachelot A, Chabbert-Buffet N, Salenave S, et al. Anti-androgen treatments. Annales D'endocrinologie. 2010;71:19–24.
11. Jordan VC. The science of selective estrogen receptor modulators: concept to clinical practice. Clin Cancer Res. 2006;12:5010–3.
12. Becker Y. Molecular immunological approaches to biotherapy of human cancers—a review, hypothesis and implications. Anticancer Res. 2006;26:1113–34.
13. Young A, Rowett L, Kerr D. Cancer biotherapy: an introductory guide. Oxford/New York: Oxford University Press; 2006.
14. Housman G, Byler S, Heerboth S, et al. Drug resistance in cancer: an overview. Cancers. 2014;6:1769–92.
15. Malek E, Jagannathan S, Driscoll JJ. Correlation of long non-coding RNA expression with metastasis, drug resistance and clinical outcome in cancer. Oncotarget. 2014;5:8027–38.
16. Zheng T, Wang J, Chen X, Liu L. Role of microRNA in anticancer drug resistance. Int J Cancer. 2010;126:2–10.
17. Hanahan D, Weinberg RA. Hallmarks of cancer: the next generation. Cell. 2011; 144:646–74.
18. Thorpe LM, Yuzugullu H, Zhao JJ. PI3K in cancer: divergent roles of isoforms, modes of activation and therapeutic targeting. Nat Rev Cancer. 2015;15:7–24.
19. Arcaroli JJ, Quackenbush KS, Powell RW, et al. Common PIK3CA mutants and a novel 3′ UTR mutation are associated with increased sensitivity to saracatinib. Clin Cancer Res. 2012;18:2704–14.
20. Meng F, Henson R, Lang M, et al. Involvement of human micro-RNA in growth and response to chemotherapy in human cholangiocarcinoma cell lines. Gastroenterology. 2006;130: 2113–29.

21. Garofalo M, Di Leva G, Romano G, et al. MiR-221&222 regulate TRAIL resistance and enhance tumorigenicity through PTEN and TIMP3 down-regulation. Cancer Cell. 2009;16:498–509.
22. Yang H, Kong W, He L, et al. MicroRNA expression profiling in human ovarian cancer: miR-214 induces cell survival and cisplatin resistance by targeting PTEN. Cancer Res. 2008;68:425–33.
23. Hamano R, Miyata H, Yamasaki M, et al. Overexpression of miR-200c induces chemoresistance in esophageal cancers mediated through activation of the Akt signaling pathway. Clin Cancer Res. 2011;17:3029–38.
24. Huang X, Taeb S, Jahangiri S, et al. miRNA-95 mediates radioresistance in tumors by targeting the sphingolipid phosphatase SGPP1. Cancer Res. 2013;73:6972–86.
25. Fornari F, Milazzo M, Chieco P, et al. MiR-199a-3p regulates mTOR and c-Met to influence the doxorubicin sensitivity of human hepatocarcinoma cells. Cancer Res. 2010;70:5184–93.
26. Weidhaas JB, Babar I, Nallur SM, et al. MicroRNAs as potential agents to alter resistance to cytotoxic anticancer therapy. Cancer Res. 2007;67:11111–6.
27. Deng Y, Bai H, Hu H. Rs11671784 G/A variation in miR-27a decreases chemo-sensitivity of bladder cancer by decreasing miR-27a and increasing the target RUNX-1 expression. Biochem Biophys Res Commun. 2015;458:321–7.
28. Li J, Wang Y, Song Y, et al. MiR-27a regulates cisplatin resistance and metastasis by targeting RKIP in human lung adenocarcinoma cells. Mol Cancer. 2014;13:193.
29. Sugimura K, Miyata H, Tanaka K, et al. Let-7 expression is a significant determinant of response to chemotherapy through the regulation of IL-6/STAT3 pathway in esophageal squamous cell carcinoma. Clin Cancer Res. 2012;18:5144–53.
30. Wang H, Sun T, Hu J, et al. MiR-33a promotes glioma-initiating cell self-renewal via PKA and NOTCH pathways. J Clin Invest. 2014;124:4489–502.
31. Fonte E, Apollonio B, Scarfo L, et al. In vitro sensitivity of CLL cells to fludarabine may be modulated by the stimulation of Toll-like receptors. Clin Cancer Res. 2013;19:367–79.
32. Mok TS, Lee K, Leung L. Targeting epidermal growth factor receptor in the management of lung cancer. Semin Oncol. 2014;41:101–9.
33. Garofalo M, Romano G, Leva D, et al. EGFR and MET receptor tyrosine kinase-altered microRNA expression induces tumorigenesis and gefitinib resistance in lung cancers. Nat Med. 2012;18:74–82.
34. Rai K, Takigawa N, Ito S, et al. Liposomal delivery of MicroRNA-7-expressing plasmid overcomes epidermal growth factor receptor tyrosine kinase inhibitor-resistance in lung cancer cells. Mol Cancer Ther. 2011;10:1720–7.
35. Ragusa M, Majorana A, Statello L, et al. Specific alterations of microRNA transcriptome and global network structure in colorectal carcinoma after cetuximab treatment. Mol Cancer Ther. 2010;9:3396–409.
36. Spector NL, Blackwell KL. Understanding the mechanisms behind trastuzumab therapy for human epidermal growth factor receptor 2-positive breast cancer. J Clin Oncol Off J Am Soc Clin Oncol. 2009;27:5838–47.
37. Bai WD, Ye XM, Zhang MY, et al. MiR-200c suppresses TGF-beta signaling and counteracts trastuzumab resistance and metastasis by targeting ZNF217 and ZEB1 in breast cancer. Int J Cancer. 2014;135:1356–68.
38. Ye XM, Zhu HY, Bai WD, et al. Epigenetic silencing of miR-375 induces trastuzumab resistance in HER2-positive breast cancer by targeting IGF1R. BMC Cancer. 2014;14:134.
39. Li L, Liu Y, Guo Y, et al. Regulatory MiR-148a-ACVR1/BMP circuit defines a cancer stem cell-like aggressive subtype of hepatocellular carcinoma. Hepatology. 2015;61:574–84.
40. Walker CL, Ho SM. Developmental reprogramming of cancer susceptibility. Nat Rev Cancer. 2012;12:479–86.
41. Takebe N, Miele L, Harris PJ, et al. Targeting Notch, Hedgehog, and Wnt pathways in cancer stem cells: clinical update. Nat Rev Clin Oncol. 2015;12:445–64.

42. Hwang WL, Jiang JK, Yang SH, et al. MicroRNA-146a directs the symmetric division of Snail-dominant colorectal cancer stem cells. Nat Cell Biol. 2014;16:268–80.
43. Li L, Li Z, Kong X, et al. Down-regulation of microRNA-494 via loss of SMAD4 increases FOXM1 and beta-catenin signaling in pancreatic ductal adenocarcinoma cells. Gastroenterology. 2014;147:485–97 e418.
44. Park EY, Chang E, Lee EJ, et al. Targeting of miR34a-NOTCH1 axis reduced breast cancer stemness and chemoresistance. Cancer Res. 2014;74:7573–82.
45. Munoz JL, Rodriguez-Cruz V, Ramkissoon SH, et al. Temozolomide resistance in glioblastoma occurs by miRNA-9-targeted PTCH1, independent of sonic hedgehog level. Oncotarget. 2015;6:1190–201.
46. Ahmad N, Kumar R. Steroid hormone receptors in cancer development: a target for cancer therapeutics. Cancer Lett. 2011;300:1–9.
47. Jordan VC, O'Malley BW. Selective estrogen-receptor modulators and antihormonal resistance in breast cancer. J Clin Oncol Off J Am Soc Clin Oncol. 2007;25:5815–24.
48. Miller PC, Clarke J, Koru-Sengul T, et al. A novel MAPK-microRNA signature is predictive of hormone-therapy resistance and poor outcome in ER-positive breast cancer. Clin Cancer Res. 2015;21:373–85.
49. Maillot G, Lacroix-Triki M, Pierredon S, et al. Widespread estrogen-dependent repression of micrornas involved in breast tumor cell growth. Cancer Res. 2009;69:8332–40.
50. Rao X, Di Leva G, Li M, et al. MicroRNA-221/222 confers breast cancer fulvestrant resistance by regulating multiple signaling pathways. Oncogene. 2011;30:1082–97.
51. Godinho M, Meijer D, Setyono-Han B, et al. Characterization of BCAR4, a novel oncogene causing endocrine resistance in human breast cancer cells. J Cell Physiol. 2011;226:1741–9.
52. Xue X, Yang YA, Zhang A, et al. LncRNA HOTAIR enhances ER signaling and confers tamoxifen resistance in breast cancer. Oncogene. 2015. doi:10.1038/onc.2015.340.
53. Muluhngwi P, Klinge CM. Roles for miRNAs in endocrine resistance in breast cancer. Endocr Relat Cancer. 2015;22:R279–300.
54. Ma S, Chan YP, Kwan PS, et al. MicroRNA-616 induces androgen-independent growth of prostate cancer cells by suppressing expression of tissue factor pathway inhibitor TFPI-2. Cancer Res. 2011;71:583–92.
55. Yang L, Lin C, Jin C, et al. LncRNA-dependent mechanisms of androgen-receptor-regulated gene activation programs. Nature. 2013;500:598–602.
56. Ribas J, Ni X, Haffner M, et al. MiR-21: an androgen receptor-regulated microRNA that promotes hormone-dependent and hormone-independent prostate cancer growth. Cancer Res. 2009;69:7165–9.
57. Rahman N. Realizing the promise of cancer predisposition genes. Nature. 2014;505:302–8.
58. Lwin T, Zhao X, Cheng F, et al. A microenvironment-mediated c-Myc/miR-548m/HDAC6 amplification loop in non-Hodgkin B cell lymphomas. J Clin Invest. 2013;123:4612–26.
59. Jeon HM, Sohn YW, Oh SY, et al. ID4 imparts chemoresistance and cancer stemness to glioma cells by derepressing miR-9*-mediated suppression of SOX2. Cancer Res. 2011;71:3410–21.
60. Zhang L, Pickard K, Jenei V, et al. MiR-153 supports colorectal cancer progression via pleiotropic effects that enhance invasion and chemotherapeutic resistance. Cancer Res. 2013;73:6435–47.
61. Williams GH, Stoeber K. The cell cycle and cancer. J Pathol. 2012;226:352–64.
62. Pouliot LM, Chen YC, Bai J, et al. Cisplatin sensitivity mediated by WEE1 and CHK1 is mediated by miR-155 and the miR-15 family. Cancer Res. 2012;72:5945–55.
63. Salerno E, Scaglione BJ, Coffman FD, et al. Correcting miR-15a/16 genetic defect in New Zealand Black mouse model of CLL enhances drug sensitivity. Mol Cancer Ther. 2009;8:2684–92.
64. Fornari F, Gramantieri L, Giovannini C, et al. MiR-122/cyclin G1 interaction modulates p53 activity and affects doxorubicin sensitivity of human hepatocarcinoma cells. Cancer Res. 2009;69:5761–7.

65. Mohammad RM, Muqbil I, Lowe L, et al. Broad targeting of resistance to apoptosis in cancer. Seminars in Cancer Biology. 2015;35(Supplement):S78–103.
66. Huang G, Nishimoto K, Zhou Z, et al. MiR-20a encoded by the miR-17-92 cluster increases the metastatic potential of osteosarcoma cells by regulating Fas expression. Cancer Res. 2012;72:908–16.
67. Razumilava N, Bronk SF, Smoot RL, et al. MiR-25 targets TNF-related apoptosis inducing ligand (TRAIL) death receptor-4 and promotes apoptosis resistance in cholangiocarcinoma. Hepatology. 2012;55:465–75.
68. Corsten MF, Miranda R, Kasmieh R, et al. MicroRNA-21 knockdown disrupts glioma growth in vivo and displays synergistic cytotoxicity with neural precursor cell delivered S-TRAIL in human gliomas. Cancer Res. 2007;67:8994–9000.
69. Incoronato M, Garofalo M, Urso L, et al. MiR-212 increases tumor necrosis factor-related apoptosis-inducing ligand sensitivity in non-small cell lung cancer by targeting the antiapoptotic protein PED. Cancer Res. 2010;70:3638–46.
70. Weng H, Huang H, Dong B, et al. Inhibition of miR-17 and miR-20a by oridonin triggers apoptosis and reverses chemoresistance by derepressing BIM-S. Cancer Res. 2014;74:4409–19.
71. Dai B, Meng J, Peyton M, et al. STAT3 mediates resistance to MEK inhibitor through microRNA miR-17. Cancer Res. 2011;71:3658–68.
72. Chen Y, Jacamo R, Konopleva M, et al. CXCR4 down-regulation of let-7a drives chemoresistance in acute myeloid leukemia. J Clin Invest. 2013;123:2395–407.
73. Lam LT, Lu X, Zhang H, et al. A microRNA screen to identify modulators of sensitivity to BCL2 inhibitor ABT-263 (navitoclax). Mol Cancer Ther. 2010;9:2943–50.
74. Braconi C, Valeri N, Gasparini P, et al. Hepatitis C virus proteins modulate microRNA expression and chemosensitivity in malignant hepatocytes. Clin Cancer Res. 2010;16:957–66.
75. Swift LH, Golsteyn RM. Genotoxic anti-cancer agents and their relationship to DNA damage, mitosis, and checkpoint adaptation in proliferating cancer cells. Int J Mol Sci. 2014;15:3403–31.
76. Curtin NJ. DNA repair dysregulation from cancer driver to therapeutic target. Nat Rev Cancer. 2012;12:801–17.
77. Zhang P, Wang L, Rodriguez-Aguayo C, et al. MiR-205 acts as a tumour radiosensitizer by targeting ZEB1 and Ubc13. Nat Commun. 2014;5:5671.
78. Galluzzi L, Morselli E, Vitale I, et al. MiR-181a and miR-630 regulate cisplatin-induced cancer cell death. Cancer Res. 2010;70:1793–803.
79. Prensner JR, Chen W, Iyer MK, et al. PCAT-1, a long non-coding RNA, regulates BRCA2 and controls homologous recombination in cancer. Cancer Res. 2014;74:1651–60.
80. Streppel MM, Pai S, Campbell NR, et al. MicroRNA 223 is up-regulated in the multistep progression of Barrett's esophagus and modulates sensitivity to chemotherapy by targeting PARP1. Clin Cancer Res. 2013;19:4067–78.
81. Wang Y, Huang JW, Calses P, et al. MiR-96 down-regulates REV1 and RAD51 to promote cellular sensitivity to cisplatin and PARP inhibition. Cancer Res. 2012;72:4037–46.
82. Duffy MJ, Synnott NC, McGowan PM, et al. P53 as a target for the treatment of cancer. Cancer Treat Rev. 2014;40:1153–60.
83. Ory B, Ramsey MR, Wilson C, et al. A microRNA-dependent program controls p53-independent survival and chemosensitivity in human and murine squamous cell carcinoma. J Clin Invest. 2011;121:809–20.
84. Ma S, Tang KH, Chan YP, et al. MiR-130b Promotes CD133(+) liver tumor-initiating cell growth and self-renewal via tumor protein 53-induced nuclear protein 1. Cell Stem Cell. 2010;7:694–707.
85. Koster R, di Pietro A, Timmer-Bosscha H, et al. Cytoplasmic p21 expression levels determine cisplatin resistance in human testicular cancer. J Clin Invest. 2010;120:3594–605.
86. Sabharwal SS, Schumacker PT. Mitochondrial ROS in cancer: initiators, amplifiers or an Achilles' heel? Nat Rev Cancer. 2014;14:709–21.

87. Reuter S, Gupta SC, Chaturvedi MM, Aggarwal BB. Oxidative stress, inflammation, and cancer: how are they linked? Free Radic Biol Med. 2010;49:1603–16.
88. Drayton RM, Dudziec E, Peter S, et al. Reduced expression of miRNA-27a modulates cisplatin resistance in bladder cancer by targeting the cystine/glutamate exchanger SLC7A11. Clin Cancer Res. 2014;20:1990–2000.
89. Murray-Stewart T, Hanigan CL, Woster PM, et al. Histone deacetylase inhibition overcomes drug resistance through a miRNA-dependent mechanism. Mol Cancer Ther. 2013;12: 2088–99.
90. Salvador MA, Wicinski J, Cabaud O, et al. The histone deacetylase inhibitor abexinostat induces cancer stem cells differentiation in breast cancer with low Xist expression. Clin Cancer Res. 2013;19:6520–31.
91. Shang Y, Cai X, Fan D. Roles of epithelial-mesenchymal transition in cancer drug resistance. Curr Cancer Drug Targets. 2013;13:915–29.
92. Izumchenko E, Chang X, Michailidi C, et al. The TGFbeta-miR200-MIG6 pathway orchestrates the EMT-associated kinase switch that induces resistance to EGFR inhibitors. Cancer Res. 2014;74:3995–4005.
93. Mirzoeva OK, Collisson EA, Schaefer PM, et al. Subtype-specific MEK-PI3 kinase feedback as a therapeutic target in pancreatic adenocarcinoma. Mol Cancer Ther. 2013;12:2213–25.
94. Adam L, Zhong M, Choi W, et al. MiR-200 expression regulates epithelial-to-mesenchymal transition in bladder cancer cells and reverses resistance to epidermal growth factor receptor therapy. Clin Cancer Res. 2009;15:5060–72.
95. Ali S, Ahmad A, Banerjee S, et al. Gemcitabine sensitivity can be induced in pancreatic cancer cells through modulation of miR-200 and miR-21 expression by curcumin or its analogue CDF. Cancer Res. 2010;70:3606–17.
96. Cochrane DR, Spoelstra NS, Howe EN, et al. MicroRNA-200c mitigates invasiveness and restores sensitivity to microtubule-targeting chemotherapeutic agents. Mol Cancer Ther. 2009;8:1055–66.
97. Li Y, VandenBoom TG, Kong D, et al. Up-regulation of miR-200 and let-7 by natural agents leads to the reversal of epithelial-to-mesenchymal transition in gemcitabine-resistant pancreatic cancer cells. Cancer Res. 2009;69:6704–12.
98. Kitamura K, Seike M, Okano T, et al. MiR-134/487b/655 cluster regulates TGF-beta-induced epithelial-mesenchymal transition and drug resistance to gefitinib by targeting MAGI2 in lung adenocarcinoma cells. Mol Cancer Ther. 2014;13:444–53.
99. Giles KM, Kalinowski FC, Candy PA, et al. Axl mediates acquired resistance of head and neck cancer cells to the epidermal growth factor receptor inhibitor erlotinib. Mol Cancer Ther. 2013;12:2541–58.
100. Bockhorn J, Dalton R, Nwachukwu C, et al. MicroRNA-30c inhibits human breast tumour chemotherapy resistance by regulating TWF1 and IL-11. Nat Commun. 2013;4:1393.
101. Galluzzi L, Pietrocola F, Bravo-San Pedro JM, et al. Autophagy in malignant transformation and cancer progression. The EMBO J. 2015;34:856–80.
102. Wang P, Zhang J, Zhang L, et al. MicroRNA 23b regulates autophagy associated with radioresistance of pancreatic cancer cells. Gastroenterology. 2013;145:1133–43 e1112.
103. Haenisch S, Werk AN, Cascorbi I. MicroRNAs and their relevance to ABC transporters. Br J Clin Pharmacol. 2014;77:587–96.
104. Wu CP, Hsieh CH, Wu YS. The emergence of drug transporter-mediated multidrug resistance to cancer chemotherapy. Mol Pharm. 2011;8:1996–2011.
105. Choi YH, Yu AM. ABC transporters in multidrug resistance and pharmacokinetics, and strategies for drug development. Curr Pharm Des. 2014;20:793–807.
106. Borel F, Han R, Visser A, et al. Adenosine triphosphate-binding cassette transporter genes up-regulation in untreated hepatocellular carcinoma is mediated by cellular microRNAs. Hepatology. 2012;55:821–32.
107. Jaiswal R, Gong J, Sambasivam S, et al. Microparticle-associated nucleic acids mediate trait dominance in cancer. FASEB J. 2012;26:420–9.

108. Anreddy N, Gupta P, Kathawala RJ, et al. Tyrosine kinase inhibitors as reversal agents for ABC transporter mediated drug resistance. Molecules. 2014;19:13848–77.
109. Kovalchuk O, Filkowski J, Meservy J, et al. Involvement of microRNA-451 in resistance of the MCF-7 breast cancer cells to chemotherapeutic drug doxorubicin. Mol Cancer Ther. 2008;7:2152–9.
110. Doyle LA, Yang W, Rishi AK, et al. H19 gene overexpression in atypical multidrug-resistant cells associated with expression of a 95-kilodalton membrane glycoprotein. Cancer Res. 1996;56:2904–7.
111. Tsang WP, Kwok TT. Riboregulator H19 induction of MDR1-associated drug resistance in human hepatocellular carcinoma cells. Oncogene. 2007;26:4877–81.
112. To KK, Robey RW, Knutsen T, et al. Escape from hsa-miR-519c enables drug-resistant cells to maintain high expression of ABCG2. Mol Cancer Ther. 2009;8:2959–68.
113. Molina-Arcas M, Casado FJ, Pastor-Anglada M. Nucleoside transporter proteins. Curr Vasc Pharmacol. 2009;7:426–34.
114. Bhutia YD, Hung SW, Patel B, et al. CNT1 expression influences proliferation and chemosensitivity in drug-resistant pancreatic cancer cells. Cancer Res. 2011;71:1825–35.
115. Dahabreh IJ, Linardou H, Siannis F, et al. Somatic EGFR mutation and gene copy gain as predictive biomarkers for response to tyrosine kinase inhibitors in non-small cell lung cancer. Clin Cancer Res. 2010;16:291–303.
116. Nahta R, Yu D, Hung MC, et al. Mechanisms of disease: understanding resistance to HER2-targeted therapy in human breast cancer. Nat Clin Pract Oncol. 2006;3:269–80.
117. Lahdaoui F, Delpu Y, Vincent A, et al. MiR-219-1-3p is a negative regulator of the mucin MUC4 expression and is a tumor suppressor in pancreatic cancer. Oncogene. 2015;34:780–8.
118. Sachdeva M, Mo YY. MicroRNA-145 suppresses cell invasion and metastasis by directly targeting mucin 1. Cancer Res. 2010;70:378–87.
119. Gomez GG, Wykosky J, Zanca C, et al. Therapeutic resistance in cancer: microRNA regulation of EGFR signaling networks. Cancer Biol Med. 2013;10:192–205.
120. Boni V, Bitarte N, Cristobal I, et al. MiR-192/miR-215 influence 5-fluorouracil resistance through cell cycle-mediated mechanisms complementary to its post-transcriptional thymidylate synthase regulation. Mol Cancer Ther. 2010;9:2265–75.
121. Holohan C, Van Schaeybroeck S, Longley DB, Johnston PG. Cancer drug resistance: an evolving paradigm. Nat Rev Cancer. 2013;13:714–26.
122. Offer SM, Butterfield GL, Jerde CR, et al. MicroRNAs miR-27a and miR-27b directly regulate liver dihydropyrimidine dehydrogenase expression through two conserved binding sites. Mol Cancer Ther. 2014;13:742–51.
123. Persson H, Kvist A, Vallon-Christersson J, et al. The non-coding RNA of the multidrug resistance-linked vault particle encodes multiple regulatory small RNAs. Nat Cell Biol. 2009;11:1268–71.
124. Blonska M, Agarwal NK, Vega F. Shaping of the tumor microenvironment: stromal cells and vessels. Semin Cancer Biol. 2015;34:3–13.
125. Kohlhapp FJ, Mitra AK, Lengyel E, Peter ME. MicroRNAs as mediators and communicators between cancer cells and the tumor microenvironment. Oncogene. 2015;34:5857–68.
126. Tanaka K, Miyata H, Sugimura K, et al. MiR-27 is associated with chemoresistance in esophageal cancer through transformation of normal fibroblasts to cancer-associated fibroblasts. Carcinogenesis. 2015;36:894–903.
127. Zhao J, Cao Y, Lei Z, et al. Selective depletion of CD4+CD25+Foxp3+ regulatory T cells by low-dose cyclophosphamide is explained by reduced intracellular ATP levels. Cancer Res. 2010;70:4850–8.
128. Kong YW, Ferland-McCollough D, Jackson TJ, et al. MicroRNAs in cancer management. Lancet Oncol. 2012;13:e249–58.
129. Zhu Y, Yu F, Jiao Y, et al. Reduced miR-128 in breast tumor-initiating cells induces chemotherapeutic resistance via Bmi-1 and ABCC5. Clin Cancer Res. 2011;17:7105–15.

130. Giovannetti E, Funel N, Peters GJ, et al. MicroRNA-21 in pancreatic cancer: correlation with clinical outcome and pharmacologic aspects underlying its role in the modulation of gemcitabine activity. Cancer Res. 2010;70:4528–38.
131. Ziliak D, Gamazon ER, Lacroix B, et al. Genetic variation that predicts platinum sensitivity reveals the role of miR-193b* in chemotherapeutic susceptibility. Mol Cancer Ther. 2012;11:2054–61.
132. Shi W, Gerster K, Alajez NM, et al. MicroRNA-301 mediates proliferation and invasion in human breast cancer. Cancer Res. 2011;71:2926–37.
133. Eto K, Iwatsuki M, Watanabe M, et al. The sensitivity of gastric cancer to trastuzumab is regulated by the miR-223/FBXW7 pathway. Int J Cancer. 2015;136:1537–45.
134. Liu Q, Paroo Z. Biochemical principles of small RNA pathways. Annu Rev Biochem. 2010;79:295–319.
135. Huang KC, Rao PH, Lau CC, et al. Relationship of XIST expression and responses of ovarian cancer to chemotherapy. Mol Cancer Ther. 2002;1:769–76.
136. Kaboli PJ, Rahmat A, Ismail P, Ling KH. MicroRNA-based therapy and breast cancer: a comprehensive review of novel therapeutic strategies from diagnosis to treatment. Pharmacol Res. 2015;97:104–21.
137. Wang Z, Li Y, Ahmad A, et al. Targeting Notch signaling pathway to overcome drug resistance for cancer therapy. Biochim Biophys Acta. 2010;1806:258–67.

Chapter 11
Noncoding RNAs Regulating Cancer Signaling Network

Jing Hu, Geoffrey J. Markowitz, and Xiaofan Wang

Abstract The cellular signaling network plays a fundamental role during development and disease, especially cancer progression. By deregulating signaling pathways, cancer cells acquire hallmarks of the disease including uncontrolled proliferation, evasion from cell death, activation of angiogenesis, invasion, and metastasis. Noncoding RNAs make substantial contributions to regulating signal transduction in cancer, thereby promoting or suppressing different biological processes during tumorigenesis. This chapter provides an overview on the regulatory functions of noncoding RNAs in the signaling network in cancer cells. It summarizes examples of noncoding RNAs that act as oncogenes or tumor-suppressing genes involved in key signal pathways as well as signal crosstalk in cancer cells.

Keywords MicroRNA • Noncoding RNA • Signal transduction • Cancer

11.1 Introduction

The cellular signaling network, comprised of a variety of potent signaling pathways which interact and inform cell functionality, is highly conserved throughout evolution and tightly controlled in normal cells to instruct key biological processes for development and maintained homeostasis of both cell numbers and tissue functions. During neoplastic transformation and tumor progression, cancer cells acquire fundamental traits including sustained proliferative signaling, immortalized

J. Hu • G.J. Markowitz
Departments of Pharmacology and Cancer Biology, Duke University Medical Center, Durham, NC 27710, USA

X. Wang (✉)
Departments of Pharmacology and Cancer Biology, Duke University Medical Center, Durham, NC 27710, USA

Duke University School of Medicine, C218 LSRC, Box 3813, Durham, NC 27710, USA
e-mail: xiao.fan.wang@duke.edu

E. Song (ed.), *The Long and Short Non-coding RNAs in Cancer Biology*,
Advances in Experimental Medicine and Biology 927,
DOI 10.1007/978-981-10-1498-7_11

replication, evasion from cellular senescence and cell death, induction of angiogenesis, and activation of invasion and metastasis. By hijacking and deregulating the signaling network, cancer cells obtain these hallmarks and become master of their own fates [1].

Recent studies have revealed essential roles for microRNAs and other noncoding RNAs in regulating gene expression during development and disease, particularly during cancer progression. The multigene regulatory features of these noncoding RNAs enable them to remodel signaling pathways, facilitating or repressing signal transmission to downstream effectors in an effective and diverse manner. Table 11.1 shows a representative, noncomprehensive list of microRNAs and noncoding RNAs that function as positive or negative regulators of key signaling pathways, thereby affecting tumorigenesis. This chapter provides an overview of the regulatory functions of noncoding RNAs in the signaling network in cancer cells. It summarizes examples of noncoding RNAs that act as oncogenes or tumor-suppressing genes modulating signal transduction in cancer cells. Since Chaps. 12 and 13 will further describe the involvement of noncoding RNAs in NFκB, c-myc, and p53 signaling, this chapter focuses on the general role of noncoding RNAs in signal transduction, especially on signaling pathways other than NFκB, c-myc, and p53.

11.2 Receptor Tyrosine Kinase Signaling

Receptor tyrosine kinases (RTKs) are a subclass of cell-surface receptors of growth factors with an intrinsic tyrosine kinase activity which play crucial roles during development and disease progression, including tumorigenesis. The activation of RTK signaling is carefully controlled in normal cells, while aberrant expression or activation of RTKs in cancer cells triggers multiple signal cascades that regulate critical cellular processes, including cell proliferation, survival, metabolism, migration, and angiogenesis, consequently driving the transformation and progression of cancer.

Since the discovery of the first RTK, epidermal growth-factor receptor (EGFR), more than 30 years ago, substantial advances have been made in characterizing other RTKs and their corresponding family members. There are 58 RTKs in humans, which share similar molecular structures: an extracellular ligand-binding domain, a transmembrane region, and a cytoplasmic region that contains a tyrosine kinase domain as well as additional regulatory domains. Upon binding to its specific ligands, RTKs dimerize, and the tyrosine kinase domain of the RTK becomes activated and phosphorylates tyrosines on downstream partners, thereby initiating key signaling cascades such as the Ras/Raf/ERK and PI3K/AKT pathways [2]. The RTK families, including EGFR, fibroblast growth-factor receptor (FGFR), platelet-derived growth factors (PDGFR), insulin-like growth-factor 1 receptor (IGF1R), hepatocyte growth-factor receptor (HGFR, MET), etc., are frequently mutated or abnormally expressed in different cancer types to promote the hallmarks of these

Table 11.1 MicroRNAs targeting signaling components in cancer

microRNA/lncRNA	Target	Effects on tumorigenesis
Receptor tyrosine kinase		
miR-7	EGFR	Suppression
miR-34	MET, PDGFR	Suppression
miR-143/145	IGF1R, ERBB3	Suppression
Ras/Raf/MEK/ERK pathway		
miR-21	Spry2, Btg2, Pdcd4	Promotion
miR-370, miR-193b	NF1	Promotion
miR-124	Sos1	Suppression
let-7	HRas, KRas, NRas	Suppression
KRAS1P	KRas	Promotion
miR-524-5p	Braf, ERK2	Suppression
miR-7	Raf	Suppression
PI3K/AKT pathway		
miR-126	p85β	Suppression
miR-7	p110δ, mTOR, p70S6K	Suppression
miR-21, miR-214, miR-221&222	PTEN	Promotion
PTENP1	PTEN	Suppression
PTENP1-AsRNA	PTEN	Promotion
miR-375	PDK1	Suppression
miR-184	AKT2	Suppression
miR-218	Rictor	Suppression
miR-205	PTEN, PHLPP2	Promotion
TGF-β signal pathway		
miR-17/20a	TGFBR2	Promotion
miR-18a	SMAD2, SMAD4	Promotion
Mir-106b-25	SMAD7	Promotion
Wnt signal pathway		
miR-34	WNT1, WNT3, LRP6, β-catenin, LEF1, Axin2	Suppression
miR-155, miR-106b	APC	Promotion
miR-222	DKK2	Promotion
Notch signal pathway		
miR-34	Notch-1, Notch-2, Delta-like 1	Suppression
miR-146a	Numb	Promotion
Hedgehog signal pathway		
mir-125b, miR-326	Smoothened	Suppression
miR-324-5p	Gli1	Suppression
miR-378	Sufu	Promotion
mir-212	ptch1	Promotion
Hippo signal pathway		

(continued)

Table 11.1 (continued)

microRNA/lncRNA	Target	Effects on tumorigenesis
mir-135b	LATS2, β-TrCP, NDR2	Promotion
miR-375	YAP	Suppression
JAK-STAT signal pathway		
mir-155	SOCS1	Promotion
miR-135a	JAK2	Suppression

cancers [2]. For example, amplification/mutation of EGFR is often involved in breast cancer, lung cancer, and glioma [3].

A series of microRNAs have been revealed to function as important tumor suppressors through targeting RTKs, thus altering the signaling output of downstream pathways. For instance, miR-7 has been reported to induce cell-cycle arrest and cell death through suppressing EGFR and the downstream AKT and ERK pathways in multiple cancer types. Notably, processing of miR-7 is decreased in glioblastoma cell lines and patient specimens, where there is an elevated level of EGFR, highlighting the tumor-suppressive role of miR-7 in this cancer by regulation of EGFR expression and signal activation [4, 5]. Similarly, miR-34 family members impair tumorigenesis and induce apoptosis via attenuating expression of MET and PDGFR in melanoma, lung cancer, and gastric cancer [6–8]. In addition, reduction of miR-143/miR-145 expression in cancer tissue leads to accumulated expression of IGF1R and ERBB3, thereby accelerating progression of colorectal cancer and breast cancer [9, 10]. Therefore, deregulated microRNAs that target RTKs serve as vital regulators at the very first stage of signaling triggered by RTKs in cancer.

11.3 The Ras/Raf/MEK/ERK Signaling Pathway

The activity of the Ras/Raf/MEK/ERK signaling pathway is increased in approximately one third of human cancers [11]. Ligand-mediated activation of RTKs allows for GTP loading of the GTPase Ras, which activates the kinase Raf. Activated Raf proceeds to phosphorylate mitogen-activated protein-kinase kinase (MEK), which subsequently phosphorylates extracellular signal-regulated kinase (ERK), leading to transcription of genes that facilitate cell proliferation, migration, and metastasis. A series of microRNAs have been illustrated to target key components of this pathway, resulting in deregulation of signaling and tumorigenesis.

Transmission of signal through the Ras/Raf/MEK pathway starts with the activation of Ras proteins, which are members of a large superfamily of GTP-binding proteins. The activity of Ras depends on whether it has bound GTP (active state, high affinity for downstream effectors) or GDP (inactive state). The process of exchanging the bound nucleotide is catalyzed by guanine nucleotide exchange factors (GEFs) and GTPase-activating proteins (GAPs). Ras GEFs facilitate the dissociation of GDP from Ras, allowing its binding of GTP to yield its activation. In response to upstream stimulation of RTKs, SOS1, one of the Ras GEFs, is recruited to the membrane through binding to the adaptor protein growth-factor-receptor-bound protein 2 (Grb2). As a consequence of the close proximity between SOS1 and Ras, GTP loading of Ras is enhanced, resulting in activation of Ras. Conversely, Ras GAP proteins accelerate the hydrolysis of GTP by Ras to insure that Ras is rapidly inactivated after stimulation. Therefore, the balance between GEF and GAP activity determines the GTP status and thereby activity of Ras [12].

A number of microRNAs serve as functional tumor suppressors and oncomirs via modulating activities of these Ras GEFs and GAPs. MiR-21 is one of the master oncomirs that is upregulated in multiple cancer types to enhance their progression. MiR-21 has been shown to promote colon cancer cell migration and growth via suppressing sprouty 2 (Spry2), which is a negative regulator of Ras activity through impairing the recruitment of Grb2-SOS1 by RTKs [13, 14]. A separate report suggests that miR-21 boosts cell-cycle progression in laryngeal carcinoma via targeting Btg2, which has been demonstrated to bind Ras and reduce its loading of GTP [15, 16]. MiR-21 also promotes neoplastic transformation through targeting Pdcd4, a negative regulator of the transcription factor AP-1 that is induced by Ras signaling to promote cell proliferation and invasion [17, 18]. Consistently, using transgenic mouse models with overexpression or deletion of miR-21, Hatley and colleagues have elucidated that miR-21 mediates KRas-dependent lung tumorigenesis by targeting Spry1, Spry2, Btg2, and PDCD4 together to promote Ras/MEK/ERK signaling activity [19]. This observation also indicates a crucial role of miR-21 in facilitating tumorigenesis robustly through regulating multiple nodes in the Ras/ERK pathway collectively, from initial activation to ultimate transcription stimulation. Interestingly, expression of miR-21 is induced by Ras signal, acting in a positive feedback loop to boost the Ras pathway [19]. Besides the aforementioned factors, other GEFs and GAPs are affected by microRNAs as well. For instance, miR-370 and miR-193b have been identified to promote acute myeloid leukemia and neck squamous cell carcinoma, respectively, by targeting NF1, a GAP which negatively regulates Ras activity and is frequently deleted in cancer [20, 21]. In contrast, miR-124 has been illustrated to suppress Sos1 and impair Ras signaling, thereby functioning as a tumor suppressor in glioblastoma [22].

In addition to the GEFs/GAPs catalyzing the process of nucleotide exchange, Ras activity can be modulated by changes in the protein itself. Mutations of Ras genes are frequently observed in cancer, with 20 % of human tumors carrying activating point mutations in Ras which result in a constitutively activated GTP-bound form [12]. Ras gene amplification and elevated expression are also common molecular alterations in lung cancer, pancreatic cancer, and gastric cancer [23–25]. Expression of Ras proteins can also be regulated by miRNAs. For example, the let-7 family of miRNAs serves as essential tumor suppressors that inhibit Ras expression and decrease cancer cell proliferation [26–29]. Johnson et al. have reported that KRas, HRas, and NRas harbor multiple binding sites complimentary to let-7 in their 3'UTRs [26]. Importantly, expression of the let-7 family and Ras proteins is negatively correlated, with lung cancer tissues demonstrating decreased let-7 expression and elevated Ras expression compared to normal tissues [26]. An SNP in the let-7 complementary site in the KRas 3'UTR, disrupting let-7's regulation of KRas, has been further shown to increase the likelihood of non-small cell lung cancer [30]. In addition to miRNAs, long-noncoding RNAs (lncRNAs) can also serve as competing endogenous RNAs (ceRNAs) for KRas to regulate its expression. Poliseno et al. have identified a long-noncoding pseudogene of KRas, KRAS1P, which contains regions highly homologous to KRas, with conserved seed sequence matches for KRas-targeting miRNAs in the 3'UTR. KRAS1P thereby exerts tumor-promoting functions by competition for microRNA binding and resultant regulation of KRas abundance. Poliseno et al. have also observed amplification of KRAS1P in different tumors and a positive correlation between KRas and KRAS1P expression [31]. Taken together, expression level of Ras is susceptible to regulation by both miRNAs and other noncoding RNAs.

Activated Ras proteins will stimulate downstream effectors through a cascade of phosphorylation events on the Raf, MEK, and ERK proteins. Deregulation of these components leads to alterations in the output of this signaling cascade. MiR-524-5p has been shown to suppress melanoma cell proliferation, migration, and tumor formation by targeting multiple downstream effectors in the Ras/Raf/ERK pathway, including both Braf and ERK2. Interestingly, activation of the Ras/Raf/ERK pathway by growth-factor stimulation or overexpression of the constitutively active mutant V600E Braf results in decreased expression of miR-524-5p, suggesting a feedback loop affecting this signal regulation [32]. Besides miR-524-5p, other miRNAs such as miR-7 have also been implicated in targeting Raf in glioblastoma and adrenocortical carcinoma to dampen the downstream Ras signal [33, 34].

Collectively, miRNAs possess a great potential to modulate one or multiple molecules in the Ras/Raf/ERK pathway, resulting in positive or negative regulation of this signaling circuit.

11.4 The PI3K/AKT Signaling Pathway

Activated by both RTKs and Ras, phosphatidylinositol-3-kinase (PI3K) activates its downstream signaling cascade through generation of the lipid second messenger phosphatidylinositol 3,4,5-trisphosphate (PIP_3) that subsequently activates AKT and multiple downstream effectors, leading to enhanced cell survival, cell-cycle progression, and cell growth [35]. The PI3K/AKT pathway is aberrantly activated in many cancers, with mutations in the PI3K pathway components accounting for around 30 % of human cancers [36]. MicroRNAs also serve as essential regulators of this pathway.

PI3K is a large family that contains three classes. The Class I PI3K is involved in oncogenesis, which is composed of heterodimers between a p85 regulatory subunit and a p110 catalytic subunit. In response to activated RTKs or Ras, p85-p110 is recruited to the cell membrane and phosphorylated by kinases, so that it proceeds to convert phosphatidylinositol 4,5-bisphosphate (PIP_2) to PIP_3 [35]. Multiple microRNAs have been reported to target the different subunits of PI3K, thereby repressing the signal pathway. For instance, miR-126, which is frequently lost in colon cancers, functions as a tumor suppressor through opposing p85β and consequently reducing the phosphorylated AKT levels substantially [37]. Fang et al. have also shown that miR-7 targets p110δ, mTOR, and p70S6K simultaneously in hepatocellular carcinoma (HCC), leading to inhibition of cancer cell proliferation, migration and tumor metastasis, which provides another example of one single miRNA targeting multiple nodes in the same pathway to impair the signal effectively. Consistently, compared to paired normal tissues, miR-7 expression is downregulated, while levels of p85β, AKT, and mTOR are upregulated in HCC tissues, suggesting a clinical association among miR-7 and the PI3K pathway components [38].

As a vital second messenger, the level of PIP_3 is stringently controlled in cells via tight regulation of PI3K activation and actions of PIP_3 phosphatases which convert PIP_3 back to PIP_2. Analogous to the regulation of GTP- or GDP-bound status of Ras by GEFs/GAPs, PI3Ks and PIP_3 phosphatases mediate levels of PIP_3 to affect the activity of this signaling pathway. PTEN is a PIP_3 phosphatase that functions as a tumor suppressor and is implicated broadly in modulating the development of various cancer types through negatively regulating PI3K/AKT signal [35]. Mutations in PTEN frequently occur in CNS, colorectal, skin, prostate cancers, and so forth [39]. In addition, monoallelic loss of PTEN contributes to tumor growth, and the dosage of PTEN expression correlates with the severity of cancers, indicating that PTEN is functionally haploinsufficient [40]. MicroRNAs and other noncoding RNAs make substantial contributions to the fine-tuned regulation of PTEN's expression level and resultant effects on cancer progression. First, a number of miRNAs are involved in suppression of PTEN expression, resulting in enhancement of PI3K/AKT signaling activity and acceleration of tumorigenesis. For instance, as a master oncomir, miR-21 represses PTEN expression in HCC to promote tumor cell proliferation and migration [41]. Mir-214, miR-221, and miR-222 have been shown to induce cell survival and promote drug resistance through targeting PTEN and boosting AKT activation in ovarian cancer and lung cancer [42, 43]. In addition, both noncoding and protein-coding RNA transcripts serve as ceRNAs to regulate PTEN expression. Analogously to KRAS1P, Poliseno et al. also have identified PTENP1, a long-noncoding pseudogene of PTEN that suppresses tumor formation by competing for the binding of microRNAs which regulate the amount of PTEN. Loss of PTENP1 in prostate cancer tissues and a positive correlation between PTEN and PTENP1 expression have also been observed [31]. Similarly, protein-coding genes, such as VAPA, CNOT6L, and Zeb2, are further elucidated to display concordant expression patterns with PTEN and function as decoys for PTEN-targeting microRNAs in colon carcinoma and melanoma cells. Attenuation of expression of these ceRNAs represses PTEN and activates the PI3K/AKT pathway in a microRNA-dependent, protein-coding-independent manner [44, 45]. In addition to ceRNAs, Johnsson et al. have characterized another lncRNA, PTENP1-encoded antisense RNA (PTENP1-AsRNA), which regulates PTEN transcription and mRNA stability. Interestingly, they illustrated that the α isoform of PTENP1-AsRNA localized to the

PTEN promoter and epigenetically modulated PTEN expression through recruitment of DNMT3 and EZH2. In contrast, the β isoform interacted with the PTENP1 transcript, destabilizing that mRNA and repressing its microRNA sponge activity, thereby suppressing PTEN's protein level. Blockage of PTENP1-AsRNA induced cell-cycle arrest through disruption of the PI3K/AKT pathway [46]. Taken together, regulation of PTEN expression occurs at multiple levels by noncoding RNAs during tumorigenesis, indicating the pivotal role of this tumor suppressor in modulating PI3K/AKT signaling.

Upon activation of PI3K, the resultant accumulated PIP3 will bind and recruit AKT to the cell membrane, where AKT undergoes two phosphorylation events by the 3-phosphoinositide-dependent protein kinases (PDKs) to become active. The phosphatases PHLPP and PHLPP2, in contrast, act to terminate AKT signaling by directly dephosphorylating and inactivating AKT [47]. Activated AKT subsequently phosphorylates and targets multiple downstream effectors, such as mTOR, BAD, and FOXO, ultimately accelerating cell growth, survival, and proliferation [35, 36]. Numerous miRNAs play tumor-suppressive roles by targeting these effectors of PI3K signaling at different steps. For example, miR-375 targets PDK1, consequently inhibiting AKT phosphorylation and gastric carcinoma cell survival; miR-184 suppresses neuroblastoma cell survival through repressing AKT2, and miR-218 targets the mTOR component Rictor to exert its tumor-suppressive functions in oral squamous cell carcinoma (OSCC) [48–50]. In the meantime, by targeting negative regulators, miRNAs may also act as oncomirs to boost AKT signaling. MiR-205 has been identified as a miRNA that is highly expressed in multiple subtypes of non-small cell lung cancer and activates both AKT/mTOR and AKT/FOXO3a through targeting PTEN and PHLPP2 [51].

In sum, miRNAs together with lncRNAs modulate PI3K/AKT signaling activity in diverse manners, ultimately altering multiple downstream biological aspects of oncogenesis.

11.5 The TGF-β Signaling Pathway

TGF-β signaling is commonly deregulated in human cancers and plays numerous varied roles during tumor progression. The TGF-β pathway exerts tumor-suppressive effects in the early stage of tumorigenesis through inhibiting cell survival and cell-cycle progression. Mutations of the pathway's effector components, such as Smad4, frequently occur in cancers, especially colon cancer. Paradoxically, in the later stages of tumor development, TGF-β signaling promotes cancer progression by inducing epithelial–mesenchymal transition (EMT), enhancing cell invasion, as well as restructuring the tumor microenvironment to promote cancer cell dissemination, metastasis, and evasion of immune surveillance [52, 53]. TGF-β signaling initiates from binding of ligands to the TGF-β type II receptor (TGFBR2), which triggers the formation of the heteromeric active receptor complexes that phosphorylate and transactivate TGF-β type I receptor (TGFBRI). The activated TGFBRI subsequently phosphorylates receptor-regulated Smads (R-Smad), which further form a complex with the common partner Smad, SMAD4 (Co-smad). This complex translocates to the nucleus and binds to target promoters to regulate expression of different subsets of genes in a cell-type- and cell-context-dependent manner [53].

MicroRNAs function as both negative and positive regulators of the TGF-β pathway. Mestdagh et al. have showed that the oncogenic miR-17-92 cluster suppresses multiple key effectors of TGF-β signaling, antagonizing the tumor-suppressive effects of TGF-β pathway and accelerating cell proliferation in neuroblastoma. Interestingly, they demonstrated that different miRNAs within the same cluster targeted different components of TGF-β pathway. MiR-17/20a repressed TGFBR2, whereas miR-18a inhibited expression of SMAD2 and SMAD4. Therefore, activation of the miR-17-92 cluster resulted in potent impairment of TGF-β signal transduction and downstream gene expression. A significant negative correlation between expression levels of miR-17-92 and TGF-β pathway components was also reported in primary neuroblastoma tumors [54]. Meanwhile, inhibitory Smads (I-Smads), such as SMAD7, have been shown to antagonize TGF-β signaling through binding to TGFBRI to interfere with phosphorylation of SMAD2/3, and through recruiting E3 ubiquitin ligases for degradation of TGFBRI. The miR-106b-25 cluster has been implicated in elevation of TGFBRI levels and activation of TGF-β signaling by targeting SMAD7, thereby inducing EMT and the tumor-initiating cell phenotype in breast cancer cells [55]. In addition, miR-106b-25 has also been reported to overcome TGF-β signaling-mediated growth suppression though targeting p21 and BIM [56], indicating the tumor-promoting roles of this miRNA cluster by modulating the output of the TGF-β pathway at multiple levels to promote tumorigenesis.

In addition to actively modulating the TGF-β signaling pathway, expression of miRNAs and lncRNAs is closely controlled by TGF-β signaling itself, strengthening the cellular responses to TGF-β. For example, miR-205 and miR-200 family members cooperatively target transcriptional repressors of E-cadherin, ZEB1, and ZEB2, to inhibit EMT [57]. Interestingly, in response to TGF-β, the miR-200 primary transcript is repressed by the accumulated ZEB1 in mesenchymal cells [58], reinforcing the TGF-β-induced EMT process. Meanwhile, an lncRNA, lncRNA-ATB, has been illustrated to be upregulated by TGF-β and subsequently boost expression of ZEB1 and ZEB2 by competitively binding with the miR-200 family to promote EMT and metastasis of HCC [59]. These feedback loops among miRNAs, lncRNA, and the TGF-β pathway ultimately strengthen the signal output. Besides transcriptional regulation, TGF-β signaling directly regulates microRNA biogenesis at the posttranscriptional level through enhancing Drosha's cleavage of a specific group of miRNAs [60].

Taken together, noncoding RNAs participate in modulation of TGF-β signaling throughout different stages of tumor progression, leading to context-dependent alterations in the TGF-β response. At the same time, TGF-β signaling actively regulates miRNA and lncRNA expression, further intensifying cellular responses.

11.6 The Wnt/β-Catenin Signaling Pathway

Wnt signaling is crucial for embryonic development and tissue regeneration. Aberrant expression of Wnt signaling components occurs in many cancer types, prominently in colon cancer, to induce tumor progression. The Wnt signaling cascade starts with Wnt ligand binding it its receptor, Frizzled (Fz), and its coreceptor, low-density lipoprotein receptor-related protein 6 (LRP6) or LRP5. On one hand, natural inhibitors such as Frizzled-related proteins (sFRPs), Wnt inhibitory protein (WIF), and Dickkopf (DKK) antagonize signal initiation by inhibiting interactions between Wnt and its receptors. On the other hand, agonists, including Norrin and R-spondins, act through the Fz/LRP complex to boost the signal. In response to Wnt ligand, LRP5 or LRP6 is phosphorylated by CK1α and GSK3β and recruits Dishevelled (DVL) to the plasma membrane to interact with Fz receptor subsequently mediating translocation of Axin to the membrane. This inactivates a destruction complex, consisting of APC and Axins as well as casein kinase 1α (CK1α) and GSK3β, which serves to mediate degradation of β-catenin. Regulation of β-catenin stability is the key event in Wnt signal transduction. In the absence of Wnt stimulation, cytoplasmic β-catenin interacts with this destruction complex, and CK1α and GSK3β phosphorylate β-catenin, targeting it for proteasome-dependent degradation. Wnt ligand stimulation-mediated disruption of this complex results in accumulation of β-catenin, which enters the nucleus and binds with TCF transcription factors to stimulate expression of genes that modulate multiple biological processes including cell proliferation, survival, stem cell self-renewal, and differentiation [61, 62].

MicroRNAs exert their tumor-suppressive or tumor-promoting functions by targeting these positive or negative regulators of the Wnt pathway. For instance, as a tumor-suppressive miRNA, miR-34 is transactivated by p53 and suppresses transcriptional activity of the β-catenin/TCF complex, thereby repressing colorectal cancer progression. Multiple Wnt pathway components have been identified as targets of miR-34, including WNT1, WNT3, LRP6, β-catenin, and LEF1 [63]. In addition, miR-34-mediated repression of Axin2 leads to increased nuclear GSK3β and decreased Snail in colorectal cancer cells [64]. These findings have revealed a key role for miR-34 in antagonizing Wnt signaling activity and colon cancer development. Meanwhile, oncogenic miRNAs, such as miR-155 and miR-106b, are upregulated in HCC and target APC directly, resulting in accumulation of β-catenin, which accelerates cell proliferation and inhibits apoptosis in HCC [65, 66]. Leading to a similar effect, miR-222 promotes glioma cell proliferation via targeting DKK2 and potentiating the Wnt/β-catenin signal [67].

11.7 The Notch Signaling Pathway

The Notch signaling pathway has a simple framework that is critical during developmental processes, including cell proliferation, apoptosis, specification of cell fate, maintenance of stem cell populations, and differentiation. Dysfunction of the Notch signaling pathway is also involved in cancer progression, in a context-dependent manner. For example, hyperactivation of Notch signaling promotes leukemia and breast cancer progression, while it induces differentiation and represses skin and lung cancer development [68]. Unlike other signaling cascades, both Notch ligands and receptors are transmembrane proteins, thereby mainly stimulating signal transduction between neighboring cells. The Notch ligands, Delta and Jagged, bind to and promote two cleavages of Notch receptors by an ADAM-family metalloprotease and the γ-secretase complex, resulting in the release of the Notch intracellular domain (NICD), which is the active effector of the signaling pathway. NICD further translocates to the nucleus, associates with the DNA-binding protein CSL (CBF1/Su(H)/Lag-1), displaces co-repressors (Co-R), and recruits co-activators (Co-A) to stimulate downstream gene transcription [68–70].

As a master tumor suppressor, miR-34 has been reported to target multiple Notch signaling components, such as Notch-1, Notch-2, and Delta-like 1, thereby dampening Notch signaling activity and impairing maintenance of cancer stem cells in different cancer types including glioblastoma, medulloblastoma, gastric cancer, colon cancer, and pancreatic cancer [71–75]. This suggests a broad role for miR-34 in modulation of cancer stem cell characteristics via suppression of Notch signaling. In contrast, as an antagonist of Notch signaling, Numb has been shown to interact with Notch and promote its degradation [70]. Through directly targeting Numb to activate Notch signaling, miR-146a accelerates initiation and progression of melanoma [76].

11.8 The Hedgehog Signaling Pathway

Hedgehog signaling plays pivotal roles in controlling cell growth and pattern formation during embryonic development. Mutations or altered activation of this signaling pathway also contribute to tumorigenesis in various cancer types, such as basal cell carcinoma and medulloblastoma. One crucial event in this signal transduction is altering the balance between the active (Gli^A) and repressive (Gli^R) forms of the transcription factor Gli. In the absence of the ligand Hedgehog (Hh), the transmembrane Hh receptor Patched (Ptch1) inhibits the GPCR-like protein Smoothened (Smo) from translocating to the cell membrane to become active. In this context, the cytoplasmic Glis are phosphorylated by proteins such as CK1α and GSK3β, and proteolytically processed to generate Gli^R. This process suppresses expression of Hh target genes. In the meantime, suppressor of fused (Sufu) impedes Gli^A function through inhibition of its nuclear translocation and transcriptional activation. Upon

Hh ligand stimulation, Ptch1 is internalized, thereby failing to repress the activity of Smo. In turn, Smo blocks Gli^{R} production and promotes Gli^{A} generation, which stimulates expression of Hh target genes [77–79].

A number of microRNAs have been reported to suppress Hedgehog pathway activity. For example, Ferretti et al. have identified that miR-125b, miR-326, and miR-324-5p are downregulated in medulloblastoma tissue with high Hh signaling. Further characterization has revealed that mir-125b and miR-326 both target Smoothened, whereas miR-324-5p targets Gli1. Downregulation of these miRNAs leads to high expression of Hh target genes that accelerate tumor cell proliferation [80]. Conversely, by suppressing the inhibitor of Hedgehog signal, Sufu, miR-378 enhances glioma cell survival, tumor proliferation and angiogenesis [81]. Similarly, miR-212 displays tumor-promoting properties via targeting Ptch1 in non-small cell lung cancer [82].

11.9 Other Signaling Pathways

Besides the aforementioned signaling circuits which are well-studied in cancer, other pathways also make tremendous contributions during tumorigenesis. One of these is the Hippo pathway, which regulates organ size in multiple species. Deregulation of this pathway is implicated in a broad range of human tumors, including lung, colorectal, ovarian, and liver cancers. Central to this pathway is a core kinase cassette that consists of mammalian STE20-like protein kinase 1 (MST1) and MST2, large tumor suppressor 1 (LATS1) and LATS2, as well as the adaptor proteins Salvador homologue 1 (SAV1), MOB kinase activator 1A (MOB1A), and MOB1B. LATS1 and LATS2 phosphorylate the oncoproteins Yes-associated protein (YAP) and transcriptional co-activator with PDZ-binding motif (TAZ), repressing their activity by sequestering them in the cytoplasm and stimulating protein degradation. However, without phosphorylation, YAP and TAZ associate with different transcription factors, such as TEADs and SMADs, to stimulate gene expression and promote cell proliferation and survival [83, 84]. MicroRNAs positively and negatively control Hippo activity as well, leading to suppression or promotion of oncogenesis. For instance, miR-135b is highly expressed in non-small cell lung cancer and represses multiple targets in the Hippo pathway, including LATS2, F-box/WD repeat-containing protein 1A (β-TrCP), and nuclear Dbf2-related kinase 2 (NDR2), resulting in accumulation of YAP/TAZ in the nucleus and promotion of lung cancer metastasis [85]. Meanwhile, miR-375 targets YAP directly and inhibits HCC cell proliferation and invasion [86]. Interestingly, nuclear YAP is also involved in suppressing miRNA biogenesis directly through sequestering p72 from the Drosha microprocessor, which is partially responsible for the enhancement of cancer progression [87].

Another potent cascade is the JAK-STAT signaling pathway, which mainly mediates cytokine-dependent inflammation and immunity. Tumorigenesis and inflammation occur concurrently. This inflammatory microenvironment can be modulated to promote or hinder tumor progression. Deregulation of JAK-STAT signaling alters the inflammatory microenvironment to promote evasion of immune surveillance and enhance tumor-promoting immune responses. Mechanistically, different cytokines bind with their corresponding cytokine receptors, activating Janus kinase (JAK) family kinases, which are the receptor-associated tyrosine kinases. Subsequently, signal transducer and activator of transcription (STAT) proteins are phosphorylated by JAK, form homodimers or heterodimers, and translocate to the nucleus to stimulate downstream gene expression, of which many are cytokines and growth factors [88]. As an oncomir in breast cancer, expression of miR-155 is upregulated and negatively correlated with expression of suppressor of cytokine signaling 1 (SOCS1), an antagonist of the JAK-STAT pathway which inhibits STAT phosphorylation by JAK. By targeting SOCS1, miR-155 enhances phosphorylation of STAT3 and boosts breast tumor formation. Notably, expression of miR-155 is induced by multiple cytokines including IL6 and IFN-γ, indicating that it may act as a bridge between inflammation and cancer [89]. Meanwhile, tumor-suppressive miRNAs, such as miR-135a, inhibit gastric cancer proliferation by targeting JAK2 and attenuating p-STAT3 activation [90].

11.10 MiRNAs in the Crosstalk Between Signaling Pathways

MicroRNAs possess the potential to target multiple genes simultaneously. We have previously mentioned several microRNAs, such as miR-34, miR-21, miR-524-5p, and miR-205, which can suppress components of the same signaling pathway, so as to potently regulate that pathway's activity during tumorigenesis. Not surprisingly, microRNAs may also exploit their regulatory capacity to modulate components of several signaling pathways, resulting in alterations in multiple key pathways integral to cancer progression. For example, the Wnt and Ras/ERK signaling pathways play prominent roles in bladder cancer progression. Through targeting β-catenin and MEK1, which belong to the Wnt and Ras pathways respectively, miR-1826 can suppress the activity of these two pathways to inhibit bladder cancer cell proliferation [91]. Similarly, during prostate cancer development, TGF-β signaling suppresses while Wnt signaling promotes tumorigenesis. MiR-183 exerts its tumor-promoting function by targeting SMAD4 and DKK3, thereby dampening TGF-β signaling but promoting Wnt signaling [92].

Besides modulating the activities of different pathways independently, microRNA also helps to connect distinct signaling circuits. Plenty of evidences have demonstrated that microRNAs can act as mediators of signaling crosstalk. For instance, miR-216a has been shown to link TGF-β signaling with the PI3K/AKT pathway. Wu et al. utilized a diethylnitrosamine (DEN)-induced rat hepatocarcinogenesis model and demonstrated that TGF-β treatment of liver progenitor cells boost expression of miR-216a, which in turn target PTEN to enhance AKT signaling. The resultant enhanced AKT activation ultimately induced enhanced self-renewal, chemoresistance, and tumorigenecity of the progenitor cells [93].

11.11 Conclusion

In summary, in this chapter, we have reviewed the roles of microRNAs and other noncoding RNAs in regulating signal transduction during cancer progression. Noncoding RNAs can function as either suppressors or promoters of signaling cascades. First, microRNAs may suppress signal transduction through targeting effectors or positive regulators of signaling pathway. For example, let-7 inhibits Ras to weaken Ras/ERK pathway activity (see Fig. 11.1a). In contrast, by repressing negative modulators, miRNA may boost signaling activity. For instance, miR-21 targets PTEN to stimulate the PI3K/AKT pathway (see Fig. 11.1b). Although the effect of a miRNA on a single target's gene expression is subtle, because of their unique regulatory features, miRNAs may amplify/dampen a specific signal potently through targeting multiple components in one pathway simultaneously (see Fig. 11.1a). Similarly, miRNAs may maximize their regulatory capacity by targeting components from different pathways, leading to modulation of activities of those pathways cooperatively to impact tumorigenic processes (see Fig. 11.1d). Besides actively controlling signal transduction, expression of microRNAs/noncoding RNAs can be influenced by the pathway they are modulating, resulting in feedback loops between microRNAs and their target signaling pathways (see Fig. 11.1c). These feedback loops can be exploited to convert a transient signal into a long-lasting cellular response, especially beneficial during transitions between cellular states, such as

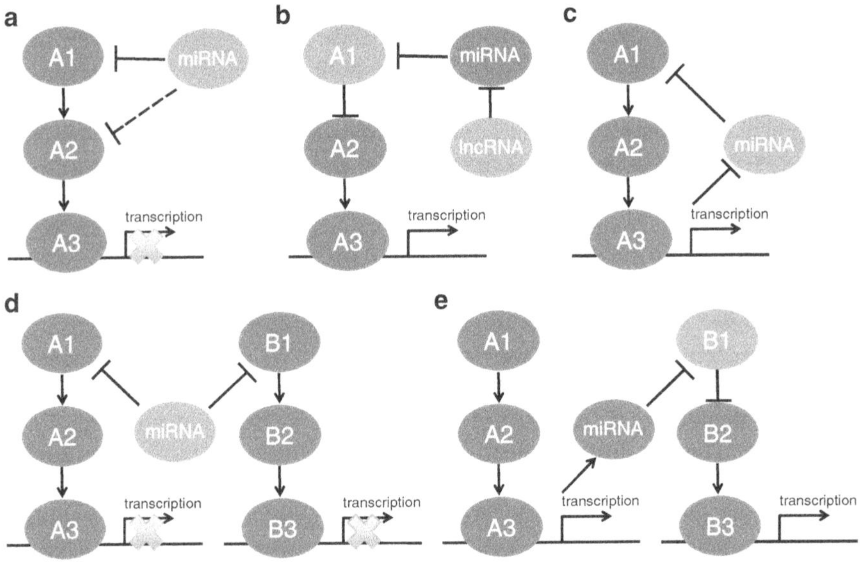

Fig. 11.1 Noncoding RNAs can function as either suppressors or promoters of signaling cascades. (**a**) miRNA acts as suppressor of a signaling pathway; (**b**) miRNA acts as promoter of a signaling pathway; (**c**) feedback loop between miRNA and its targeting signaling pathway; (**d**) miRNA targeting multiple pathways simultaneously; (**e**) miRNA mediating signal crosstalk between two pathways

EMT. Once induction of one state is initiated by the signaling cue, that signal transduction can be self-sustained and reinforced to complete the transition. Finally, signaling pathways are highly interconnected and microRNAs/noncoding RNAs contribute to connecting multiple signaling circuits. A microRNA that is downstream of one signaling cascade may modulate activity of another pathway, mediating the crosstalk between these two pathways (see Fig. 11.1e). Therefore, the regulation of microRNAs/noncoding RNAs on signal transduction is very diverse, having multiple context-dependent methods and outputs. Utilizing these methodologies, microRNAs make significant contributions to tumor formation and progression by aiding to tightly control signal transduction in cancer cells.

References

1. Hanahan D, Weinberg RA. Hallmarks of cancer: the next generation. Cell. 2011;144(5):646–74. doi:10.1016/j.cell.2011.02.013.
2. Gschwind A, Fischer OM, Ullrich A. Timeline-the discovery of receptor tyrosine kinases: targets for cancer therapy. Nat Rev Cancer. 2004;4(5):361–70. doi:10.1038/nrc1360.
3. Normanno N, De Luca A, Bianco C, et al. Epidermal growth factor receptor (EGFR) signaling in cancer. Gene. 2006;366(1):2–16. doi:10.1016/j.gene.2005.10.018.
4. Webster RJ, Giles KM, Price KJ, et al. Regulation of epidermal growth factor receptor signaling in human cancer cells by microRNA-7. J Biol Chem. 2009;284(9):5731–41. doi:10.1074/jbc.M804280200.
5. Kefas B, Godlewski J, Comeau L, et al. MicroRNA-7 inhibits the epidermal growth factor receptor and the Akt pathway and is down-regulated in glioblastoma. Cancer Res. 2008;68(10):3566–72. doi:10.1158/0008-5472.Can-07-6639.
6. Garofalo M, Jeon YJ, Nuovo GJ, et al. MiR-34a/c-dependent PDGFR-alpha/beta downregulation inhibits tumorigenesis and enhances TRAIL-induced apoptosis in lung cancer. PLoS ONE. 2013;8(6):e67581. doi:10.1371/journal.pone.0067581.
7. Migliore C, Petrelli A, Ghiso E, et al. MicroRNAs impair MET-mediated invasive growth. Cancer Res. 2008;68(24):10128–36. doi:10.1158/0008-5472.CAN-08-2148.
8. Peng Y, Guo JJ, Liu YM. Wu XL MicroRNA-34A inhibits the growth, invasion and metastasis of gastric cancer by targeting PDGFR and MET expression. Biosci Rep. 2014;34(3):247–56. doi:10.1042/BSR20140020.
9. Su J, Liang H, Yao W, et al. MiR-143 and MiR-145 regulate IGF1R to suppress cell proliferation in colorectal cancer. PLoS ONE. 2014;9(12):e114420. doi:10.1371/journal.pone.0114420.
10. Yan X, Chen X, Liang H, et al. MiR-143 and miR-145 synergistically regulate ERBB3 to suppress cell proliferation and invasion in breast cancer. Mol Cancer. 2014;13:220. doi:10.1186/1476-4598-13-220.
11. Dhillon AS, Hagan S, Rath O, Kolch W. MAP kinase signalling pathways in cancer. Oncogene. 2007;26(22):3279–90. doi:10.1038/sj.onc.1210421.
12. Downward J. Targeting RAS signalling pathways in cancer therapy. Nat Rev Cancer. 2003;3(1):11–22. doi:10.1038/nrc969.
13. Sayed D, Rane S, Lypowy J, et al. MicroRNA-21 targets Sprouty2 and promotes cellular outgrowths. Mol Biol Cell. 2008;19(8):3272–82. doi:10.1091/mbc.E08-02-0159.
14. Hanafusa H, Torii S, Yasunaga T, Nishida E. Sprouty1 and Sprouty2 provide a control mechanism for the Ras/MAPK signalling pathway. Nat Cell Biol. 2002;4(11):850–8. doi:10.1038/ncb867.
15. Liu M, Wu H, Liu T, et al. Regulation of the cell cycle gene, BTG2, by miR-21 in human laryngeal carcinoma. Cell Res. 2009;19(7):828–37. doi:10.1038/cr.2009.72.

16. Buganim Y, Solomon H, Rais Y, et al. p53 regulates the Ras circuit to inhibit the expression of a cancer-related gene signature by various molecular pathways. Cancer Res. 2010;70(6):2274–84. doi:10.1158/0008-5472.Can-09-2661.
17. Talotta F, Cimmino A, Matarazzo MR, et al. An autoregulatory loop mediated by miR-21 and PDCD4 controls the AP-1 activity in RAS transformation. Oncogene. 2009;28(1):73–84. doi:10.1038/onc.2008.370.
18. Lu Z, Liu M, Stribinskis V, et al. MicroRNA-21 promotes cell transformation by targeting the programmed cell death 4 gene. Oncogene. 2008;27(31):4373–9. doi:10.1038/onc.2008.72.
19. Hatley ME, Patrick DM, Garcia MR, et al. Modulation of K-Ras-dependent lung tumorigenesis by microRNA-21. Cancer Cell. 2010;18(3):282–93. doi:10.1016/j.ccr.2010.08.013.
20. Lenarduzzi M, Hui ABY, Alajez NM et al. MicroRNA-193b enhances tumor progression via down regulation of neurofibromin 1. Plos ONE. 2013;8(1). doi:ARTN e53765.
21. Garcia-Orti L, Cristobal I, Cirauqui C et al. Integration of SNP and mRNA arrays with microRNA profiling reveals that miR-370 is up-regulated and targets NF1 in acute myeloid leukemia. Plos ONE. 2012;7(10). doi:ARTN e47717.
22. Lv ZH, Yang LZ. MiR-124 inhibits the growth of glioblastoma through the down-regulation of SOS1. Mol Med Rep. 2013;8(2):345–9. doi:10.3892/mmr.2013.1561.
23. Singh A, Greninger P, Rhodes D, et al. A gene expression signature associated with "K-Ras Addiction" reveals regulators of EMT and tumor cell survival. Cancer Cell. 2009;15(6):489–500. doi:10.1016/j.ccr.2009.03.022.
24. Wagner PL, Stiedl AC, Wilbertz T, et al. Frequency and clinicopathologic correlates of KRAS amplification in non-small cell lung carcinoma. Lung Cancer. 2011;74(1):118–23. doi:10.1016/j.lungcan.2011.01.029.
25. Mita H, Toyota M, Aoki F et al. A novel method, digital genome scanning detects KRAS gene amplification in gastric cancers: involvement of overexpressed wild-type KRAS in downstream signaling and cancer cell growth. BMC Cancer. 2009;9. doi:Artn 198.
26. Johnson SM, Grosshans H, Shingara J, et al. RAS is regulated by the let-7 MicroRNA family. Cell. 2005;120(5):635–47. doi:10.1016/j.cell.2005.01.014.
27. Akao Y, Nakagawa Y, Naoe T. Let-7 microRNA functions as a potential growth suppressor in human colon cancer cells. Biol Pharm Bull. 2006;29(5):903–6. doi:10.1248/bpb.29.903.
28. Yu F, Yao H, Zhu PC, et al. Let-7 regulates self renewal and tumorigenicity of breast cancer cells. Cell. 2007;131(6):1109–23. doi:10.1016/j.cell.2007.10.054.
29. Kumar MS, Erkeland SJ, Pester RE, et al. Suppression of non-small cell lung tumor development by the let-7 microRNA family. Proc Natl Acad Sci U S A. 2008;105(10):3903–8. doi:10.1073/pnas.0712321105.
30. Chin LJ, Ratner E, Leng SG, et al. A SNP in a let-7 microRNA complementary site in the KRAS 3' untranslated region increases non-small cell lung cancer risk. Cancer Res. 2008;68(20):8535–40. doi:10.1158/0008-5472.Can-08-2129.
31. Poliseno L, Salmena L, Zhang JW, et al. A coding-independent function of gene and pseudogene mRNAs regulates tumour biology. Nature. 2010;465(7301):1033–8. doi:10.1038/nature09144.
32. Liu SM, Lu J, Lee HC, et al. MiR-524-5p suppresses the growth of oncogenic BRAF melanoma by targeting BRAF and ERK2. Oncotarget. 2014;5(19):9444–59.
33. Liu Z, Jiang Z, Huang J, et al. MiR-7 inhibits glioblastoma growth by simultaneously interfering with the PI3K/ATK and Raf/MEK/ERK pathways. Int J Oncol. 2014;44(5):1571–80. doi:10.3892/ijo.2014.2322.
34. Glover AR, Zhao JT, Gill AJ et al. MicroRNA-7 as a tumor suppressor and novel therapeutic for adrenocortical carcinoma. Oncotarget. 2015; doi:10.18632/oncotarget.5383.
35. Vivanco I, Sawyers CL. The phosphatidylinositol 3-Kinase AKT pathway in human cancer. Nat Rev Cancer. 2002;2(7):489–501. doi:10.1038/nrc839.
36. Shaw RJ, Cantley LC. Ras, PI(3)K and mTOR signalling controls tumour cell growth. Nature. 2006;441(7092):424–30. doi:10.1038/nature04869.

37. Guo CG, Sah JF, Beard L, et al. The non-coding RNA, miR-126, suppresses the growth of neoplastic cells by targeting phosphatidylinositol 3-kinase signaling and is frequently lost in colon cancers. Gene Chromosomes Cancer. 2008;47(11):939–46. doi:10.1002/gcc.20596.
38. Fang Y, Xue JL, Shen Q, et al. MicroRNA-7 inhibits tumor growth and metastasis by targeting the phosphoinositide 3-kinase/Akt pathway in hepatocellular carcinoma. Hepatology. 2012;55(6):1852–62. doi:10.1002/hep.25576.
39. Yuan TL, Cantley LC. PI3K pathway alterations in cancer: variations on a theme. Oncogene. 2008;27(41):5497–510. doi:10.1038/onc.2008.245.
40. Alimonti A, Carracedo A, Clohessy JG, et al. Subtle variations in Pten dose determine cancer susceptibility. Nat Genet. 2010;42(5):454–8. doi:10.1038/ng.556.
41. Meng FY, Henson R, Wehbe-Janek H, et al. MicroRNA-21 regulates expression of the PTEN tumor suppressor gene in human hepatocellular cancer. Gastroenterology. 2007;133(2):647–58. doi:10.1053/j.gastro.2007.05.022.
42. Yang H, Kong W, He L, et al. MicroRNA expression profiling in human ovarian cancer: miR-214 induces cell survival and cisplatin resistance by targeting PTEN. Cancer Res. 2008;68(2):425–33. doi:10.1158/0008-5472.Can-07-2488.
43. Garofalo M, Di Leva G, Romano G, et al. miR-221&222 regulate TRAIL resistance and enhance tumorigenicity through PTEN and TIMP3 down-regulation. Cancer Cell. 2009;16(6):498–509. doi:10.1016/j.ccr.2009.10.014.
44. Tay Y, Kats L, Salmena L, et al. Coding-independent regulation of the tumor suppressor PTEN by competing endogenous mRNAs. Cell. 2011;147(2):344–57. doi:10.1016/j.cell.2011.09.029.
45. Karreth FA, Tay Y, Pema D et al. In vivo identification of tumor-suppressive PTEN ceRNAs in an oncogenic BRAF-induced mouse model of melanoma. Cell. 2011;147(4):948–948. doi:10.1016/j.cell.2011.10.032.
46. Johnsson P, Ackley A, Vidarsdottir L, et al. A pseudogene long-non-coding-RNA network regulates PTEN transcription and translation in human cells. Nat Struct Mol Biol. 2013;20(4):440–6. doi:10.1038/nsmb.2516.
47. Brognard J, Sierecki E, Gao TY, Newton AC. PHLPP and a second isoform, PHLPP2, differentially attenuate the amplitude of Akt signaling by regulating distinct Akt isoforms. Mol Cell. 2007;25(6):917–31. doi:10.1016/j.molcel.2007.02.017.
48. Tsukamoto Y, Nakada C, Noguchi T, et al. MicroRNA-375 is down-regulated in gastric carcinomas and regulates cell survival by targeting PDK1 and 14-3-3 zeta. Cancer Res. 2010;70(6):2339–49. doi:10.1158/0008-5472.Can-09-2777.
49. Foley NH, Bray IM, Tivnan A et al. MicroRNA-184 inhibits neuroblastoma cell survival through targeting the serine/threonine kinase AKT2. Mol Cancer. 2010;9. doi:Artn 83.
50. Uesugi A, Kozaki K, Tsuruta T, et al. The tumor suppressive microRNA miR-218 targets the mTOR component Rictor and inhibits AKT phosphorylation in oral cancer. Cancer Res. 2011;71(17):5765–78. doi:10.1158/0008-5472.Can-11-0368.
51. Cai JC, Fang LS, Huang YB, et al. MiR-205 targets PTEN and PHLPP2 to augment AKT signaling and drive malignant phenotypes in non-small cell lung cancer. Cancer Res. 2013;73(17):5402–15. doi:10.1158/0008-5472.Can-13-0297.
52. Massague J. TGFbeta in cancer. Cell. 2008;134(2):215–30. doi:10.1016/j.cell.2008.07.001.
53. Ikushima H, Miyazono K. TGFbeta signalling: a complex web in cancer progression. Nat Rev Cancer. 2010;10(6):415–24. doi:10.1038/nrc2853.
54. Mestdagh P, Bostrom AK, Impens F, et al. The miR-17-92 microRNA cluster regulates multiple components of the TGF-beta pathway in neuroblastoma. Mol Cell. 2010;40(5):762–73. doi:10.1016/j.molcel.2010.11.038.
55. Smith AL, Iwanaga R, Drasin DJ, et al. The miR-106b-25 cluster targets Smad7, activates TGF-beta signaling, and induces EMT and tumor initiating cell characteristics downstream of Six1 in human breast cancer. Oncogene. 2012;31(50):5162–71. doi:10.1038/onc.2012.11.
56. Petrocca F, Visone R, Onelli MR, et al. E2F1-regulated microRNAs impair TGFbeta-dependent cell-cycle arrest and apoptosis in gastric cancer. Cancer Cell. 2008;13(3):272–86. doi:10.1016/j.ccr.2008.02.013.

57. Gregory PA, Bert AG, Paterson EL, et al. The miR-200 family and miR-205 regulate epithelial to mesenchymal transition by targeting ZEB1 and SIP1. Nat Cell Biol. 2008;10(5):593–601. doi:10.1038/ncb1722.
58. Bracken CP, Gregory PA, Kolesnikoff N, et al. A double-negative feedback loop between ZEB1-SIP1 and the microRNA-200 family regulates epithelial-mesenchymal transition. Cancer Res. 2008;68(19):7846–54. doi:10.1158/0008-5472.Can-08-1942.
59. Yuan JH, Yang F, Wang F, et al. A long non-coding RNA activated by TGF-beta promotes the invasion-metastasis cascade in hepatocellular carcinoma. Cancer Cell. 2014;25(5):666–81. doi:10.1016/j.ccr.2014.03.010.
60. Davis BN, Hilyard AC, Lagna G, Hata A. SMAD proteins control DROSHA-mediated microRNA maturation. Nature. 2008;454(7200):56–61. doi:10.1038/nature07086.
61. Klaus A, Birchmeier W. Wnt signalling and its impact on development and cancer. Nat Rev Cancer. 2008;8(5):387–98. doi:10.1038/nrc2389.
62. Clevers H, Nusse R. Wnt/beta-catenin signaling and disease. Cell. 2012;149(6):1192–205. doi:10.1016/j.cell.2012.05.012.
63. Kim NH, Kim HS, Kim NG et al. p53 and microRNA-34 are suppressors of canonical Wnt signaling. Sci Signal. 2011;4(197). doi:ARTN ra71.
64. Kim NH, Cha YH, Kang SE, et al. P53 regulates nuclear GSK-3 levels through miR-34-mediated Axin2 suppression in colorectal cancer cells. Cell Cycle. 2013;12(10):1578–87. doi:10.4161/cc.24739.
65. Zhang Y, Wei W, Cheng N, et al. Hepatitis C virus-induced up-regulation of microRNA-155 promotes hepatocarcinogenesis by activating Wnt signaling. Hepatology. 2012;56(5):1631–40. doi:10.1002/hep.25849.
66. Shen G, Jia H, Tai Q, et al. MiR-106b down-regulates adenomatous polyposis coli and promotes cell proliferation in human hepatocellular carcinoma. Carcinogenesis. 2013;34(1):211–9. doi:10.1093/carcin/bgs320.
67. Li Q, Shen K, Zhao Y, et al. MicroRNA-222 promotes tumorigenesis via targeting DKK2 and activating the Wnt/beta-catenin signaling pathway. FEBS Lett. 2013;587(12):1742–8. doi:10.1016/j.febslet.2013.04.002.
68. Radtke F, Raj K. The role of Notch in tumorigenesis: oncogene or tumour suppressor? Nat Rev Cancer. 2003;3(10):756–67. doi:10.1038/nrc1186.
69. Bray SJ. Notch signalling: a simple pathway becomes complex. Nat Rev Mol Cell Biol. 2006;7(9):678–89. doi:10.1038/nrm2009.
70. Kopan R, Ilagan MX. The canonical Notch signaling pathway: unfolding the activation mechanism. Cell. 2009;137(2):216–33. doi:10.1016/j.cell.2009.03.045.
71. Ji Q, Hao X, Meng Y, et al. Restoration of tumor suppressor miR-34 inhibits human p53-mutant gastric cancer tumorspheres. BMC Cancer. 2008;8:266. doi:10.1186/1471-2407-8-266.
72. Li YQ, Guessous F, Zhang Y, et al. MicroRNA-34a inhibits glioblastoma growth by targeting multiple oncogenes. Cancer Res. 2009;69(19):7569–76. doi:10.1158/0008-5472.Can-09-0529.
73. Ji Q, Hao X, Zhang M, et al. MicroRNA miR-34 inhibits human pancreatic cancer tumor-initiating cells. PLoS ONE. 2009;4(8):e6816. doi:10.1371/journal.pone.0006816.
74. de Antonellis P, Medaglia C, Cusanelli E, et al. MiR-34a targeting of Notch ligand delta-like 1 impairs CD15+/CD133+ tumor-propagating cells and supports neural differentiation in medulloblastoma. PLoS ONE. 2011;6(9):e24584. doi:10.1371/journal.pone.0024584.
75. Bu PC, Chen KY, Chen JH, et al. A microRNA miR-34a-regulated bimodal switch targets Notch in colon cancer stem cells. Cell Stem Cell. 2013;12(5):602–15. doi:10.1016/j.stem.2013.03.002.
76. Forloni M, Dogra SK, Dong Y, et al. MiR-146a promotes the initiation and progression of melanoma by activating Notch signaling. Elife. 2014;3:e01460. doi:10.7554/eLife.01460.
77. Rubin LL, de Sauvage FJ. Targeting the Hedgehog pathway in cancer. Nat Rev Drug Discov. 2006;5(12):1026–33. doi:10.1038/nrd2086.

78. Jiang J, Hui CC. Hedgehog signaling in development and cancer. Dev Cell. 2008;15(6):801–12. doi:10.1016/j.devcel.2008.11.010.
79. Merchant AA, Matsui W. Targeting Hedgehog-a cancer stem cell pathway. Clin Cancer Res. 2010;16(12):3130–40. doi:10.1158/1078-0432.CCR-09-2846.
80. Ferretti E, De Smaele E, Miele E, et al. Concerted microRNA control of Hedgehog signalling in cerebellar neuronal progenitor and tumour cells. EMBO J. 2008;27(19):2616–27. doi:10.1038/emboj.2008.172.
81. Lee DY, Deng ZQ, Wang CH, Yang BB. MicroRNA-378 promotes cell survival, tumor growth, and angiogenesis by targeting SuFu and Fus-1 expression. Proc Natl Acad Sci U S A. 2007;104(51):20350–5. doi:10.1073/pnas.0706901104.
82. Li Y, Zhang DQ, Chen CW, et al. MicroRNA-212 displays tumor-promoting properties in non-small cell lung cancer cells and targets the hedgehog pathway receptor PTCH1. Mol Biol Cell. 2012;23(8):1423–34. doi:10.1091/mbc.E11-09-0777.
83. Pan DJ. The Hippo signaling pathway in development and cancer. Dev Cell. 2010;19(4):491–505. doi:10.1016/j.devcel.2010.09.011.
84. Harvey KF, Zhang XM, Thomas DM. The Hippo pathway and human cancer. Nat Rev Cancer. 2013;13(4):246–57. doi:10.1038/nrc3458.
85. Lin CW, Chang YL, Chang YC et al. MicroRNA-135b promotes lung cancer metastasis by regulating multiple targets in the Hippo pathway and LZTS1. Nat Commun 2013;4. doi:ARTN 1877.
86. Liu AM, Poon RTP, Luk JM. MicroRNA-375 targets Hippo-signaling effector YAP in liver cancer and inhibits tumor properties. Biochem Bioph Res Commun. 2010;394(3):623–7. doi:10.1016/j.bbrc.2010.03.036.
87. Mori M, Triboulet R, Mohseni M, et al. Hippo signaling regulates microprocessor and links cell-density-dependent miRNA biogenesis to cancer. Cell. 2014;156(5):893–906. doi:10.1016/j.cell.2013.12.043.
88. Yu H, Pardoll D, Jove R. STATs in cancer inflammation and immunity: a leading role for STAT3. Nat Rev Cancer. 2009;9(11):798–809. doi:10.1038/nrc2734.
89. Jiang SA, Zhang HW, Lu MH, et al. MicroRNA-155 functions as an OncomiR in breast cancer by targeting the suppressor of cytokine signaling 1 gene. Cancer Res. 2010;70(8):3119–27. doi:10.1158/0008-5472.Can-09-4250.
90. Wu H, Huang M, Cao P, et al. MiR-135a targets JAK2 and inhibits gastric cancer cell proliferation. Cancer Biol Ther. 2012;13(5):281–8. doi:10.4161/cbt.13.5.18943.
91. Hirata H, Hinoda Y, Ueno K, et al. MicroRNA-1826 targets VEGFC, beta-catenin (CTNNB1) and MEK1 (MAP2K1) in human bladder cancer. Carcinogenesis. 2012;33(1):41–8. doi:10.1093/carcin/bgr239.
92. Ueno K, Hirata H, Shahryari V, et al. microRNA-183 is an oncogene targeting Dkk-3 and SMAD4 in prostate cancer. Brit J Cancer. 2013;108(8):1659–67. doi:10.1038/bjc.2013.125.
93. Wu K, Ding J, Chen C, et al. Hepatic transforming growth factor beta gives rise to tumor-initiating cells and promotes liver cancer development. Hepatology. 2012;56(6):2255–67. doi:10.1002/hep.26007.

Chapter 12
Noncoding RNAs Regulating NF-κB Signaling

Mengfeng Li and Hongyu Guan

Abstract As transcription factors that regulate expression of a variety of genes essential for diverse physiological and pathological processes, nuclear factor kappa B (NF-κB) family molecules play important roles in the development and progression of malignant tumor, and constitutive activation of NF-κB has been evidenced in various types of tumor tissues. Underlying its pathologic role, deregulated expression and/or transactivating activity of NF-κB usually involves multiple layers of molecular mechanisms. Noncoding RNAs, including microRNAs (miRNAs) and long noncoding RNAs (lncRNAs), are known to modulate expression and biological functions of regulatory proteins in a variety of cancer contexts. In this chapter, the regulatory role of miRNAs and lncRNAs in NF-κB signaling in malignant diseases will be discussed.

Keywords NF-κB • miRNAs • lncRNAs • Noncoding RNA • Tumorigenesis • Therapeutics

12.1 Introduction

The NF-κB transcription factors consist of five Rel family members, i.e., NF-κB1/p105, NF-κB2/p100, RelA/p65, RelB, and c-Rel [1], all of which contain a conserved Rel homology domain for dimerization, DNA binding, and nuclear translocation [2]. All Rel members are capable of binding DNA, but only RelA/p65, RelB, and c-Rel carry transactivating domains that are responsible for interacting with basal transcription factors and cofactors [3]. NF-κB members can form homodimeric and heterodimeric complexes that directly regulate genes involved in a wide variety of biological processes, such as the immune response, inflammation, cell growth, cell adhesion, and differentiation [4]. As deregulation of these biological processes is usually associated with malignant transformation of cells and

M. Li (✉) • H. Guan
Zhongshan School of Medicine, Sun Yat-sen University,
74 Zhongshan Road II, Guangzhou, China
e-mail: limf@mail.sysu.edu.cn

E. Song (ed.), *The Long and Short Non-coding RNAs in Cancer Biology*,
Advances in Experimental Medicine and Biology 927,
DOI 10.1007/978-981-10-1498-7_12

dissemination of tumors, NF-κB represents a family of molecules that play important roles in multiple aspects of tumorigenesis and cancer progression, and constitutive activation of NF-κB indeed has been found in many cancer types [5]. Understanding how the expression and the biological functions of NF-κB family members are deregulated in cancer has been a topic of high interest, which is expected to facilitate the development of new, effective anticancer strategies.

Noncoding RNAs, including microRNAs (miRNAs) and long noncoding RNAs (lncRNAs), play fundamental roles in modulating gene expression and protein functions. miRNAs represent a class of endogenous noncoding small RNAs capable of modifying the expression level of genes by base pairing with the 3'-untranslated region of target mRNAs [6]. Numerous miRNAs have been found to be involved in the initiation, development, and progression of cancer, and interestingly, a significant number of these miRNAs are regulator of NF-κB expression or signaling. Meanwhile, lncRNAs are endogenous cellular RNAs longer than 200 bases, as defined by widely accepted criteria [7]. Unlike miRNAs, lncRNAs can regulate gene expression at multiple levels through various mechanisms distinct among different lncRNAs [8]. Studies have suggested a role of lncRNAs in regulating NF-κB signaling in tumorigenesis.

In this chapter, the significance of noncoding miRNAs and lncRNAs in modulating NF-κB activity, tumor development, and progression and potential implications of these noncoding RNAs as interventional targets in human cancer will be summarized.

12.2 Significance of NF-κB Signaling in Cancer Biology

NF-κB was identified by Sen and Baltimore in 1986 as a transcription factor in the nucleus of mature lymphocytes that binds the enhancer of immunoglobulin κ-light chain [9]. In mammals, NF-κB family consists of five Rel family members, i.e., NF-κB1/p105, NF-κB2/p100, RelA/p65, RelB, and c-Rel [1]. All these Rel family members contain a conserved region termed Rel homology domain, which is key to the dimerization, DNA binding, and nuclear translocation of Rel proteins [2]. While all Rel members are capable of binding DNA, only RelA/p65, RelB, and c-Rel contain transactivating domains in their carboxyl termini, which are responsible for interacting with basal transcription factors and cofactors [3]. It has been well demonstrated that NF-κB is ubiquitously expressed in the cytoplasm of various types of cells. When activated by specific stimuli, such as inflammatory signals, NF-κB factors translocate to the nucleus, where it transcriptionally regulates genes involved in immune response, inflammation, cell growth, adhesion, and differentiation. To become transcriptionally active, NF-κB members form homodimeric and heterodimeric complexes, among which p50/RelA (p50/p65) heterodimer is found to be the most abundant and active form in mammalian cell [4].

Given its significant role involved in cellular growth, survival, invasion, angiogenesis, and transformation, NF-κB has been implicated in the initiation, development, metastasis, and chemoresistance of human malignant diseases. Constitutive activation of NF-κB has been observed in a number of cancer types. Of particular note, the functional consequences of deregulated activation of NF-κB are closely correlated with nearly all hallmarks of cancer identified thus far [10].

12.2.1 NF-κB Signaling Links Inflammation and Cancer

Inflammation represents a fundamental defense mechanism in the innate immunity. However, uncontrolled chronic inflammation can lead to biological scenarios characteristic of cellular stress and unstability, such as over-recruitment of inflammatory factors and accumulation of DNA damage. Notably, such stressful conditions are usually underlying development of cancer [11]. Indeed, nearly 20 % of human cancers are associated with chronic inflammatory diseases [12]. For example, helicobacter pylori infection in gastric cancer, HBV or HCV infection in hepatocellular cancer, and inflammatory bowel disease in colorectal cancer are common causes of chronic inflammation related to carcinogenesis [13–15]. Thus, inflammation has been recognized as one of the contributory factors associated with the pathogenesis of human cancer. In this context, accumulating evidence has implicated that NF-κB activation, a featured master regulator of inflammatory responses, acts as a signaling link between inflammation and cancer at multiple levels. It is of note that NF-κB overactivation has been evidenced both in cells that become malignant and in stromal cells surrounding tumor cells and constituting intratumoral microenvironment. At tumor site with activation of NF-κB, the pro-inflammatory cytokines directly contribute to the establishment of microenvironment permissive for initiation and development of cancer [16]. Tumor stromal cells, which include inflammatory cells such as macrophages, neutrophils, lymphocytes, mast cells, and dendritic cells, produce inflammatory cytokines, angiogenic factors, growth factors, proteases, and other inflammatory mediators [17]. Inflammatory cytokines (TNF-α, IL-1, etc.) secreted by these microenvironmental cells have been found to act on premalignant cells prone to malignant transformation, which in turn induce expression of genes associated with cellular proliferation, cell survival, and angiogenesis [18, 19]. Proteases that degrades the extracellular matrix facilitates tumor cell migration, invasion, and dissemination, which are key to cancer metastasis [20]. Furthermore, in stromal cells, activation of NF-κB induces production of reactive oxygen species (ROS), which may increase DNA damage in the premalignant cells and in turn trigger NF-κB activation [21].

As described above, a major contribution of NF-κB activation to tumorigenesis can be associated with the establishment of inflammatory microenvironment. Indeed, constitutive NF-κB activity has been demonstrated in significant numbers of human cancers [22]. Suppression of NF-κB in tumor cell model might lead to tumor suppression, suggesting that the NF-κB signaling pathway could be a promising anticancer target [23].

12.2.2 NF-κB Signaling Promotes Cell Proliferation and Inhibits Cell Apoptosis

Cancer is essentially a disease which involves unlimited cell proliferation and aberrantly excessive cell survival [24]. Deregulated expression and/or overactivation of cell cycle-driven proteins contribute to this process. The essential role of NF-κB in cell proliferation is largely due to its effects on regulating expression of cyclins and proto-oncogene [25]. Studies in both human cancer cells and murine embryonic fibroblasts by Hinz et al. has demonstrated that NF-κB contributes to the G1-to-S phase transition and that cyclin D1 may be a target gene of NF-κB transactivation [26]. Lately a study carried out by Guttridge et al. has identified cyclin D1 as a direct transcriptional target of NF-κB [27]. Meanwhile, expression of antiapoptotic genes, such as FLIP, c-IAP1/2, XIAP, and members of the Bcl-2 family can be modulated by NF-κB activity [28, 29], suggesting that NF-κB leads cancer cells to escaping apoptosis. Collectively, evidence has shown that NF-κB contributes to tumorigenesis through promoting cell proliferation, as well as inhibiting apoptosis.

12.2.3 NF-κB Stimulates Tumor Cell Invasion and Metastasis

NF-κB signaling stimulates tumor cell invasion and metastasis at multiple levels. As an early event in the process of metastasis, epithelial-mesenchymal transition (EMT) has been found to be associated with NF-κB signaling. For example, Twist1 and Snail, two key regulators in EMT, are transactivated by NF-κB [30, 31]. Moreover, although the underlying mechanisms still remain unclear, an involvement of NF-κB in cancer cell extravasation and colonization in distal organs has been evidently shown [32]. Notably, Parthenolide, a natural inhibitor of NF-κB, suppresses lung colonization of a highly metastatic murine osteosarcoma cell line LM8 [33]. Furthermore, NF-κB regulates many adhesion proteins, including integrins and their receptors [34]. In addition, matrix metalloproteinases (MMPs), a family of zinc-dependent endopeptidases, contribute to many steps of cancer progression by proteolytic interaction with multiple substrates. Farina et al. have showed that MMP9 expression is transcriptionally regulated by NF-κB [35]. Intriguingly, elevated level of NF-κB in pre-metastatic niche facilitates creating favorable environment for tumor cell seeding [36].

12.2.4 *NF-κB Modulates Tumor Angiogenesis*

Angiogenesis is a hallmark of cancer and is required for tumor growth. Vascular endothelial growth factor (VEGF) is a key angiogenic factor in the process of tumor angiogenesis [37]. The expression of VEGF is regulated by hypoxia-inducible factor-1α (HIF-1α) under hypoxic conditions [38]. Intriguingly, NF-κB activation promotes expression of VEGF, thereby contributing to the process of angiogenesis [39]. Indeed, it has been demonstrated that inhibition of NF-κB results in suppression of VEGF and angiogenesis [40]. Moreover, other important pro-angiogenic factors, such as TNF, bFGF, MCP-1, MMPs, and IL-8, are upregulated by NF-κB. For example, oncogene Bmi-1 plays an important role in glioma angiogenesis, and further molecular mechanistic study has revealed that NF-κB/VEGF mediates the positive effects of Bmi-1 on tumor progression [41]. In vivo study of a thyroid cancer model has shown that NF-κB-dependent angiogenesis may also be mediated by IL-8 secretion [42]. MMPs discussed above also play functional roles in tumor-promoting angiogenesis and are regulated by NF-κB [43]. These studies underscore the importance of NF-κB activation in tumor angiogenesis.

12.2.5 *NF-κB Is Involved in Cancer Metabolism*

It has been suggested that NF-κB may be involved in regulation of cell metabolism in malignant diseases, which are also supported by several recent studies. Mauro et al. provided evidence for the role of NF-κB in metabolic adaption in cancer and normal cells. Their data indicate that NF-κB remodels energy metabolism networks by modulating the balance between the glycolysis utilization and mitochondrial respiration [44]. Study by Kawauchi et al. demonstrated that glycolysis activates IKK-NF-κB via a positive feedback loop and that p53 deficiency-induced hyperactivation of this loop plays an important role in oncogene Ras-induced cell transformation [45]. In the absence of p53, NF-κB family member RelA can transport into the mitochondria and suppress the expression of mitochondrial genes, consumption of oxygen, and levels of cellular ATP, suggesting that RelA regulates the function of mitochondria and thereby influences the metabolism and energy production in tumor cells [46].

12.3 The Molecular Process of NF-κB Signaling: Mechanism and Regulation

As a family of transcriptional factors associated with many physiological and pathological processes, the activities of NF-κB are tightly controlled by both classical (canonical) and alternative (noncanonical) regulatory pathways [47]. When regulated by the canonical NF-κB pathway, stimuli, such as viral/bacterial infections and pro-inflammatory cytokines, activate the IKK complex and subsequently the two catalytic subunits, IKKα (IKK1) and IKKβ (IKK2), which then form the IKK complex together with a regulatory subunit IKKγ (NEMO) [48]. This complex phosphorylates the N-terminal serines 32 and 36 of the inhibitor of NF-κB (IκB) [49], followed by ubiquitylation and proteasomal degradation of the phosphorylated IκB, thereby releasing the NF-κB dimers (mainly p65/p50 and p50/c-Rel) to enter the nucleus and activate the transcription of their target genes [50]. Thus, the ubiquitination and proteasome-mediated degradation of IκB plays crucial roles in regulating NF-κB activity. Meanwhile, the alternative NF-κB activation pathway relies on the upstream kinase NF-κB-inducing kinase (NIK) and its activation of the IKKα homodimers in an IKKβ- and IKKγ-independent manner, leading to phosphorylation of p100. Ubiquitin-mediated proteasomal degradation of the phosphorylated p100 generates p52 and forms the p52/RelB dimer that translocates to the nucleus and transactivates target gene expression [51].

As discussed above, ubiquitin modification is essential for activation of NF-κB signaling [52]. Dysregulated ubiquitin conjugation/deconjugation in NF-κB signaling was shown in various types of human cancers [53]. Ubiquitin conjugation acts as a positive regulator in NF-κB signaling and has been found to be involved in multiple steps along the activating cascade [54]. Furthermore, deubiquitinating enzymes, including CYLD, cellular zinc finger anti-NF-κB (Cezanne), and A20, are responsible for the ubiquitin conjugation and therefore function as negative NF-κB regulators [55, 56].

Importantly, the activity of NF-κB signaling is modulated at multiple levels. The expression quantities of key regulatory proteins in the canonical as well as alternative pathways are important parameters that contribute to determining the intensity of NF-κB activity. As the biological significance of noncoding RNAs is gradually recognized, their roles in modulating NF-κB signaling is becoming increasingly appreciated. In the following sections, implications of noncoding RNAs in the constitutive activation of NF-κB will be summarized.

12.4 Noncoding RNAs Involved in Regulation of NF-κB Signaling in Cancer

12.4.1 MicroRNAs (miRNAs) in NF-κB Signaling

MiRNAs, a class of endogenous noncoding small RNAs, are involved in modulation of many biological processes by base pairing, usually imperfectly, with the 3'-untranslated region of a target mRNA, resulting in posttranscriptional inhibition and/or mRNA cleavage [6]. More than 1000 human miRNAs have been identified thus far. As each miRNA is predicted to regulate a few hundreds of protein-coding genes, current estimates suggest that over 45 000 sites in human genome are targeted by miRNAs, resulting in altered expression of ~60 % of genes [57]. Numerous studies have demonstrated the important roles of miRNAs in development and progression of cancer, in which aberrant activation of NF-κB often occurs. Naturally, accumulating evidence has suggested that a number of miRNAs and their target genes in the NF-κB signaling cascades can be critical to tumorigenesis and the progression of cancer.

12.4.1.1 MiR-30e* Inhibits NF-κB Inhibitor IκB

Biological evidence has established that NF-κB activation is tightly controlled by its natural inhibitors, IκBs, which bind and sequester NF-κB in the cytoplasm until nuclear translocation is induced by signal-triggered IκB degradation. However, how cancer cells override the NF-κB-IκB negative feedback loop, through which NF-κB promotes expression of its own inhibitors, IκBs, that reversely acts to retain NF-κB in the cytoplasm as an inactive form, remains unclear. One of our previous studies found that miR-30e* directly targets IκBα 3'-UTR, suppresses IκBα expression, and causes deregulated hyperactivation of NF-κB and expression of NF-κB-regulated genes (e.g., MMPs and VEGF-C), promoting invasiveness of glioma cells and angiogenesis [58]. More importantly, such effects of miR-30e* on the aggressiveness of glioma are clinically relevant as upregulated miR-30e* is found in human glioma tissue and highly correlates with clinical progression of the disease and poor survival of patients. Hence, miR-30e* mediates a novel epigenetic mechanism that disrupts the NF-κB-IκBα loop. This finding provides new insights in developing therapeutic intervention against both tumor invasion and neovessel formation in gliomas by inhibiting the expression of miR-30e*.

12.4.1.2 MiR-138 Suppresses NF-κB Activity by Inhibiting Ubiquitination of TRAF2 and RIP1

Esophageal squamous cell carcinoma (ESCC) is one of the most aggressive malignancies of the gastrointestinal tract [59]. Constitutive abnormal activation of NF-κB signaling is involved in ESCC progression. To reveal the possible role of miRNAs in NF-κB activation and ESCC progression, Gong et al. [60] demonstrated that miR-138 is downregulated in esophageal squamous cell carcinoma and is inversely associated with malignant progression of the disease. They also found that inhibition of miR-138 enhances K63-linked poly-ubiquitination of TRAF2 and RIP1, intermediaries of NF-κB signaling, and thereby prolongs activation of NF-κB. These data indicate a novel mechanism for abnormal NF-κB activation in ESCCs and may provide a novel target for treatment of ESCCs.

12.4.1.3 MiR-486 Suppresses SYLD, ITCH, and TNIPs

It has been well demonstrated that the ubiquitin proteasome system is essential for regulating NF-κB signaling via degradation of IκBs [52]. In addition to IκBs, however, several key proteins in the NF-κB signaling regulatory network are also modulated or interact with ubiquitin [54]. Intriguingly, dysregulated ubiquitin conjugation/deconjugation is fairly frequent in human cancers. For example, CYLD, A20, and Cezanne are deubiquitinases that negatively regulate NF-κB activation [61]. To elucidate the underlying mechanism of the simultaneous disruption of these regulators in cancers, Song et al. [62] demonstrated that miR-486 directly suppresses the expression of CYLD, Cezanne, as well as several A20 regulators, ITCH, TNIP-1, TNIP-2, and TNIP-3 in glioma cells. Consequently, miR-486 results in increased ubiquitin conjugations, aberrant NF-κB activation, and promotion of glioma aggressiveness.

12.4.1.4 MiR-182 Links NF-κB with TGF-*β* Signaling

Ubiquitin conjugation is a vital modification for NF-κB signaling and occurs at nearly every step of NF-κB signaling cascades, serving as a positive modulator in activation of NF-κB signaling [63]. In association with this context, the TGF-*β*/Smad signaling has been found to promote tumor progression, and of note, NF-κB can act as an oncogenic mediator of TGF-*β* signaling in cancer cells. To further explore the cross-talk between TGF-*β* signaling and NF-κB signaling, Song et al. [64] found that miR-182 is overexpressed in glioma and directly inhibits the expression of CYLD, a negative regulator of NF-κB signaling. The suppression of CYLD enhances the ubiquitin conjugation of NF-κB signaling components, thereby promoting the aggressive phenotype of glioma cells. Importantly, TGF-*β*-induced miR-182 contributes to constitutive activation of NF-κB signaling. Moreover, there are significant correlations among miR-182 level, TGF-*β* hyperactivation and activation of NF-κB in clinical glioma specimens. At the molecular lever, MiR-182 was also found to target gene transcription elongation factor A-like 7 (TCEAL7), an NF-κB negative regulator [65] in endometrial cancer cells [66]. The expression of miR-182 is upregulated in endometrial cancer, and loss function of miR-182 suppresses the cell proliferation. These data indicate a novel mechanism for constitutive activation of NF-κB in glioma.

12.4.1.5 MiR-892b Inhibits NF-κB Signaling via Targeting TRAF2, TAKa, and TAB3

In breast cancer, Jiang et al. [67] found that miR-892b is significantly downregulated and associated with clinical outcome of patients with the disease. Ectopic overexpression of miR-892 resulted in suppression of tumor growth, metastasis, and angiogenesis, whereas silencing miR-892b promoted these phenotypes. In further

Table 12.1 MiRNAs and NF-κB in cancer cells

MiRNA	Effects on NF-κB	Cancer type	Targets	Reference
miR-125b	↑	Glioma	Tnfaip3 and Nkiras2	[68]
miR-141	↑	Ovarian cancer	KEAP1	[69]
miR-146a/b	↓	Breast cancer	Traf6 and Irak1	[70]
miR-17~92	↑	Lymphomas	Cyld, A20, Rnf11, Tax1bp1, and Traf3	[71]
miR-181b	↑	Pancreatic cancer	Cyld	[72]
miR-19	↑	Leukemia	Cyld	[73]
miR-196a	↑	Pancreatic cancer	IκBα	[74]
miR-199a	↓	Ovarian cancer	IKKβ	[75]
miR-200b	↓	Breast cancer	IKKβ	[76]
miR-218	↓	Gastric cancer	ECOP	[77]
miR-221/222	↑	Colorectal cancer	RelA	[78]
miR-26b	↓	Hepatocellular carcinoma	Tak1 and Tab3	[79]
miR-29b	↓	Multiple myeloma	Socs-1	[80]
miR-30c-2-3p	↓	Breast cancer	Tradd	[81]
miR-301a	↑	Pancreatic cancer	Nkrf	[82]
miR-31	↓	Leukemia	NIK	[83]
miR-342-3p	↓	Hepatocellular carcinoma	NF-κB, IKKγ, Tab2, and Tab3	[84]
miR-362	↑	Gastric cancer	Cyld	[85]
miR-372	↑	Gastric cancer	Tnfip1	[86]
miR-448	↓	Breast cancer	Satb1	[87]
miR-451	↓	Hepatocellular carcinoma	IKKβ	[88]
miR-491	↓	Hepatocellular carcinoma	GIP-1	[89]
miR-500	↑	Gastric cancer	Taxibp1, and Ptud7b	[90]
miR-520b	↓	Breast cancer	Hbxip and IL-8	[91]
miR-520e	↓	Hepatocellular carcinoma	NIK	[92]
miR-657	↓	Hepatocellular carcinoma	Tle1	[93]
miR-7	↓	Hepatocellular carcinoma	RelA and IKKε	[94]
miR-708	↓	Leukemia	IKKβ	[95]
miR-9	↓	Melanoma	NF-κB1	[96]

mechanistic studies, the authors found that miR-892b directly targets TRAF2, TAK1, and TAB3, thereby inhibiting NF-κB signaling.

In summary, an increasing body of evidence suggests a significant involvement of miRNAs in the modulation of NF-κB activation in various cancer types. In addition to the miRNAs introduced above, numerous other miRNAs have also been shown to contribute to NF-κB signaling associate with cancer biology (see Table 12.1). Of note, these miRNAs are found to target expression of gene regulators in the NF-κB signaling network, therefore underscoring the importance of post-transcriptional regulation of miRNAs in modulating NF-κB signaling. Nevertheless, although this layer of regulation has been widely investigated over the past years, how the complex regulatory networks control miRNAs and NF-κB signaling remains elucidated. A better understanding of whether the miRNA-based regulatory mechanism is truly essential for the oncogenic role of NF-κB signaling will benefit future development of miRNA-targeted anticancer strategies.

12.4.2 Long Noncoding RNAs (lncRNAs) in NF-κB Signaling

Long noncoding RNAs (lncRNAs) represent a class of endogenous cellular RNAs longer than 200 bases in size, which neither possess the capacity of encoding proteins nor exert biological functions of being translated into proteins. Numerous lncRNAs have been identified to be key players in various physiological and pathological processes, including tumor formation and metastasis [7]. Unlike miRNAs, which generally display similar lengths (mostly 20–25 nt) and interspecies conservation in sequences and alter gene expression mainly through two most common mechanisms, i.e., mRNA degradation and translational suppression, lncRNAs, whose lengths are highly variable, are far less conserved among species. Furthermore, it is essentially noteworthy that distinct molecular mechanisms are utilized by different lncRNAs to mediate their biological functions [8]. In this context, various examples of mechanistic basis upon which lncRNAs contribute to alterations of gene expression or gene functions have been reported. To illustrate a few, lncRNAs can function as scaffolds for the assembly of other molecular components, thereby forming a complex to execute specific functions [97]. They can also guide ribonucleoprotein complexes to specific loci and eventually modulate gene expression either in an in cis or in trans manner [98]. Moreover, lncRNAs can act as decoys to bind transcriptional factors, splicing proteins or miRNAs, thus changing the expression of their affected genes [99, 100]. In addition, they are able to modify epigenetic regulation via interacting with chromatin-modifying complexes [101]. While the abovementioned modes of lncRNA action have been widely recognized in the field, understanding additional molecular and biological scenarios in which lncRNAs function as cancer-promoting or cancer-suppressive molecules will help address the complexity of cancer biology and develop more potentially effective anticancer strategies.

As recent advances in sequencing technology would facilitate identification of additional bioactive lncRNAs and their roles in biological events key to the pathogenesis of human disease, it is of great significance to explore whether, and how, lncRNAs are involved in modulating NF-κB signaling in the context of cancer biology. Interestingly, several recent studies have demonstrated that lncRNAs indeed are an important class of molecular modifiers of the signaling network, and typical examples of interactions between lncRNAs and NF-κB signaling are summarized below.

12.4.2.1 NF-κB Interacting lncRNA (NKILA)

The most direct connection between a cancer-related NF-κB activation and lncRNA overexpression is derived from a study on the inhibition of IκB, a pivotal negative regulator of NF-κB signaling. IκB binds to and sequesters NF-κB in the cytoplasm until specific signals stimulate the cells and subsequently induce the phosphorylation of IKK, which in turn phosphorylates IκB and leads to its degradation. The removal of IκB liberates NF-κB and subsequently leads to NF-κB translocation into the nucleus, resulting in transcriptional activation of target genes. Given that the stimuli-induced activation of IKK persists far longer than activation of NF-κB [102], in combination with notion that a basal activity of IKK is sufficient to phosphorylate NF-κB-bound IκB [102], it appears that other factors might be involved in protecting NF-κB-bound IκB from being phosphorylated by IKK. Liu et al. identified a lncRNA, designated NF-κB interacting lncRNA (NKILA), during a search for inflammatory cytokine-upregulated lncRNAs in breast cancer cells. NKILA inhibits IκB phosphorylation by directly masking IKK phosphorylation motifs of IκB, thereby suppressing IκB phosphorylation and the subsequent activation of NF-κB. Further study revealed that NKILA binds p65 at two distinct sites in the presence of IκBα, thus forming a stable NKILA/NF-κB/IκBα complex. Moreover, in inflammatory stimuli-stimulated breast epithelial cells, NKILA is key to the inhibition of NF-κB overactivation. The results of this research highlight the important negative feedback regulatory role of NKILA in NF-κB signaling [103].

12.4.2.2 P50-Associated COX-2 Extragenic RNA (PACER)

To study novel regulatory mechanisms of COX-2 transcription in cells prone to tumorigenesis, Emerson et al. found a nuclear antisense long noncoding RNA (p50-associated COX-2 extragenic RNA), which is expressed in the upstream region of COX-2. PACER physically interacts and sequesters NF-κB p50 from binding the COX-2 promoter, which facilitates the exchange of p50/p50-repressive homodimers for p65/p50 heterodimers, recruitment of p300 histone acetyltransferase, increased chromatin accessibility, and assembly of RNA polymerase II initiation complexes, suggesting that PACER constitutes a new layer of COX-2 transcriptional modulation in cancer and inflammation [104].

12.4.3 BRAF-Activated Noncoding RNA (BANCR)

The study by Zhang et al. [105] showed that the expression of BRAF-activated noncoding RNA (BANCR) was significantly elevated in gastric cancer tissue and cell lines. Silencing BANCR retarded gastric cancer cell growth and enhanced cell apoptosis. Mechanistically, BANCR decreases NF-κB1 (p50/105). Moreover, miR-9 is involved in the regulation of NF-κB1 by BANCR.

So far, only several lncRNAs have been reported to have essential roles in regulating NF-κB signaling in cancer cells. However, with regard to the roles of lncRNAs in controlling expression of genes at all levels, discovery of interplays between lncRNAs and NF-κB signaling is emerging [106]. It is likely that more lncRNAs will be found to directly influence the key components of NF-κB signaling in cancer cells. Moreover, understanding the roles of lncRNAs in microenvironmental stromal cells, in which lncRNAs activate NF-κB signaling and influence the tumor-associated microenvironment, might represent a new direction in the investigation of the association between lncRNAs and NF-κB signaling. In summary, further understanding of the biological significance and functioning of lncRNAs in NF-κB signaling and cancer will shed light into the disease etiology, as well as future therapeutic strategies.

12.4.4 Other Noncoding RNAs in Modulation of Cellular Signal Transduction and Potential Significance in NF-κB Signaling

While miRNAs and lncRNAs have been widely recognized as important modulators of cellular signal transduction, several other classes of noncoding RNAs, including circular RNAs (circRNAs) and PIWI-interacting RNAs (piRNAs), also participate in regulating tumor-associated signaling pathways. For example, Hansen et al. revealed that a circRNA sponge for miR-7 (ciRS-7), one circRNA with more than 70 conventional miR-7 binding sites, can efficiently tether miR-7, leading to suppression of miR-7 and enhanced levels of miR-7-targeted genes and their associated signaling pathways [107, 108]. Another circRNA, the sex-determining region Y (Sry) found in the testis acts as a miR-138 sponge [107]. Although little is known about the function of piRNA in carcinogenesis, research by Fu et al. explored the role of piRNA-021285 (piR-021285) in the breast cancer, and they have found that piR-021285 can induce methylation in breast cancer-associated genes [109]. Taken together, these data on other noncoding RNAs than miRNA and lncRNA warrant further investigation of their potential roles in modulating NF-κB signaling.

12.5 Potential Therapeutic Implication and Limitation

Accumulating evidence has demonstrated that targeting NF-κB signaling in cancer cells may represent a new strategy of preventing cancer development or halting cancer progression. Moreover, NF-κB overactivation can be associated with resistance to chemotherapy and escape from apoptosis due to the well-recognized effects of chemotherapy drugs on activation of NF-κB. For example, treatment of Hela cells with SN38 (7-ethyl-10-hydroxycamptothecin) and doxorubicin leads to NF-κB activation and suppression of NF-κB activity by IκBα-mut super-repressor resensitizes Hela cell to drug-induced apoptosis [110]. Given the important roles of NF-κB in tumorigenesis and resistance toward chemotherapeutic agents, together with the findings that NF-κB can be regulated, fairly precisely, by miRNAs in a variety cancer cells, implicating therapeutic possibilities of targeting NF-κB in cancer using miRNAs as targets or therapeutic agents.

Currently, the most commonly tested in vivo approach to miRNA-based therapy is the inhibition of the function of a miRNA using anti-miR oligonucleotides [111]. In order to efficiently silence dysregulated miRNAs in vivo, the anti-miR oligonucleotides need to be chemically modified to improve their biostability and binding affinity. Endogenous mature miRNAs can be sequestered by anti-miRs and thereby releasing the target genes. Thus, anti-miRs approach can be designed for inhibiting miRNAs which activate NF-κB in malignant cells. Krützfeldt et al. showed that chemically modified antisense oligonucleotides, antagomir, can efficiently and specifically silence endogenous miRNAs in mice. Intravenous injection of antagomirs targeting miRNAs leads to significant suppression of corresponding miRNA levels in many tissues, and the effects can be efficient, specific, and long lasting [112]. MiR-21, one of the most widely recognized oncomiRs, directly targets the *Pten* gene [113], a negative regulator of Akt signaling. As Akt activation can be linked to NF-κB signaling, it will be of interest to investigate whether targeting miR-21 could attenuate NF-κB signals. In this context, while studies have focused on targeting miR-21 by specific, synthetic oligonucleotides, in vivo tests are needed to validate the efficacy [114, 115]. In addition to the aforementioned antagomirs, miRNA sponges that carry multiple complementary binding sites of a specific miRNA or miRNA family have been shown to suppress expression of specific miRNAs in in vitro as well as animal experiments, their effectiveness in human cancers needs further preclinical and clinical investigation [116].

Meanwhile, restoring miRNA function can be of therapeutic potential against diseases in which downregulation of a specific miRNA plays a pivotal role in the pathogenesis, including cancer. Approaches to such purposes include use of synthetic double-stranded miRNAs with chemical modifications or viral vector-based overexpression [117]. In the context of antagonizing NF-κB signaling, pharmacological administration of a miRNA that inhibits NF-κB activities may be of therapeutic significance against NF-κB stimulated cancer progression and chemoresistance. Of note, one specific miRNA might be able to target multiple components of the NF-κB signaling pathway, as demonstrated in previous studies, including our own. For instance, miR-7 can directly target RelA and IKKε and inhibit NF-κB signaling in gastric cancer [94], and therefore re-expressing miR-7 may provide advantages in targeting multiple modulators of NF-κB signaling simultaneously.

It is of particular note that in a double-stranded miRNA mimic, the guide strand has sequence identical to the target miRNA, and the complementary strand can be modified with chemical groups to improve cellular uptake. Moreover, chemical modifications of the complementary strand, such as 5'-O-methylation, can prevent the loading of RISC [118]. In addition to chemically modified miRNA mimics, viral-vectors delivery systems, including lentivirus, adenovirus, and adeno-associated viruses (AAV), can also be used to express a given miRNA [119, 120]. Several in vivo replacement therapies have been tested and shown promising results [121]. For example, systemic treatment of a mouse model of non-small cell lung cancer with miR-34a and let-7 mimics results in a significant suppression of tumorigenesis [122]. Development of safe, efficacious delivery technologies will be key to the eventual success of miRNA-based intervention of NF-κB signaling in the clinic.

References

1. Baud V, Karin M. Is NF-kappaB a good target for cancer therapy? Hopes and pitfalls. Nat Rev Drug Discov. 2009;8(1):33–40.
2. Verma IM, Stevenson JK, Schwarz EM, Van Antwerp D, Miyamoto S. Rel/NF-kappa B/I kappa B family: intimate tales of association and dissociation. Genes Dev. 1995;9(22):2723–35.
3. O'Shea JM, Perkins ND. Regulation of the RelA (p65) transactivation domain. Biochem Soc Trans. 2008;36(Pt 4):603–8.
4. Conti L, Hiscott J, Papacchini M, Roulston A, Wainberg MA, Belardelli F, Gessani S. Induction of relA(p65) and I kappa B alpha subunit expression during differentiation of human peripheral blood monocytes to macrophages. Cell Growth Differ. 1997;8(4):435–42.
5. Hoesel B, Schmid JA. The complexity of NF-kappaB signaling in inflammation and cancer. Mol Cancer. 2013;12:86.
6. Bartel DP. MicroRNAs: genomics, biogenesis, mechanism, and function. Cell. 2004;116(2):281–97.
7. Gardini A, Shiekhattar R. The many faces of long noncoding RNAs. FEBS J. 2015;282(9):1647–57.
8. Quinodoz S, Guttman M. Long noncoding RNAs: an emerging link between gene regulation and nuclear organization. Trends Cell Biol. 2014;24(11):651–63.
9. Sen R, Baltimore D. Inducibility of kappa immunoglobulin enhancer-binding protein Nf-kappa B by a posttranslational mechanism. Cell. 1986;47(6):921–8.
10. Hanahan D, Weinberg RA. Hallmarks of cancer: the next generation. Cell. 2011;144(5):646–74.
11. Grivennikov SI, Greten FR, Karin M. Immunity, inflammation, and cancer. Cell. 2010;140(6):883–99.
12. Kundu JK, Surh YJ. Inflammation: gearing the journey to cancer. Mutat Res. 2008;659(1–2):15–30.

13. Hartung ML, Gruber DC, Koch KN, Gruter L, Rehrauer H, Tegtmeyer N, Backert S, Muller A. H. pylori-induced DNA strand breaks are introduced by nucleotide excision repair endonucleases and promote NF-kappaB target gene expression. Cell Rep. 2015;13(1):70–9.
14. Yang D, Zuo C, Wang X, Meng X, Xue B, Liu N, Yu R, Qin Y, Gao Y, Wang Q, Hu J, Wang L, Zhou Z, Liu B, Tan D, Guan Y, et al. Complete replication of hepatitis B virus and hepatitis C virus in a newly developed hepatoma cell line. Proc Natl Acad Sci U S A. 2014;111(13):E1264–73.
15. Herszenyi L, Barabas L, Miheller P, Tulassay Z. Colorectal cancer in patients with inflammatory bowel disease: the true impact of the risk. Dig Dis. 2015;33(1):52–7.
16. Terzic J, Grivennikov S, Karin E, Karin M. Inflammation and colon cancer. Gastroenterology. 2010;138(6):2101–14. e2105.
17. Blouin S, Basle MF, Chappard D. Interactions between microenvironment and cancer cells in two animal models of bone metastasis. Br J Cancer. 2008;98(4):809–15.
18. Pikarsky E, Porat RM, Stein I, Abramovitch R, Amit S, Kasem S, Gutkovich-Pyest E, Urieli-Shoval S, Galun E, Ben-Neriah Y. NF-kappaB functions as a tumour promoter in inflammation-associated cancer. Nature. 2004;431(7007):461–6.
19. Mantovani A, Allavena P, Sica A, Balkwill F. Cancer-related inflammation. Nature. 2008;454(7203):436–44.
20. Wong GS, Rustgi AK. Matricellular proteins: priming the tumour microenvironment for cancer development and metastasis. Br J Cancer. 2013;108(4):755–61.
21. Radisky DC, Levy DD, Littlepage LE, Liu H, Nelson CM, Fata JE, Leake D, Godden EL, Albertson DG, Nieto MA, Werb Z, Bissell MJ. Rac1b and reactive oxygen species mediate MMP-3-induced EMT and genomic instability. Nature. 2005;436(7047):123–7.
22. Xiao G, Fu J. NF-kappaB and cancer: a paradigm of Yin-Yang. Am J Cancer Res. 2011;1(2):192–221.
23. Luo JL, Kamata H, Karin M. IKK/NF-kappaB signaling: balancing life and death-a new approach to cancer therapy. J Clin Invest. 2005;115(10):2625–32.
24. Shaw RJ, Cantley LC. Ras, PI(3)K and mTOR signalling controls tumour cell growth. Nature. 2006;441(7092):424–30.
25. Ledoux AC, Perkins ND. NF-kappaB and the cell cycle. Biochem Soc Trans. 2014;42(1):76–81.
26. Hinz M, Krappmann D, Eichten A, Heder A, Scheidereit C, Strauss M. NF-kappaB function in growth control: regulation of cyclin D1 expression and G0/G1-to-S-phase transition. Mol Cell Biol. 1999;19(4):2690–8.
27. Guttridge DC, Albanese C, Reuther JY, Pestell RG, Baldwin Jr AS. NF-kappaB controls cell growth and differentiation through transcriptional regulation of cyclin D1. Mol Cell Biol. 1999;19(8):5785–99.
28. Bortul R, Tazzari PL, Cappellini A, Tabellini G, Billi AM, Bareggi R, Manzoli L, Cocco L, Martelli AM. Constitutively active Akt1 protects HL60 leukemia cells from TRAIL-induced apoptosis through a mechanism involving NF-kappaB activation and cFLIP(L) up-regulation. Leukemia. 2003;17(2):379–89.
29. Guseva NV, Taghiyev AF, Sturm MT, Rokhlin OW, Cohen MB. Tumor necrosis factor-related apoptosis-inducing ligand-mediated activation of mitochondria-associated nuclear factor-kappaB in prostatic carcinoma cell lines. Mol Cancer Res. 2004;2(10):574–84.
30. Pham CG, Bubici C, Zazzeroni F, Knabb JR, Papa S, Kuntzen C, Franzoso G. Upregulation of Twist-1 by NF-kappaB blocks cytotoxicity induced by chemotherapeutic drugs. Mol Cell Biol. 2007;27(11):3920–35.
31. Julien S, Puig I, Caretti E, Bonaventure J, Nelles L, van Roy F, Dargemont C, de Herreros AG, Bellacosa A, Larue L. Activation of NF-kappaB by Akt upregulates snail expression and induces epithelium mesenchyme transition. Oncogene. 2007;26(53):7445–56.
32. Tomonaga M, Hashimoto N, Tokunaga F, Onishi M, Myoui A, Yoshikawa H, Iwai K. Activation of nuclear factor-kappa B by linear ubiquitin chain assembly complex contributes to lung metastasis of osteosarcoma cells. Int J Oncol. 2012;40(2):409–17.

33. Kishida Y, Yoshikawa H, Myoui A. Parthenolide, a natural inhibitor of nuclear factor-kappaB, inhibits lung colonization of murine osteosarcoma cells. Clin Cancer Res. 2007;13(1):59–67.
34. Wang JH, Manning BJ, Wu QD, Blankson S, Bouchier-Hayes D, Redmond HP. Endotoxin/lipopolysaccharide activates NF-kappa B and enhances tumor cell adhesion and invasion through a beta 1 integrin-dependent mechanism. J Immunol. 2003;170(2):795–804.
35. Farina AR, Tacconelli A, Vacca A, Maroder M, Gulino A, Mackay AR. Transcriptional up-regulation of matrix metalloproteinase-9 expression during spontaneous epithelial to neuroblast phenotype conversion by SK-N-SH neuroblastoma cells, involved in enhanced invasivity, depends upon GT-box and nuclear factor kappaB elements. Cell Growth Differ. 1999;10(5):353–67.
36. Lorusso G, Ruegg C. New insights into the mechanisms of organ-specific breast cancer metastasis. Semin Cancer Biol. 2012;22(3):226–33.
37. Wong HL, Jin G, Cao R, Zhang S, Cao Y, Zhou Z. MT1-MMP sheds LYVE-1 on lymphatic endothelial cells and suppresses VEGF-C production to inhibit lymphangiogenesis. Nat Commun. 2016;7:10824.
38. Oh SY, Seok JY, Choi YS, Lee SH, Bae JS, Lee YM. The histone methyltransferase inhibitor BIX01294 inhibits HIF-1alpha stability and angiogenesis. Mol Cells. 2015;38(6):528–34.
39. Noort AR, van Zoest KP, Weijers EM, Koolwijk P, Maracle CX, Novack DV, Siemerink MJ, Schlingemann RO, Tak PP, Tas SW. NF-kappaB-inducing kinase is a key regulator of inflammation-induced and tumour-associated angiogenesis. J Pathol. 2014;234(3):375–85.
40. Varinska L, Gal P, Mojzisova G, Mirossay L, Mojzis J. Soy and breast cancer: focus on angiogenesis. Int J Mol Sci. 2015;16(5):11728–49.
41. Jiang L, Song L, Wu J, Yang Y, Zhu X, Hu B, Cheng SY, Li M. Bmi-1 promotes glioma angiogenesis by activating NF-kappaB signaling. PLoS ONE. 2013;8(1):e55527.
42. Bauerle KT, Schweppe RE, Lund G, Kotnis G, Deep G, Agarwal R, Pozdeyev N, Wood WM, Haugen BR. Nuclear factor kappaB-dependent regulation of angiogenesis, and metastasis in an in vivo model of thyroid cancer is associated with secreted interleukin-8. J Clin Endocrinol Metab. 2014;99(8):E1436–44.
43. Kim A, Im M, Yim NH, Ma JY. Reduction of metastatic and angiogenic potency of malignant cancer by Eupatorium fortunei via suppression of MMP-9 activity and VEGF production. Sci Rep. 2014;4:6994.
44. Mauro C, Leow SC, Anso E, Rocha S, Thotakura AK, Tornatore L, Moretti M, De Smaele E, Beg AA, Tergaonkar V, Chandel NS, Franzoso G. NF-kappaB controls energy homeostasis and metabolic adaptation by upregulating mitochondrial respiration. Nat Cell Biol. 2011;13(10):1272–9.
45. Kawauchi K, Araki K, Tobiume K, Tanaka N. p53 regulates glucose metabolism through an IKK-NF-kappaB pathway and inhibits cell transformation. Nat Cell Biol. 2008;10(5):611–8.
46. Johnson RF, Witzel II, Perkins ND. p53-dependent regulation of mitochondrial energy production by the RelA subunit of NF-kappaB. Cancer Res. 2011;71(16):5588–97.
47. Sun SC. Non-canonical NF-kappaB signaling pathway. Cell Res. 2011;21(1):71–85.
48. Yamamoto Y, Gaynor RB. IkappaB kinases: key regulators of the NF-kappaB pathway. Trends Biochem Sci. 2004;29(2):72–9.
49. Chen ZJ, Parent L, Maniatis T. Site-specific phosphorylation of IkappaBalpha by a novel ubiquitination-dependent protein kinase activity. Cell. 1996;84(6):853–62.
50. Karin M, Ben-Neriah Y. Phosphorylation meets ubiquitination: the control of NF-[kappa]B activity. Annu Rev Immunol. 2000;18:621–63.
51. Hayden MS, Ghosh S. Signaling to NF-kappaB. Genes Dev. 2004;18(18):2195–224.
52. Wertz IE, O'Rourke KM, Zhou H, Eby M, Aravind L, Seshagiri S, Wu P, Wiesmann C, Baker R, Boone DL, Ma A, Koonin EV, Dixit VM. De-ubiquitination and ubiquitin ligase domains of A20 downregulate NF-kappaB signalling. Nature. 2004;430(7000):694–9.

53. Harhaj EW, Dixit VM. Deubiquitinases in the regulation of NF-kappaB signaling. Cell Res. 2011;21(1):22–39.
54. Liu S, Chen ZJ. Expanding role of ubiquitination in NF-kappaB signaling. Cell Res. 2011;21(1):6–21.
55. Duwel M, Hadian K, Krappmann D. Ubiquitin conjugation and deconjugation in NF-kappaB signaling. Subcell Biochem. 2010;54:88–99.
56. Brummelkamp TR, Nijman SM, Dirac AM, Bernards R. Loss of the cylindromatosis tumour suppressor inhibits apoptosis by activating NF-kappaB. Nature. 2003;424(6950):797–801.
57. Niu T, Liu N, Zhao M, Xie G, Zhang L, Li J, Pei YF, Shen H, Fu X, He H, Lu S, Chen XD, Tan LJ, Yang TL, Guo Y, Leo PJ, et al. Identification of a novel FGFRL1 MicroRNA target site polymorphism for bone mineral density in meta-analyses of genome-wide association studies. Hum Mol Genet. 2015;24(16):4710–27.
58. Jiang L, Lin C, Song L, Wu J, Chen B, Ying Z, Fang L, Yan X, He M, Li J, Li M. MicroRNA-30e* promotes human glioma cell invasiveness in an orthotopic xenotransplantation model by disrupting the NF-kappaB/IkappaBalpha negative feedback loop. J Clin Invest. 2012;122(1):33–47.
59. Enzinger PC, Mayer RJ. Esophageal cancer. N Engl J Med. 2003;349(23):2241–52.
60. Gong H, Song L, Lin C, Liu A, Lin X, Wu J, Li M, Li J. Downregulation of miR-138 sustains NF-kappaB activation and promotes lipid raft formation in esophageal squamous cell carcinoma. Clin Cancer Res. 2013;19(5):1083–93.
61. Wertz IE, Dixit VM. Signaling to NF-kappaB: regulation by ubiquitination. Cold Spring Harb Perspect Biol. 2010;2(3):a003350.
62. Song L, Lin C, Gong H, Wang C, Liu L, Wu J, Tao S, Hu B, Cheng SY, Li M, Li J. miR-486 sustains NF-kappaB activity by disrupting multiple NF-kappaB-negative feedback loops. Cell Res. 2013;23(2):274–89.
63. Iwai K, Fujita H, Sasaki Y. Linear ubiquitin chains: NF-kappaB signalling, cell death and beyond. Nat Rev Mol Cell Biol. 2014;15(8):503–8.
64. Song L, Liu L, Wu Z, Li Y, Ying Z, Lin C, Wu J, Hu B, Cheng SY, Li M, Li J. TGF-beta induces miR-182 to sustain NF-kappaB activation in glioma subsets. J Clin Invest. 2012;122(10):3563–78.
65. Rattan R, Narita K, Chien J, Maguire JL, Shridhar R, Giri S, Shridhar V. TCEAL7, a putative tumor suppressor gene, negatively regulates NF-kappaB pathway. Oncogene. 2010;29(9):1362–73.
66. Guo Y, Liao Y, Jia C, Ren J, Wang J, Li T. MicroRNA-182 promotes tumor cell growth by targeting transcription elongation factor A-like 7 in endometrial carcinoma. Cell Physiol Biochem. 2013;32(3):581–90.
67. Jiang L, Yu L, Zhang X, Lei F, Wang L, Liu X, Wu S, Zhu J, Wu G, Cao L, Liu A, Song L, Li J. miR-892b silencing activates NF-kappaB and promotes aggressiveness in breast cancer. Cancer Res. 2016;76(5):1101–11.
68. Haemmig S, Baumgartner U, Gluck A, Zbinden S, Tschan MP, Kappeler A, Mariani L, Vajtai I, Vassella E. miR-125b controls apoptosis and temozolomide resistance by targeting TNFAIP3 and NKIRAS2 in glioblastomas. Cell Death Dis. 2014;5:e1279.
69. van Jaarsveld MT, Helleman J, Boersma AW, van Kuijk PF, van Ijcken WF, Despierre E, Vergote I, Mathijssen RH, Berns EM, Verweij J, Pothof J, Wiemer EA. miR-141 regulates KEAP1 and modulates cisplatin sensitivity in ovarian cancer cells. Oncogene. 2013;32(36):4284–93.
70. Bhaumik D, Scott GK, Schokrpur S, Patil CK, Campisi J, Benz CC. Expression of microRNA-146 suppresses NF-kappaB activity with reduction of metastatic potential in breast cancer cells. Oncogene. 2008;27(42):5643–7.
71. Jin HY, Oda H, Lai M, Skalsky RL, Bethel K, Shepherd J, Kang SG, Liu WH, Sabouri-Ghomi M, Cullen BR, Rajewsky K, Xiao C. MicroRNA-17~92 plays a causative role in lymphomagenesis by coordinating multiple oncogenic pathways. EMBO J. 2013;32(17):2377–91.

72. Takiuchi D, Eguchi H, Nagano H, Iwagami Y, Tomimaru Y, Wada H, Kawamoto K, Kobayashi S, Marubashi S, Tanemura M, Mori M, Doki Y. Involvement of microRNA-181b in the gemcitabine resistance of pancreatic cancer cells. Pancreatology. 2013;13(5):517–23.
73. Ye H, Liu X, Lv M, Wu Y, Kuang S, Gong J, Yuan P, Zhong Z, Li Q, Jia H, Sun J, Chen Z, Guo AY. MicroRNA and transcription factor co-regulatory network analysis reveals miR-19 inhibits CYLD in T-cell acute lymphoblastic leukemia. Nucleic Acids Res. 2012;40(12):5201–14.
74. Huang F, Tang J, Zhuang X, Zhuang Y, Cheng W, Chen W, Yao H, Zhang S. MiR-196a promotes pancreatic cancer progression by targeting nuclear factor kappa-B-inhibitor alpha. PLoS ONE. 2014;9(2):e87897.
75. Chen R, Alvero AB, Silasi DA, Kelly MG, Fest S, Visintin I, Leiser A, Schwartz PE, Rutherford T, Mor G. Regulation of IKKbeta by miR-199a affects NF-kappaB activity in ovarian cancer cells. Oncogene. 2008;27(34):4712–23.
76. Wu H, Wang G, Wang Z, An S, Ye P, Luo S A negative feedback loop between miR-200b and the NF-kappaB pathway via IKBKB/IKK-beta in breast cancer cells. FEBS J, 2015. doi: 10.1111/febs.13543. [Epub ahead of print]
77. Gao C, Zhang Z, Liu W, Xiao S, Gu W, Lu H. Reduced microRNA-218 expression is associated with high nuclear factor kappa B activation in gastric cancer. Cancer. 2010;116(1):41–9.
78. Liu S, Sun X, Wang M, Hou Y, Zhan Y, Jiang Y, Liu Z, Cao X, Chen P, Liu Z, Chen X, Tao Y, Xu C, Mao J, Cheng C, Li C, et al. A microRNA 221- and 222-mediated feedback loop maintains constitutive activation of NFkappaB and STAT3 in colorectal cancer cells. Gastroenterology, 2014;147(4):847–859 e811
79. Zhao N, Wang R, Zhou L, Zhu Y, Gong J, Zhuang SM. MicroRNA-26b suppresses the NF-kappaB signaling and enhances the chemosensitivity of hepatocellular carcinoma cells by targeting TAK1 and TAB3. Mol Cancer. 2014;13:35.
80. Mott JL, Kurita S, Cazanave SC, Bronk SF, Werneburg NW, Fernandez-Zapico ME. Transcriptional suppression of mir-29b-1/mir-29a promoter by c-Myc, hedgehog, and NF-kappaB. J Cell Biochem. 2010;110(5):1155–64.
81. Shukla K, Sharma AK, Ward A, Will R, Hielscher T, Balwierz A, Breunig C, Munstermann E, Konig R, Keklikoglou I, Wiemann S. MicroRNA-30c-2-3p negatively regulates NF-kappaB signaling and cell cycle progression through downregulation of TRADD and CCNE1 in breast cancer. Mol Oncol. 2015;9(6):1106–19.
82. Lu Z, Li Y, Takwi A, Li B, Zhang J, Conklin DJ, Young KH, Martin R, Li Y. miR-301a as an NF-kappaB activator in pancreatic cancer cells. EMBO J. 2011;30(1):57–67.
83. Yamagishi M, Nakano K, Miyake A, Yamochi T, Kagami Y, Tsutsumi A, Matsuda Y, Sato-Otsubo A, Muto S, Utsunomiya A, Yamaguchi K, Uchimaru K, Ogawa S, Watanabe T. Polycomb-mediated loss of miR-31 activates NIK-dependent NF-kappaB pathway in adult T cell leukemia and other cancers. Cancer Cell. 2012;21(1):121–35.
84. Zhao L, Zhang Y. miR-342-3p affects hepatocellular carcinoma cell proliferation via regulating NF-kappaB pathway. Biochem Biophys Res Commun. 2015;457(3):370–7.
85. Xia JT, Chen LZ, Jian WH, Wang KB, Yang YZ, He WL, He YL, Chen D, Li W. MicroRNA-362 induces cell proliferation and apoptosis resistance in gastric cancer by activation of NF-kappaB signaling. J Transl Med. 2014;12:33.
86. Zhou C, Li X, Zhang X, Liu X, Tan Z, Yang C, Zhang J. microRNA-372 maintains oncogene characteristics by targeting TNFAIP1 and affects NFkappaB signaling in human gastric carcinoma cells. Int J Oncol. 2013;42(2):635–42.
87. Li QQ, Chen ZQ, Cao XX, Xu JD, Xu JW, Chen YY, Wang WJ, Chen Q, Tang F, Liu XP, Xu ZD. Involvement of NF-kappaB/miR-448 regulatory feedback loop in chemotherapy-induced epithelial-mesenchymal transition of breast cancer cells. Cell Death Differ. 2011;18(1):16–25.

88. Li HP, Zeng XC, Zhang B, Long JT, Zhou B, Tan GS, Zeng WX, Chen W, Yang JY. miR-451 inhibits cell proliferation in human hepatocellular carcinoma through direct suppression of IKK-beta. Carcinogenesis. 2013;34(11):2443–51.
89. Yang X, Ye J, Yan H, Tang Z, Shen J, Zhang J, Yang L. MiR-491 attenuates cancer stem cells-like properties of hepatocellular carcinoma by inhibition of GIT-1/NF-kappaB-mediated EMT. Tumour Biol. 2015;37(1):201–9.
90. Zhang L, Ding Y, Yuan Z, Liu J, Sun J, Lei F, Wu S, Li S, Zhang D. MicroRNA-500 sustains nuclear factor-kappaB activation and induces gastric cancer cell proliferation and resistance to apoptosis. Oncotarget. 2015;6(4):2483–95.
91. Hu N, Zhang J, Cui W, Kong G, Zhang S, Yue L, Bai X, Zhang Z, Zhang W, Zhang X, Ye L. miR-520b regulates migration of breast cancer cells by targeting hepatitis B X-interacting protein and interleukin-8. J Biol Chem. 2011;286(15):13714–22.
92. Zhang S, Shan C, Kong G, Du Y, Ye L, Zhang X. MicroRNA-520e suppresses growth of hepatoma cells by targeting the NF-kappaB-inducing kinase (NIK). Oncogene. 2012;31(31):3607–20.
93. Zhang L, Yang L, Liu X, Chen W, Chang L, Chen L, Loera S, Chu P, Huang WC, Liu YR, Yen Y. MicroRNA-657 promotes tumorigenesis in hepatocellular carcinoma by targeting transducin-like enhancer protein 1 through nuclear factor kappa B pathways. Hepatology. 2013;57(5):1919–30.
94. Zhao XD, Lu YY, Guo H, Xie HH, He LJ, Shen GF, Zhou JF, Li T, Hu SJ, Zhou L, Han YN, Liang SL, Wang X, Wu KC, Shi YQ, Nie YZ, et al. MicroRNA-7/NF-kappaB signaling regulatory feedback circuit regulates gastric carcinogenesis. J Cell Biol. 2015;210(4):613–27.
95. Baer C, Oakes CC, Ruppert AS, Claus R, Kim-Wanner SZ, Mertens D, Zenz T, Stilgenbauer S, Byrd JC, Plass C. Epigenetic silencing of miR-708 enhances NF-kappaB signaling in chronic lymphocytic leukemia. Int J Cancer. 2015;137(6):1352–61.
96. Liu N, Sun Q, Chen J, Li J, Zeng Y, Zhai S, Li P, Wang B, Wang X. MicroRNA-9 suppresses uveal melanoma cell migration and invasion through the NF-kappaB1 pathway. Oncol Rep. 2012;28(3):961–8.
97. Yoon JH, Abdelmohsen K, Kim J, Yang X, Martindale JL, Tominaga-Yamanaka K, White EJ, Orjalo AV, Rinn JL, Kreft SG, Wilson GM, Gorospe M. Scaffold function of long non-coding RNA HOTAIR in protein ubiquitination. Nat Commun. 2013;4:2939.
98. Hung T, Chang HY. Long noncoding RNA in genome regulation: prospects and mechanisms. RNA Biol. 2010;7(5):582–5.
99. Puvvula PK, Desetty RD, Pineau P, Marchio A, Moon A, Dejean A, Bischof O. Long noncoding RNA PANDA and scaffold-attachment-factor SAFA control senescence entry and exit. Nat Commun. 2014;5:5323.
100. Pfeiffer V, Lingner J. TERRA promotes telomere shortening through exonuclease 1-mediated resection of chromosome ends. PLoS Genet. 2012;8(6):e1002747.
101. Mercer TR, Dinger ME, Mattick JS. Long non-coding RNAs: insights into functions. Nat Rev Genet. 2009;10(3):155–9.
102. O'Dea EL, Kearns JD, Hoffmann A. UV as an amplifier rather than inducer of NF-kappaB activity. Mol Cell. 2008;30(5):632–41.
103. Liu B, Sun L, Liu Q, Gong C, Yao Y, Lv X, Lin L, Yao H, Su F, Li D, Zeng M, Song E. A cytoplasmic NF-kappaB interacting long noncoding RNA blocks IkappaB phosphorylation and suppresses breast cancer metastasis. Cancer Cell. 2015;27(3):370–81.
104. Krawczyk M, Emerson BM. p50-associated COX-2 extragenic RNA (PACER) activates COX-2 gene expression by occluding repressive NF-kappaB complexes. eLife. 2014;3:e01776.
105. Zhang ZX, Liu ZQ, Jiang B, Lu XY, Ning XF, Yuan CT, Wang AL. BRAF activated non-coding RNA (BANCR) promoting gastric cancer cells proliferation via regulation of NF-kappaB1. Biochem Biophys Res Commun. 2015;465(2):225–31.
106. Schmitz SU, Grote P, Herrmann BG. Mechanisms of long noncoding RNA function in development and disease. Cell Mol Life Sci, 2016. doi: 10.1007/s00018-016-2174-5. [Epub ahead of print]

107. Hansen TB, Jensen TI, Clausen BH, Bramsen JB, Finsen B, Damgaard CK, Kjems J. Natural RNA circles function as efficient microRNA sponges. Nature. 2013;495(7441):384–8.
108. Hansen TB, Kjems J, Damgaard CK. Circular RNA and miR-7 in cancer. Cancer Res. 2013;73(18):5609–12.
109. Fu A, Jacobs DI, Hoffman AE, Zheng T, Zhu Y. PIWI-interacting RNA 021285 is involved in breast tumorigenesis possibly by remodeling the cancer epigenome. Carcinogenesis. 2015;36(10):1094–102.
110. Bottero V, Busuttil V, Loubat A, Magne N, Fischel JL, Milano G, Peyron JF. Activation of nuclear factor kappaB through the IKK complex by the topoisomerase poisons SN38 and doxorubicin: a brake to apoptosis in HeLa human carcinoma cells. Cancer Res. 2001;61(21):7785–91.
111. van Rooij E, Purcell AL, Levin AA. Developing microRNA therapeutics. Circ Res. 2012;110(3):496–507.
112. Krutzfeldt J, Rajewsky N, Braich R, Rajeev KG, Tuschl T, Manoharan M, Stoffel M. Silencing of microRNAs in vivo with 'antagomirs'. Nature. 2005;438(7068):685–9.
113. Iliopoulos D, Jaeger SA, Hirsch HA, Bulyk ML, Struhl K. STAT3 activation of miR-21 and miR-181b-1 via PTEN and CYLD are part of the epigenetic switch linking inflammation to cancer. Mol Cell. 2010;39(4):493–506.
114. Davis S, Lollo B, Freier S, Esau C. Improved targeting of miRNA with antisense oligonucleotides. Nucleic Acids Res. 2006;34(8):2294–304.
115. Zhang J, Jiao J, Cermelli S, Muir K, Jung KH, Zou R, Rashid A, Gagea M, Zabludoff S, Kalluri R, Beretta L. miR-21 inhibition reduces liver fibrosis and prevents tumor development by inducing apoptosis of CD24+ progenitor cells. Cancer Res. 2015;75(9):1859–67.
116. Ma L, Young J, Prabhala H, Pan E, Mestdagh P, Muth D, Teruya-Feldstein J, Reinhardt F, Onder TT, Valastyan S, Westermann F, Speleman F, Vandesompele J, Weinberg RA. miR-9, a MYC/MYCN-activated microRNA, regulates E-cadherin and cancer metastasis. Nat Cell Biol. 2010;12(3):247–56.
117. Thorsen SB, Obad S, Jensen NF, Stenvang J, Kauppinen S. The therapeutic potential of microRNAs in cancer. Cancer J. 2012;18(3):275–84.
118. Chen PY, Weinmann L, Gaidatzis D, Pei Y, Zavolan M, Tuschl T, Meister G. Strand-specific 5'-O-methylation of siRNA duplexes controls guide strand selection and targeting specificity. RNA. 2008;14(2):263–74.
119. Miyazaki Y, Adachi H, Katsuno M, Minamiyama M, Jiang YM, Huang Z, Doi H, Matsumoto S, Kondo N, Iida M, Tohnai G, Tanaka F, Muramatsu S, Sobue G. Viral delivery of miR-196a ameliorates the SBMA phenotype via the silencing of CELF2. Nat Med. 2012;18(7):1136–41.
120. Xie J, Burt DR, Gao G. Adeno-associated virus-mediated microRNA delivery and therapeutics. Semin Liver Dis. 2015;35(1):81–8.
121. Pichler M, Calin GA. MicroRNAs in cancer: from developmental genes in worms to their clinical application in patients. Br J Cancer. 2015;113(4):569–73.
122. Trang P, Wiggins JF, Daige CL, et al. Systemic delivery of tumor suppressor microRNA mimics using a neutral lipid emulsion inhibits lung tumors in mice. Mol Ther. 2011;19(6):1116–22.

Chapter 13
Noncoding RNAs Regulating p53 and c-Myc Signaling

Yide Mei and Mian Wu

Abstract p53 is one of the most important tumor suppressors and is known to play critical roles in the process of tumor development. Similarly, as an important proto-oncogenes, c-Myc is activated in over half of human cancers. Both p53 and c-Myc participate in almost every crucial decision of almost every cell. Therefore, it is utmost important to gain a better understanding of how they affect multiple cellular processes. The physiological and pathologic patterns of p53 and c-Myc regulations are modulated by a large number of cis-elements and transfactors (RNAs and proteins). These elements and factors are composed of a complicated network of intracellular and extracellular pathways. How the noncoding RNAs are involved in their regulations has not been comprehensively reviewed. In this chapter, we will list and describe recently published important noncoding RNAs including microRNAs and long noncoding RNAs, which act as effectors and regulators for both p53 and c-Myc regulation. The purpose of this chapter is to provide a recent progress of noncoding RNA in the regulation of p53 and c-Myc on network of cellular signaling and its potential implications in both basic science and clinical application.

Keywords p53 • miRNA • c-Myc • Noncoding RNA

13.1 Introduction

The p53 tumor suppressor maintains the normal cell growth and genomic stability in check by either imposing the cell cycle arrest or inducing apoptosis in response to DNA damage or other types of cellular stress [1, 2]. The importance of p53 as a tumor suppressor is highlighted by the fact that p53 is inactivated in more than half of all human cancers. p53 has been implicated in the regulation of various cellular

Y. Mei • M. Wu (✉)
CAS Key Laboratory of Innate Immunity and Chronic Disease, CAS Center for Excellence in Molecular Cell Science, Innovation Center for Cell Signaling Network, School of Life Sciences and Medical Center, University of Science and Technology of China, Hefei 230027, China
e-mail: meiyide@ustc.edu.cn; wumian@ustc.edu.cn

E. Song (ed.), *The Long and Short Non-coding RNAs in Cancer Biology*,
Advances in Experimental Medicine and Biology 927,
DOI 10.1007/978-981-10-1498-7_13

processes, including cell proliferation, cell senescence, and cell metabolism. The majority of the p53 downstream effectors are mediated by its intrinsic nature as a master transcriptional factor. Similarly, the importance of c-Myc as an important oncogene is highlighted by the fact that c-Myc is activated in more than half of human cancers [3]. Many mechanisms involved in c-Myc activation during tumorigenesis include gene amplification, chromosomal rearrangement, and point mutations in the coding sequence. It has been recognized that c-Myc has the potent transforming ability to promote tumorigenesis [4]. Therefore, it is not surprising that c-Myc expression is under the tight control in normal cells [5]. For instance, c-Myc is an immediate early gene and its transcription is controlled at the level of initiation in response to a range of growth stimuli. In addition, c-Myc mRNA is highly unstable, and the export and translation of c-Myc mRNA are also tightly controlled in the cell. Furthermore, c-Myc is a short-lived protein, and its protein stability is regulated by multiple E3 ubiquitin ligases, among which SCF (Skp-Cullin-F-box)-Fbw7 (F-box and WD repeat domain-containing 7) is the best-characterized E3 ubiquitin ligase for c-Myc [6]. Once these control mechanisms are compromised, c-Myc becomes dysregulated, thus leading to the tumor initiation and progression. As emerging roles of microRNAs and long noncoding RNAs in cancer research have opened exciting new areas in p53 and c-Myc investigations, it is necessary to have a review on recent progress in the regulation of noncoding RNA in networks for physiological or pathological signaling pathways including the tumor suppressor p53 and the oncogene c-Myc.

13.2 Noncoding RNAs Regulating p53 Signaling

Due to its potent antiproliferative effect, p53 is under extraordinarily tight regulations involving mRNA stability, protein translation, and posttranslational modifications in healthy cells. The best-known negative regulator of p53 is Mdm2, which is the principle E3 ubiquitin ligase for p53 and targets p53 for rapid degradation through the ubiquitin-proteasome pathway [7]. Upon cellular stress, p53 is stabilized and binds to a consensus response element (p53 RE) in the promoter of p53 target genes, leading to their transcriptional activation. Among them, p21 is a well-known cell cycle regulator. In addition to the protein-coding genes, noncoding RNAs are emerging as a novel class of p53 target genes. Also, noncoding RNAs are able to regulate p53 function and serve as p53 regulators [8]. Due to the importance of p53 in cancer initiation and development, it is conceivable that the p53-related noncoding RNAs are critical players in the regulation of tumorigenesis via controlling the p53 signaling pathway, and also these noncoding RNAs may represent as potential diagnostic markers and therapeutic targets for cancer treatment. Here, we will overview the role of noncoding RNAs, particularly lncRNAs and miRNAs, in the regulation of p53 signaling network in cancer.

13.2.1 Long Noncoding RNAs Regulating p53 Signaling

Increasing evidence suggests that lncRNAs play an important role in the regulation of various cellular processes. Dysregulation of lncRNAs has been implicated in a number of human diseases, including cancer [9]. LncRNAs can function as oncogenic and tumor-suppressive genes. In particular, a strong correlation between lncRNAs and the p53 pathway has been pointed out in a number of models both in vitro and in vivo [10]. Some lncRNAs are involved in the regulation of p53 expression, whereas some others are p53 targets and direct effectors of p53 signaling. For the sake of simplicity, we have classified the p53-related lncRNAs as "p53 effectors" and "p53 regulators" in this review.

13.2.1.1 Long Noncoding RNAs as Effectors of p53 Signaling

Given that p53 is a master transcriptional factor, several high-throughput technologies including lncRNA-specific microarray, genome-wide ChIP-seq and RNA-seq analyses have been applied to identify novel p53-responsive lncRNAs [11, 12]. Dozens of lncRNAs have been uncovered as bona fide p53 transcriptional targets. Here, we present a few representative examples of p53-regulated lncRNAs that act as effectors of p53 signaling in cancer.

LincRNA-p21

By performing a tiling microarray analysis, Huarte et al. have identified a set of p53-responsive lncRNAs [13]. Among them, lincRNA-p21 was induced in mouse cells where p53 was selectively activated. LincRNA-p21 is located approximately 15 kb upstream of CDKN1A gene (p21) mapped to mouse chromosome 17 and human chromosome 6 [14] and serves as a repressor in the p53 pathway because knockdown of lincRNA-p21 affects the expression of hundreds of target genes that are normally repressed by p53. LincRNA-p21 functions through interaction with hnRNP-K and is required for p53-triggered apoptosis. By using a lincRNA-p21 conditional knockout mouse model, it has been recently shown that lincRNA-p21 influences the p53 pathway by predominantly acting in cis to activate expression of its neighboring gene, CDKN1A (p21) [15]. Together, these findings demonstrate lincRNA-p21 as an important mediator of p53 activity.

Despite the limited sequence homology, the human counterpart of mouse lincRNA-p21 has been identified and appears to be downregulated in multiple tumor types [16]. Unlike mouse lincRNA-p21 that is predominantly localized in the nucleus [13], human lincRNA-p21 is mainly localized in the cytosol [17, 18]. The human lincRNA-p21 is able to interact with the RNA-binding protein HuR, through which lincRNA-p21 regulates target mRNA translation [17]. Also, the human lincRNA-p21 is a hypoxia-responsive lncRNA and is essential for hypoxia-enhanced glycolysis [18]. p53 is known to be a reprogramming blocking factor, and recently, Bao X et al. have reported that p53-regulated lincRNA-p21 prevents somatic cell reprogramming by sustaining H3K9me3 and CpG methylation at pluripotency gene promoters [19].

PANDA

By using an ultrahigh density array, PANDA (p21-associated ncRNA DNA damage activated) has been identified as a p53-regulated lncRNA that is involved in cell cycle progression and apoptosis [20]. PANDA is located 5 kb upstream of the CDKN1A gene promoter. p53 can bind to a single p53 RE located between CDKN1A and PANDA divergent promoters [20, 21]. As a result, PANDA expression is induced upon DNA damage in a p53-dependent manner. Upon induction, PANDA interacts with and sequesters the transcriptional factor NF-YA to limit expression of proapoptotic genes such as APAF1, BIK, and FAS. Suppression of PANDA strongly sensitizes human fibroblasts to DNA damage-induced apoptosis [20]. These findings indicate that PANDA is a p53 downstream effector in response to DNA damage to inhibit apoptosis. However, since PANDA-sequestered NF-YA can control the expression of both proapoptotic and anti-apoptotic genes depending on the cellular context, it is not surprising that PANDA also shows the antiproliferative activity in addition to the abovementioned anti-apoptotic function. For instance, low expression of PANDA has been linked to the progression of non-small cell lung carcinoma (NSCLC). Downregulation of PANDA allows NF-YA to upregulate anti-apoptotic gene Bcl2 expression and promotes cell survival [21]. Altogether, these findings suggest an intriguing role of PANDA as a p53 target in orchestrating the delicate balance between apoptosis and cell survival in response to different cellular stresses.

Loc285194

Loc285194 is an lncRNA located at chromosome 3q13.31. This locus harbors frequent focal copy number alterations (CNAs) and loss of heterozygosity (LOH) in primary osteosarcoma samples [22], indicating loc285194 is a tumor suppressor. In support of this, downregulation of loc285194 has been found in several cancer types [23, 24]. Loc285194 is a p53 transcriptional target [23]. Ectopic expression of loc285194 inhibits tumor cell growth both in vitro and in vivo. Mechanistically, loc285194 exerts its tumor-suppressive function via acting as a molecular sponge for miR-211, which has been known to promote cell growth. These findings suggest that loc285194 is a p53-regulated tumor suppressor.

TUG1

TUG1 (taurine upregulated gene 1) was originally identified as an important regulator of murine retina differentiation [25]. More recently, TUG1 has been characterized as a p53 transcriptional target [26]. TUG1 controls gene expression via acting as a molecular scaffold to recruit the chromatin-modifying complexes PRC1 or PRC2 [27, 28], indicating an intriguing possibility that p53 may regulate a specific set of protein-coding gene expression via TUG1. Functionally, TUG1 has been shown to regulate cell proliferation in several tumor types. Downregulation of TUG1 is associated with poor prognosis of non-small cell lung cancer, whereas in esophageal squamous cell carcinoma, upregulation of TUG1 promotes cell proliferation [26, 29]. This discrepancy may be explained by lncRNA tissue specificity or tumor heterogeneity (e.g., p53 status).

p53-Related eRNAs

eRNAs are a class of noncoding RNAs transcribed from the DNA sequence of enhancer regions [30]. A recent study has shown that genome-wide p53 binding occurs predominantly within enhancer regions in both human and mouse model system [31]. This data strongly suggest that p53 may regulate enhancer activity. The ability of p53 to modulate enhancer activity provides an additional layer of complexity to the p53 network. Intriguingly, several eRNAs, including DUSP4, PAPPA, and IER5, are expressed in a p53-dependent manner. Such p53-regulated eRNAs are required for efficient transcriptional enhancement of interacting target genes and induction of p53-dependent cell cycle arrest [32]. Together, these findings suggest that p53 may be able to regulate a specific set of target gene expression via eRNAs.

13.2.1.2 Long Noncoding RNAs as Regulators of p53 Signaling

Although gene expression is regulated at both the transcriptional and posttranscriptional levels and lncRNAs are capable of modulating gene transcription, the evidence of lncRNA-mediated transcriptional regulation of p53 is still lacking. Here, we will discuss several lncRNAs that have been reported to regulate p53 expression at the posttranscriptional level.

MALAT1

Metastasis-associated lung adenocarcinoma transcript 1 (MALAT1) was originally reported to be upregulated in lung cancer [33] and subsequently shown to be overexpressed in several other cancer types including breast cancer and colon cancer [34]. Ectopic expression of MALAT1 enhances cell proliferation in vitro and promotes tumor formation in nude mice. Evidence from MALAT1 knockout model reveals that MALAT1 promotes metastasis of lung cancer through regulating expression of metastasis-related genes [35]. These studies strongly suggest that MALAT1 possess an oncogenic activity. Of great interest, MALAT1 serves as a repressor of p53 [36]. Depletion of MALAT1 in human fibroblasts activates double-stranded DNA damage response resulting in the induction of p53 and its down-stream target genes. Correlated with p53 activation, MALAT1-depleted human fibroblasts exhibit the phenotype of G1 cell cycle arrest [36]. However, it is still unclear that the observed defects in cell cycle progression are due to the deregulation of p53. Therefore, it would be interesting and important to determine whether MALAT1 exerts its oncogenic activity through regulating the p53 pathway in the future.

MEG3

As an imprinted gene, maternally expressed gene 3 (MEG3) has been found to be downregulated in various types of human cancers [34]. Ectopic expression of MEG3 markedly inhibits growth of human cancer cells [37]. These studies suggest that MEG3 functions as a tumor suppressor. Of note, MEG3 is able to active p53 and stimulate p53-mediated gene expression [38]. To induce p53 expression, MEG3 appears to act through an indirect mechanism by suppressing Mdm2 levels and attenuating the inhibitory effect of Mdm2 on p53. However, it is still unknown how MEG3 downregulates Mdm2 expression. Intriguingly, MEG3-stimulated p53 transcription is selective, since MEG3 enhances expression of growth differentiation factor 15 (GDF15) by promoting p53 binding to its promoter, whereas expression of other p53 targets, like p21, is unaffected [38]. Therefore, it would be interesting to investigate whether and how MEG3 directs p53 binding to its specific target gene promoters.

Wrap53

WD repeat containing, antisense to p53 (Wrap53) is a natural antisense transcript of p53 [39]. Wrap53 is located upstream of the p53 gene on the opposite strand. Wrap53 gene is transcribed as three different isoforms, α, β, and γ, but only the α form, containing a complementary sequence to the first exon of p53, is able to regulate p53 expression [39]. The other two isoforms β and γ lacking this sequence fail to affect p53 levels. Functionally, knockdown of Wrap53 abrogates p53 induction in response to DNA damage, whereas ectopic expression of Wrap53 potentiates p53-dependent apoptosis [39]. Since Wrap53 and p53 mRNAs are able to form an RNA-RNA duplex, it has been speculated that this RNA-RNA interaction is required to stabilize the p53 transcript. Interestingly, a recent study has shown that CCCTC-binding factor (CTCF) is able to physically interact with Wrap53 and thereafter affect p53 levels [40], reinforcing the important role of Wrap53 in the regulation of p53 expression.

ROR

Regulator of reprogramming (ROR) was originally identified as a promoting factor of somatic cell reprogramming via attenuating p53-dependent apoptosis [41]. Further studies have demonstrated that ROR plays a special role in the p53 signaling network, since ROR not only regulates p53 protein expression but ROR itself is also regulated by p53 [42]. ROR represses p53 translation through a direct interaction with heterogeneous nuclear ribonucleoprotein I (hnRNP I). A 28-base ROR sequence with hnRNP I-binding ability is essential and sufficient for p53 repression. Functionally, ROR inhibits p53-mediated cell cycle arrest and apoptosis. On the other side, p53 binds to the p53-responsive element (p53 RE) in the ROR promoter and activates ROR expression. These findings suggest an existence of a unique autoregulatory feedback loop between p53 and ROR, through which p53 expression is delicately controlled in response to various cellular stresses.

In summary, the detailed regulations for long noncoding RNA as either p53 effectors or regulators are far more complicated than we have anticipated, and the listed examples above are just one aspect of these multifaceted investigations. Examples listed in the contents above are outlined in Fig. 13.1.

13.2.2 MiRNAs Regulating p53 Signaling

MiRNAs are small noncoding RNAs that regulate gene expression at the posttranscriptional level [43]. Dysregulation of miRNAs has been linked to a variety of human diseases including cancer [44]. Over the last decade, a growing number of miRNAs have been involved in the regulation of p53 signaling pathway [45, 46]. These noncoding RNAs join the p53 network either as effectors or regulators. More interestingly, some of the p53-regulated miRNAs are involved in complex feedback loops, through which miRNAs either amplify or fine-tune p53 signaling in response to different cellular stresses.

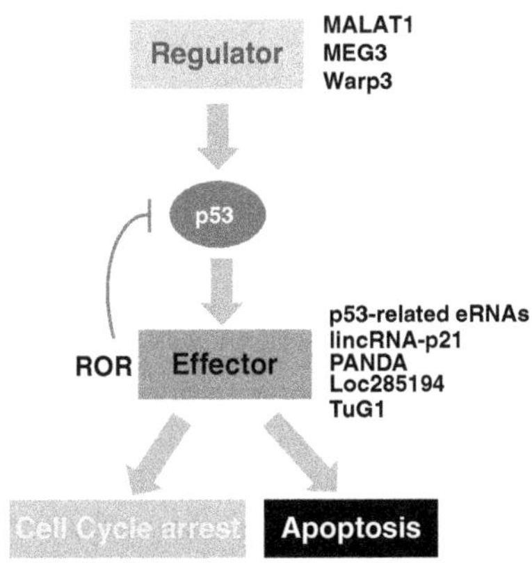

Fig. 13.1 Long noncoding RNAs regulating p53 signaling

13.2.2.1 MiRNAs as Effectors of p53 Signaling

Since miR-34 was identified as the first p53 target miRNA gene, more than dozens of p53-regulated miRNAs have been discovered. We here focus on the miRNAs that act as effectors of p53 signaling in cancer. MiR-34 is able to induce p53-dependent apoptosis, cell cycle arrest, and senescence [47–52]. MiR-34 acts to induce cell cycle arrest by suppressing expression of a batch of cell cycle-related factors, including E2F3, cyclin E2, CDK4/6, and c-Myc. In addition, miR-34 can promote apoptosis in response to p53 activation through inhibiting expression of a number of anti-apoptotic proteins, including Bcl2 and DcR3. Besides, miR-34 is also involved in the regulation of other cellular processes, such as cell metabolism, epithelial to mesenchymal transition (EMT), and angiogenesis, all of which are of central importance to cancer cell biology. For instance, miR-34 regulates metabolic processes such as glycolysis and lipid metabolism through targeting LDHA, Sirt1, ASCL1, and ASCL4 [53]. Given the strong tumor-suppressive ability of miR-34, it is not surprising that miR-34 is downregulated in various cancer types [47, 48]. However, in contrast to p53-deficient mice, miR-34 knockout mice do not display increased susceptibility to spontaneous, irradiation-induced, or c-Myc-initiated tumorigenesis [54], indicating that miR-34 alone is not sufficient to mediate the potent tumor-suppressive function of p53.

Other miRNAs acting as p53 effectors include miR-15a/16, miR-107, miR-205, miR-145, miR-192/215, and miR-200 family members. Soon after the discovery of miR-34, miR-192/215 were found to induce cell cycle arrest in response to p53 activation by targeting several G1 and G2 checkpoint proteins, including CDC7, Cul5, and LMNB2 [55]. MiR205 was also shown to reduce cell cycle progression by targeting E2F1 in response to p53 activation [56].

As the p53 transcriptional target, miR-15a/16 has been implicated in targeting genes involved in various p53 signaling pathways, such as apoptosis, cell cycle progression, cell proliferation, migration, and invasion. MiR-15a/16 deletion has been found in several tumor types, including non-small cell lung cancer and prostate cancer [57]. The miR-200 family members have been associated with p53-regulated signaling pathways and also play an important role in suppressing cancer metastasis through direct targeting of genes including ZEB1/2, SOX2, and VEGF [58, 59]. MiR-107 contributes to the function of p53 in the regulation of angiogenesis and hypoxic signaling through the targeting of HIF-1β. MiR-107 is also able to induce G1 cell cycle arrest through suppression of CDK6 and RBL2 expression. Functionally, ectopic expression of miR-107 inhibits both tumor growth and angiogenesis in mouse colon cancer models [60, 61]. p53 directly binds to the p53 RE in the miR-145 promoter and induces its expression. Upon induction, miR-145 targets and negatively regulates expression of several cell cycle regulators including c-Myc and CDK4/6 [62], thereby leading to the inhibition of tumor cell proliferation.

Interestingly, some miRNAs are transcriptionally suppressed rather than induced by p53. For instance, the miR-17-92 cluster can be transcriptionally repressed by p53 in response to hypoxia, sensitizing cells to hypoxia-induced apoptosis [63]. In addition to regulating miRNA expression transcriptionally, p53 can also control miRNA expression posttranscriptionally by modulating miRNA processing and maturation [64]. p53 has been shown to interact with the Drosha complex and promote the cleavage of pri-miRNAs to pre-miRNAs. It has also been reported to affect miRNA target gene selection via the regulation of the RNA-binding protein RBM38 [65]. Furthermore, although most p53-regulated miRNAs function as tumor suppressors, some potentially act as oncogenes. MiR-194 is transcriptionally upregulated by p53 and is able to target thrombospondin-1, leading to increased tumor angiogenesis [66]. In addition, the anti-apoptotic miR-149* has been shown to suppress expression of GSK-3α in response to p53 activation, resulting in increased expression of Mcl-1 and consequent resistance of melanoma cells to apoptosis [67].

13.2.2.2 MiRNAs as Regulators of p53 Signaling

In addition to being the p53 targets, miRNAs are able to regulate p53 expression. miRNAs contribute to the tight control of p53 by either directly interacting with the 3'-UTR of p53 mRNA or indirectly downregulating p53 regulators.

Those miRNAs that bind p53 directly and function in a p53-repressive manner include miR-125b, miR-504, miR-33, miR-1285, miR-30d, miR-25, and miR-380 [68]. Due to the strongly tumor-suppressive function of p53, these p53-repressive miRNAs may be clinically relevant oncogenes. For example, miR-125b negatively regulates p53 expression by binding to the 3'-UTR of p53, resulting in decreased apoptosis [69]. In contrast, knockdown of miR-125b increases p53 protein levels and induces apoptosis. MiR-125b also targets many other genes involved in the p53 signaling pathway. By using a gain- and/or loss-of-function screen for miR-125 targets, in humans, mice, and zebra fish, miR-125b has been found to directly control at least 20 genes in the p53 network. Among them are modulators of apoptosis, such as Puma, Igfbp3, and Bak, and also several cell cycle regulators, including Cdc25C and cyclin C [70]. In colorectal cancers, elevated expression of miR-125b is associated with increased tumor size and invasion and correlated with poor prognosis and decreased survival. Also, the miR-125b gene is inherently activated by a chromosomal translocation t(11;14)(q24;q32) in human B-cell precursor acute lymphoblastic leukemia (BCP-ALL). Eu/miR-125b transgenic mice with miR-125b overexpression develop lethal B-cell malignancies with clonal proliferation [71]. These studies indicate miR-125b as a potential oncogene. However, miR-125b has also been shown to function as a tumor suppressor in breast cancer [72], suggesting that miR-125b may exert its function in a context-dependent fashion.

MiR-504, miR-33, and miR-1285 downregulate p53 levels through two seed match sequences in the 3'-UTR of p53 [73]. Ectopic expression of these miRNAs attenuates p53-mediated cell cycle arrest and apoptosis and promotes tumorigenesis in colon cancer models. MiR-30d and miR-25 are also able to decrease p53 levels by directly binding to its 3'-UTR [74], leading to the impaired downstream effects of p53, such as senescence, apoptosis, and cell cycle arrest. It has been observed that miR-30d and miR-25 are upregulated in multiple myelomas, which exhibit a concomitant downregulation of p53 expression. Additionally, the miR-30d gene is amplified in more than 30% of multiple types of human solid tumors (n = 1283), and enhanced expression of miR-30d is associated significantly with poor clinical outcomes in ovarian cancer patients [75]. MiR-380 is found to repress p53 levels via a conserved sequence in the p53 3'-UTR in neuroblastomas commonly harboring wild-type p53. Neuroblastomas with elevated miR-380 expression have showed decreased patient survival. MiR-380 overexpression cooperates with H-Ras oncoprotein in transformation, blocks oncogene-induced senescence, and promotes tumor formation in mice [76]. Furthermore, in vivo delivery of a miR-380 antagonist decreases tumor size in an orthotopic mouse model of neuroblastoma. Intriguingly, a recent study has found hundreds of novel somatic mutations in the 3'-UTR of p53 from B-cell lymphoma patients, and the seed match binding sites of 8 out of 11 p53-targeting miRNAs are disrupted by these mutations [77]. Altogether, these studies demonstrate the physiological importance of miRNAs in suppressing p53 tumor-suppressive function.

In addition to abovementioned miRNAs that directly repress p53, a number of miRNAs have been discovered to activate p53 by directly repressing Mdm2, such as miR-192, miR-194, miR-605, miR-25, miR-32, miR-143, miR-145, miR-660, and miR-661 [68]. Almost all of these miRNAs are able to inhibit cancer cell proliferation via promoting p53-mediated apoptosis, senescence, and/or cell cycle arrest. Some of them are also capable of repressing the migration and invasion of cancer cells to inhibit cancer metastasis. Other than directly repressing Mdm2, miR-122 indirectly decreases Mdm2 activity via the downregulation of cyclin G1 and subsequent inhibition of the recruitment of the PP2A phosphatase to Mdm2, resulting in increased p53 levels and activity. In addition to suppressing Mdm2, several miRNAs can also target other p53 regulators, such as Sirt1 and HDAC1, and thus control p53 activity [45]. For instance, miR-34a and miR-449 have been shown to target Sirt1, leading to increased p53 acetylation and p53-induced apoptosis. Interestingly, some of the miRNAs that positively regulate p53 activity, such as miR-192, miR-194, miR-215, miR-605, miR-143, and miR-145, are also the transcriptional targets of p53, indicating the existence of a positive feedback loop that amplifies the p53 response to cellular stress.

13.2.2.3 Feedback Loops Involved in MiRNA-Regulated p53 Signaling

It has long been accepted that p53 and Mdm2 form a negative feedback loop, where p53 positively regulates Mdm2 by activating its transcription and Mdm2 negatively regulates p53 by promoting its ubiquitination and degradation. This feedback regulation has been recognized as a key mechanism in determining the cellular outcome in response to p53 activation. Interestingly, the feedback regulation can also be achieved by some p53-regulated miRNAs, either positively or negatively [73]. For instance, p53-induced miRNAs miR-192, miR-194, and miR-215 directly inhibit Mdm2 expression and protect p53 from Mdm2-mediated degradation [78, 79]. The combined ectopic expression of these miRNAs greatly enhances the therapeutic effectiveness of Mdm2 inhibitor MI-219 to treat multiple myeloma [79]. These miRNAs have been found to be downregulated in several cancer types, such as colorectal cancer and renal cell carcinoma (RCC). Also, miR-215 expression correlates positively with survival of RCC patients. In agreement with the identification of Mdm2 as the direct target of miR-192, miR-194, and miR-215, specimens from RCC patients exhibit an inversely correlated expression of mdm2 and these three miRNAs [80]. A similar positive feedback loop has been recently described for miR-605, which is induced by p53 and is able to negatively regulate Mdm2 expression [81]. MiR-143 and miR-145, which belong to the same cluster, have been found to directly target Mdm2, and both miRNAs are posttranscriptionally upregulated by p53. These two miRNAs are downregulated in head and neck squamous cell carcinoma, while Mdm2 is upregulated in these tumors [82].

Some p53-regulated miRNAs can also modulate p53 levels by controlling the p53 regulators other than Mdm2. For example, as a p53-inducible miRNA, miR-29 targets and inhibits the expression of Cdc42 and the p85α regulatory subunit of PI3K [83], both of which are p53 negative regulators. In addition, upon DNA damage, miR-29 induced transcriptionally by p53 can upregulate p53 expression by targeting Ppm1d phosphatase [84], which is a negative regulator of p53. An additional example of a positive feedback loop is the signaling pathway involving p53, miR-34a, and Sirt1 [85]. In response to cellular stress, p53 induces the expression of miR-34a, which in turn increases p53 acetylation by targeting Sirt1. The resultant increase in p53 activity amplifies p53-mediated tumor-suppressive signaling to accelerate apoptosis, senescence, and cell cycle arrest.

Unlike the positive feedback loops discussed above, there are a few examples of negative feedback loops between p53 and miRNA. For instance, in glioblastoma, miR-25 and miR-32 are downregulated by p53. These two miRNAs can directly target Mdm2. Downregulation of Mdm2 by these miRNAs leads to p53 accumulation with subsequent cell cycle arrest, cell proliferation inhibition, and impaired tumor formation [86].

More and more p53-regulated or to-be-regulated microRNAs are emerging as important players in various aspects of biological or pathological processes. The abovementioned microRNAs involved in p53 signaling are depicted in Fig. 13.2.

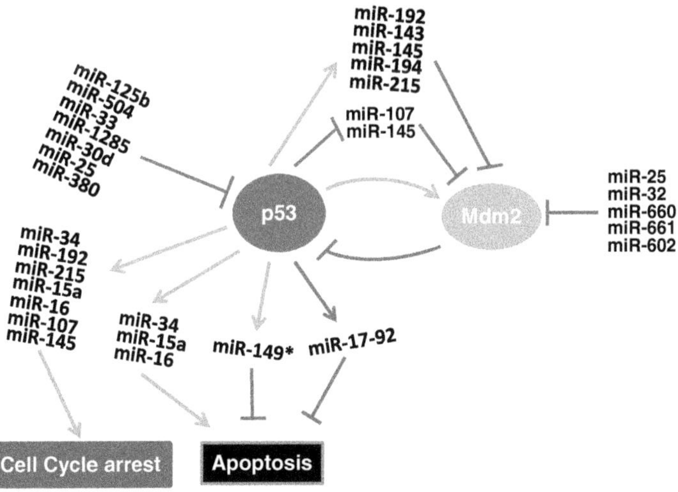

Fig. 13.2 MiRNAs regulating p53 signaling

13.3 Noncoding RNAs Regulating c-Myc Signaling

As a master transcriptional factor, c-Myc has been shown to bind to approximately 10–15% of genes in the genome [87]. Recent studies also suggest that c-Myc may function as a global amplifier of already active promoters [88]. It has been shown that by modulating expression of a variety of protein-coding genes, c-Myc regulates various cellular processes, including cell growth, cell differentiation, cell cycle, cell metabolism, and cell transformation. Increasing evidence suggests that noncoding RNAs, particularly long noncoding RNAs and miRNAs, play a critical role in the regulation of the c-Myc signaling pathway [89, 90]. Long noncoding RNAs and miRNAs join the c-Myc network as effectors or regulators. Given the strong tumor-promoting effect of c-Myc, these noncoding RNAs may be critical regulators of c-Myc-initiated tumorigenesis. Also, these c-Myc-related noncoding RNAs may represent as potential diagnostic markers and therapeutic targets for cancer.

13.3.1 Long Noncoding RNAs Regulating c-Myc Signaling

A significant association between c-Myc and lncRNAs has been observed in a wide range of cancers. It has been shown that many newly identified lncRNAs are c-Myc downstream target genes [89]. These c-Myc-responsive lncRNAs are able to regulate cancer cell proliferation and invasion [91]. Also, some lncRNAs have been reported to regulate c-Myc expression and its function. Here, we will discuss the functional role of the c-Myc-lncRNA network in tumorigenesis.

13.3.1.1 Long Noncoding RNAs as Effectors of c-Myc Signaling

By using lncRNA-specific microarray and RNA-seq analyses, a great number of lncRNAs have been recently identified as direct c-Myc downstream targets [92]. Some of these lncRNAs are indeed oncogenic molecules. Here, we will discuss the functional role of several lncRNAs as effectors in mediating c-Myc-initiated tumorigenesis.

H19

The H19 gene, located in the H19/insulin-like growth factor 2 (IGF2) locus, is subjected to genomic imprinting that leads to differential allelic expression of H19 from maternal allele and IGF2 from the paternal allele [93]. Aberrant expression of H19 has been linked to diverse human cancers. Although H19 was initially described as a tumor suppressor, more recent studies suggest H19 as an oncogene because H19 is reactivated in various human cancers, such as breast, lung, and cervical cancers [34].

c-Myc has been shown to induce H19 expression via binding to the H19 promoter [94]. Knockdown of H19 significantly decreases clonogenicity and anchorage-independent growth of breast and lung cancer cells. Moreover, c-Myc and H19 expression exhibits strong positive correlation in primary breast and lung carcinomas [94]. These findings indicate c-Myc-induced H19 is important for the regulation of tumorigenesis.

CCAT1

Colorectal cancer-associated transcript 1 (CCAT1) is a highly effective biomarker for CRC, and its upregulation is evident through all stages of CRC [95]. In addition, CCAT1 is also upregulated in other cancer types, such as gastric carcinoma. The expression of c-Myc and CCAT1 shows a strong association in gastric carcinoma [96]. c-Myc is able to enhance CCAT1 expression by directly binding to its promoter region. In both gastric and colon cancer cells, overexpression of CCAT1 promotes cell proliferation and invasion [96, 97]. These studies suggest that c-Myc-activated CCAT1 may contribute to both gastric and colon cancer formation.

MINCR

Myc-induced long noncoding RNA (MINCR) has been recently identified as a c-Myc-inducible lncRNA [98]. MINCR has a strong correlation with c-Myc expression in cancer. Knockdown of MINCR decreases the binding of c-Myc to the promoters of selective cell cycle genes, leading to reduced expression of these genes and resultant cell proliferation inhibition. These findings suggest MINCR as an important player in c-Myc-regulated transcriptional network.

BCYRN1

Brain cytoplasmic RNA 1 (BCYRN1) was originally identified as a 200-nt-long noncoding RNA (brain cytoplasmic 200, lncRNA-BC200) that was selectively expressed in the primate nervous system. It was later found to be overexpressed in cancers from the breast, cervix, esophagus, lung, ovary, parotid, and tongue [99]. The evidences from a recent study have demonstrated that BCYRN1 is also upregulated in non-small cell lung cancer [100]. c-Myc is shown to bind to the promoter region of BCYRN1 gene and increase BCYRN1 expression. Functionally, BCYN1 is necessary for c-Myc-promoted cell migration and invasion, indicating c-Myc-activated BCYN1 may be an oncogenic molecule.

13.3.1.2 Long Noncoding RNAs as Regulators of c-Myc Signaling

Emerging evidence suggests that lncRNAs are able to regulate gene expression at different levels, such as chromatin remodeling, transcription, and posttranscriptional processing. Several lncRNAs have recently been shown to regulate c-Myc expression at multiple levels, thus acting as regulators of c-Myc signaling in cancer.

CCAT1-L

Like the abovementioned CCAT1, its alternative splicing isoform CCAT1-L is also found to be significantly upregulated in CRC tissue samples compared to their normal tissue samples. In addition, CCAT1-L is expressed in several CRC-derived cell lines but not in non-CRC cell lines. The CCAT1-L transcript is encoded within the enhancer region 515 kb upstream of the c-Myc gene. Interestingly, CCAT1-L is able to promote c-Myc transcription via establishing an intra-chromosome looping between the Myc promoter and its upstream enhancer element [101]. Knockdown of CCAT1-L reduces long-distance interaction between the c-Myc promoter and its enhancers, resulting in the reduction of c-Myc transcription, suggesting that CCAT1-L functions in cis to regulate c-Myc expression.

GHET1

Gastric carcinoma high expressed transcript 1 (GHET1) was originally identified as a lncRNA that was overexpressed in gastric carcinoma [102]. It was later found to be also upregulated in bladder cancer [103]. Overexpression of GHET1 is closely related to increased tumor size, enhanced tumor invasion, and poor survival in patients. Ectopic expression of GHET1 promotes cancer cell proliferation, whereas knockdown of GHET1 has the opposite effect. Mechanistically, GHET1 enhances the interaction between c-Myc mRNA and insulin-like growth factor 2 mRNA-binding protein 1 (IGF2BP1), thereby resulting in the increased stability of c-Myc mRNA and expression. Knockdown of c-Myc suppresses the ability of GHET1 to promote cancer cell proliferation. Besides, the expression of GHET1 and c-Myc is strongly correlated in gastric carcinoma tissues. Altogether, these findings suggest that GHET1 promotes tumorigenesis via increasing c-Myc mRNA stability.

GAS5

Growth arrest-specific 5 (GAS5) has been implicated in the regulation of multiple cellular processes, including apoptosis, cell cycle arrest, and cell proliferation [104]. Low expression of GAS5 is associated with a poor prognosis in head and neck squamous cell carcinoma. Also, GAS5 is considered as a potential diagnostic marker and therapeutic target for non-small cell lung cancer. Molecular mechanisms of GAS5 action include riborepression of certain steroid hormone receptors and sequestration of several miRNAs [104]. A recent study has showed that GAS5 binds to c-Myc mRNA and suppresses c-Myc translation via cooperating with the eukaryotic translation initiation factor 4E (eIF4E) [105], suggesting that GAS5 may exert its tumor-suppressive function through repressing c-Myc expression.

PCAT-1

Prostate cancer-associated ncRNA transcript 1 (PCAT-1) has been shown to be upregulated in prostate cancer. In addition, PCAT-1 is implicated as a prognostic biomarker for colorectal cancer metastasis and poor patient survival [106, 107]. Ectopic expression of PCAT-1 promotes prostate cell proliferation. This PCAT-1-enhanced proliferation is dependent on c-Myc, as knockdown of c-Myc reverses the effect of PCAT-1 on cell proliferation. Mechanistically, PCAT-1 posttranscriptionally regulates c-Myc expression by abrogating the downregulation of c-Myc by miR-34a [108]. These findings indicate an oncogenic role of PCAT-1 in prostate cancer proliferation through c-Myc.

PVT1

Plasmacytoma variant translocation 1 (PVT1) is transcribed from approximately 100–500 kb downstream of the c-Myc gene locus within the chromosomal region 8q24.21. It has been shown that overexpression of PVT1 is significantly associated with increased metastasis and poor prognosis in many cancers, including breast, colon, gastric, and ovarian cancers [109, 110]. By analyzing available databases, it has been found that >97% of tumors with amplified 8q24 region show a co-gain of c-Myc and PVT1. This PVT1 and c-Myc protein co-amplification is also confirmed in a panel of eight human primary tumors [111]. Interestingly, knockdown of PVT1 reduces c-Myc protein levels, leading to inhibited cell proliferation and impaired tumor formation. Mechanistically, PVT1 positively regulates c-Myc protein expression via reducing its phosphorylation at threonine 58 (Thr58) and protecting it from proteasome-dependent degradation [111]. These findings suggest that PVT1 is an important regulator of tumorigenesis by controlling c-Myc protein stability.

PCGEM

Prostate cancer gene expression marker 1 (PCGEM) is a prostate tissue-specific lncRNA and highly associated with prostate cancer [112]. Overexpression of PCGEM has been found in over 80% of patient prostate tumor specimens. Ectopic expression of PCGEM is able to promote cell proliferation, increase colony formation, and confer resistance to doxorubicin-induced apoptosis [113]. These findings

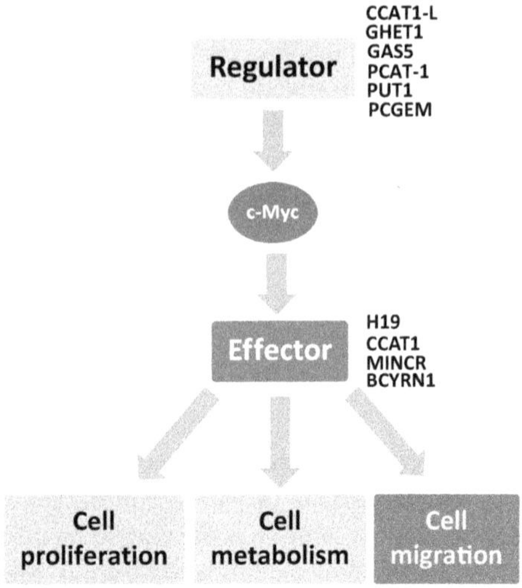

Fig. 13.3 Long noncoding RNAs regulating c-Myc signaling

highly suggest the oncogenic role of PCGEM in prostate cancer. Interestingly, it has been recently shown that PCGEM1 functions as a critical regulator of cell metabolism that facilitates the biosynthesis of cellular building materials, thus providing growth advantages for cancer cells [114]. PCGEM1 regulates metabolic reprogramming predominantly by enhancing activation of c-Myc and controlling c-Myc-dependent expression of multiple genes involved in the key metabolic pathways. The PCGEM-mediated c-Myc activation involves the direct binding of PCGEM to c-Myc that facilitates the recruitment of c-Myc to the chromatin target sites [114]. Altogether, these findings uncover PCGEM as a critical regulator of metabolic reprogramming of prostate cancer cells by being a coactivator of c-Myc.

In conclusion, long noncoding RNAs involved in regulating c-Myc signaling are emerging as novel strategies in studying gene regulation which will benefit both basic and clinical investigations. The pathways utilized by several long noncoding RNAs are briefly sketched in Fig. 13.3.

13.3.2 MiRNAs Regulating c-Myc Signaling

MiRNAs have been widely implicated as components of both tumor-suppressive and oncogenic pathways. In particular, miRNAs have been linked to the c-Myc signaling pathway. c-Myc is able to regulate a number of miRNAs, which contribute to all key c-Myc-driven phenotypes, including apoptosis, cell cycle progression, metabolism, angiogenesis, and metastasis [90, 115]. Moreover, the expression of c-Myc itself is subjected to the regulation by miRNAs, leading to sustained c-Myc activity and the corresponding c-Myc downstream pathway. Here, we will discuss how miRNAs mediate and regulate c-Myc functions in cancer.

13.3.2.1 MiRNAs as Effectors of c-Myc Signaling

c-Myc can either induce or repress miRNA expression. miRNAs that are induced by c-Myc include miR-17-92, miR-378, and miR-22. MiR-17-92 is identified as the first c-Myc-induced miRNA cluster [116, 117]. The miR-17-92 cluster encodes six distinct miRNAs, including miR-17, miR-18a, miR-19a, miR-20a, miR-19b-1, and miR-92a-1. The c-Myc-stimulated miR-17-92 cluster has been shown to mediate c-Myc-regulated cellular processes, such as cell cycle, apoptosis, and metabolism. For instance, miR-17 and miR-20a are able to directly target p21 and inhibit its expression, leading to accelerated cell cycle progression. Besides, the E2F1 family of transcription factors that regulate cell cycle progression is also targeted by miR-17 and miR-20a. The miR-17-92 cluster also inhibits apoptosis by targeting Bim, PTEN, PP2A, and AMPK [118], all of which are positive regulator of apoptosis. Furthermore, by suppressing expression of PP2A and AMPK, miR-19a/b stimulates the Akt and mTOR signaling and thereafter promotes aerobic glycolysis [119].

In agreement with its oncogenic activity, the miR-17-92 cluster is frequently activated in multiple solid tumors and B-cell lymphomas. Transgenic expression of the miR-17-92 cluster, under the control of the IgH promoter, results in B lymphoma development in mice [120]. In addition, conditional knockout of miR-17-92 in c-Myc-initiated lymphomas reduces their tumorigenicity. Furthermore, ectopic expression of the miR-17-92 cluster in PTEN deletion-induced retinoblastomas, as well as in granule neuron progenitors and colorectal colonocytes, greatly enhances tumorigenesis [121–123]. Altogether, these findings suggest that the miR-17-92 cluster represents a bona fide oncogene and an important mediator of c-Myc-driven tumorigenesis.

Another c-Myc-induced miRNA, miR-378, has been recently shown to act as an oncogenic miRNA [124]. MiR-378 is able to cooperate with activated Ras or HER2 to promote cellular transformation. MiR-378 exerts this oncogenic effect by targeting and inhibiting the antiproliferative BTG family member TOB2. In addition, miR-22 is also an oncogenic c-Myc-induced miRNA. MiR-22 triggers epithelial-mesenchymal transition (EMT), enhances invasiveness, and promotes metastasis in mouse xenografts. It also exerts its metastatic potential by silencing miR-200 gene through direct inhibition of methylcytosine dioxygenase ten-eleven translocation (TET) proteins [125].

In addition to activating miRNA expression, c-Myc is also able to repress the expression of a number of miRNAs. These c-Myc-repressed miRNAs include let7, miR-23a/b, miR-15a/16, miR-34, and miR-26. Let-7 is repressed by c-Myc in an unconventional fashion. c-Myc stimulates the expression of Lin28 and Lin28b, which bind the let-7 pre-miRNA stem loop, and thereby prevents processing by Drosha and Dicer [126]. Inhibition of let-7 leads to increased proliferation, whereas overexpression of let-7 results in cell cycle arrest. This let-7-regulated phenotype involves in the direct suppression of high-mobility group protein A2 (HMGA2) and several positive cell cycle regulators, such as CDK6 and CDC25A [127, 128]. In addition to let-7, miR-23a and miR-23b are also repressed by c-Myc. By repressing miR-23a and miR-23b, c-Myc increases the expression of GLS-1, a key enzyme responsible for the conversion of glutamine to glutamate, which serves as a substrate in the TCA cycle for the energy production [129], thereby leading to accelerated cancer cell proliferation.

The miR-15a/16 cluster is located in an intronic region of the DLEU2 gene, which is directly repressed by c-Myc [130]. Deregulation of these miRNAs has been implicated in multiple human cancers. Loss of miR-15a/16 expression is sufficient to develop chronic lymphatic leukemia (CLL) in mice [131]. By directly targeting several oncogenic factors, such as Bcl2 and cyclin D1, miR-15a/16 induces apoptosis and cell cycle arrest, thereby exerting its tumor-suppressive function [132, 133].

As mentioned earlier, miRNA-34 is positively regulated by p53 and is important for the tumor-suppressive function of p53. It has been also shown that miR-34 is repressed by c-Myc [134]. Therefore, the regulation of miR-34 may serve as an important platform of the antagonism between c-Myc and p53. By repressing miR-34 expression, c-Myc antagonizes several functions ascribed to miR-34, such as inhibition of cell cycle progression and induction of apoptosis and senescence, thereby contributing to tumor initiation and progression [115]. Interestingly, in certain contexts, miR-34 shows an oncogenic activity instead of tumor-suppressive function. For example, in B lymphoid cells with enhanced c-Myc expression, ectopic expression of miR-34, which does not directly target p53, significantly decreases p53 protein levels [135]. This effect is mediated by downregulation of c-Myc, which stimulates p53 expression via the ARF-Mdm2 axis. As a result, in cells with the intact c-Myc-ARF-Mdm2-p53 pathway, miR-34 is able to inhibit p53-dependent apoptosis. Thus, in certain tumors with upregulated c-Myc expression, miR-34a may serve as a potential therapeutic target.

MiR-26 is another c-Myc-repressed miRNA. Ectopic expression of miR-26 has been shown to induce cell cycle arrest in hepatocellular carcinoma cells (HCC) by directly targeting repression of several positive cell cycle regulators, such as CCND2 and CCNE2 [136]. These results suggest that c-Myc may contribute to HCC through repressing miR-26 expression. In support of this, restoration of miR-26 expression shows the therapeutic efficacy in a c-Myc-driven mouse model of HCC [137]. This efficacy is likely due to the cell cycle arrest caused by miR-26. Besides repressing miR-26, c-Myc may also promote HCC tumorigenesis through a miRNA-mediated positive feedback loop comprising of miR-148a, miR-363, and the ubiquitin-specific protease 28 (Usp28) [138]. c-Myc is able to directly bind to the promoters of miR-148a and miR-363 and suppress their expression. MiR-148a directly targets c-Myc and inhibits its expression, while miR-363 destabilizes c-Myc protein via targeting and inhibiting Usp28. As a result, ectopic expression of miR-148a and miR-363 promotes HCC tumorigenesis, whereas inhibition of these miRNAs has an opposite effect. Taken together, these studies demonstrate miRNA as an important class of noncoding RNA that is involved in the regulation of c-Myc-driven tumorigenesis.

13.3.2.2 MiRNAs as Regulators of c-Myc Signaling

The interaction of c-Myc and miRNAs is mutual, as a number of miRNAs have been described that regulate c-Myc expression. For example, miR-33b is a negative regulator of c-Myc through direct binding to the 3'-UTR of c-Myc mRNA [139]. Restored expression of miR-33b in a cell line without endogenous miR-33b decreases c-Myc levels, reduces anchorage-independent growth, and attenuates tumor formation in nude mice. MiR-375 is able to indirectly repress c-Myc expression by targeting the cancerous inhibitor of PP2A (CIP2A) [140], a guardian of c-Myc protein. As expected, ectopic expression of miR-375 in oral cancer cells decreases c-Myc protein levels and reduces cell proliferation, colony formation, migration, and invasion.

As discussed above, p53-induced miR-34 represses c-Myc expression. Several other p53-induced miRNAs can also target c-Myc. For instance, as a p53-induced miRNA, miR-145 is able to directly bind to the 3'-UTR of c-Myc mRNA [62]. Ectopic expression of miR-145 silences the expression of c-Myc, whereas knockdown of miR-145 enhances its expression. Furthermore, miR-145-mediated c-Myc silencing accounts at least in part for the miR-145-mediated tumor growth inhibition.

As a c-Myc-repressed miRNA, let-7 has been shown to directly repress c-Myc [141]. Ectopic expression of let-7 reduces both mRNA and protein levels of c-Myc, thus reverting c-Myc-induced growth in Burkitt lymphoma cells. Interestingly, two different miRNAs, miR-196b and miR-184, are able to concomitantly suppress c-Myc and Bcl2 expression, leading to the inhibition of cell proliferation and survival [142, 143]. Similarly, overexpression of miR-449c inhibits tumor cell migration and invasion via direct targeting of c-Myc. MiR-449 also achieves its tumor-suppressive function by targeting other factors such as E2F1.

Interestingly, miR-24, which is upregulated during terminal differentiation of multiple lines, can repress c-Myc expression via "seedless" 3'-UTR microRNA recognition element [144]. MiR-135b targets and inhibits c-Myc expression in osteosarcoma cells [145]. Inhibition of miR-135b accelerates cell proliferation, migration, and invasion, whereas forced expression of miR-135b has the opposite effect. Moreover, ectopic expression of c-Myc recovers miR-135b-inhibited cell proliferation and invasion, suggesting that miR-135b may function as a tumor suppressor via targeting c-Myc. It has been recently shown that ribosomal protein L11 regulates c-Myc mRNA turnover [146]. The unique protein-RNA complex formed by L11, Ago2, RISC, and miR-24 binds to the 3'-UTR of c-Myc mRNA and promotes its degradation. Altogether, these findings suggest that miRNA represents another important layer of the complexity of c-Myc regulation.

More microRNAs involved in c-Myc signaling still await to be characterized, and we have every reason to believe that by unveiling the underlying mechanism of those microRNAs in regulating c-Myc signaling, it will help us to gain a better understanding in noncoding RNA research field. Regulation of microRNAs in c-Myc signaling in this chapter can be seen in Fig. 13.4.

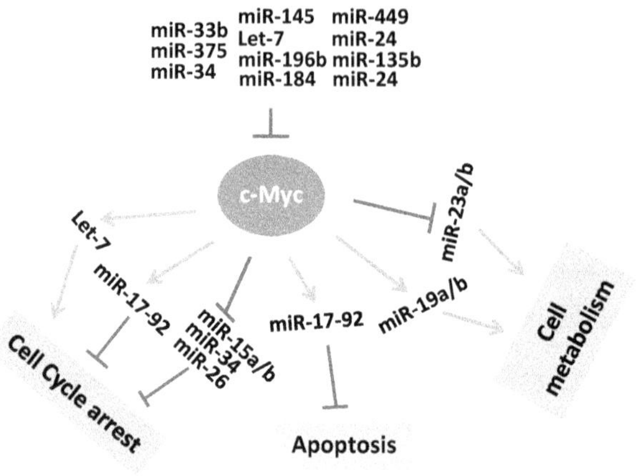

Fig. 13.4 MiRNAs regulating c-Myc signaling

13.4 Conclusions

With the advancement of high-throughput sequencing technologies, it has been recognized that the number of noncoding genes largely exceeds that of protein-coding genes. This has led to one of the most significant shifts in our understandings of gene regulatory networks. Given that RNA is an evolutionary predecessor of proteins, noncoding RNA may represent an ancient mechanism by which gene expression is finely controlled. Here, we reviewed recent findings regarding the critical role of noncoding RNAs, especially miRNAs and lncRNAs, in regulating p53 and c-Myc signaling. Given the importance of p53 and c-Myc in the regulation of tumorigenesis, it is not surprising that some p53- and c-Myc-related noncoding RNAs are involved in tumor initiation and/or progression, and also these noncoding RNAs may represent as potential cancer biomarkers and targets for cancer treatment. However, there are still some critical issues that remain to be addressed. For example, by acting as master transcription factors, both p53 and c-Myc are able to regulate a variety of noncoding RNA target gene expressions. Also, p53 and c-Myc signaling are subjected to the tight regulation by noncoding RNAs. It remains unclear how these complex noncoding RNA networks are integrated into a mechanical system for regulating p53 and c-Myc functions in cancer. In addition, although dysregulation of several p53- and c-Myc-regulated noncoding RNAs has been implicated in tumorigenesis, the underlying mechanisms are largely unknown. Nevertheless, we believe that the noncoding RNA field is a rich landscape waiting to be further unveiled and functional exploration of noncoding RNA world will open unexpected possibilities, not only for biological but also for translational research.

References

1. Kruiswijk F, Labuschagne CF, Vousden KH. p53 in survival, death and metabolic health: a lifeguard with a licence to kill. Nat Rev Mol Cell Biol. 2015;16:393–405.
2. Levine AJ, Oren M. The first 30 years of p53: growing ever more complex. Nat Rev Cancer. 2009;9:749–58.
3. Vita M, Henriksson M. The Myc oncoprotein as a therapeutic target for human cancer. Semin Cancer Biol. 2006;16:318–30.
4. Dang CV. MYC on the path to cancer. Cell. 2012;149:22–35.
5. Dang CV. c-Myc target genes involved in cell growth, apoptosis, and metabolism. Mol Cell Biol. 1999;19:1–11.
6. Farrell AS, Sears RC. MYC degradation. Cold Spring Harb Perspect Med. 2014;4, pii: a014365.
7. Brooks CL, Gu W. Dynamics in the p53-Mdm2 ubiquitination pathway. Cell Cycle. 2004;3:895–9.
8. Li Y, Lu H. Non-coding RNAs: 'our turn' to join the p53 network. J Mol Cell Biol. 2014;6:179–80.
9. Batista PJ, Chang HY. Long non-coding RNAs: cellular address codes in development and disease. Cell. 2013;152:1298–307.
10. Zhang A, Xu M, Mo YY. Role of the lncRNA-p53 regulatory network in cancer. J Mol Cell Biol. 2014;6:181–91.
11. Idogawa M, Ohashi T, Sasaki Y, Maruyama R, Kashima L, Suzuki H, Tokino T. Identification and analysis of large intergenic non-coding RNAs regulated by p53 family members through a genome-wide analysis of p53-binding sites. Hum Mol Genet. 2014;23:2847–57.
12. Sanchez Y, Segura V, Marin-Bejar O, Athie A, Marchese FP, Gonzalez J, Bujanda L, Guo S, Matheu A, Huarte M. Genome-wide analysis of the human p53 transcriptional network unveils a lncRNA tumour suppressor signature. Nat Commun. 2014;5:5812.
13. Huarte M, Guttman M, Feldser D, Garber M, Koziol MJ, Kenzelmann-Broz D, Khalil AM, Zuk O, Amit I, Rabani M, Attardi LD, Regev A, Lander ES, Jacks T, Rinn JL. A large intergenic non-coding RNA induced by p53 mediates global gene repression in the p53 response. Cell. 2010;142:409–19.
14. Tang SS, Zheng BY, Xiong XD. LincRNA-p21: implications in human diseases. Int J Mol Sci. 2015;16:18732–40.
15. Dimitrova N, Zamudio JR, Jong RM, Soukup D, Resnick R, Sarma K, Ward AJ, Raj A, Lee JT, Sharp PA, Jacks T. LincRNA-p21 activates p21 in cis to promote Polycomb target gene expression and to enforce the G1/S checkpoint. Mol Cell. 2014;54:777–90.
16. Grossi E, Sanchez Y, Huarte M. Expanding the p53 regulatory network: lncRNAs take up the challenge. Biochim Biophys Acta. 2015;1859(1):200–8.
17. Yoon JH, Abdelmohsen K, Srikantan S, Yang X, Martindale JL, De S, Huarte M, Zhan M, Becker KG, Gorospe M. LincRNA-p21 suppresses target mRNA translation. Mol Cell. 2012;47:648–55.
18. Yang F, Zhang H, Mei Y, Wu M. Reciprocal regulation of HIF-1alpha and lincRNA-p21 modulates the Warburg effect. Mol Cell. 2014;53:88–100.
19. Bao X, Wu H, Zhu X, Guo X, Hutchins AP, Luo Z, Song H, Chen Y, Lai K, Yin M, Xu L, Zhou L, Chen J, Wang D, Qin B, Frampton J, Tse HF, Pei D, Wang H, Zhang B, Esteban MA. The p53-induced lincRNA-p21 derails somatic cell reprogramming by sustaining H3K9me3 and CpG methylation at pluripotency gene promoters. Cell Res. 2015;25:80–92.
20. Hung T, Wang Y, Lin MF, Koegel AK, Kotake Y, Grant GD, Horlings HM, Shah N, Umbricht C, Wang P, Kong B, Langerod A, Borresen-Dale AL, Kim SK, van de Vijver M, Sukumar S, Whitfield ML, Kellis M, Xiong Y, Wong DJ, Chang HY. Extensive and coordinated transcription of non-coding RNAs within cell-cycle promoters. Nat Genet. 2011;43:621–9.

21. Han L, Zhang EB, Yin DD, Kong R, Xu TP, Chen WM, Xia R, Shu YQ, De W. Low expression of long non-coding RNA PANDAR predicts a poor prognosis of non-small cell lung cancer and affects cell apoptosis by regulating Bcl-2. Cell Death Dis. 2015;6:e1665.
22. Pasic I, Shlien A, Durbin AD, Stavropoulos DJ, Baskin B, Ray PN, Novokmet A, Malkin D. Recurrent focal copy-number changes and loss of heterozygosity implicate two noncoding RNAs and one tumor suppressor gene at chromosome 3q13.31 in osteosarcoma. Cancer Res. 2010;70:160–71.
23. Liu Q, Huang J, Zhou N, Zhang Z, Zhang A, Lu Z, Wu F, Mo YY. LncRNA loc285194 is a p53-regulated tumor suppressor. Nucleic Acids Res. 2013;41:4976–87.
24. Qi P, Xu MD, Ni SJ, Huang D, Wei P, Tan C, Zhou XY, Du X. Low expression of LOC285194 is associated with poor prognosis in colorectal cancer. J Transl Med. 2013;11:122.
25. Young TL, Matsuda T, Cepko CL. The non-coding RNA taurine upregulated gene 1 is required for differentiation of the murine retina. Curr Biol. 2005;15:501–12.
26. Zhang EB, Yin DD, Sun M, Kong R, Liu XH, You LH, Han L, Xia R, Wang KM, Yang JS, De W, Shu YQ, Wang ZX. P53-regulated long non-coding RNA TUG1 affects cell proliferation in human non-small cell lung cancer, partly through epigenetically regulating HOXB7 expression. Cell Death Dis. 2014;5:e1243.
27. Yang L, Lin C, Liu W, Zhang J, Ohgi KA, Grinstein JD, Dorrestein PC, Rosenfeld MG. ncRNA- and Pc2 methylation-dependent gene relocation between nuclear structures mediates gene activation programs. Cell. 2011;147:773–88.
28. Khalil AM, Guttman M, Huarte M, Garber M, Raj A, Rivea Morales D, Thomas K, Presser A, Bernstein BE, van Oudenaarden A, Regev A, Lander ES, Rinn JL. Many human large intergenic non-coding RNAs associate with chromatin-modifying complexes and affect gene expression. Proc Natl Acad Sci U S A. 2009;106:11667–72.
29. Xu Y, Wang J, Qiu M, Xu L, Li M, Jiang F, Yin R. Upregulation of the long non-coding RNA TUG1 promotes proliferation and migration of esophageal squamous cell carcinoma. Tumour Biol. 2015;36:1643–51.
30. Natoli G, Andrau JC. Non-coding transcription at enhancers: general principles and functional models. Annu Rev Genet. 2012;46:1–19.
31. Younger ST, Kenzelmann-Broz D, Jung H, Attardi LD, Rinn JL. Integrative genomic analysis reveals widespread enhancer regulation by p53 in response to DNA damage. Nucleic Acids Res. 2015;43:4447–62.
32. Melo CA, Drost J, Wijchers PJ, van de Werken H, de Wit E, Oude Vrielink JA, Elkon R, Melo SA, Leveille N, Kalluri R, de Laat W, Agami R. eRNAs are required for p53-dependent enhancer activity and gene transcription. Mol Cell. 2013;49:524–35.
33. Ji P, Diederichs S, Wang W, Boing S, Metzger R, Schneider PM, Tidow N, Brandt B, Buerger H, Bulk E, Thomas M, Berdel WE, Serve H, Muller-Tidow C. MALAT-1, a novel non-coding RNA, and thymosin beta4 predict metastasis and survival in early-stage non-small cell lung cancer. Oncogene. 2003;22:8031–41.
34. Gibb EA, Brown CJ, Lam WL. The functional role of long non-coding RNA in human carcinomas. Mol Cancer. 2011;10:38.
35. Gutschner T, Hammerle M, Eissmann M, Hsu J, Kim Y, Hung G, Revenko A, Arun G, Stentrup M, Gross M, Zornig M, MacLeod AR, Spector DL, Diederichs S. The non-coding RNA MALAT1 is a critical regulator of the metastasis phenotype of lung cancer cells. Cancer Res. 2013;73:1180–9.
36. Tripathi V, Shen Z, Chakraborty A, Giri S, Freier SM, Wu X, Zhang Y, Gorospe M, Prasanth SG, Lal A, Prasanth KV. Long non-coding RNA MALAT1 controls cell cycle progression by regulating the expression of oncogenic transcription factor B-MYB. PLoS Genet. 2013;9:e1003368.
37. Zhang X, Zhou Y, Mehta KR, Danila DC, Scolavino S, Johnson SR, Klibanski A. A pituitary-derived MEG3 isoform functions as a growth suppressor in tumor cells. J Clin Endocrinol Metab. 2003;88:5119–26.

38. Zhou Y, Zhong Y, Wang Y, Zhang X, Batista DL, Gejman R, Ansell PJ, Zhao J, Weng C, Klibanski A. Activation of p53 by MEG3 non-coding RNA. J Biol Chem. 2007;282:24731–42.
39. Mahmoudi S, Henriksson S, Corcoran M, Mendez-Vidal C, Wiman KG, Farnebo M. Wrap53, a natural p53 antisense transcript required for p53 induction upon DNA damage. Mol Cell. 2009;33:462–71.
40. Saldana-Meyer R, Gonzalez-Buendia E, Guerrero G, Narendra V, Bonasio R, Recillas-Targa F, Reinberg D. CTCF regulates the human p53 gene through direct interaction with its natural antisense transcript, Wrap53. Genes Dev. 2014;28:723–34.
41. Loewer S, Cabili MN, Guttman M, Loh YH, Thomas K, Park IH, Garber M, Curran M, Onder T, Agarwal S, Manos PD, Datta S, Lander ES, Schlaeger TM, Daley GQ, Rinn JL. Large intergenic non-coding RNA-RoR modulates reprogramming of human induced pluripotent stem cells. Nat Genet. 2010;42:1113–7.
42. Zhang A, Zhou N, Huang J, Liu Q, Fukuda K, Ma D, Lu Z, Bai C, Watabe K, Mo YY. The human long non-coding RNA-RoR is a p53 repressor in response to DNA damage. Cell Res. 2013;23:340–50.
43. Bartel DP. MicroRNAs: genomics, biogenesis, mechanism, and function. Cell. 2004;116:281–97.
44. Iorio MV, Croce CM. MicroRNA dysregulation in cancer: diagnostics, monitoring and therapeutics. EMBO Mol Med. 2012;4:143–59.
45. Liao JM, Cao B, Zhou X, Lu H. New insights into p53 functions through its target microRNAs. J Mol Cell Biol. 2014;6:206–13.
46. Krell J, Frampton AE, Colombo T, Gall TM, De Giorgio A, Harding V, Stebbing J, Castellano L. The p53 miRNA interactome and its potential role in the cancer clinic. Epigenomics. 2013;5:417–28.
47. Bommer GT, Gerin I, Feng Y, Kaczorowski AJ, Kuick R, Love RE, Zhai Y, Giordano TJ, Qin ZS, Moore BB, MacDougald OA, Cho KR, Fearon ER. p53-mediated activation of miRNA34 candidate tumor-suppressor genes. Curr Biol. 2007;17:1298–307.
48. Chang TC, Wentzel EA, Kent OA, Ramachandran K, Mullendore M, Lee KH, Feldmann G, Yamakuchi M, Ferlito M, Lowenstein CJ, Arking DE, Beer MA, Maitra A, Mendell JT. Transactivation of miR-34a by p53 broadly influences gene expression and promotes apoptosis. Mol Cell. 2007;26:745–52.
49. Corney DC, Flesken-Nikitin A, Godwin AK, Wang W, Nikitin AY. MicroRNA-34b and MicroRNA-34c are targets of p53 and cooperate in control of cell proliferation and adhesion-independent growth. Cancer Res. 2007;67:8433–8.
50. He L, He X, Lim LP, de Stanchina E, Xuan Z, Liang Y, Xue W, Zender L, Magnus J, Ridzon D, Jackson AL, Linsley PS, Chen C, Lowe SW, Cleary MA, Hannon GJ. A microRNA component of the p53 tumour suppressor network. Nature. 2007;447:1130–4.
51. Raver-Shapira N, Marciano E, Meiri E, Spector Y, Rosenfeld N, Moskovits N, Bentwich Z, Oren M. Transcriptional activation of miR-34a contributes to p53-mediated apoptosis. Mol Cell. 2007;26:731–43.
52. Tazawa H, Tsuchiya N, Izumiya M, Nakagama H. Tumor-suppressive miR-34a induces senescence-like growth arrest through modulation of the E2F pathway in human colon cancer cells. Proc Natl Acad Sci U S A. 2007;104:15472–7.
53. Yamakuchi M, Ferlito M, Lowenstein CJ. miR-34a repression of SIRT1 regulates apoptosis. Proc Natl Acad Sci U S A. 2008;105:13421–6.
54. Concepcion CP, Han YC, Mu P, Bonetti C, Yao E, D'Andrea A, Vidigal JA, Maughan WP, Ogrodowski P, Ventura A. Intact p53-dependent responses in miR-34-deficient mice. PLoS Genet. 2012;8:e1002797.
55. Georges SA, Biery MC, Kim SY, Schelter JM, Guo J, Chang AN, Jackson AL, Carleton MO, Linsley PS, Cleary MA, Chau BN. Coordinated regulation of cell cycle transcripts by p53-Inducible microRNAs, miR-192 and miR-215. Cancer Res. 2008;68:10105–12.

56. Piovan C, Palmieri D, Di Leva G, Braccioli L, Casalini P, Nuovo G, Tortoreto M, Sasso M, Plantamura I, Triulzi T, Taccioli C, Tagliabue E, Iorio MV, Croce CM. Oncosuppressive role of p53-induced miR-205 in triple negative breast cancer. Mol Oncol. 2012;6:458–72.
57. Calin GA, Dumitru CD, Shimizu M, Bichi R, Zupo S, Noch E, Aldler H, Rattan S, Keating M, Rai K, Rassenti L, Kipps T, Negrini M, Bullrich F, Croce CM. Frequent deletions and down-regulation of micro- RNA genes miR15 and miR16 at 13q14 in chronic lymphocytic leukemia. Proc Natl Acad Sci U S A. 2002;99:15524–9.
58. Brabletz S, Bajdak K, Meidhof S, Burk U, Niedermann G, Firat E, Wellner U, Dimmler A, Faller G, Schubert J, Brabletz T. The ZEB1/miR-200 feedback loop controls Notch signalling in cancer cells. EMBO J. 2011;30:770–82.
59. Gregory PA, Bert AG, Paterson EL, Barry SC, Tsykin A, Farshid G, Vadas MA, Khew-Goodall Y, Goodall GJ. The miR-200 family and miR-205 regulate epithelial to mesenchymal transition by targeting ZEB1 and SIP1. Nat Cell Biol. 2008;10:593–601.
60. Bohlig L, Friedrich M, Engeland K. p53 activates the PANK1/miRNA-107 gene leading to downregulation of CDK6 and p130 cell cycle proteins. Nucleic Acids Res. 2011;39:440–53.
61. Yamakuchi M, Lotterman CD, Bao C, Hruban RH, Karim B, Mendell JT, Huso D, Lowenstein CJ. P53-induced microRNA-107 inhibits HIF-1 and tumor angiogenesis. Proc Natl Acad Sci U S A. 2010;107:6334–9.
62. Sachdeva M, Zhu S, Wu F, Wu H, Walia V, Kumar S, Elble R, Watabe K, Mo YY. p53 represses c-Myc through induction of the tumor suppressor miR-145. Proc Natl Acad Sci U S A. 2009;106:3207–12.
63. Yan HL, Xue G, Mei Q, Wang YZ, Ding FX, Liu MF, Lu MH, Tang Y, Yu HY, Sun SH. Repression of the miR-17-92 cluster by p53 has an important function in hypoxia-induced apoptosis. EMBO J. 2009;28:2719–32.
64. Suzuki HI, Yamagata K, Sugimoto K, Iwamoto T, Kato S, Miyazono K. Modulation of microRNA processing by p53. Nature. 2009;460:529–33.
65. Leveille N, Elkon R, Davalos V, Manoharan V, Hollingworth D, Oude Vrielink J, le Sage C, Melo CA, Horlings HM, Wesseling J, Ule J, Esteller M, Ramos A, Agami R. Selective inhibition of microRNA accessibility by RBM38 is required for p53 activity. Nat Commun. 2011;2:513.
66. Sundaram P, Hultine S, Smith LM, Dews M, Fox JL, Biyashev D, Schelter JM, Huang Q, Cleary MA, Volpert OV, Thomas-Tikhonenko A. p53-responsive miR-194 inhibits thrombospondin-1 and promotes angiogenesis in colon cancers. Cancer Res. 2011;71:7490–501.
67. Jin L, Hu WL, Jiang CC, Wang JX, Han CC, Chu P, Zhang LJ, Thorne RF, Wilmott J, Scolyer RA, Hersey P, Zhang XD, Wu M. MicroRNA-149*, a p53-responsive microRNA, functions as an oncogenic regulator in human melanoma. Proc Natl Acad Sci U S A. 2011;108:15840–5.
68. Zhang C, Liu J, Wang X, Feng Z. The regulation of the p53/MDM2 feedback loop by microRNAs. RNA Dis. 2015;2:e502.
69. Sun YM, Lin KY, Chen YQ. Diverse functions of miR-125 family in different cell contexts. J Hematol Oncol. 2013;6:6.
70. Le MT, Shyh-Chang N, Khaw SL, Chin L, Teh C, Tay J, O'Day E, Korzh V, Yang H, Lal A, Lieberman J, Lodish HF, Lim B. Conserved regulation of p53 network dosage by microRNA-125b occurs through evolving miRNA-target gene pairs. PLoS Genet. 2011;7:e1002242.
71. Enomoto Y, Kitaura J, Hatakeyama K, Watanuki J, Akasaka T, Kato N, Shimanuki M, Nishimura K, Takahashi M, Taniwaki M, Haferlach C, Siebert R, Dyer MJ, Asou N, Aburatani H, Nakakuma H, Kitamura T, Sonoki T. Emu/miR-125b transgenic mice develop lethal B-cell malignancies. Leukemia. 2011;25:1849–56.
72. Zhang Y, Yan LX, Wu QN, Du ZM, Chen J, Liao DZ, Huang MY, Hou JH, Wu QL, Zeng MS, Huang WL, Zeng YX, Shao JY. miR-125b is methylated and functions as a tumor suppressor by regulating the ETS1 proto-oncogene in human invasive breast cancer. Cancer Res. 2011;71:3552–62.

73. Hermeking H. MicroRNAs in the p53 network: micromanagement of tumour suppression. Nat Rev Cancer. 2012;12:613–26.
74. Kumar M, Lu Z, Takwi AA, Chen W, Callander NS, Ramos KS, Young KH, Li Y. Negative regulation of the tumor suppressor p53 gene by microRNAs. Oncogene. 2011;30:843–53.
75. Li N, Kaur S, Greshock J, Lassus H, Zhong X, Wang Y, Leminen A, Shao Z, Hu X, Liang S, Katsaros D, Huang Q, Butzow R, Weber BL, Coukos G, Zhang L. A combined array-based comparative genomic hybridization and functional library screening approach identifies mir-30d as an oncomir in cancer. Cancer Res. 2012;72:154–64.
76. Swarbrick A, Woods SL, Shaw A, Balakrishnan A, Phua Y, Nguyen A, Chanthery Y, Lim L, Ashton LJ, Judson RL, Huskey N, Blelloch R, Haber M, Norris MD, Lengyel P, Hackett CS, Preiss T, Chetcuti A, Sullivan CS, Marcusson EG, Weiss W, L'Etoile N, Goga A. miR-380-5p represses p53 to control cellular survival and is associated with poor outcome in MYCN-amplified neuroblastoma. Nat Med. 2010;16:1134–40.
77. Li Y, Gordon MW, Xu-Monette ZY, Visco C, Tzankov A, Zou D, Qiu L, Montes-Moreno S, Dybkaer K, Orazi A, Zu Y, Bhagat G, Richards KL, Hsi ED, Choi WW, van Krieken JH, Huang Q, Ai W, Ponzoni M, Ferreri AJ, Winter JN, Go RS, Piris MA, Moller MB, Wu L, Wang M, Ramos KS, Medeiros LJ, Young KH. Single nucleotide variation in the TP53 3' untranslated region in diffuse large B-cell lymphoma treated with rituximab-CHOP: a report from the International DLBCL Rituximab-CHOP Consortium Program. Blood. 2013;121:4529–40.
78. Braun CJ, Zhang X, Savelyeva I, Wolff S, Moll UM, Schepeler T, Orntoft TF, Andersen CL, Dobbelstein M. p53-Responsive micrornas 192 and 215 are capable of inducing cell cycle arrest. Cancer Res. 2008;68:10094–104.
79. Pichiorri F, Suh SS, Rocci A, De Luca L, Taccioli C, Santhanam R, Zhou W, Benson Jr DM, Hofmainster C, Alder H, Garofalo M, Di Leva G, Volinia S, Lin HJ, Perrotti D, Kuehl M, Aqeilan RI, Palumbo A, Croce CM. Downregulation of p53-inducible microRNAs 192, 194, and 215 impairs the p53/MDM2 autoregulatory loop in multiple myeloma development. Cancer Cell. 2010;18:367–81.
80. Hoffman Y, Pilpel Y, Oren M. microRNAs and Alu elements in the p53-Mdm2-Mdm4 regulatory network. J Mol Cell Biol. 2014;6:192–7.
81. Xiao J, Lin H, Luo X, Wang Z. miR-605 joins p53 network to form a p53:miR-605:Mdm2 positive feedback loop in response to stress. EMBO J. 2011;30:524–32.
82. Zhang J, Sun Q, Zhang Z, Ge S, Han ZG, Chen WT. Loss of microRNA-143/145 disturbs cellular growth and apoptosis of human epithelial cancers by impairing the MDM2-p53 feedback loop. Oncogene. 2013;32:61–9.
83. Park SY, Lee JH, Ha M, Nam JW, Kim VN. miR-29 miRNAs activate p53 by targeting p85 alpha and CDC42. Nat Struct Mol Biol. 2009;16:23–9.
84. Ugalde AP, Ramsay AJ, de la Rosa J, Varela I, Marino G, Cadinanos J, Lu J, Freije JM, Lopez-Otin C. Aging and chronic DNA damage response activate a regulatory pathway involving miR-29 and p53. EMBO J. 2011;30:2219–32.
85. Yamakuchi M, Lowenstein CJ. MiR-34, SIRT1 and p53: the feedback loop. Cell Cycle. 2009;8:712–5.
86. Suh SS, Yoo JY, Nuovo GJ, Jeon YJ, Kim S, Lee TJ, Kim T, Bakacs A, Alder H, Kaur B, Aqeilan RI, Pichiorri F, Croce CM. MicroRNAs/TP53 feedback circuitry in glioblastoma multiforme. Proc Natl Acad Sci U S A. 2012;109:5316–21.
87. Meyer N, Penn LZ. Reflecting on 25 years with MYC. Nat Rev Cancer. 2008;8:976–90.
88. Nie Z, Hu G, Wei G, Cui K, Yamane A, Resch W, Wang R, Green DR, Tessarollo L, Casellas R, Zhao K, Levens D. c-Myc is a universal amplifier of expressed genes in lymphocytes and embryonic stem cells. Cell. 2012;151:68–79.
89. Deng K, Guo X, Wang H, Xia J. The lncRNA-MYC regulatory network in cancer. Tumour Biol. 2014;35:9497–503.
90. Psathas JN, Thomas-Tikhonenko A. MYC and the art of microRNA maintenance. Cold Spring Harb Perspect Med. 2014;4, pii: a014175.

91. Kim T, Cui R, Jeon YJ, Fadda P, Alder H, Croce CM. MYC-repressed long non-coding RNAs antagonize MYC-induced cell proliferation and cell cycle progression. Oncotarget. 2015;6:18780–9.
92. Hart JR, Roberts TC, Weinberg MS, Morris KV, Vogt PK. MYC regulates the non-coding transcriptome. Oncotarget. 2014;5:12543–54.
93. Reik W, Walter J. Genomic imprinting: parental influence on the genome. Nat Rev Genet. 2001;2:21–32.
94. Barsyte-Lovejoy D, Lau SK, Boutros PC, Khosravi F, Jurisica I, Andrulis IL, Tsao MS, Penn LZ. The c-Myc oncogene directly induces the H19 non-coding RNA by allele-specific binding to potentiate tumorigenesis. Cancer Res. 2006;66:5330–7.
95. Nissan A, Stojadinovic A, Mitrani-Rosenbaum S, Halle D, Grinbaum R, Roistacher M, Bochem A, Dayanc BE, Ritter G, Gomceli I, Bostanci EB, Akoglu M, Chen YT, Old LJ, Gure AO. Colon cancer associated transcript-1: a novel RNA expressed in malignant and pre-malignant human tissues. Int J Cancer. 2012;130:1598–606.
96. Yang F, Xue X, Bi J, Zheng L, Zhi K, Gu Y, Fang G. Long non-coding RNA CCAT1, which could be activated by c-Myc, promotes the progression of gastric carcinoma. J Cancer Res Clin Oncol. 2013;139:437–45.
97. He X, Tan X, Wang X, Jin H, Liu L, Ma L, Yu H, Fan Z. C-Myc-activated long non-coding RNA CCAT1 promotes colon cancer cell proliferation and invasion. Tumour Biol. 2014;35:12181–8.
98. Doose G, Haake A, Bernhart SH, Lopez C, Duggimpudi S, Wojciech F, Bergmann AK, Borkhardt A, Burkhardt B, Claviez A, Dimitrova L, Haas S, Hoell JI, Hummel M, Karsch D, Klapper W, Kleo K, Kretzmer H, Kreuz M, Kuppers R, Lawerenz C, Lenze D, Loeffler M, Mantovani-Loffler L, Moller P, Ott G, Richter J, Rohde M, Rosenstiel P, Rosenwald A, Schilhabel M, Schneider M, Scholz I, Stilgenbauer S, Stunnenberg HG, Szczepanowski M, Trumper L, Weniger MA, Hoffmann S, Siebert R, Iaccarino I. MINCR is a MYC-induced lncRNA able to modulate MYC's transcriptional network in Burkitt lymphoma cells. Proc Natl Acad Sci U S A. 2015;112:E5261–70.
99. Chen W, Bocker W, Brosius J, Tiedge H. Expression of neural BC200 RNA in human tumours. J Pathol. 1997;183:345–51.
100. Hu T, Lu YR. BCYRN1, a c-MYC-activated long non-coding RNA, regulates cell metastasis of non-small-cell lung cancer. Cancer Cell Int. 2015;15:36.
101. Xiang JF, Yin QF, Chen T, Zhang Y, Zhang XO, Wu Z, Zhang S, Wang HB, Ge J, Lu X, Yang L, Chen LL. Human colorectal cancer-specific CCAT1-L lncRNA regulates long-range chromatin interactions at the MYC locus. Cell Res. 2014;24:513–31.
102. Yang F, Xue X, Zheng L, Bi J, Zhou Y, Zhi K, Gu Y, Fang G. Long non-coding RNA GHET1 promotes gastric carcinoma cell proliferation by increasing c-Myc mRNA stability. FEBS J. 2014;281:802–13.
103. Li LJ, Zhu JL, Bao WS, Chen DK, Huang WW, Weng ZL. Long non-coding RNA GHET1 promotes the development of bladder cancer. Int J Clin Exp Pathol. 2014;7:7196–205.
104. Pickard MR, Williams GT. Molecular and cellular mechanisms of action of tumour suppressor GAS5 lncRNA. Genes (Basel). 2015;6:484–99.
105. Hu G, Lou Z, Gupta M. The long non-coding RNA GAS5 cooperates with the eukaryotic translation initiation factor 4E to regulate c-Myc translation. Plos ONE. 2014;9:e107016.
106. Prensner JR, Iyer MK, Balbin OA, Dhanasekaran SM, Cao Q, Brenner JC, Laxman B, Asangani IA, Grasso CS, Kominsky HD, Cao X, Jing X, Wang X, Siddiqui J, Wei JT, Robinson D, Iyer HK, Palanisamy N, Maher CA, Chinnaiyan AM. Transcriptome sequencing across a prostate cancer cohort identifies PCAT-1, an unannotated lincRNA implicated in disease progression. Nat Biotechnol. 2011;29:742–9.
107. Ge X, Chen Y, Liao X, Liu D, Li F, Ruan H, Jia W. Overexpression of long non-coding RNA PCAT-1 is a novel biomarker of poor prognosis in patients with colorectal cancer. Med Oncol. 2013;30:588.

108. Prensner JR, Chen W, Han S, Iyer MK, Cao Q, Kothari V, Evans JR, Knudsen KE, Paulsen MT, Ljungman M, Lawrence TS, Chinnaiyan AM, Feng FY. The long non-coding RNA PCAT-1 promotes prostate cancer cell proliferation through cMyc. Neoplasia. 2014;16:900–8.
109. Colombo T, Farina L, Macino G, Paci P. PVT1: a rising star among oncogenic long non-coding RNAs. Biomed Res Int. 2015;2015:304208.
110. Johnsson P, Morris KV. Expanding the functional role of long non-coding RNAs. Cell Res. 2014;24:1284–5.
111. Tseng YY, Moriarity BS, Gong W, Akiyama R, Tiwari A, Kawakami H, Ronning P, Reuland B, Guenther K, Beadnell TC, Essig J, Otto GM, O'Sullivan MG, Largaespada DA, Schwertfeger KL, Marahrens Y, Kawakami Y, Bagchi A. PVT1 dependence in cancer with MYC copy-number increase. Nature. 2014;512:82–6.
112. Srikantan V, Zou Z, Petrovics G, Xu L, Augustus M, Davis L, Livezey JR, Connell T, Sesterhenn IA, Yoshino K, Buzard GS, Mostofi FK, McLeod DG, Moul JW, Srivastava S. PCGEM1, a prostate-specific gene, is overexpressed in prostate cancer. Proc Natl Acad Sci U S A. 2000;97:12216–21.
113. Petrovics G, Zhang W, Makarem M, Street JP, Connelly R, Sun L, Sesterhenn IA, Srikantan V, Moul JW, Srivastava S. Elevated expression of PCGEM1, a prostate-specific gene with cell growth-promoting function, is associated with high-risk prostate cancer patients. Oncogene. 2004;23:605–11.
114. Hung CL, Wang LY, Yu YL, Chen HW, Srivastava S, Petrovics G, Kung HJ. A long non-coding RNA connects c-Myc to tumor metabolism. Proc Natl Acad Sci U S A. 2014;111:18697–702.
115. Jackstadt R, Hermeking H. MicroRNAs as regulators and mediators of c-MYC function. Biochim Biophys Acta. 2015;1849:544–53.
116. O'Donnell KA, Wentzel EA, Zeller KI, Dang CV, Mendell JT. c-Myc-regulated microRNAs modulate E2F1 expression. Nature. 2005;435:839–43.
117. He L, Thomson JM, Hemann MT, Hernando-Monge E, Mu D, Goodson S, Powers S, Cordon-Cardo C, Lowe SW, Hannon GJ, Hammond SM. A microRNA polycistron as a potential human oncogene. Nature. 2005;435:828–33.
118. Mavrakis KJ, Wolfe AL, Oricchio E, Palomero T, de Keersmaecker K, McJunkin K, Zuber J, James T, Khan AA, Leslie CS, Parker JS, Paddison PJ, Tam W, Ferrando A, Wendel HG. Genome-wide RNA-mediated interference screen identifies miR-19 targets in Notch-induced T-cell acute lymphoblastic leukaemia. Nat Cell Biol. 2010;12:372–9.
119. Mu P, Han YC, Betel D, Yao E, Squatrito M, Ogrodowski P, de Stanchina E, D'Andrea A, Sander C, Ventura A. Genetic dissection of the miR-17~92 cluster of microRNAs in Myc-induced B-cell lymphomas. Genes Dev. 2009;23:2806–11.
120. Sandhu SK, Fassan M, Volinia S, Lovat F, Balatti V, Pekarsky Y, Croce CM. B-cell malignancies in microRNA Emu-miR-17~92 transgenic mice. Proc Natl Acad Sci U S A. 2013;110:18208–13.
121. Dews M, Homayouni A, Yu D, Murphy D, Sevignani C, Wentzel E, Furth EE, Lee WM, Enders GH, Mendell JT, Thomas-Tikhonenko A. Augmentation of tumor angiogenesis by a Myc-activated microRNA cluster. Nat Genet. 2006;38:1060–5.
122. Conkrite K, Sundby M, Mukai S, Thomson JM, Mu D, Hammond SM, MacPherson D. miR-17~92 cooperates with RB pathway mutations to promote retinoblastoma. Genes Dev. 2011;25:1734–45.
123. Uziel T, Karginov FV, Xie S, Parker JS, Wang YD, Gajjar A, He L, Ellison D, Gilbertson RJ, Hannon G, Roussel MF. The miR-17~92 cluster collaborates with the Sonic Hedgehog pathway in medulloblastoma. Proc Natl Acad Sci U S A. 2009;106:2812–7.
124. Feng M, Li Z, Aau M, Wong CH, Yang X, Yu Q. Myc/miR-378/TOB2/cyclin D1 functional module regulates oncogenic transformation. Oncogene. 2011;30:2242–51.
125. Song SJ, Poliseno L, Song MS, Ala U, Webster K, Ng C, Beringer G, Brikbak NJ, Yuan X, Cantley LC, Richardson AL, Pandolfi PP. MicroRNA-antagonism regulates breast cancer

stemness and metastasis via TET-family-dependent chromatin remodeling. Cell. 2013;154:311–24.
126. Chang TC, Zeitels LR, Hwang HW, Chivukula RR, Wentzel EA, Dews M, Jung J, Gao P, Dang CV, Beer MA, Thomas-Tikhonenko A, Mendell JT. Lin-28B transactivation is necessary for Myc-mediated let-7 repression and proliferation. Proc Natl Acad Sci U S A. 2009;106:3384–9.
127. Lee YS, Dutta A. The tumor suppressor microRNA let-7 represses the HMGA2 oncogene. Genes Dev. 2007;21:1025–30.
128. Johnson CD, Esquela-Kerscher A, Stefani G, Byrom M, Kelnar K, Ovcharenko D, Wilson M, Wang X, Shelton J, Shingara J, Chin L, Brown D, Slack FJ. The let-7 microRNA represses cell proliferation pathways in human cells. Cancer Res. 2007;67:7713–22.
129. Gao P, Tchernyshyov I, Chang TC, Lee YS, Kita K, Ochi T, Zeller KI, De Marzo AM, Van Eyk JE, Mendell JT, Dang CV. c-Myc suppression of miR-23a/b enhances mitochondrial glutaminase expression and glutamine metabolism. Nature. 2009;458:762–5.
130. Rissland OS, Hong SJ, Bartel DP. MicroRNA destabilization enables dynamic regulation of the miR-16 family in response to cell-cycle changes. Mol Cell. 2011;43:993–1004.
131. Klein U, Lia M, Crespo M, Siegel R, Shen Q, Mo T, Ambesi-Impiombato A, Califano A, Migliazza A, Bhagat G, Dalla-Favera R. The DLEU2/miR-15a/16-1 cluster controls B cell proliferation and its deletion leads to chronic lymphocytic leukemia. Cancer Cell. 2010;17:28–40.
132. Bonci D, Coppola V, Musumeci M, Addario A, Giuffrida R, Memeo L, D'Urso L, Pagliuca A, Biffoni M, Labbaye C, Bartucci M, Muto G, Peschle C, De Maria R. The miR-15a-miR-16-1 cluster controls prostate cancer by targeting multiple oncogenic activities. Nat Med. 2008;14:1271–7.
133. Cimmino A, Calin GA, Fabbri M, Iorio MV, Ferracin M, Shimizu M, Wojcik SE, Aqeilan RI, Zupo S, Dono M, Rassenti L, Alder H, Volinia S, Liu CG, Kipps TJ, Negrini M, Croce CM. miR-15 and miR-16 induce apoptosis by targeting BCL2. Proc Natl Acad Sci U S A. 2005;102:13944–9.
134. Chang TC, Yu D, Lee YS, Wentzel EA, Arking DE, West KM, Dang CV, Thomas-Tikhonenko A, Mendell JT. Widespread microRNA repression by Myc contributes to tumorigenesis. Nat Genet. 2008;40:43–50.
135. Sotillo E, Laver T, Mellert H, Schelter JM, Cleary MA, McMahon S, Thomas-Tikhonenko A. Myc overexpression brings out unexpected antiapoptotic effects of miR-34a. Oncogene. 2011;30:2587–94.
136. Yang X, Liang L, Zhang XF, Jia HL, Qin Y, Zhu XC, Gao XM, Qiao P, Zheng Y, Sheng YY, Wei JW, Zhou HJ, Ren N, Ye QH, Dong QZ, Qin LX. MicroRNA-26a suppresses tumor growth and metastasis of human hepatocellular carcinoma by targeting interleukin-6-Stat3 pathway. Hepatology. 2013;58:158–70.
137. Kota J, Chivukula RR, O'Donnell KA, Wentzel EA, Montgomery CL, Hwang HW, Chang TC, Vivekanandan P, Torbenson M, Clark KR, Mendell JR, Mendell JT. Therapeutic microRNA delivery suppresses tumorigenesis in a murine liver cancer model. Cell. 2009;137:1005–17.
138. Han H, Sun D, Li W, Shen H, Zhu Y, Li C, Chen Y, Lu L, Zhang J, Tian Y, Li Y. A c-Myc-microRNA functional feedback loop affects hepatocarcinogenesis. Hepatology. 2013;57:2378–89.
139. Takwi AA, Li Y, Becker Buscaglia LE, Zhang J, Choudhury S, Park AK, Liu M, Young KH, Park WY, Martin RC. A statin-regulated microRNA represses human c-Myc expression and function. EMBO Mol Med. 2012;4:896–909.
140. Jung HM, Patel RS, Phillips BL, Wang H, Cohen DM, Reinhold WC, Chang LJ, Yang LJ, Chan EK. Tumor suppressor miR-375 regulates MYC expression via repression of CIP2A coding sequence through multiple miRNA-mRNA interactions. Mol Biol Cell. 2013;24(1638–1648):S1631–7.

141. Sampson VB, Rong NH, Han J, Yang Q, Aris V, Soteropoulos P, Petrelli NJ, Dunn SP, Krueger LJ. MicroRNA let-7a down-regulates MYC and reverts MYC-induced growth in Burkitt lymphoma cells. Cancer Res. 2007;67:9762–70.
142. Abe W, Nasu K, Nakada C, Kawano Y, Moriyama M, Narahara H. miR-196b targets c-myc and Bcl-2 expression, inhibits proliferation and induces apoptosis in endometriotic stromal cells. Hum Reprod. 2013;28:750–61.
143. Zhen Y, Liu Z, Yang H, Yu X, Wu Q, Hua S, Long X, Jiang Q, Song Y, Cheng C, Wang H, Zhao M, Fu Q, Lyu X, Chen Y, Fan Y, Liu Y, Li X, Fang W. Tumor suppressor PDCD4 modulates miR-184-mediated direct suppression of C-MYC and BCL2 blocking cell growth and survival in nasopharyngeal carcinoma. Cell Death Dis. 2013;4:e872.
144. Lal A, Navarro F, Maher CA, Maliszewski LE, Yan N, O'Day E, Chowdhury D, Dykxhoorn DM, Tsai P, Hofmann O, Becker KG, Gorospe M, Hide W, Lieberman J. miR-24 Inhibits cell proliferation by targeting E2F2, MYC, and other cell-cycle genes via binding to "seedless" 3'UTR microRNA recognition elements. Mol Cell. 2009;35:610–25.
145. Liu Z, Zhang G, Li J, Liu J, Lv P. The tumor-suppressive microRNA-135b targets c-myc in osteosarcoma. PLoS ONE. 2014;9:e102621.
146. Challagundla KB, Sun XX, Zhang X, DeVine T, Zhang Q, Sears RC, Dai MS. Ribosomal protein L11 recruits miR-24/miRISC to repress c-Myc expression in response to ribosomal stress. Mol Cell Biol. 2011;31:4007–21.

Chapter 14
Viral Noncoding RNAs in Cancer Biology

Zhi Li, Shujun Fu, and Lun-Quan Sun

Abstract Over 12 % of all human cancers are caused by oncoviruses, primarily including Epstein–Barr virus (EBV), high-risk human papillomaviruses (HPVs), hepatitis B and C viruses (HBV and HCV, respectively), and Kaposi's sarcoma herpesvirus (KSHV). In addition to viral oncoproteins, a variety of noncoding RNAs (ncRNAs) produced by oncoviruses have been recognized as important cofactors that contribute to the oncogenic events. In this chapter, we will focus on the recent understanding of the long and short noncoding RNAs, as well as microRNAs of the viruses, and discuss their roles in the biology of multistep oncogenesis mediated by established human oncoviruses.

Keywords Oncoviruses • Noncoding RNA • MicroRNAs • Cancer

14.1 Introduction

Tumor viruses can be classified into two groups based on their genetic material, i.e., DNA or RNA. Cancer-causing DNA tumor viruses and RNA retroviruses have been extensively investigated for their roles in oncogenesis (see Table 14.1). Despite great efforts being made to uncover the oncogenic mechanisms of the virus-encoded proteins, it is until recently that the importance of noncoding RNAs of the viruses has been realized by researchers, which is largely due to the development of sequencing technology. Although viral genomes are limited in size, any space expended on a ncRNA is rationed. Thus, viral ncRNAs are expected to play important roles in enhancing the viral replication or counteracting the defenses that the

Z. Li • S. Fu • L.-Q. Sun (✉)
Center for Molecular Medicine of Xiangya Hospital, Collaborative Innovation Center for Cancer Medicine, Central South University, Changsha, Hunan 410008, China

Key Laboratory of Molecular Radiation Oncology, Hunan, China

Research Center for Technologies of Nucleic Acid-Based Diagnostics and Therapeutics, Hunan, China
e-mail: lunquansun@csu.edu.cn

E. Song (ed.), *The Long and Short Non-coding RNAs in Cancer Biology*,
Advances in Experimental Medicine and Biology 927,
DOI 10.1007/978-981-10-1498-7_14

Table 14.1 Representative human oncogenic viruses and key oncoproteins

Group	Examples	Viral oncoproteins	Tumor types
DNA virus			
Hepadnaviridae	HBV	HBx	Hepatocellular carcinoma
Herpesviridae	EBV	LMP1, BARF-1	Lymphoma, nasopharyngeal carcinoma, gastric cancer
	KSHV	vGPCR	Kaposi sarcoma, body cavity lymphomas
	HVS	–	T-cell leukemias and lymphomas
Papovaviridae	Merkel cell	T antigens	Merkel cell carcinoma
	Polyomavirus	–	–
Papillomaviridae	HPV	E6, E7	Cervical and anal cancer
	16,18,31,45	–	Oral cancer
RNA virus			
Flaviviridae	Hepatitis C	–	Hepatocellular carcinoma
	Virus	–	–

host raises against viral infection. Indeed, accumulating evidence argues for their participation in the process of carcinogenesis [1].

In mammalian cells, ncRNAs can be grouped into small noncoding RNAs (sncRNAs), mid-size noncoding RNAs (mid-size ncRNAs), and long noncoding RNAs (lncRNAs) according to their length. Within these groups, there can be additional subclasses of the moieties [2], such as PIWI-interacting RNAs (piRNAs), small nucleolar RNAs (snoRNAs), transcribed ultra-conserved regions (T-UCRs), and large intergenic noncoding RNAs (lincRNAs), which act as key elements for cellular homeostasis [3]. In the case of the ncRNAs transcribed by human tumor viruses, microRNA (miRNAs) and lncRNAs are the major species of ncRNAs that are produced at different stages of the viral replication and latency and impact on immortalization, transformation, and malignancy of host cells. Thus, they will be the focus of this chapter. In addition, we will also pay special attention to some unique noncoding transcripts from human viruses such as sub-genomic flavivirus RNA (sfRNA) and stable intronic sequence (sis) RNAs, which are characteristically produced during virus infection and lytic replication, as they have been unveiled to facilitate virus–host interaction and may shed a light on the missing details regarding the oncogenic process of tumor viruses.

14.2 Virally Transcribed Long Noncoding RNAs

It has been well recognized that tumor viruses encode some abundant RNA molecules without coding capacity, but the biological significance of these ncRNAs remains to be molecularly defined. Among those, there are six well-characterized lncRNAs that have been explored extensively, such as EBV-encoded RNAs (EBERs), KSHV PAN RNA (polyadenylated nuclear RNA), herpesvirus saimiri

Table 14.2 Virally encoded lncRNAs and their features

Length	Name	Virus	Oncogenic characteristics	Reference
EBER1:167 nt EBER2:172 nt	EBERs	EBV	EBERs are involved in lymphoma, NPC, and gastric carcinoma partly through TLR3- and RIG-I-mediated inflammation	[9–31]
75–143 nt	HSURs	HVS	HSURs are involved in transformation of T cells, and regulation of miR-27 by HSUR1 may represent such a case	[32–43]
~2700 nt	lncRNA β2.7	HCMV	β2.7 may be oncogenic through protecting cells from apoptosis via interaction with complex I	[44–46]
1060 nt	PAN RNA	KSHV	PAN RNA may promote Kaposi's sarcoma via facilitating EZH2 chromosomes location	[47–60]
520 nt	sfRNA	*Flaviviridae*	XRN1 repression by 5′ UTR- or 3′ UTR -localized sfRNA may contribute to carcinogenesis via stabilizing cellular mRNAs of oncogenes	[61–72]
~160	VA I and II	Human adenovirus	Possibly not involved in human cancers	[8]
2000 nt	LAT	HSV-1	Not investigated	[74–76]
ebv-sisRNA1:81 nt ebv-sisRNA2:2971 nt	ebv-sis RNAs	EBV –	Not investigated	[78, 79]

small nuclear RNAs, and virus-associated RNA I and II (VA I and II), encoded by adenovirus (see Table 14.2) [4–8]. In the process of establishing the viral infection, the host cells also produce an array of ncRNAs, which function in coordination with viral ncRNAs to promote oncogenesis.

14.2.1 EBV-Encoded RNAs (EBERs)

EBERs are the most abundant EBV transcripts (about 10^7 copies per cell) during latent infection by EBV in a variety of cells. Owing to its expression abundance and universal existence in all of the three forms of latent infection, EBERs have been under intensive studies since they were discovered by Lerner for the first time [9].

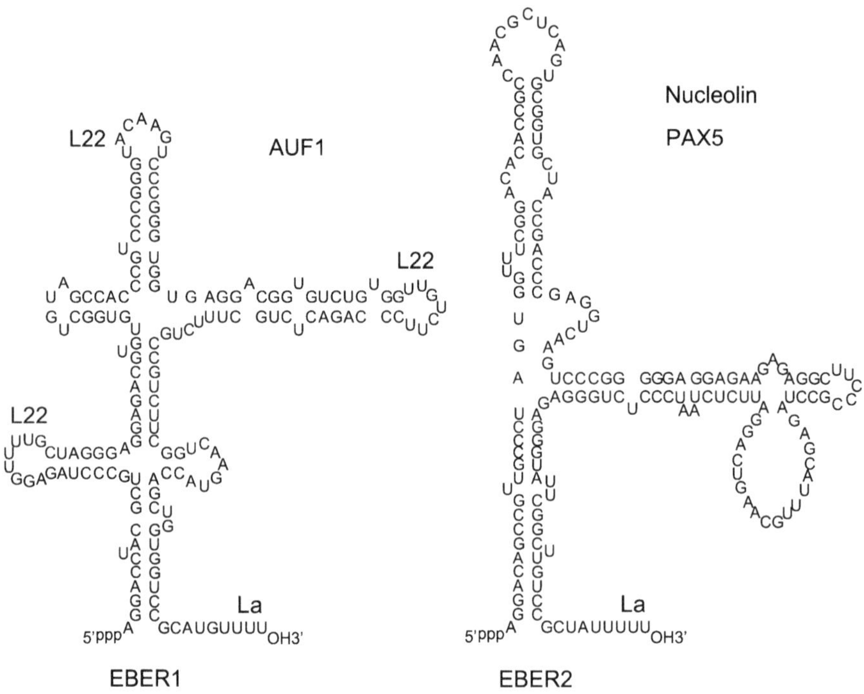

Fig. 14.1 EBERs' secondary structures and binding sites for associated proteins

EBERs are encoded by the right-hand 1000 base pairs of the EcoRI J fragment of EBV DNA [10]. EBER1 is 166 (167) nucleotides long and EBER2 is 172 ± 1 nucleotides long with the heterogeneity resides at the 3′ termini (see Fig. 14.1). Although Schwemmle et al. have claimed that both RNAs are found in the cytoplasm as well as in the nuclei of interphase cells [11], most researches suggest that EBERs are confined to the nucleus [12], which imply EBERs may largely fulfill their function in the nucleus.

In early studies, EBERs were first found to interact with La antigen [9]. The significance of EBER–protein interaction, however, cannot be overemphasized in the EBER-involved cellular function and oncogenic process. As mentioned above, La was identified to be complexed with EBERs, and together they could form the antigen recognized by systemic lupus erythematosus (SLE) anti-La antiserum. Moreover, EBERs were reported to be released in the complex with La, but the pathological significance remains to be clarified [13]. Furthermore, PKR was reported to be complexed with EBERs, which prevented the induction of interferon (IFN), PKR-mediated protein translation inhibition, and apoptosis [14, 15]. L22, a ribosomal composing protein, was also confirmed to be complexed with EBER1, which could translocate L22 from cytoplasm to nucleoplasm and thus might result in a depletion of the protein from ribosomes [16]. Additionally, EBER1 was reported to form complex with AUF1/hnRNP D, which indicated EBER1 might disturb the normal homeostasis between AUF1 and AU-rich elements (AREs)-containing mRNAs or compete with other AUF1-interacting targets in cells latently infected by EBV [17]. More recently, EBER2 was shown to base pairs with a nascent RNA transcribed from TRs of EBV genome, which recruited PAX5 to the TRs. Consequently, EBER–PAX5 complex decreased latent membrane protein 1 (LMP1) and LMP2A transcription and promoted following viral lytic replication [18], though this transcription regulation of LMP1 and LMP2A did not coincide with their unchanged protein level in LCLs [19].

Owing to a high expression level of EBERs and the close association between EBV and lymphoma and nasopharyngeal carcinoma (NPC) and gastric cancer, great efforts have been made to explore their possible involvement in the process of oncogenesis. With regard to the association of EBERs with lymphoma, Yajima et al. showed that the presence of EBERs in Akata cells was required for the cells to be more malignant and apoptosis resistant [20, 21]. In support of this conclusion, a transgenic mouse model expressing EBER1 suggested that EBER1 promoted lymphoid hyperplasia and lymphoma in cooperation with c-Myc [22]. However, discrepancy existed with a previous study, which indicated that EBERs were not essential for B lymphocytes transformation into LCLs (lymphoblastoid cells) and EBV replication in LCLs [23]. Later, observations based on BAC-EBV system again contradicted with Yajima's results, which showed that EBV could transform primary B cells into LCLs at the same frequency in the presence or absence of EBERs [24]. In consideration of the reliability of BAC-EBV system, EBERs may not account for the transformation of lymphocytes by EBV, especially for B95-8 stain background. In the case of NPC, it was reported that EBER expression could confer an apoptotic-resistant phenotype in immortalized nasopharyngeal epithelial cells. The EBER-expressing NP69 cells (an immortalized nasopharyngeal cell line) attained a higher growth rate compared to the cells transfected with control plasmid (pcDNA3). However, the EBER-expressing NP69 cells did not form colonies in soft agar and were non-tumorigenic in nude mice [25]. In line with these observations, the NPC-KT cells highly expressing EBERs did not show any growth advantages or anti-apoptosis ability over the mock transfected NPC-KT counterparts [26], in spite of the previous observations that EBV infection induced IGF-1 expression in NPC cell lines and that the secreted IGF-1 acted as an autocrine growth factor [27]. The evidence of EBER involvement in the oncogenesis of lymphoma and NPC is still obscure, possibly owing to variation of EBV strains and different genetic backgrounds of the cell lines used in the studies.

The double-strand secondary structures of EBERs were reported to trigger their recognition by Toll-like receptor 3 (TLR3) and RIG-I, which consequently led to lymphocyte pathogenesis or lymphoma, respectively [13, 28–31]. In line with these studies, we thoroughly examined the involvement of TLR3 and RIG-I in EBER-mediated inflammation and associated pathological effect on NPC [30, 31]. Inspired by the correlation of EBERs and RIG-I expression, we verified that EBERs could trigger the RIG-I-dependent NFκB and IRF3 signaling pathways and subsequently promoted NPC progress. Furthermore, we confirmed that EBER-triggered inflammatory cytokines potentiated TAM chemoattraction and polarization via RIG-I

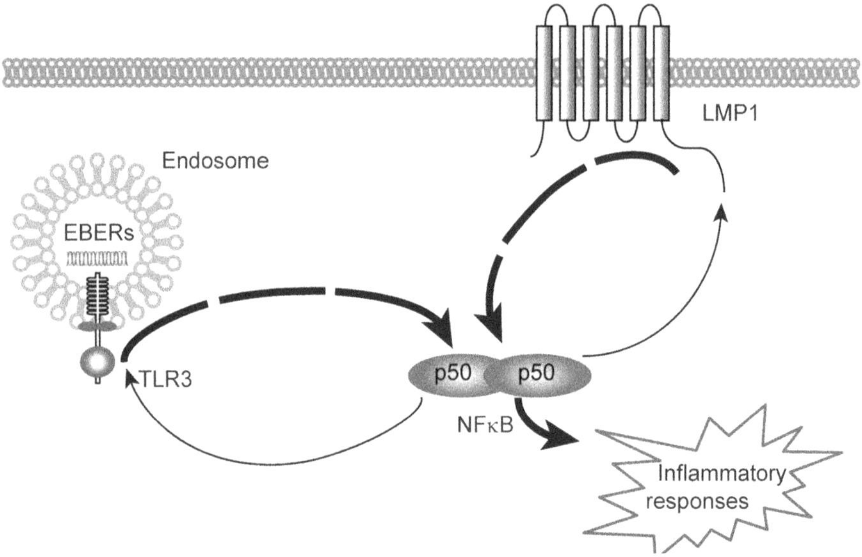

Fig. 14.2 Schematic description of a proposed model for EBER–LMP1 interactive feedforward loop in promoting cancer-related inflammation. EBERs and LMP1 can both trigger moderate inflammatory response through NFκB as a key node. In addition to the role as an inflammation mediator, NFκB can transcriptionally up-regulate EBERs and LMP1. As a result, EBERs promote LMP1 transcription via NFκB which in turn transcriptionally induces more EBERs expression via NFκB likewise. Consequently, EBERs and LMP1 synergistically generate overwhelming signals to promote NFκB-mediated cancer-related inflammation

[30]. Also, we demonstrated that EBERs could induce inflammatory response in NPC cells through TLR3, mainly featured by high level of TNFα production. Intriguingly, EBERs and EBV latent membrane protein 1 (LMP1) could form a positive regulatory loop with NFκB as a key node that amplified the inflammatory signals in EBV-infected epithelial cells [13], which for the first time demonstrated that cancer-promoting inflammation could be triggered and amplified through the ncRNA and protein components of EBV in a coordinated way (see Fig. 14.2).

14.2.2 Herpesvirus Saimiri U-Rich RNAs (HSURs)

Herpesvirus saimiri (HVS) is a γ-herpesvirus that causes fatal T-cell leukemia and lymphomas in new world primates and transforms human primary T cells in vitro [32, 33]. HVS genome harbors seven ncRNA genes that encode for seven Sm-class small nuclear RNAs (snRNAs) called herpesvirus saimiri U-rich RNAs (HSURs), with HSURs ranking among the most abundantly expressed RNAs in HVS-transformed T cells [7, 34]. HSURs share common features with snRNAs assembled into Sm snRNPs, including 5′ trimethyl caps, Sm protein-binding sites, and

3′-terminal stem loops. HSURs1, 2, and 5 have highly conserved 5′-end sequences containing the AUUUA pentamer characteristic of AU-rich elements (AREs) that regulate the stability of many host mRNAs, including those encoding most proto-oncogenes and cytokines. It has been shown that expression levels of the HSURs are regulated by cellular ARE-binding proteins, such as hnRNPD and HuR [35, 36]. Microarray and Northern analyses reveal that HSUR1 and 2 expression correlates with significant increases in a small number of host mRNAs which are involved in the activation of virally transformed T cells during latency [37].

HSUR1 and HSUR2 have been recently reported to possess the capacity to associate with three host miRNAs, namely, miR-142-3p, miR-27, and miR-16, in marmoset T cells transformed by HVS. The abundance of miR-27 is dramatically lowered in transformed cells in the presence of HSUR1, with consequent effects on expression of miR-27 target genes. This viral strategy illustrates the use of a ncRNA to manipulate host cell gene expression via the miRNA pathway. Of note, rather than function as a miRNA sponge, HSUR1-specific regulation of miRNA is proposed to mimic host splicing snRNA genes, which bind target miRNA for degradation [38, 39]. In agreement, murine cytomegalovirus (MCMV) uses a similar antisense RNA-based mechanism to target miR-27 for degradation [40, 41]. Because miR-27 is a repressor of T-cell activation [42], degradation of miR-27 mediated by HSUR1 promotes activation and presumably proliferation of HVS-infected host T cells. Moreover, since HSURs are involved in transformation of T cells by HSV [43], the accurate role and mechanism of this unique regulation of miR-27 by HSUR1 in T-cell lymphoma remain to be ascertained.

14.2.3 LncRNA β2.7

Human cytomegalovirus (HCMV) is a highly multifaceted host-specific beta-herpesvirus that is regarded as asymptomatic or mildly pathogenic virus in immunocompetent host. The involvement of HCMV in late inflammatory complications underscores its possible role in inflammatory diseases and cancer. HCMV encodes β2.7, a noncoding RNA that accounts for a large fraction (~20 %) of the transcripts that are expressed during lytic infection [44]. Recent report has indicated that HCMV infection protects cells from rotenone-induced apoptosis, which is mediated by β2.7. During infection, β2.7 RNA interacted with the complex I and prevented the relocalization of the essential subunit of the gene product associated with retinoid/interferon-induced mortality-19, in response to apoptotic stimuli. Complex I targeting by a viral RNA represents a refined strategy to modulate the metabolic viability of the infected host cell [45]. Given the association of anti-apoptosis ability with oncogenesis, it's rational to propose and validate the possible role of β2.7 in the process of carcinogenesis. Intriguingly, ribosome profiling and mass spectrometry reveal that β2.7 is also translated into small peptides and therefore has a coding function [46].

14.2.4 *KSHV-Encoded Polyadenylated Nuclear RNA (PAN RNA)*

Kaposi's sarcoma-associated herpesvirus (KSHV) is an oncogenic γ-herpesvirus, the causative agent of Kaposi's sarcoma and body cavity lymphomas [47]. Kaposi's sarcoma is a multifocal neovascular tumor of proliferating spindle endothelial cells latently infected by KSHV, which encodes a viral long ncRNA known as polyadenylated nuclear RNA (PAN RNA) [48]. Although assumed to be primarily a lytic transcript, recent observations have showed that PAN RNA is highly abundant within virions and is already the most abundant transcript in a few hours postinfection [49]. Thus PAN RNA is expressed during all phases of the viral life cycle and has the potential to influence all of them. Expression of PAN RNA during KSHV lytic infection was regulated by major viral trans-activator K-RTA through the interaction of a specific DNA interaction domain referred to as RTA-responsive element (RRE) [50]. Furthermore, PAN RNA has been recently found to interact with the KSHV latency-associated nuclear antigen (LANA) [51], and this interaction is thought to serve to sequester LANA away from viral DNA episomes to facilitate lytic reactivation.

Transcribed by RNA Pol II and 5′ capped and polyadenylated at the 3′ end, PAN RNA remained in the nucleus where it accumulated to extremely high levels per infected cell (~500,000 copies/cell), and this high copy number was attributable to two features of the ~1.1-kb PAN transcript: MTA/ORF57 binding near the 5′ end and a triple-helical stabilization element (ENE) upstream of the 3′ polyA tail adenovirus-associated RNAs VAI and VAII [52–54]. However, insertion of the minimal MRE along with the ENE showed no added increase in mRNA accumulation [55], indicating that there might be other cis-elements or motifs within PAN RNA that contributed to its extremely high abundance.

The functional studies revealed that depletion of PAN RNA using RNase H-targeting antisense oligonucleotides led to impairment of the KSHV late gene expression and decrease of the amount of infectious virus [56]. Subsequent studies showed that PAN RNA also interacted with histones H1 and H2A, mitochondrial and cellular single-stranded binding proteins, interferon regulatory factor 4 (IRF4), and KSHV ORF59 [57]. Using recombinant KSHV BAC with a large deletion in the PAN RNA gene indicated that PAN RNA was required for full activation of KSHV gene expression and virus production [58]. RNA-chromatin immunoprecipitation (RNA-CHIP) assays showed that PAN RNA interacted with demethylases JMJD3 and UTX and the histone methyltransferase MLL2. Consistent with the interaction with demethylases, expression of PAN RNA resulted in a decrease of the repressive H3K27me3 mark at the ORF50 promoter, and thus it could be postulated that PAN RNA mediated latency or reactivation through its interaction with different chromatin-modifying complexes.

There has always been interest in the speculation about the coding potential of PAN RNA. The most recent development in the coding potential of PAN RNA has come from a study that uses mRNA-Seq, along with ribosome footprinting (Ribo-

Seq) and DNA-Seq, to compile a comprehensive analysis of viral gene expression [59], which suggests PAN RNA represents up to 92 % of the total viral mRNA-Seq reads and the ribosome-protected RNA corresponding to the small PAN peptides represents up to 1.7 % of the total cycloheximide Ribo-Seq reads. Thus, these data strongly indicate that in addition to its documented functions as a "noncoding" RNA, PAN may also be a presumptive coding RNA.

Furthermore, RNA-Seq analysis of the cell lines that express PAN RNA has showed that transcription of the genes involving expression of proteins participated in cell cycle, immune response, and inflammation is dysregulated. Expression of PAN RNA in various cell types results in an enhanced growth phenotype, higher cell densities, and increased survival compared to control cells. Also, PAN RNA expression mediates a decrease in the production of inflammatory cytokines [49]. Especially, ChIRP-Seq has showed that PAN RNA occupies much of the KSHV genome during lytic infection, including the PAN RNA gene promoter itself, and that PAN RNA also interacts with protein components that comprise the polycomb repression complex 2 (PRC2) [49]. EZH2 is a component of the epigenetic regulator PRC2 that suppresses gene expression. Elevated expression of EZH2 is common in human cancers and is associated with tumor progression and poor prognosis. Since KSHV-mediated upregulation of EZH2 is required for the induction of ephrin-B2, an essential proangiogenic factor that drives endothelial cell tubule formation, it's conceivable that PAN RNA may promote Kaposi's sarcoma via facilitating EZH2 chromosomes location [60]. Collectively, PAN RNA may contribute to KSHV-associated Kaposi's sarcoma by enhancing proliferation and apoptosis resistance, probably partly through interaction with the regulatory complexes that modifies the epigenetic status of the host cells.

14.2.5 Noncoding Sub-genomic Flavivirus RNA (sfRNA)

Arthropod-borne members of the *Flaviviridae* (e.g., West Nile virus (WNV) and the dengue viruses (DENV)) have been shown to contain a conserved structure at the beginning of their 3′ untranslated regions (UTRs), namely, sfRNA, which is a product of incomplete degradation of genomic RNA by the cellular 5′–3′ exoribonuclease XRN1 [61]. The 3′ UTRs folded into an interesting three-helix junction that stalled the cellular XRN1 enzyme as it tried to degrade flaviviral transcripts [62]. The stalling of XRN1 at this structure resulted in the accumulation of large amounts of a short 3′ UTR-derived sub-genomic flavivirus RNA (sfRNA) during infection [63].

The contribution of each of the stem-loop (SL) and pseudoknot (PK) structures in facilitating XRN1 stalling has been investigated for West Nile (WNV) and yellow fever virus (YFV) [63–65]. Exploring 3′ UTR secondary structures as well as tertiary interactions indicated secondary structures SL-IV and dumbbell 1 (DB1) downstream of SL-II were able to prevent further degradation of gRNA when the SL-II structure was deleted, leading to production of sfRNA2 and sfRNA3, respectively. The authors also showed that a number of PK interactions, in particular PK1 stabilizing SL-II and PK3 stabilizing DB1, were required for protection of gRNA from nuclease degradation and production of sfRNA [64].

Fluorescence in situ hybridization (FISH) assay using probes to specifically detect the WNV 3′ UTR (sfRNA) has demonstrated that sfRNA forms punctuate foci which co-localizes variably with XRN1 (a marker of PBs) and with the stress granule (SG) marker protein eIF3ŋ [61]. Localization of sfRNA to PBs is consistent with the generation of this RNA species, while the underlying significance of the co-localization of WNV sfRNA to SG remains to be unveiled.

sfRNAs have been reported to impact on flavivirus replication, cytopathicity, and pathogenicity. Decreased replication in mammalian and mosquito cell lines was observed for WNV and YFV mutants deficient in sfRNA [63–65], while preliminary evidence suggested sfRNA might play a negative role in JEV viral RNA replication and/or translation [66]. WNV mutants that produced less abundant, truncated sfRNAs were significantly less pronounced in Vero cells, leading to marked reduction in size or complete absence of viral plaques [63]. In addition, pseudoknot (PK) interactions play a vital role in the production of nuclease-resistant sfRNA, which is essential for viral cytopathicity in cells and pathogenicity in mice [64, 65]. Transfection of minus-sense sfRNA into JEV-infected cells, in order to counter the effects of plus-sense sfRNA, resulted in higher levels of antigenome, suggesting that the presence of the sfRNA inhibited antigenome synthesis. Trans-acting effect of sfRNA on JEV translation was further confirmed using a reporter mRNA containing the luciferase gene fused to partial coding regions of JEV and flanked by the respective JEV UTRs. sfRNA-inhibited JEV translation was also demonstrated in the in vivo and in vitro translation assays [66].

Mechanistically, WNV sfRNAs are critical for viral cytopathicity via counteracting type I interferon (IFN) effect. Utilizing mice and cells that were deficient in IFN response, replication of the sfRNA mutant WNV could be rescued in the mice and cells lacking interferon regulatory factor 3 (IRF-3) and IRF-7 and in the mice lacking the type I alpha/beta interferon receptor (IFNAR), suggesting a contribution of sfRNA in overcoming the antiviral response mediated by type I IFN. This has been confirmed by a series of studies. For example, the mutant virus replication was rescued in the presence of IFNAR-neutralizing antibodies; greater sensitivity of mutant virus replication to IFN-α pretreatment could be achieved; partial rescue of its infectivity in cells deficient in RNase L and direct effects of transfected sfRNA on rescuing replication of unrelated Semliki Forest virus (SFV) in cells pretreated with IFN-α were also observed. All of the results confirm that sfRNA is involved in flavivirus pathogenesis via its contribution to viral evasion of the type I interferon response [67]. Moreover, in PKR−/− or RNase L−/− MEFs, no observable rescue of FL-IRAΔCS3 (which is unable to transcribe sfRNA) replication was shown, compared to that observed via infection in wild-type MEFs. Interestingly, transfection of in vitro transcribed sfRNA may inhibit IRF-3 phosphorylation in JEV-infected cells, which suggests this inhibition of IRF-3 activation may lead to a decrease in IFN-β transcription [68].

Growing evidence has indicated that production of viral noncoding RNAs in RNA virus-infected cells can serve as one of the strategies for RNA viruses to avoid the cellular RNA decay machinery. In the case of flaviviruses, stalling of the XRN1 enzyme in the process of generating sfRNA resulted in the repression of its activity. Cells infected with dengue or Kunjin viruses accumulate uncapped mRNAs and decay intermediates normally targeted by XRN1. XRN1 repression also resulted in the increased overall stability of cellular mRNAs in flavivirus-infected cells [69]. Disruption of the cellular RNA decay machinery can benefit for the virus, including increased stability of viral transcripts as well as interference with the regulation of cellular gene expression that may influence the ability of the cells to mount an effective innate immune response.

RNA interference (RNAi) is the predominant antiviral response against invading RNA viruses in insects and plants. Enhanced flavivirus replication in mosquitoes with depleted RNAi factors suggests an important biological role for cellular RNAi in restricting virus replication. Recent investigation has demonstrated a novel role for sfRNA, i.e., as a nucleic acid-based regulator of RNAi pathways, a strategy that may be conserved among flaviviruses. The authors first established that flaviviral RNA replication suppressed siRNA-induced gene silencing in WNV and DENV replicon-expressing cells. Next, they showed that none of the WNV-encoded proteins displayed RNA silencing suppressor (RSS) activity in mammalian and insect cells and in plants by using robust RNAi suppressor assays. In contrast, they found that sfRNA efficiently suppressed siRNA- and miRNA-induced RNAi pathways in both mammalian and insect cells. Interference with human Dicer processing of dsRNA in vitro suggested that sfRNA acted as a decoy molecule upstream of the RNA-induced silencing complex (RISC). As a result from this, less (antiviral) siRNA was produced when sfRNA was present, which was in line with the observation that sfRNA enhanced the replication of a heterologous arbovirus in mosquito cells [70]. In addition to counteracting the RNAi pathways targeting virus itself, flaviviruses took advantage of RNAi to facilitate their replication in infected cells. Intriguingly, WNV sfRNA was deemed the likely source of KUN-miR-1, the first flavivirus-derived miRNA, in WNV-infected mosquito cells. Transcription of WNV (KUN) pre-miRNA (3′ SL) in mosquito cells either from plasmid or Semliki Forest virus (SFV) RNA replicon resulted in the production of mature KUN-miR-1. Silencing of Dicer-1 but not Dicer-2 led to a reduction in the miRNA levels. Further, when a synthetic inhibitor of KUN-miR-1 was transfected into mosquito cells, replication of viral RNA was significantly reduced. In addition, a host mRNA target for KUN-miR-1 in mosquito cells was determined to be zinc finger transcription factor GATA4. Depletion of GATA4 mRNA by RNA silencing led to a significant reduction in virus RNA replication, while a KUN-miR-1 RNA mimic enhanced replication of a mutant WNV (KUN) virus producing reduced amounts of KUN-miR-1, suggesting that GATA4 induction via KUN-miR-1 played an important role in virus replication [71].

Hepatitis C virus (HCV) is a positive-sense RNA virus of the *Hepacivirus* genus within the *Flaviviridae* that chronically infects approximately 130–150 million people worldwide. Chronic HCV infection causes acute liver dysfunction and cirrhosis and is associated with the development of hepatocellular carcinoma (HCC). Interestingly, HCV and other non-arthropod-associated members of the *Flaviviridae* such as the economically important bovine viral diarrhea virus (BVDV) do not generate a sfRNA-like molecule from their 3′ UTRs. Instead, HCV and BVDV contain structured regions in their 5′ UTR near or including the IRES region that both stall and repress XRN1. XRN1 repression by the 5′ UTRs of these viruses can be demonstrated both in biochemical assays and in living cells. Interestingly, HCV or BVDV repression of XRN1 is associated with a dramatic repression of the major 5′–3′ decay pathway and a large increase in the stability and abundance of numerous normally short-lived cellular mRNAs. Of note, the mRNAs of many cellular oncogenes and angiogenic factors are significantly stabilized and increased in abundance during HCV infection [72]. Therefore, XRN1 repression is a highly conserved and important facet of infections by disparate members of the *Flaviviridae*, which under specific circumstances may contribute to oncogenic process of liver cancers.

14.2.6 Intronic Long Noncoding RNAs

The pre-mRNA splicing machinery generates a number of non-polyadenylated noncoding RNAs (intronic long noncoding RNAs), such as snoRNA, sno-lncRNA, circular intronic RNA (ciRNA), and circular RNA (circRNA) [73]. During the past few decades, significant progress has been made with respect to the identification and function validation of the virus-encoded intronic long noncoding RNAs.

The ncRNAs composed of stable introns were first discovered in the α-herpesvirus HSV-1, which produced an RNA species in high abundance—the latency-associated transcript gene (LAT) ncRNA [74]. Experiments in mice and rabbits have shown that LAT plays a critical role in enhancing the reactivation phenotype. This appeared to be a function of LAT anti-apoptosis activity, since wild-type levels of reactivation could be restored to a LAT(−) virus by inserting one of several different alternative anti-apoptosis genes in place of LAT. LAT might also contribute to latency/reactivation via several immune evasion mechanisms, such as an increase of HVEM expression (herpes virus entry mediator, a member of the tumor necrosis family), which could act as a switch to decrease T-cell function [75]. Interestingly, the excised intronic LAT persisted and accumulated to high levels in infected cells, and the functional form of the LAT ciRNA appeared to be the branched lariat. Expression of the HSV-1 LAT ciRNA maintained infection by inhibiting apoptosis of neuronal cells and by silencing viral lytic gene expression through alteration of heterochromatin structures at viral promoters [76]. In addition, human and murine CMVs could also encode ciRNAs with a similar structure to HSV-1 LAT ciRNA, which persisted stably with the lariat structure after splicing [77].

In EBV-infected cells, analysis of the data from small RNA-Seq, and ribosome-depleted RNA-Seq identified both the short and long introns as stable intronic sequence RNAs (sisRNAs): ebv-sisRNA-1 (81 nt) and ebv-sisRNA-2 (2791 nt), respectively [78, 79]. ebv-sisRNA-1 is estimated to be present at the 21 % level of

EBER1 or roughly on a par with EBER2, which is estimated to be 25 % as abundant as EBER1. The ebv-sisRNA-1 differs from the LAT in two major respects: the LAT is much larger (>2 kb), and the functional form of the LAT is believed to be the lariat-intron splicing intermediate, whereas the ebv-sisRNA-1 is likely to be a linear molecule. Thus, ebv-sisRNA-1 may be referred to ebv-snoRNA-1, which is composed of stem-loop configuration rather than branched lariat. Indeed, the short hairpin of ebv-sisRNA-1 presents a conserved U-rich motif in the loop region, while the downstream sequence is unstructured and contains a completely conserved CA-rich region, both of which motifs confer potential protein interaction sites for ebv-sisRNA-1.

A notable feature of ebv-sisRNA-2 is the presence of predicted conserved and thermodynamically stable RNA secondary structure [78], which encompasses a massive (586 nt) hairpin that is extensively conserved in structure, but not sequence, between divergent lymphocryptoviruses [39]. Furthermore, v-snoRNA1, a canonical C/D box snoRNA encoded by EBV, has been identified. This genetic element displays all hallmark sequence motifs of a canonical C/D box snoRNA, namely, C/C′ and D/D′ boxes. The nucleolar localization of v-snoRNA1 was verified by in situ hybridization of EBV-infected cells. V-snoRNA1 was also confirmed to bind to the three canonical snoRNA proteins, fibrillarin, Nop56, and Nop58. The C-box motif of v-snoRNA1 was shown to be crucial for the stability of the viral snoRNA. Its selective deletion in the viral genome led to a complete downregulation of v-snoRNA1 expression levels within EBV-infected B cells. V-snoRNA1 might serve as a miRNA-like precursor, which was processed into 24 nt sized RNA species, designated as v-snoRNA124pp. A potential target site of v-snoRNA124pp was identified within the 3′ UTR of BALF5 mRNA which encoded the viral DNA polymerase, indicative of the probable involvement of v-snoRNA1 in EBV lytic replication transition. Of note, v-snoRNA1 was found to be expressed in all investigated EBV-positive cell lines, including lymphoblastoid cell lines (LCL), which implied v-snoRNA1 expression might contribute to lymphoma incidence [80]. Thus far, there have been no studies showing this class of sisRNAs is involved in the EBV-associated carcinomas of epithelial origin.

14.3 Virus-Encoded MicroRNAs

MicroRNA (miRNA) is a small noncoding RNA, mainly consisting of 20–23 nucleotides with partial or total homology to target sequence [81]. It promotes target mRNA degradation or negatively regulates gene expression through binding to the 3′-untranslated region [82]. As a vital modulator, miRNA regulates numerous cellular processes including apoptosis and tumorigenesis. In virus-positive cancers, many viral miRNAs are expressed, which suggests a critical role of these miRNAs in viral latency and oncogenesis. The representative examples of viral miRNAs involved in various cancers are listed in Table 14.3.

Table 14.3 Overview of viral miRNAs and their target genes

Virus	miRNA	Target genes	Function	Reference
EBV	miR-BART5	PUMA	Inhibit apoptosis	[89]
	miR-BART cluster 2	NDRG1	Promote metastasis	[90]
	miR-BART cluster 1	LMP1	Reduce the sensitivity to cisplatin	[91]
	miR-BART9	E-cadherin	Promote migration and invasion	[93]
	miR-BART15-3p	BRUCE	Inhibit cell proliferation and promote apoptosis	[94]
	miR-BART3	DICE1	Promote cell growth and transformation	[95]
KSHV	MiR-K12-7 and MiR-K12-9	ORF50	Inhibit lytic replication	[104, 105]
	MiR-K12-4	Rbl2	Inhibit lytic replication	[106]
	MiR-K12-1	IκBα	Promote NFκB-dependent viral latency and cell cycle	[97]
		p21		[107]
	MiR-K12-3 and MiR-K12-7	C/EBPβ	Promote IL-6, IL-10 expression	[114]
HPV	HPV16 encoded	CYP26B1	Inhibit apoptosis	–
miRNAs	–	PBRM1	Allow expansion of the HPV-infected cell population	[119]
MCV	MCV-miR-M1-5p	PIK3CD, PS ME3, RUNX1	Regulate host immune response and viral proliferation	[121]

14.3.1 EBV-Encoded miRNA

EBV is the first human virus to be shown expressing miRNAs. To date, 25 EBV miRNA precursors and 48 mature miRNAs have been identified [83, 84]. They mainly locate in two regions of the viral genome, BamHI fragment H rightward open reading frame 1 (BHRF1) and BamHI A rightward transcript (BART). The BHRF1 region encodes three miRNA precursors and four mature miRNAs, while BART region generates two clusters of miRNAs with 22 miRNA precursors and 44 mature miRNAs [85].

Studies on EBV-encoded miRNAs have revealed that the viral miRNAs play important roles in EBV-associated malignancies. By bioinformatic analysis and biological function validation, the potential targets of the miRNAs were identified in their effects on oncogenesis [86, 87]. The p53 up-regulated modulator of apoptosis (PUMA), a pro-apoptotic protein belonging to Bcl-2 family, was found to be a target of miR-BART5 [88]. In EBV-infected epithelial cells, miR-BART5 had a relatively high level, which was negatively correlated with PUMA expression. When treated with Adriamycin, a DNA damage agent, miR-BART5, could protect host cells from apoptosis [89]. The entire BART miRNA cluster 2 was identified to be responsible for the downregulation of N-myc downstream-regulated gene 1 (NDRG1), an epithelial differentiation marker and metastasis suppressor [90]. EBV latent membrane protein 1 (LMP1) is a virus protein that pays a key role in oncogenesis of nasopharyngeal carcinoma (NPC). On one hand, MiR-BART cluster 1 was identified to suppress LMP1 expression and could reduce the sensitivity of the NPC cells to cisplatin [91]. On the other hand, other group validated miR-BART9 could positively regulate LMP1 in two nasal NK/T-cell lymphoma (NKTCL) cell lines to maintain the EBV latency program [92]. Meanwhile, miR-BART9 was highly expressed in NPC tissues. Depleting endogenous miR-BART9 could inhibit the migration and invasion of EBV-positive NPC cells via targeting E-cadherin [93]. Cell growth assay and annexin V staining showed that miR-BART15-3p could inhibit cell proliferation and promote apoptosis [94]. Integrator complex subunit 6 (DICE1/INTS6) is a tumor suppressor which is often inactivated in cancers. It has also been identified as a target of miR-BART3. Overexpression miR-BART3 suppressed the DICE1 activity and stimulated cell proliferation. Consistent with the low expression of DICE1 in EBV-positive NPC tumors, it is suggested that EBV-encoded miR-BART3 promote cell growth and transformation via targeting tumor suppressor DICE1 [95].

14.3.2 KSHV-Encoded MiRNA

Kaposi's sarcoma-associated herpesvirus (KSHV) is another etiologic agent that can cause cancer. Its lifecycle involves two distinct phases: latent and lytic. The oncogenic potential of KSHV is greatly dependent on the genes expressed during latency and the setting of immune escape. KSHV can encode its own miRNAs, which are believed to promote latent viral persistence, immune suppression, and tumor progression [96–99]. Thus far, 12 KSHV miRNA precursors and 18 mature miRNAs have been identified [100–102]. All these miRNAs locate in the KSHV latency-associated region (KLAR) [103].

Studies have indicated KSHV miRNAs work as a switch of latent–lytic regulation. MiR-K12-7 and miR-K12-9 suppressed the transcription of ORF50 which encoded the replication and transcription activator (RTA), a critical factor for the initiation of lytic replication [104, 105]. MiR-K12-4 repressed retinoblastoma (Rb)-like protein 2 (Rbl2), a repressor of DNA methyl transferases (DNMT) that methylated the RTA promoter and inhibited its expression [106]. MiR-K12-1 could target both IκBα to promote NFκB-dependent viral latency and cell survival [97] and p21 to attenuate cell cycle arrest [107]. MiR-K12-5, MiR-K12-9, MiR-K12-10, MiR-K12-11, and MiR-K12-12 were shown to facilitate virus lytic replication [108–110]. Several factors secreted by KSHV-positive cells were associated with cancer pathogenesis [111], such as IL-6 and IL-10 [112, 113]. MiR-K12-3 and MiR-K12-7 were identified to repress C/EBPβ, which served as a transcriptional repressor of IL-6 and IL-10 [114]. MiR-K12-10 repressed the tumor necrosis factor-like weak inducer of apoptosis receptor (TWEAKR), resulting in reduced expression of IL-8, and the cells transfected with miR-K12-10 were more resistant to TWEAK-induced apoptosis [115].

14.3.3 HPV-Encoded MicroRNA

Human papillomaviruses (HPV) is one of the well-characterized human viruses that directly link to cervical cancer [116, 117]. HPV infection is limited to the basal cell layer and requires host cell to enter M phase [118]. The papillomavirus-encoded miRNAs in human cervical cancer and cell lines were first identified and validated by Qian et al. [119]. They established small RNA (sRNA) libraries from two HPV-immortalized cell lines and ten formalin-fixed paraffin-embedded HPV-positive cervical epithelium samples. The libraries were then sequenced with SOLiD 4 technology. Nine putative papillomavirus microRNAs were predicted and four were successfully validated: two were encoded by HPV16, one by HPV38, and the other by HPV68. Cellular target analysis of HPV16-encoded miRNAs indicated they had pathological effects on cell cycle, cell migration, and cancer development [119].

14.3.4 MCV-Encoded MicroRNA

Merkel cell polyomavirus (MCV) is present in the majority of merkel cell carcinomas (MCC), an aggressive neuroendocrine skin cancer. Extensive effort has been made to figure out the role of MCV-encoded miRNAs in pathogenesis. Seo et al. computationally predicted potential pre-miRNAs and then cloned them to an expression vector for screening. At least, one candidate MR17 was pursued for its function validation [120]. Other groups also identified a mature miRNA MCV-miR-M1-5p after sequencing 30 million sRNAs from seven MCC tumors and one perilesional sample. These miRNA-targeted genes were related to immune evasion and viral replication [121].

Together, it is evident that all the viral miRNAs discussed above are involved in the regulation of virus latency, immune evasion, or oncogenesis (see Table 14.3). Therefore, silencing specific viral miRNAs would be a worthy strategy to therapeutically intervene viral oncogenesis and cancer progression [122, 123].

14.4 Future Perspective

Owing to the development of sequencing technologies such as massive parallel sequencing and the evolved bioinformatics analysis methodologies, large amounts of noncoding RNAs transcribed by oncogenic virus have been identified. While this progress has deepen our knowledge of the category and complexity of noncoding

RNAs during the process of cell pathogenesis and oncogenesis, much work needs to be done to elucidate the roles of viral noncoding oncogenic RNAs in oncogenesis and other disease settings.

Great care should be taken when some viral noncoding RNAs are to be annotated as "pathogenic" or even "oncogenic." Given the unique structures of viral noncoding RNAs such as special 5′ caps, one may not investigate the candidate noncoding RNAs by simply overexpressing in vitro transcribed RNAs, which could produce RNA structures differing from the real modifications of RNAs in vivo. Likewise, the complicated secondary structures along with the nuclear localization of some ncRNAs such as sisRNAs make it almost impossible to knockdown the expression with routine siRNA or shRNA strategies. Instead, appropriate antisense oligonucleotides, specifically devised ncRNAs knockout strategy, and appropriate control viruses should be employed for the respective viral ncRNA investigations.

Large ncRNAs tend to be flexible modular scaffolds, which means RNAs contain the discrete domains that interact with specific protein complexes. This will bring specific regulatory components into proximity with each other to result in the formation of a unique functional complex. Thus, understanding these principles will require the identification of the sites of the RNA–protein interactions and the exact RNA-binding proteins in vivo. In addition, the way in which large ncRNAs localize to their target genes is unknown but could involve direct RNA–DNA interactions or interactions with nascent RNAs. With the development of methodologies such as RNA-chromatin immunoprecipitation (RNA-CHIP) and capture hybridization analysis of RNA targets (CHART), resolution of the molecular interaction details regarding the mechanism of viral ncRNAs is possible. By truly understanding this modular RNA code, it may be possible to devise accurate pharmaceuticals targeting oncogenic virus ncRNAs to reduce the health burden of cancer patients.

Acknowledgments This work was partly supported by the National Natural Science Foundation of China (81172188, 91129709), the National Key Basic Research Program, and the National Key Technologies R&D Program of the Ministry of Science and Technology, China (2013CB967203, 2013BAI01B07).

References

1. Zheng ZM. Viral oncogenes, non-coding RNAs, and RNA splicing in human tumor viruses. Int J Biol Sci. 2010;6(7):730–55.
2. Fatica A, Bozzoni I. Long non-coding RNAs: new players in cell differentiation and development. Nat Rev Genet. 2014;15(1):7–21.
3. Esteller M. Non-coding RNAs in human disease. Nat Rev Genet. 2011;12(12):861–74.
4. Iwakiri D. Epstein-Barr virus-encoded RNAs: key molecules in viral pathogenesis. Cancers (Basel). 2014;6(3):1615–30.
5. Sun R, Lin SF, Gradoville L, Miller G. Polyadenylylated nuclear RNA encoded by Kaposi sarcoma-associated herpesvirus. Proc Natl Acad Sci U S A. 1996;21:11883–8.
6. Zhong W, Ganem D. Characterization of ribonucleoprotein complexes containing an abundant polyadenylated nuclear RNA encoded by Kaposi's sarcoma-associated herpesvirus (human herpesvirus 8). J Virol. 1997;71(2):1207–12.

7. Lee SI, Murthy SC, Trimble JJ, et al. Four novel U RNAs are encoded by a herpesvirus. Cell. 1988;54(5):599–607.
8. Mathews MB, Shenk T. Adenovirus virus-associated RNA and translation control. J Virol. 1991;65(11):5657–62.
9. Lerner MR, Andrews NC, Miller G, Steitz JA. Two small RNAs encoded by Epstein-Barr virus and complexed with protein are precipitated by antibodies from patients with systemic lupus erythematosus. Proc Natl Acad Sci U S A. 1981;78(2):805–9.
10. Rosa MD, Gottlieb E, Lerner MR, Steitz JA. Striking similarities are exhibited by two small Epstein-Barr virus-encoded ribonucleic acid and the adenovirus-associated ribonucleic acids VAI and VAII. Mol Cell Biol. 1981;1(9):785–96.
11. Schwemmle M, Clemens MJ, Hilse K, et al. Localization of Epstein-Barr virus-encoded RNAs EBER-1 and EBER-2 in interphase and mitotic Burkitt lymphoma cells. Proc Natl Acad Sci U S A. 1992;89(21):10292–6.
12. Fok V, Friend K, Steitz JA. Epstein-Barr virus non-coding RNAs are confined to the nucleus, whereas their partner, the human La protein, undergoes nucleocytoplasmic shuttling. J Cell Biol. 2006;173(3):319–25.
13. Iwakiri D, Zhou L, Samanta M, et al. Epstein-Barr virus (EBV)-encoded small RNA is released from EBV-infected cells and activates signaling from Toll-like receptor 3. J Exp Med. 2009;206(10):2091–9.
14. Clarke PA, Schwemmle M, Schickinger J, et al. Binding of Epstein-Barr virus small RNA EBER-1 to the double-stranded RNA-activated protein kinase DAI. Nucleic Acids Res. 1991;19(2):243–8.
15. Ruf IK, Lackey KA, Warudkar S, Sample JT. Protection from interferon-induced apoptosis by Epstein-Barr virus small RNAs is not mediated by inhibition of PKR. J Virol. 2005;79(23):14562–9.
16. Houmani JL, Davis CI, Ruf IK. Growth-promoting properties of Epstein-Barr virus EBER-1 RNA correlate with ribosomal protein L22 binding. J Virol. 2009;83(19):9844–53.
17. Lee N, Pimienta G, Steitz JA. AUF1/hnRNP D is a novel protein partner of the EBER1 non-coding RNA of Epstein-Barr virus. RNA. 2012;18(11):2073–82.
18. Lee N, Moss WN, Yario TA, Steitz JA. EBV non-coding RNA binds nascent RNA to drive host PAX5 to viral DNA. Cell. 2014;160(4):607–18.
19. Gregorovic G, Boulden EA, Bosshard R, et al. Epstein-Barr viruses deficient in EBER RNAs give higher LMP2 RNA expression in lymphoblastoid cell lines and efficiently establish persistent infection in humanized mice. J Virol. 2015;89(22):11711–4.
20. Komano J, Maruo S, Kurozumi K, et al. Oncogenic role of Epstein-Barr virus-encoded RNAs in Burkitt's lymphoma cell line Akata. J Virol. 1997;73(12):9827–31.
21. Yajima M, Kanda T, Takada K. Critical role of Epstein-Barr Virus (EBV)-encoded RNA in efficient EBV-induced B-lymphocyte growth transformation. J Virol. 2005;79(7):4298–307.
22. Repellin CE, Tsimbouri PM, Philbey AW, Wilson JB. Lymphoid hyperplasia and lymphoma in transgenic mice expressing the small non-coding RNA, EBER1 of Epstein-Barr virus. PLoS ONE. 2010;5(2), e9092.
23. Swaminathan S, Tomkinson B, Kieff E. Recombinant Epstein-Barr virus with small RNA (EBER) genes deleted transforms lymphocytes and replicates in vitro. Proc Natl Acad Sci U S A. 1991;88(4):1546–50.
24. Gregorovic G, Bosshard R, Karstegl CE, et al. Cellular gene expression that correlates with EBER expression in Epstein-Barr Virus-infected lymphoblastoid cell lines. J Virol. 2011;85(7):3535–45.
25. Wong HL, Wang X, Chang RC, et al. Stable expression of EBERs in immortalized nasopharyngeal epithelial cells confers resistance to apoptotic stress. Mol Carcinog. 2005;44(2):92–101.
26. Yoshizaki T, Endo K, Ren Q, et al. Oncogenic role of Epstein-Barr virus-encoded small RNAs (EBERs) in nasopharyngeal carcinoma. Auris Nasus Larynx. 2007;34(1):73–8.
27. Iwakiri D, Sheen TS, Chen JY, et al. Epstein-Barr virus-encoded small RNA induces insulin-like growth factor 1 and supports growth of nasopharyngeal carcinoma-derived cell lines. Oncogene. 2005;24(10):1767–73.

28. Samanta M, Iwakiri D, Kanda T, et al. EB virus-encoded RNAs are recognized by RIG-I and activate signaling to induce type I IFN. EMBO. 2006;25(18):4207–14.
29. Samanta M, Iwakiri D, Takada K. Epstein-Barr virus-encoded small RNA induces IL-10 through RIG-I-mediated IRF-3 signaling. Oncogene. 2008;27(30):4150–60.
30. Duan Y, Li Z, Cheng S, et al. Nasopharyngeal carcinoma progression is mediated by EBER-triggered inflammation via the RIG-I pathway. Cancer Lett. 2015;361(1):67–74.
31. Li Z, Duan Y, Cheng S, Chen Y, et al. EBV-encoded RNA via TLR3 induces inflammation in nasopharyngeal carcinoma. Oncotarget. 2015;6(27):24291–303.
32. Albrecht JC, Fleckenstein B. Nucleotide sequence of HSUR 6 and HSUR 7, two small RNAs of herpesvirus saimiri. Nucleic Acids Res. 1992;20(7):1810.
33. Ensser A, Fleckenstein B. T-cell transformation and oncogenesis by gamma2-herpesviruses. Adv Cancer Res. 2005;93:91–128.
34. Murthy S, Kamine J, Desrosiers RC. Viral-encoded small RNAs in herpes virus saimiri induced tumors. EMBO J. 1986;5(7):1625–32.
35. Myer VE, Lee SI, Steitz JA. Viral small nuclear ribonucleoproteins bind a protein implicated in messenger RNA destabilization. Proc Natl Acad Sci U S A. 1992;89(4):1296–300.
36. Cook HL, Mischo HE, Steitz JA. The Herpesvirus saimiri small nuclear RNAs recruit AU-rich element-binding proteins but do not alter host AU-rich element-containing mRNA levels in virally transformed T cells. Mol Cell Biol. 2004;24(10):4522–33.
37. Cook HL, Lytle JR, Mischo HE, et al. Small nuclear RNAs encoded by Herpesvirus saimiri up-regulate the expression of genes linked to T cell activation in virally transformed T cells. Curr Biol. 2005;15(10):974–9.
38. Cazalla D, Yario T, Steitz JA. Down-regulation of a host microRNA by a Herpesvirus saimiri non-coding RNA. Science. 2010;328(5985):1563–6.
39. Tycowski KT, Guo YE, Lee N, et al. Viral non-coding RNAs: more surprises. Genes Dev. 2015;29(6):567–84.
40. Libri V, Helwak A, Miesen P, et al. Murine cytomegalovirus encodes a miR-27 inhibitor disguised as a target. Proc Natl Acad Sci U S A. 2012;109(1):279–84.
41. Marcinowski L, Tanguy M, Krmpotic A, et al. Degradation of cellular mir-27 by a novel, highly abundant viral transcript is important for efficient virus replication in vivo. PLoS Pathog. 2012;8(2), e1002510.
42. Guo YE, Riley KJ, Iwasaki A, Steitz JA. Alternative capture of non-coding RNAs or protein-coding genes by herpesviruses to alter host T cell function. Mol Cell. 2014;54(1):67–79.
43. Murthy SC, Trimble JJ, Desrosiers RC. Deletion mutants of herpesvirus saimiri define an open reading frame necessary for transformation. J Virol. 1989;63(8):3307–14.
44. Greenaway PJ, Wilkinson GW. Nucleotide sequence of the most abundantly transcribed early gene of human cytomegalovirus strain AD169. Virus Res. 1987;7(1):17–31.
45. Reeves MB, Davies AA, McSharry BP, et al. Complex I binding by a virally encoded RNA regulates mitochondria-induced cell death. Science. 2007;316(5829):1345–8.
46. Stern-Ginossar N, Weisburd B, Michalski A, et al. Decoding human cytomegalovirus. Science. 2012;338(6110):1088–93.
47. Ganem D. KSHV and the pathogenesis of Kaposi sarcoma: listening to human biology and medicine. J Clin Invest. 2010;120(4):939–49.
48. Rossetto CC, Pari GS. PAN's Labyrinth: molecular biology of Kaposi's sarcoma-associated herpesvirus (KSHV) PAN RNA, a multifunctional long non-coding RNA. Viruses. 2014;6(11):4212–26.
49. Rossetto CC, Tarrant-Elorza M, Verma S, et al. Regulation of viral and cellular gene expression by Kaposi's sarcoma-associated herpesvirus polyadenylated nuclear RNA. J Virol. 2013;87(10):5540–53.
50. Song MJ, Brown HJ, Wu TT, Sun R. Transcription activation of polyadenylated nuclear rna by rta in human herpesvirus 8/Kaposi's sarcoma-associated herpesvirus. J Virol. 2001;75(7):3129–40.
51. Campbell M, Kim KY, Chang PC, et al. A lytic viral long non-coding RNA modulates the function of a latent protein. J Virol. 2014;88(3):1843–8.

52. Sahin BB, Patel D, Conrad NK. Kaposi's sarcoma-associated herpesvirus ORF57 protein binds and protects a nuclear non-coding RNA from cellular RNA decay pathways. PLoS Pathog. 2010;6(3), e1000799.
53. Conrad NK, Shu MD, Uyhazi KE, Steitz JA. Mutational analysis of a viral RNA element that counteracts rapid RNA decay by interaction with the polyadenylate tail. Proc Natl Acad Sci U S A. 2007;104(25):10412–7.
54. Mitton-Fry RM, DeGregorio SJ, et al. Poly(A) tail recognition by a viral RNA element through assembly of a triple helix. Science. 2010;330(6008):1244–7.
55. Massimelli MJ, Kang JG, Majerciak V, et al. Stability of a long non-coding viral RNA depends on a 9-nt core element at the RNA 5′ end to interact with viral ORF57 and cellular PABPC1. Int J Biol Sci. 2011;7(8):1145–60.
56. Borah S, Darricarrere N, Darnell A, et al. A viral nuclear non-coding RNA binds re-localized poly(A) binding protein and is required for late KSHV gene expression. PLoS Pathog. 2011;7(10), e1002300.
57. Rossetto CC, Pari GS. Kaposi's sarcoma-associated herpesvirus non-coding polyadenylated nuclear RNA interacts with virus- and host cell-encoded proteins and suppresses expression of genes involved in immune modulation. J Virol. 2011;85(24):13290–7.
58. Rossetto CC, Pari G. KSHV PAN RNA associates with demethylases UTX and JMJD3 to activate lytic replication through a physical interaction with the virus genome. PLoS Pathog. 2012;8(5), e1002680.
59. Arias C, Weisburd B, Stern-Ginossar N, et al. KSHV 2.0: a comprehensive annotation of the Kaposi's sarcoma-associated herpesvirus genome using next-generation sequencing reveals novel genomic and functional features. PLoS Pathog. 2014;10(1), e1003847.
60. He M, Zhang W, Bakken T, et al. Cancer angiogenesis induced by Kaposi sarcoma-associated herpesvirus is mediated by EZH2. Cancer Res. 2012;72(14):3582–92.
61. Roby JA, Pijlman GP, Wilusz J, Khromykh AA. Non-coding subgenomic flavivirus RNA: multiple functions in West Nile virus pathogenesis and modulation of host responses. Viruses. 2014;6(2):404–27.
62. Chapman EG, Costantino DA, Rabe JL, et al. The structural basis of pathogenic subgenomic flavivirus RNA (sfRNA) production. Science. 2014;344(6181):307–10.
63. Pijlman GP, Funk A, Kondratieva N, et al. A highly structured, nuclease-resistant, non-coding RNA produced by flaviviruses is required for pathogenicity. Cell Host Microbe. 2008;4(6):579–91.
64. Funk A, Truong K, Nagasaki T, et al. RNA structures required for production of subgenomic flavivirus RNA. J Virol. 2010;84(21):11407–17.
65. Silva PA, Pereira CF, Dalebout TJ, et al. An RNA pseudoknot is required for production of yellow fever virus subgenomic RNA by the host nuclease XRN1. J Virol. 2010;84(21):11395–406.
66. Fan YH, Nadar M, Chen CC, et al. Small non-coding RNA modulates Japanese encephalitis virus replication and translation in trans. Virol J. 2011;8:492.
67. Schuessler A, Funk A, Lazear HM, et al. West Nile virus non-coding subgenomic RNA contributes to viral evasion of the type I interferon-mediated antiviral response. J Virol. 2012;86(10):5708–18.
68. Chang RY, Hsu TW, Chen YL, et al. Japanese encephalitis virus non-coding RNA inhibits activation of interferon by blocking nuclear translocation of interferon regulatory factor 3. Vet Microbiol. 2013;166(1–2):11–21.
69. Moon SL, Anderson JR, Kumagai Y, et al. A non-coding RNA produced by arthropod-borne flaviviruses inhibits the cellular exoribonuclease XRN1 and alters host mRNA stability. RNA. 2012;18(11):2029–40.
70. Schnettler E, Sterken MG, Leung JY, et al. Non-coding flavivirus RNA displays RNA interference suppressor activity in insect and Mammalian cells. J Virol. 2012;86(24):13486–500.
71. Hussain M, Torres S, Schnettler E, et al. West Nile virus encodes a microRNA-like small RNA in the 3′ untranslated region which up-regulates GATA4 mRNA and facilitates virus replication in mosquito cells. Nucleic Acids Res. 2012;40(5):2210–23.

72. Moon SL, Blackinton JG, Anderson JR, et al. XRN1 stalling in the 5′ UTR of Hepatitis C virus and Bovine Viral Diarrhea virus is associated with dysregulated host mRNA stability. PLoS Pathog. 2015;11(3), e1004708.
73. Wilusz JE. Long non-coding RNAs: re-writing dogmas of RNA processing and stability. Biochim Biophys Acta. 2015;1859(1):128–38.
74. Stevens JG, Wagner EK, Devi-Rao GB, et al. RNA complementary to a herpesvirus alpha gene mRNA is prominent in latently infected neurons. Science. 1987;235(4792):1056–9.
75. Allen SJ, Rhode-Kurnow A, Mott KR, et al. Interactions between herpesvirus entry mediator (TNFRSF14) and latency-associated transcript during herpes simplex virus 1 latency. J Virol. 2014;88(4):1961–71.
76. Cliffe AR, Garber DA, Knipe DM. Transcription of the herpes simplex virus latency-associated transcript promotes the formation of facultative heterochromatin on lytic promoters. J Virol. 2009;83(16):8182–90.
77. Schwarz TM, Kulesza CA. Stability determinants of murine cytomegalovirus long non-coding RNA7.2. J Virol. 2014;88(19):11630–3.
78. Moss WN, Steitz JA. Genome-wide analyses of Epstein-Barr virus reveal conserved RNA structures and a novel stable intronic sequence RNA. BMC Genomics. 2013;14:543.
79. Moss WN, Lee N, Pimienta G, Steitz JA. RNA families in Epstein-Barr virus. RNA Biol. 2014;11(1):10–7.
80. Hutzinger R, Feederle R, Mrazek J, et al. Expression and processing of a small nucleolar RNA from the Epstein-Barr virus genome. PLoS Pathog. 2009;5(8), e1000547.
81. Zeng Z, Huang H, Huang L, et al. Regulation network and expression profiles of Epstein-Barr virus-encoded microRNAs and their potential target host genes in nasopharyngeal carcinomas. Sci China Life Sci. 2014;57(3):315–26.
82. Bartel DP. MicroRNAs: genomics, biogenesis, mechanism, and function. Cell. 2004;116(2):281–97.
83. Xie Y-J, Long Z-F, He X-S. Involvement of EBV-encoded BART-miRNAs and dysregulated cellular miRNAs in nasopharyngeal carcinoma genesis. Asian Pac J Cancer Prev. 2013;14(10):5637–44.
84. Lo AK, Dawson CW, Jin DY, Lo KW. The pathological roles of BART miRNAs in nasopharyngeal carcinoma. J Pathol. 2012;227(4):392–403.
85. Marquitz AR, Raab-Traub N. The role of miRNAs and EBV BARTs in NPC. Semin Cancer Biol. 2012;22(2):166–72.
86. Witkos TM, Koscianska E, Krzyzosiak WJ. Practical aspects of microRNA target prediction. Curr Mol Med. 2011;11(2):93–109.
87. Kuzembayeva M, Hayes M, Sugden B. Multiple functions are mediated by the miRNAs of Epstein-Barr virus. Curr Opin Virol. 2014;7:61–5.
88. Wong AM, Kong KL, Tsang JW, et al. Profiling of Epstein-Barr virus-encoded microRNAs in nasopharyngeal carcinoma reveals potential biomarkers and oncomirs. Cancer. 2012;118(3):698–710.
89. Choy EY, Siu KL, Kok KH, et al. An Epstein-Barr virus-encoded microRNA targets PUMA to promote host cell survival. J Exp Med. 2008;205(11):2551–60.
90. Kanda T, Miyata M, Kano M, et al. Clustered microRNAs of the Epstein-Barr virus cooperatively down-regulate an epithelial cell-specific metastasis suppressor. J Virol. 2015;89(5):2684–97.
91. Lo AK, To KF, Lo KW, et al. Modulation of LMP1 protein expression by EBV-encoded microRNAs. Proc Natl Acad Sci U S A. 2007;104(41):16164–9.
92. Ramakrishnan R, Donahue H, Garcia D, et al. Epstein-Barr virus BART9 miRNA modulates LMP1 levels and affects growth rate of nasal NK T cell lymphomas. PLoS ONE. 2011;6(11), e27271.
93. Hsu CY, Yi YH, Chang KP, et al. The Epstein-Barr virus-encoded microRNA MiR-BART9 promotes tumor metastasis by targeting E-cadherin in nasopharyngeal carcinoma. PLoS Pathog. 2014;10(2), e1003974.

94. Choi H, Lee H, Kim SR, et al. Epstein-Barr virus-encoded microRNA BART15-3p promotes cell apoptosis partially by targeting BRUCE. J Virol. 2013;87(14):8135–44.
95. Lei T, Yuen KS, Xu R, et al. Targeting of DICE1 tumor suppressor by Epstein-Barr virus-encoded miR-BART3* microRNA in nasopharyngeal carcinoma. Int J Cancer. 2013;133(1):79–87.
96. Hansen A, Henderson S, Lagos D, et al. KSHV-encoded miRNAs target MAF to induce endothelial cell reprogramming. Genes Dev. 2010;24(2):195–205.
97. Lei X, Bai Z, Ye F, et al. Regulation of NF-kappaB inhibitor IkappaBalpha and viral replication by a KSHV microRNA. Nat Cell Biol. 2010;12(2):193–9.
98. Samols MA, Skalsky RL, Maldonado AM, et al. Identification of cellular genes targeted by KSHV-encoded microRNAs. PLoS Pathog. 2007;3(5), e65.
99. Gottwein E, Corcoran DL, Mukherjee N, et al. Viral microRNA targetome of KSHV-infected primary effusion lymphoma cell lines. Cell Host Microbe. 2011;10(5):515–26.
100. Cai X, Lu S, Zhang Z, et al. Kaposi's sarcoma-associated herpesvirus expresses an array of viral microRNAs in latently infected cells. Proc Natl Acad Sci U S A. 2005;102(15):5570–5.
101. Samols MA, Hu J, Skalsky RL, Renne R. Cloning and identification of a microRNA cluster within the latency-associated region of Kaposi's sarcoma-associated herpesvirus. J Virol. 2005;79(14):9301–5.
102. Pfeffer S, Sewer A, Lagos-Quintana M, et al. Identification of microRNAs of the herpesvirus family. Nat Methods. 2005;2(4):269–76.
103. Qin Z, Jakymiw A, Findlay V, Parsons C. KSHV-encoded microRNAs: lessons for viral cancer pathogenesis and emerging concepts. Int J Cell Biol. 2012;2012:603961.
104. Lin X, Liang D, He Z, et al. MiR-K12-7-5p encoded by Kaposi's sarcoma-associated herpesvirus stabilizes the latent state by targeting viral ORF50/RTA. PLoS ONE. 2011;6(1), e16224.
105. Bellare P, Ganem D. Regulation of KSHV lytic switch protein expression by a virus-encoded microRNA: an evolutionary adaptation that fine-tunes lytic reactivation. Cell Host Microbe. 2009;6(6):570–5.
106. Lu F, Stedman W, Yousef M, et al. Epigenetic regulation of Kaposi's sarcoma-associated herpesvirus latency by virus-encoded microRNAs that target Rta and the cellular Rbl2-DNMT pathway. J Virol. 2010;84(6):2697–706.
107. Gottwein E, Cullen BR. A human herpesvirus microRNA inhibits p21 expression and attenuates p21-mediated cell cycle arrest. J Virol. 2010;84(10):5229–37.
108. Ziegelbauer JM, Sullivan CS, Ganem D. Tandem array-based expression screens identify host mRNA targets of virus-encoded microRNAs. Nat Genet. 2009;41(1):130–4.
109. Liang D, Gao Y, Lin X, et al. A human herpesvirus miRNA attenuates interferon signaling and contributes to maintenance of viral latency by targeting IKKepsilon. Cell Res. 2011;21(5):793–806.
110. Umbach JL, Cullen BR. In-depth analysis of Kaposi's sarcoma-associated herpesvirus microRNA expression provides insights into the mammalian microRNA-processing machinery. J Virol. 2010;84(2):695–703.
111. Mesri EA, Cesarman E, Boshoff C. Kaposi's sarcoma and its associated herpesvirus. Nat Rev Cancer. 2010;10(10):707–19.
112. Jones KD, Aoki Y, Chang Y, et al. Involvement of interleukin-10 (IL-10) and viral IL-6 in the spontaneous growth of Kaposi's sarcoma herpesvirus-associated infected primary effusion lymphoma cells. Blood. 1999;94(8):2871–9.
113. Cirone M, Lucania G, Aleandri S, et al. Suppression of dendritic cell differentiation through cytokines released by Primary Effusion Lymphoma cells. Immunol Lett. 2008;120(1–2):37–41.
114. Qin Z, Kearney P, Plaisance K, Parsons CH. Pivotal advance: Kaposi's sarcoma-associated herpesvirus (KSHV)-encoded microRNA specifically induce IL-6 and IL-10 secretion by macrophages and monocytes. J Leukoc Biol. 2010;87(1):25–34.
115. Abend JR, Uldrick T, Ziegelbauer JM. Regulation of tumor necrosis factor-like weak inducer of apoptosis receptor protein (TWEAKR) expression by Kaposi's sarcoma-associated her-

pesvirus microRNA prevents TWEAK-induced apoptosis and inflammatory cytokine expression. J Virol. 2010;84(23):12139–51.
116. Walboomers JM, Jacobs MV, Manos MM, et al. Human papillomavirus is a necessary cause of invasive cervical cancer worldwide. J Pathol. 1999;189(1):12–9.
117. Cuschieri K, Brewster DH, Graham C, et al. Influence of HPV type on prognosis in patients diagnosed with invasive cervical cancer. Int J Cancer. 2014;135(11):2721–6.
118. Pyeon D, Pearce SM, Lank SM, et al. Establishment of human papillomavirus infection requires cell cycle progression. PLoS Pathog. 2009;5(2), e1000318.
119. Qian K, Pietila T, Ronty M, et al. Identification and validation of human papillomavirus encoded microRNAs. PLoS ONE. 2013;8(7), e70202.
120. Seo GJ, Chen CJ, Sullivan CS. Merkel cell polyomavirus encodes a microRNA with the ability to autoregulate viral gene expression. Virology. 2009;383(2):183–7.
121. Lee S, Paulson KG, Murchison EP, et al. Identification and validation of a novel mature microRNA encoded by the Merkel cell polyomavirus in human Merkel cell carcinomas. J Clin Virol. 2011;52(3):272–5.
122. Elmen J, Lindow M, Schutz S, et al. LNA-mediated microRNA silencing in non-human primates. Nature. 2008;452(7189):896–9.
123. Krutzfeldt J, Rajewsky N, Braich R, et al. Silencing of microRNAs in vivo with 'antagomirs'. Nature. 2005;438(7068):685–9.

Chapter 15
Noncoding RNAs in Cancer Diagnosis

Mu-Sheng Zeng

Abstract The accuracy and efficiency of tumor treatment depends mainly on early and precise diagnosis. Although histopathology is always the gold standard for cancer diagnosis, noninvasive biomarkers represent an opportunity for early detection and molecular staging of cancer. Besides the classical tumor markers, noncoding RNAs (ncRNAs) emerge to be a novel category of biomarker for cancer diagnosis since the dysregulation of ncRNAs is closely associated with the development and progression of human cancers such as liver, lung, breast, gastric, and other kinds of cancers. In this chapter, we will summarize the different types of ncRNAs in the diagnosis of major human cancers. In addition, we will introduce the recent advances in the detection and applications of circulating serum or plasma ncRNAs and non-blood fluid ncRNAs because the noninvasive body fluid-based assays are easy to examine for cancer diagnosis and monitoring.

Keywords Noncoding RNA • Cancer diagnosis • Circulating ncRNA • Exosome

15.1 Introduction

The accuracy and efficiency of tumor treatment depends mainly on early diagnosis. However, histopathology is always the gold standard for cancer diagnosis, but is usually invasive since the examined tissues are obtained by surgery or biopsy. Since the last decade, it has been shown that dysregulation of ncRNAs is closely associated with the development and progression of human cancers such as liver, lung, breast, gastric, and other kinds of cancers [1, 2].

As is traditionally accepted in central dogma, DNA is transcribed into mRNA, which in turn serves as the template for protein synthesis. However, we now know that there are only about 20,000 protein-coding genes, representing <2 % of the total genome sequence [3], whereas at least 90 % of the genome are actively transcribed

M.-S. Zeng (✉)
Sun Yat-sen University Cancer Center, 651 Dongfeng East Road, Guangzhou, China
e-mail: zengmsh@mail.sysu.edu.cn

E. Song (ed.), *The Long and Short Non-coding RNAs in Cancer Biology*,
Advances in Experimental Medicine and Biology 927,
DOI 10.1007/978-981-10-1498-7_15

into noncoding RNAs (ncRNAs), implicating that ncRNAs could have significant regulatory roles in complex organisms. It is also thought that RNAs are actually precursors to all current lives on Earth. ncRNAs are classified into housekeeping and regulatory ncRNAs, based on their abundance in distribution and biological function. Most of the cancer-associated ncRNAs are regulatory ncRNAs, which can be divided into small ncRNAs (sncRNAs), smaller than 200 nucleotides (nts) and long ncRNAs (lncRNAs, longer than 200 nts) [4, 5], according to their length. ncRNAs can also present circular structure [6], called circular RNAs (circRNAs). The small ncRNAs include microRNAs (miRNAs), Piwi-associated RNAs (piRNAs) [7], small nucleolar RNAs (snoRNA), and so on. In contrast to sncRNAs, lncRNAs are much longer and tend to have a more complex secondary structure [8]. All of these ncRNAs have the potential to act as biomarkers for cancer diagnosis.

In this chapter, we will introduce noncoding RNAs in cancer diagnosis through the following aspects: types of ncRNAs and cancer diagnosis, ncRNAs and major cancer types, circulating ncRNAs, and non-blood body fluid ncRNAs.

15.2 Types of ncRNAs and Cancer Diagnosis

Cancer is a complicated disease. Cancer diagnosis is very important and challenging, especially for early- stage cancer. Recent studies have begun to address the underlying causes of cancer mutational heterogeneity by comparing mutation rate variation to the distribution of sequence features, gene expression, and epigenetic marks along the genome [9]. Cancer diagnosis has also been focused on cancer biology, since ncRNAs play important roles in the oncogenesis and development of cancer [10]. Besides, ncRNAs can be specifically expressed in certain tissues [11] or certain cancers [12]. Thus, ncRNAs should be great potential markers in cancer diagnosis.

Recently, with the great development of next-generation sequencing, it brings in big progression in analysis of ncRNAs. In general, ncRNAs can be divided into two types in terms of their sizes, small RNAs and long noncoding RNAs (lncRNAs), which are based on the threshold of 200 nucleotides (nt) of the RNA length [13]. Small RNAs include many different types of RNAs, including microRNAs (miRNAs), small nucleolar RNAs (snoRNAs), and piwiRNAs (piRNAs) [14], which are classified by functional type. Meanwhile, lncRNAs also have different classification criteria. Ma summarized the classification methods of lncRNAs according to their different features including their (1) genome location and context, (2) exerted effects on DNA sequences, (3) mechanisms of functioning, and (4) targeting [15]. Laurent [16] summarized the classification of the great majority of lncRNAs relying on the empirical attributes originally used to detect them, such as classification based on transcript length, association with annotated protein-coding genes, association with other DNA elements of known function, protein-coding RNA resemblance, association with repeats, association with a biochemical pathway or stability, sequence and structure conservation, expression in different biological states, association with subcellular structures, and function. In this chapter, we will summarize lncRNAs to three categories: intergenic lncRNAs, sense lncRNAs, and antisense lncRNAs.

15.2.1 Small RNAs

For small RNAs, there is a large amount of evidence supporting their potential application in cancer diagnosis.

15.2.1.1 snoRNAs and Cancer Diagnosis

snoRNAs range in size from 60 to 300 nt. At least 200 snoRNAs have been identified in mammals, but many more remain elusive [17]. There are two types of snoRNAs: Box C/D and BoxH/ACA snoRNAs [18]. Box C/D snoRNAs serve as guides for the 2'-O-ribose methylation of rRNAs, whereas box H/ACA snoRNAs are guides for isomerization of uridine residues into pseudouridine. Therefore, snoRNAs were recognized as housekeeping RNAs due to their critical roles in rRNA maturation in the traditional opinion. However, recently the opinion has been changed at a certain extent because some evidence reveal a completely new and previously unrecognized role of snoRNAs in the control of cell fate and oncogenesis in various cancers. For example, Maarabouni has found that GAS5 controls apoptosis and is downregulated in breast cancer [19]; Dong used a series of assays to confirm that snoRNA U50 is a candidate tumor suppressor gene at 6q14.3 with a mutation associated with clinically significant prostate cancer [20].

15.2.1.2 piwiRNAs and Cancer Diagnosis

Piwi-interacting RNAs (piRNAs) belong to a novel class of sncRNAs, with a length of 24–33 nts, which only bind specifically with P-element-induced wimpy testis (Piwi) protein family members. Piwi proteins in humans have four homologs: PiwiL1/Hiwi, PiwiL2/Hili, PiwiL3, and PiwiL4/Hiwi2 [21]. To date, piRNAs are thought to mainly exist in the intergenic region and rarely in the gene region and repeat region. In July 2006, piRNAs were first identified as being closely associated with germ cell development [22]. Localization of Piwi and piwiRNAs has been reported mostly in the nucleus and cytoplasm of higher eukaryotes germ-line cells, and known piwiRNA sequences are believed to be located in repeat regions of the nuclear genome in germ-line cells. There have been increasing evidence suggesting that piRNAs may play a similar epigenetic silencing role in human cancers and be biomarkers for cancer diagnosis. In addition, Nam [23] has found that there are some piwiRNAs expressed in the mitochondria of mammalian cancer cells. Stefani [24] has demonstrated that the presence of piwiRNAs in cancer correlates with poorer clinical outcomes, suggesting the piwiRNA plays a functional role in cancer biology. Therefore, piwiRNAs are candidate cancer-associated genes, which can be used for cancer diagnosis.

15.2.1.3 miRNAs and Cancer Diagnosis

microRNAs (miRNAs), which were first reported in roundworms (*Caenorhabditis elegans*) [25], are short noncoding RNAs of 20–24 nucleotides that play important roles in virtually all biological pathways in mammals and other multicellular organisms. Accordingly, miRNAs influence numerous cancer-relevant processes such as proliferation, cell-cycle control, apoptosis, differentiation, migration, and metabolism.

In recent years, there are mounting cancer research focusing on the miRNAs. The first evidence for a direct link between miRNAs and human cancer came from the observation that two miRNA genes, mir-15 and mir-16, were located in a 30 kb region on chromosome 13 that had been found to be deleted in chronic lymphocytic leukemia (CLL) cases and that mir-15 and mir-16 expression was often reduced in CLL [26]. A second study [27] found that mir-143 and mir-145 expression levels were reduced in adenomatous and cancer stages of colorectal neoplasia. A number of studies have shown specific miRNAs identified in expression profiling experiments forward to carry out functional analysis. For example, mir-21 was found to be overexpressed in malignant cholangiocytes and mir-21 expression was found to downregulate the tumor suppressor PTEN in these cells [28]. However, several miRNAs have been proposed to act as tumor suppressors. One of these miRNAs is the members of the let-7 family, which have been reported to regulate expression of the RAS oncogene in *C. elegans* and in human cells. For example, let-7 is downregulated in lung cancer and its expression is negatively correlated with that of RAS [29]. Overall, many tumor-associated miRNAs have been reported and studied.

Apparently, miRNAs can be potentially useful for cancer diagnosis. Although initial miRNA studies focused on comparing its expression between normal tissues and tumors to gauge its diagnostic potential, it will be more interesting and important to correlate miRNA expression with tumor subtypes or clinical parameters. miRNA expression data can be used to build discriminators with clinical values, because miRNAs have clear advantages over mRNAs as they are more stable in vivo and in vitro [30],

15.2.2 lncRNAs

Recently, more and more researchers have focused on long noncoding RNAs in cancer study, and some lncRNAs have a better cancer diagnose potential.

15.2.2.1 Intergenic lncRNAs and Cancer Diagnosis

Long intergenic noncoding RNAs, which are expressed on a noncoding region, represent an emerging and understudied class of transcripts that plays a significant role in human cancers. The intergenic lncRNA sequence and transcription tend to turn over rapidly during evolution [31], so the intergenic lncRNAs may be species specific. White et al. [32] characterized the lncRNA landscape in lung cancer using publicly available transcriptome sequencing data from a cohort of 567 adenocarcinoma and squamous cell carcinoma tumors and found an intergenic lncRNA LCAL1 by using a functional assay, which plays a role in lung cancer cellular proliferation. Tens of thousands of intergenic lncRNAs have been identified [33]. It is no doubt that there are many intergenic lncRNAs that can be used for cancer diagnosis.

15.2.2.2 Antisense lncRNAs and Cancer Diagnosis

Antisense lncRNAs, whose sequence overlaps with the antisense strand of a protein-coding gene, can have a direct relationship with protein-coding gene expression. PANDA (promoter of CDKN1A antisense DNA damage-activated RNA), produced from the CDKN1A promoter region, shows p53-dependent induction after DNA damage and aids in cell proliferation by inhibiting apoptotic genes [34]. ANRIL, which is an antisense noncoding RNA at the INK4 locus, is an antisense transcript of the CDKN2B gene. Its abnormal expression is associated to Philadelphia-positive acute lymphoblastic leukemia [35]. HOTAIR (HOX transcript antisense intergenic RNA), which is located in the HOXC cluster and regulates human HOXD gene cluster expression in trans by epigenetic events [36], is upregulated in breast, colorectal, hepatocellular, gastrointestinal, and pancreatic carcinomas. There are many other antisense lncRNA-associated cancers, such as ANRASSF1 [37] and HIF1a antisense RNAs [38] or so.

15.2.2.3 Sense lncRNAs and Cancer Diagnosis

Many coding genes have alternative splicing subtypes, and some transcripts are noncoding RNAs. SRAs (steroid receptor RNA activators) can be translated into proteins, and the tRNA sequence can also act as a scaffold for several coactivator and repressor proteins to form complexes that regulate gene transcription in breast cancer [39]. Besides, there are some intronic lncRNAs [40] and cirRNAs [41] on the sense strand formed by splicing. For example, intronic lncRNA CAI2 contributes to advanced-stage neuroblastoma [42], and circular RNAs are enriched and stable in exosomes, which may serve as a promising biomarker for cancer diagnosis [43].

Overall, more and more lncRNAs can serve as biomarkers for cancer diagnosis. Cancer is a multifactorial, multistep, and complicated disease. An advantage in the diagnostic use of ncRNA detection versus protein-coding RNAs is that ncRNAs are effector molecules, whose expression levels may be a better indicator of the intrinsic biology of the tumor [12].

15.3 Noncoding RNAs and Major Cancer Types

15.3.1 microRNAs Acting as Diagnostic Cancer Biomarker

microRNAs (miRNAs), 21–22 nts small noncoding RNAs (ncRNAs), are involved in regulating gene expression by forming the RNA-induced silencing complex (RISC) and binding to specific sequences in the 3'-UTR of their target mRNAs to suppress translation or to induce mRNA degradation. Among the ncRNAs, miRNAs have been extensively studied in various types of cancers. Here, we have summarized their potential use in cancer diagnosis (see Fig. 15.1).

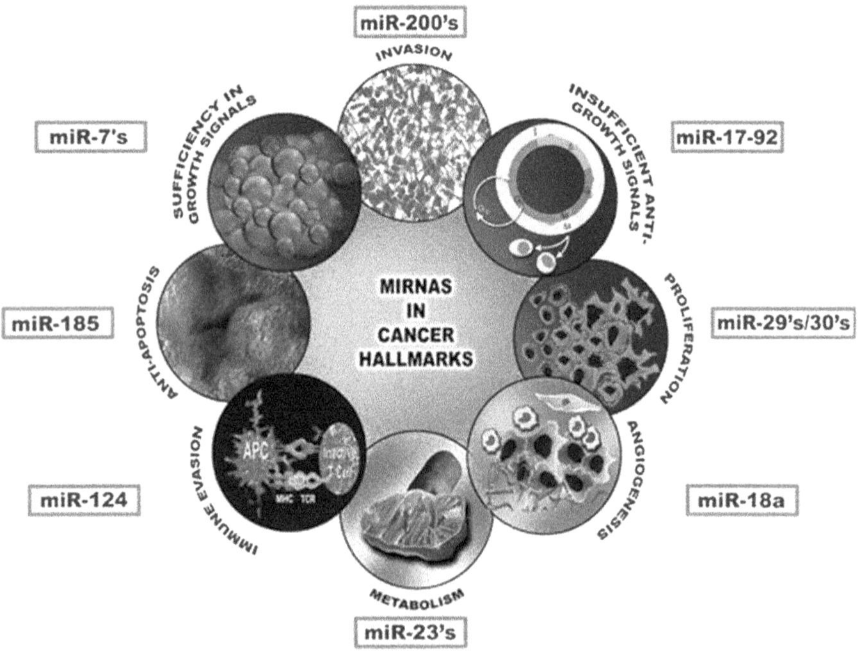

Fig. 15.1 Summary about miRNAs and diagnostic cancer. Examples of microRNAs involved in cancer hallmarks. The eight biological capabilities acquired during the multistep development of human tumors include sustaining proliferative signaling, evading growth suppressors, resisting cell death, and enabling replicative immortality. An example of a microRNA is presented. This figure is adopted from "MicroRNAome genome: A treasure for cancer diagnosis and therapy" [76]

15.3.1.1 microRNAs and Colorectal Cancer

Li, L. et al. detected the level of miRNA-29b from 400 healthy controls and patients with colon cancer and found miRNA-29b can act as diagnostic biomarker for early-stage colon cancer both in the tissue and the plasma [44]. Valeri, N. et al. found that miRNA-135b is upregulated in colon cancer [45]. These studies have shown that more than one miRNAs have the potential to be biomarkers for diagnosis in colorectal cancer.

15.3.1.2 microRNAs and Prostate Cancer

The role of miRNAs in the diagnosis, treatment, and prognosis of prostate cancer has drawn increasing attention in recent years, and the results regarding miRNAs in prostate cancer diagnosis are encouraging [46]. In 2006, Volinia et al. found that miR-106a is overexpressed in the tumor samples from 56 tumorous and seven normal tissue samples of prostate cancer [47]. The representative studies concerning miRNAs and prostate cancer are summarized in Tables 15.1 and 15.2.

15.3.1.3 microRNAs and Hepatocellular Carcinoma (HCC)

miRNAs associated with HCC development have been investigated as diagnostic biomarkers for HCC [73]. They have been shown to accurately predict the prognosis of HCC. For example, studies have demonstrated that members of the miR-200 family, miR-200a and miR-200b, are deregulated in HCC. Besides, a set of miRNAs, including miR-122, miR-192, miR-21, miR-223, miR-26a, miR-27a, and miR-801, have been shown to have high diagnostic accuracy in the early diagnosis of HBV-related HCC [74]. Lin and his colleagues showed that classifying the miRNA for HCC diagnosis was even better than AFP, since the specificities of the classifier are around 80–90 %, offering a promising diagnostic accuracy and specificity for HCC [75]. More details are summarized in Table 15.3.

15.3.2 lncRNAs Acting as Diagnostic Cancer Biomarker

Long noncoding RNAs (lncRNAs) are a major class of transcripts, longer than 200 nt and lack of protein-coding potential. Increasing evidence shows that many lncRNAs play vital roles in various cellular processes, especially in exerting influence on tumor biology. At the same time, some lncRNAs have tissue-specific expression in particular types of cancer, which makes them promising biomarkers for cancer diagnosis [94–96]. Here, we will review the application of some lncRNAs on cancer diagnosis. Table 15.4 has also summarized representative lncRNAs involved in carcinogenesis and acting as potential cancer biomarkers.

Table 15.1 miRNAs with the highest diagnostic value in prostate cancer

Sample type	Deregulated miRNAs (number of miRNAs)	miRNAs, selected as biomarkers		References
60 microdissected cellular elements of tumor tissue and 16 samples of normal tissue	Upregulated (21)	miR-32, miR-26a	↑	[48]
	Downregulated (21)	miR-181a, miR-196a, miR-25, miR-92, and let-7i		
102 PC samples and 102 normal tissue samples	Downregulated (54)	miR-205	↓	[49]
49 PC samples and 25 normal tissue samples	Downregulated (7)	miR-96-5p, miR-183-5p	↑	[50]
		miR-145-4, miR-221-5P	↓	
36 serum samples of PC patients and 12 healthy donors	Upregulated (6)	miR-20b, miR-874, miR-1274a, miR-1207-5pmiR-93, miR-106a	↑	[51]
	Downregulated (4)			
		miR-223, miR-26b, miR-30c, miR-24	↓	
78 plasma samples of PC patients and 28 healthy donors; 118 urine samples of PC patients and 17 healthy donors	Downregulated (12)	miR-107, miR-574-3p	↑	[52]
40 PC samples and 40 normal tissue samples; urine samples of 36 PC patients and of 12 healthy donors	Downregulated (12)	miR-205, miR-214	↓	[53]
Urine samples of eight patients with PC,12 BPH and ten healthy donors	Downregulated (17)	miR-1825 (only in PC)	↑	[54]
		miR-484 (in PC and BPH)	↓	
76 samples of PC and 76 normal tissue	Upregulated (5)	miR-96, miR-182, miR-183, miR-375	↑	[55]
	Downregulated (10)			
		miR-16, miR-31, miR-125b, miR-145, miR-149, miR-181b, miR-184, miR-205, miR-221, miR-222	↓	

Notes
PC: prostate cancer;
BPH: benign prostatic hyperplasia;
↑: Increased expression
↓: decreased expression
Adopted from "The role of miRNAs in the development of prostate cancer" [46]

Table 15.2 miRNAs with altered expression in cancer diseases

Tumor type	Increased expression	Decreased expression	References
Breast cancer	miR-21, miR-29b-2, miR-31	miR-125b, miR-10b, miR-155, miR-17-5p, miR-27b	[56, 57]
Ovarian cancer	miR-373, miR-200	let-7f, miR-140, miR-145, miR-199a, miR-101	[5, 58]
Endometrial cancer	miR-205	miR-505	[59, 60]
Glioblastoma	miR-221, miR-21	miR-124	[61 ,62]
Lymphoma	miR-155, miR-17-92cluster	miR-15a	[63, 64]
Colorectal cancer	miR-10a, miR-17, miR-20a, miR-24-1, miR-29b-2, miR-31	miR-143, miR-145, let-7	[65, 66]
Thyroid cancer	miR-146, miR-181b, miR-197, miR-346		[67, 68]
Hepatocellular cancer	miR-18, miR-224	miR-199a, miR-195, miR-200a, miR-125a	[69, 70]
Testicular cancer	miR-372, miR-373		[71]
Pancreatic cancer	miR-221, miR-376a, miR-301, miR-21, miR-24, miR-100, miR-103, miR-107, miR-125	miR-375	[70, 71]
Cholangiocarcinoma	miR-21, miR-141, miR-200b		[69]
Prostate cancer	let-7d, miR-195, miR-203, miR-96, miR-183	miR-128a, miR-145, miR-221	[71, 72]
Gastric cancer	miR-223, miR-21, miR-103	miR-218	[70, 71]
Lung cancer	miR-17	let-7	[69, 70]

Notes
Adopted from "The role of miRNAs in the development of prostate cancer" [46]

15.3.2.1 lncRNAs and Prostate Cancer

For example, lncRNA PCA3 (prostate cancer antigen 3) is overexpressed in prostate cancer tissue compared to non-neoplastic prostatic tissue in the same patients [97]. PCGEM1 (prostate cancer gene expression marker 1), another lncRNA gene with highly prostate cancer tissue-specific expression, is associated with high-risk groups in prostate cancer [98]. PRNCR1 (prostate cancer noncoding RNA1) is highly expressed in prostate cancer cells [99]. Antti Ylipaa et al. have found that PCAT-5 (prostate-cancer-associated noncoding RNA transcript 5) as a novel biomarker for diagnosis in ERG-regulated prostate cancer [100]. PCAT-18, exhibiting a highly specific expression pattern in prostate cancer, is specifically expressed in

Table 15.3 Examples of malignant cells or body fluid microRNA profiles with clinical significances for patients with cancer

Cancer type	No. of patients	miRNA	Role	References
Multiple cancer	1809 patients: seven different types of cancer: breast cancer, primary head and neck squamous cell carcinoma, renal cancer, soft tissue sarcoma, pediatric osteosarcoma, bladder cancer, and glioblastoma	miR-210	Predictive effect on survival of patients with studied cancer types as indexed by disease-free survival, progression-free survival, and relapse-free survival	[77]
Multiple cancer	174 patients and 39 controls: 50 breast cancers, 30 gastric cancers, 31 lung cancers, 31 esophageal cancers, and 31 colorectal cancers	miR-21	Potential broad-spectrum serum-based diagnostic marker for the detection of solid tumors	[78]
Breast cancer	120 patients and 40 controls	miRNAs: miR-10b, miR-17, miR-34a, miR-93, miR-155, miR-373	Known to be relevant for tumor development and progression; serum concentrations of deregulated miRNAs may be linked to a particular biology of breast cancer favoring progression and metastatic speed	[79]
Colon cancer	102 patients and 40 controls	miR-141	Highly correlated with TNM stage in patients with colorectal cancer; elevated levels associated with liver metastasis in patients with colorectal cancer	[80]
Colon cancer	100 patients with colorectal cancer, 37 with adenomas, and 59 controls	miR-29a	Association with TNM stage in patients with colorectal cancer	[81]
Colon cancer	103 patients with colorectal cancer and 37 controls	miR-221	Potential noninvasive molecular marker for diagnosis of colorectal cancer	[82]

(continued)

Table 15.3 (continued)

Cancer type	No. of patients	miRNA	Role	References
Colon cancer	Phase1: 12 patients with stage 1 and stage IV colorectal cancer Phase 2: 182 patients with colorectal cancer and 24 controls Phase 3: 156 matched tumor tissues from the phase 2 colorectal cancer cohort plus an independent set of 20 matched primary colorectal cancers with corresponding liver metastasis	miR-200c	Independent predictor of lymph node metastasis and tumor recurrence, emerging as an independent prognostic marker for colorectal cancer	[83]
Colon cancer	Exosome-enriched serum samples from 88 patients with primary colorectal cancer and 11 healthy controls and 29 paired samples from patients after tumor resection	let-7a, miR-21, miR-23a, miR-150, miR-223, miR-1229, and miR-1246	The serum exosomal levels of seven miRNAs were significantly higher in patients with primary colorectal cancer, even those with early-stage disease, compared with healthy controls and were significantly downregulated after surgical resection of tumors	[84]
Gastric cancer	124 patients with noncardia gastric cancer and 36 patients with cardia adenocarcinoma and 160 controls	miR-16, miR-25, miR-92a, miR-451, and miR-486-5p	Detection of the early-stage gastric cancer	[85]
Gastric cancer	104 patients and 65 controls	miR-18a	Diagnostic power (high sensitivity and specificity)	[82, 86]
HCC	101 patients, 89 controls, and 48 patients with hepatitis B	miR-21, miR-122, and miR-223	Elevated in patients with HCC or chronic hepatitis, with strong potential to serve as novel biomarker for liver injury but not specifically for HCC	[87]

(continued)

Table 15.3 (continued)

Cancer type	No. of patients	miRNA	Role	References
Lung cancer	Training set of 64 patients with adenocarcinoma; The Cancer Genome Atlas data set; 223 patients with adenocarcinoma	miR-31	Predictor of survival in a multivariate cox regression model even when checking for cancer staging; exploratory in silico analysis indicated that low expression of miR-31 is associated with excellent survival for patients with T2N0 disease	[88]
Lung cancer	25 paired NSCLC paracancerous tissues and serum,103control sera, and 201 patients with NSCLC	miR-19a	High-serum miR-19a expression may be an independent poor prognostic factor for survival in patients with NSCLC	[89]
Lung cancer	30 NSCLC patients and 75 individuals without tumor pathology (e.g., inflammatory interstitial diseases, infections, nontumor lung nodules, hemotypsis, and other diseases)	Exosomes isolated in plasma and BAL	In plasma, a higher percentage of miRNAs with increased levels compared with tumor BAL or in nontumor plasma; the data reveal differences between BAL and plasma exosome amount and miRNA content	[90]
Glioblastoma	Phase 1: 122 patients with untreated WHO grades 3 to 4 disease and 123 control serum samples Phase 2: 55 WHO grade 2, 15 WHO grade 1, 11 astrogliosis serum samples, and eight WHO grades 2 to 4 astrocytoma tumor tissues	miR-15b, miR-23a, miR-133a, miR-150, miR-197, miR-497, and miR-548-5p and the seven-miRNA panel	The seven-serum miRNA panel demonstrated a high sensitivity and specificity for the prediction of malignant astrocytomas; a marked difference in serum miRNA profile was observed between high-grade astrocytomas and normal controls	[91]

(continued)

Table 15.3 (continued)

Cancer type	No. of patients	miRNA	Role	References
Prostate cancer	Prebiopsy serum samples of 133 enrolled patients from three study centers	miR-26a-1 and miR-141	The analysis of circulating miRNAs does not appear to help identify patients with cancer undergoing prostate biopsy; however, their levels may be useful to identify patients with high-risk prostate cancer	[92]
Ovarian cancer	360 patients with epithelial ovarian cancer and 200 controls from two institutions	miR-205, let-7f	Plasma miR-205 and let-7f are biomarkers for ovarian cancer detection that complement CA 125; let-7f may be predictive of ovarian cancer prognosis	[93]

Notes
Adopted from “MicroRNAome genome: A treasure for cancer diagnosis and therapy” [76]

Table 15.4 Representative lncRNAs involved in carcinogenesis and potential cancer biomarkers

Cancer type	lncRNAs	References
Esophagus	ENST00000435885.1, XLOC-013014, ENST00000547963.1	[107]
Stomach	GCAT1SUMOIP3	[108, 109]
Colon and rectum	HOTAIR, uc. 73	[110, 111]
Liver	HULC	[112]
Lung	MALAT1, TUG1	[113, 114]
Breast	HOTAIR, UCA1	[115, 116]
Ovary	HOTAIR	[117]
Bladder	UCA1, H19, Linc-UBC1, MALAT1, GAS5	[118–120]
Prostate	PCA3, PCAT1, PCGEM1	[121–123]
Glioma	H19	[124]
Melanoma	SAMMSON	[125]
Oral cavity and nasopharynx	lnc-c22orf32-1, lnc-AL355149.1-1, lnc-ZNF674-1	[126, 127]

Notes
Adopted from “Long Noncoding RNAs as New Architects in Cancer Epigenetics, Prognostic Biomarkers, and Potential Therapeutic Targets” [128]

prostate tissue and upregulated in prostate cancer compared with other benign and malignant tissues, indicating that it can act as a novel biomarker in diagnosis [101]. The potential utility of these long noncoding RNAs in a multi-biomarker method may successfully distinguish cancerous tissues from normal tissues.

15.3.2.2 lncRNAs and the Esophagus

In esophageal squamous cell carcinoma (ESCC), several lncRNAs have been studied. The lncRNA Epist is generally highly expressed in the esophagus and down-regulated during ESCC progression, acting as a tumor suppressor [102]. HOTAIR (HOX transcript antisense RNA) is upregulated in ESCC cell lines and patient samples and exhibits oncogenic activity in ESCC [103]. Levels of linc-POU3F3 are increased in ESCC samples from patients compared with noncancerous tissues [104]. Elevated expression of lncRNA CCAT2 is associated with poor prognosis in esophageal squamous cell carcinoma [105].

15.3.2.3 lncRNAs and Hepatocellular Carcinoma

Several lncRNAs have been shown to be promising biomarkers for HCC. Li et al. detected eight lncRNAs from HCC patients and evaluated the diagnostic value and accuracy of the lncRNA profiling system [59]. They found that HULC and Linc00152 were significantly overexpressed in HCC, and the ROC values were 0.78 and 0.85, respectively. The ROC value of the combination of HULC and Linc00152 was 0.87, which is less than the AFP (0.89), suggesting the HULC and Linc00152 achieve a fine diagnostic accuracy in diagnosing HCC.

15.3.2.4 lncRNAs and Nasopharyngeal Carcinoma

Nie et al. have found that HOTAIR is of great clinical value for the diagnosis and treatment of nasopharyngeal carcinoma (NPC). NPC patients with higher HOTAIR levels have poor prognosis and short overall survival compared to those with lower HOTAIR levels using univariate and multivariate analysis [106].

15.4 Circulating ncRNAs in Cancer

ncRNAs in plasma or serum may also serve as novel noninvasive biomarkers and blood-based assays and are easy to examine for cancer diagnosis and monitoring.

15.4.1 Circulating miRNAs as Cancer Biomarkers

Recently, the prevalent discovery of miRNAs in serum or plasma has provided a new approach for diagnostic screening in blood [129]. Many studies examined circulating miRNAs in the plasma or serum of cancer patients, which shows a good promise to serve as biomarkers for diagnosis too. The spectrum of circulating miRNAs varies among different types of cancers, such as lung cancer, breast cancer, hepatocellular carcinoma, colorectal cancer, leukemia, prostate cancer, and so on.

15.4.1.1 Lung Cancer

Lung cancer represents the most common cancer worldwide with a high mortality rate due mainly to the fact that the disease generally only becomes clinically apparent at advanced stages and the lack of validated or cost-effective screening methods for early diagnosis [130]. More recently, several microRNAs (miRNAs) present in the peripheral blood have been proposed as stable and reproducible biomarkers for lung cancer diagnosis.

The first comprehensive analysis of the miRNA spectrum in the serum of non-small-cell lung cancer (NSCLC) patients was carried by Chen et al. [131] in 2008. After high-throughput miRNA sequencing, they identified miR-25 and miR-223 as NSCLC-specific miRNAs, which were upregulated in the serum of NSCLC patients. Following this study, most research aimed to identify circulating miRNAs with a diagnostic relevance in serum or plasma, but the results were highly heterogeneous, even in studies using similar biological materials. Bianchi et al. [132] used miR-197 and miR-24 as internal references for their stability in the serum, while other authors [133, 134] reported that miR-24 and miR-197 were upregulated in the serum and plasma of NSCLC patients, respectively. Several studies [78, 135–137] identified that miR-21 was one of the most frequently upregulated miRNAs in NSCLC. But among other studies [138, 139], two differentially expressed miR-15b and miR-27b were capable of distinguishing NSCLC from healthy cases with highest sensitivity and specificity. With more and more miRNAs being found, it appears that one or two miRNAs could not be used in diagnosis of specific cancers because the same single miRNA might be dysregulated in several different cancers, and in certain cases a set of several miRNAs might show the same characteristics in cancer diagnosis. Therefore, miRNAs are applied in the diagnosis of cancers by using a cluster of miRNAs and estimating their integral effects in special diagnostic models. For example, Boeri et al. [140] demonstrated that nine circulating miRNAs (miR-221, miR-660, miR-486-5p, miR28-3p, miR-197, miR-106a, miR-451, miR-140-5p, and miR-16) contributed to the malignancy and poor prognosis of lung cancer. Up to now, the circulating miRNA signatures for lung cancer diagnosis even contain a four-miRNA panel (miR-21, miR-126, miR-210, and miR-486-5p) [136], a ten-miRNA panel (miR-20a, miR-24, miR-25, miR-145, miR-152, miR-199a-5p, miR-221, miR-222, miR-223, and miR-320) [139], and so on. Therefore, the application of multiple miRNAs in a novel diagnostic model may open up new opportunities for lung cancer diagnosis.

15.4.1.2 Breast Cancer

Breast cancer is a heterogeneous disease with different molecular subtypes, biological behaviors, and responses to therapy [141]. The therapeutic management depends on accuracy diagnostic biomarkers to guide the decision and treatment choices. Recently, more and more circulating miRNAs were employed as biomarkers to differentiate breast cancer from normal.

In one of the first studies, the authors found that miR-155 was upregulated in PR-positive breast cancer compared with negative tumors and healthy controls in serum specimens [142]. In another study, Heneghan et al. [143] identified miR-195 as a cancer-specific miRNA with elevated expression in the blood of patients with breast cancer. miR-29a and miR-21 were reported to be upregulated in the serum of patients with breast cancer [144, 145]. With the growing discovery of miRNAs in the blood, Hu et al. [146] conducted miRNA profiling and demonstrated that the four-miRNA signature (miR-16, miR-25, miR-222, and miR-324-3p) could distinguish breast cancer patients from normal individuals. Furthermore, another study [147] found that miR-155 in combination with miR-145 and miR-182 could significantly increase the sensitivity and specificity of breast cancer diagnosis.

In addition, dysregulated expression of circulating miRNAs has also been shown to correlate with different breast cancer subtypes. For example, women with luminal A-like breast tumors exhibited downregulated miR-29a, miR-181a, miR-223, and miR-652 compared with healthy controls [148]. More importantly, combination of the three miRNAs (miR-29a, miR-181a, and miR-652) will get an AUC of 0.80 to differentiate luminal A-like breast cancer from controls [148]. The elevated levels of serum miR-214 [149], miR-10b, and miR-373 [150] are associated with lymph node metastasis in breast cancer patients.

Finally the altered levels of circulating miRNAs were also closely associated with breast cancer sex hormone receptor expression status and used to guide treatment choices. Heneghan et al. [143] found that the high serum levels of miR-10b were associated with the estrogen receptor status of breast cancer patients. Ruihua Zhao et al. [151] analyzed the correlation between miRNA-221 and chemosensitivity in breast cancer patients who previously received neoadjuvant chemotherapy (NAC) and found that the plasma miRNA-221 expression level was correlated with overall response rate but not with pathologic complete response. The miRNA-221 expression was negatively correlated with hormone receptor (HR) expression which was definitely predictive for chemoresistance, suggesting that patients with high miRNA-221 expression were more likely to be chemoresistant to taxane and anthracycline. Meanwhile, Rao et al. [152] showed that miRNA-221/222 overexpression confers fulvestrant resistance in breast cancer. These studies indicate that plasma miRNA-221 might be an efficient biomarker for sensitivity to NAC in breast cancer patients. Furthermore, two studies from Wang et al. [153] and Jung et al. [154] presented that circulating miRNA-125b was associated with chemotherapeutic resistance of breast cancer via directly targeting the E2F3 gene and miRNA-210 was correlated with trastuzumab resistance in patients with HER2-positive breast cancer. Overall, these studies have shown that circulating miRNAs can serve as promising markers for breast cancer diagnosis, prognosis, and therapy monitoring.

15.4.1.3 Liver Cancer

Recently, many efforts have been made to develop noninvasive serum biomarkers for the diagnosis of liver cancer. Despite remarkable advances, the reliability of biomarkers such as AFP is still debatable [155]. Indeed, the specificity of AFP is low, especially in the context of chronic liver disease. Fortunately, altered levels of circulating miRNAs were shown in patients with liver cancers. In 2010, the first study [156] on circulating miRNAs in liver cancer patients was reported, where the authors had identified 13 miRNAs that apparently expressed in the hepatitis B virus (HBV) serum compared with those in control serum. These 13 miRNAs can accurately discriminate not only HBV cases and HCV cases from controls but also HBV-positive hepatocellular carcinoma (HCC) cases from control and HBV cases. Interestingly, six miRNAs (miRNA-1, miRNA-25, miRNA-92a, miRNA-206, miRNA-375, and let-7f) were significantly upregulated in HCC patients compared to healthy controls. Moreover, a set of three miRNAs (miR-25, miR-375, and let-7f) out of the six miRNAs showed a high ability to distinguish HCC cases from healthy controls. Among these miRNAs, miR-375 alone had 100 % sensitivity and 96 % specificity in predicting HCC156. Therefore, this study has demonstrated that circulating miRNAs can serve as novel noninvasive biomarkers for HBV-positive HCC diagnosis. In another independent study, Xu et al. [157] showed that miR-21, miR-122, and miR-223 had higher serum levels in HCC and chronic hepatitis B (CHB) patients compared with healthy controls. Other studies [158] found that miR-221 increased both in tissue and in serum of HCC patients, and its expression was correlated with tumor size, cirrhosis, and tumor stages.

The high specificity of circulating miRNAs for HCC diagnosis is very important to avoid false positivity. A recent study [159] has showed that decreased serum levels of miR-16 and miR-199a can discriminate HCC patients from patients with chronic liver diseases. More importantly, the combination of miR-16 and AFP can improve diagnostic sensitivity and specificity for HCC. Additionally, combined utilization of circulating miR-15b and miR-130b can potentially distinguish HCC from HBV hepatitis patients [160]. For early diagnosis, circulating miRNAs such as miR-15b, miR-130b, and miR-16 have been identified as potential candidates [160, 161]. Especially, miR-15b and miR-130b can get an increased sensitivity for liver cancer diagnosis when combined with AFP.

15.4.1.4 Gastric Cancer

Gastric cancer (GC) is the fourth most common cancer and the third leading cause of cancer mortality worldwide. To date, there is a lack of high sensitivity or specificity markers for current GC diagnosis. Circulating miRNAs used as potential biomarkers have been extensively studied in GC, and numerous miRNAs are considered as promising candidates for GC diagnosis [162]. In 2010, Tsujiura et al. [163] first reported that four miRNAs (miR-17-5p, miR-21, miR-106a, and miR-106b) are significantly increased in plasma from a GC patient. Then another group [164] also identified that miR-17-5p and miR-106a in plasma can obviously distinguish GC patients from normal individuals. In the subsequent studies [165–167], several groups also found that miR-21 is altered in GC patients' plasma. Most importantly the combination of miR-21 with miR-223 and miR-218 can get the highest diagnostic value for GC detection in humans, regardless of tumor site, tumor stage, and pathological type [168]. Furthermore, other miRNAs (such as miR-1, miR-16, miR-18a, miR-20a, miR-25, miR-27a, miR-34a, miR-92a, miR-100, miR-103, miR-106a, miR-106b, miR-107, miR-146a, miR-148a, miR-192, miR-194, miR-196a, miR-199a-3p, miR-200c, miR-210, miR-221, miR-223, miR-376c, miR-378, miR-423-5p, miR-421, miR-451, miR-486-5p, miR-744, and miR-93) are upregulated in the circulation of GC patients, of which miR-378 and miR-199a were shown to be potential markers for the early GC diagnosis. Moreover, besides the upregulated miRNAs, many circulating miRNAs are reduced in the blood of GC patients. miR-122 and miR-195 are the two mainly downregulated circulating miRNAs in GC patients, thus suggesting that both of them were potential diagnostic markers for GC screening [169].

15.4.1.5 Colorectal Cancer

Circulating miRNAs are also widely studied in colorectal cancer (CRC) diagnosis. Firstly, Ng, E. K. et al. [170] found a differential expression of microRNAs in the plasma of patients with colorectal cancer compared to normal cohort. Then, they identified that miR-92 could serve as a potential noninvasive marker for CRC diagnosis. Moreover, another group [81] even confirmed that miR-29a and miR-92 are upregulated in the plasma of early-stage CRC. A recent study [171] also showed that serum miR-29a is a promising novel marker for early detection of colorectal liver metastasis, yielding an ROC–AUC of 80.3 %. Another characteristic miRNA profiling (upregulated miR-21 and let-7 g and downregulated miR-31, miR-181b, miR-92a, and miR-203) demonstrated that miR-92 can distinguish CRC patients from normal controls with high sensitivity and specificity [172]. Besides, miR-29a, miR-92, and miR-21 are the most widespread miRNAs in CRC patients. The serum level of miR-21 can differentiate CRC patients from controls with 90 % specificity and sensitivity and get an ROC curve area of 0.85 for CRC [173]. Moreover, miR-200c and miR-29a were reported as potential noninvasive biomarkers for CRC prognosis and predicting metastasis [174]. There are other several miRNAs, such as miR-141, miR-601, miR-760, miR-130, miR-145, miR-216, and miR-372, which were reported as potential biomarkers for CRC diagnosis [175, 176].

15.4.1.6 Hematologic Cancer

Circulating miRNAs have the closest relationship with blood cells. As early as 2004, Chen and colleagues identified that miR-181 is specifically expressed in lymphoid lineages, while miR-223 and miR-142 are expressed only in myeloid cells [177]. Until 2008, the first study on circulating miRNAs in hematologic cancer was reported in which the authors noted that miR-21, miR-155, and miR-210 were significantly upregulated in the serum of patients with diffuse large B-cell lymphoma (DLBCL) [178]. Another study examined that both miR-21 and miR-92 are potential biomarkers for DLBCL diagnosis [179]. Apart from DLBCL, circulating miRNAs were also shown to be significantly altered in other hematologic cancers. For example, miR-150 and miR-342 were shown to be promising biomarkers in acute myeloid leukemia (AML) diagnosis [180]. Chronic lymphocytic leukemia (CLL) patients showed seven upregulated plasma miRNAs (miR-150, miR-19b, miR-92a, miR-223, miR-320, miR-484, and miR-17) [181]. Furthermore, more and more miRNAs were identified as biomarkers for different hematologic cancer diagnoses. As reported, miR-155, miR-125, miR-181, miR-221/222, and miR-29a are upregulated in AML, and miR-181, miR-221/222, and miR-29a also showed their diagnostic potential in CLL [182–184]. Additionally, miR-17-92, miR-15a/16-1, and miR21 have been proposed as diagnostic/prognostic biomarkers for multiple myeloma (MM) [185]. But among the three miRNAs, the levels of miR-17-92 [186] and miR21 [187] are altered in B-cell lymphoma. Due to the similarity among different hematologic cancers, many miRNAs overlap for diagnosis in hematologic cancers. Combination with traditional pathology will get better diagnostic efficiency in hematologic cancer early screening.

15.4.2 Circulating Piwi-Interacting RNAs as Cancer Biomarkers

Up to now, most of reported piwiRNAs focus on gastric cancer [188]. It's reported that piR-823 and piR-651 are both at lower levels in circulating tumor cells (CTCs) in the peripheral blood of gastric cancer patients, compared with normal controls. Furthermore, both piR-651 and piR-823 are more sensitive than the commonly used biomarkers such as serum carcinoembryonic antigen (CEA) and carbohydrate antigen 19–9 (CA19-9) for gastric cancer, thus making them possible as early diagnosis markers for gastric cancer [189]. Although, as a drawback, most of the piwiRNAs are instable, while piR-651 and piR-823 in blood samples are relatively stable. Most piRNAs play a role in other cancer types, such as piR-Hep1, which was shown to contribute to the invasion of hepatocellular carcinomas [190]. These findings indicate that piRNAs may be promising molecular markers for the diagnosis of certain cancers.

15.4.3 Circulating snoRNAs in Cancer

Some snoRNAs exhibit altered expression patterns in a variety of human cancers and can affect cell transformation, tumorigenesis, and/or metastasis. Up to now, many reports have shown that snoRNAs are stably detectable in blood plasma and serum samples. Therefore, snoRNAs can serve as potential fluid-based novel biomarkers for cancer diagnosis. Jipei Liao et al. [191] identified that three plasma snoRNAs (snoRD33, snoRD66, and snoRD76) could significantly distinguish NSCLC patients from both normal individuals and patients with chronic obstructive pulmonary disease with 81.1 % sensitivity and 95.8 % specificity. Furthermore, snoRA42 was regarded as a potential biomarker in lung cancer diagnosis [192]. Besides, the potential for the development of snoRNAs as biomarkers is supported by a study in head and neck squamous cell carcinomas and breast cancers. Gee et al. [193] found that the expression levels of C/D box snoRNAs RNU44 and RNU43 are associated with poor prognosis in head and neck squamous cell carcinomas and breast cancers. Furthermore, Appaiah et al. [194] found that the U6/SNORD44 ratio is consistently high in breast cancer patients with or without active diseases. Taken together, snoRNAs may serve as potential biomarkers for cancer diagnosis. Now the high-throughput techniques can identify more informative snoRNAs as biomarkers for cancer diagnosis and prognosis with higher sensitivity and specificity.

15.4.4 Circulating Circular RNAs

Covalently closed circular RNA molecules (circRNAs) have recently emerged as a class of RNA isoforms with widespread and tissue-specific expression across animals, usually independent on the corresponding linear mRNAs. circRNAs are remarkably stable and sometimes highly expressed molecules, which are ideal biomarkers for cancer diagnosis. CDR1as, a circRNA reported by Memczak et al., [41] and ciRS-7 by Hansen et al. [195] contain roughly 63–70 evolutionarily conserved binding sites for microRNA-7 (miR-7) and form a complex with AGO proteins. ciRS-7 also plays an important role in Alzheimer's disease [196]. There are some controversies regarding to the quantity of circRNAs existing in plasma. Koh et al. reported only a few (less than 100) circular RNAs were detected in plasma [197], while, in another report, around 2400 circRNA candidates were detected in human whole blood and observed with RNA-Seq [198]. In addition, the overall circRNA expression level in blood is unexpectedly similar to that of neuronal tissues where circRNAs are highly abundant [199]. For example, Li et al. [200] found that the plasma level of Hsa_circ_002059, a typical circular RNA, can significantly differentiate postoperative gastric cancer patients and preoperative gastric cancer patients. Importantly, they further demonstrated that reduced expression of Hsa_circ_002059 was significantly correlated with distal metastasis. Furthermore, Qu et al. [201] employed a circular RNA microarray to explore the signature of circular RNAs in pancreatic ductal adenocarcinoma (PDAC), and they revealed that the circRNA expression signatures of PDAC are dysregulated. All of these findings indicate that circRNAs can be involved in the initiation and progression of PDAC. Circular RNAs as a novel class of RNAs with high stability may serve as prefect biomarkers for cancer diagnosis, though more studies are needed to explore this field.

15.4.5 Exosomal ncRNAs

Recent studies showed that a significant portion of circular ncRNAs in plasma or serum are carried by exosomes. Exosomes are bilayered lipid vesicles of endocytic origin with diameters 40–100 nm in size [202]. They can also be found in nearly all human body fluids, including plasma, serum, urine, saliva, breast milk, cerebrospinal fluid, chest water, and amniotic fluid [203]. Exosomes are derived from many cell types such as B-cells, T-cells, dendritic cells, neurons, and epithelial cells especially tumor cells, which contain multiple functional substances such as proteins, DNA, mRNA, miRNA, and long noncoding RNA. They play an important role in intercellular communicating by transferring some effector biomolecules such as noncoding RNAs, DNA, and protein. It has been reported that exosomes mediate intercellular communication by which viruses transfer genetic materials from infected NPC cells to neighboring cells. Gourzones et al. [204] detected BART-miRNAs in exosomes from the plasma of NPC patients. BART-miRNAs encoded by EBV play an important role in interrupting host-cell regulatory pathways and evading immune responses. A free exosome database named ExoCarta (http://www.exocarta.org) was launched in 2009 as a resource to compile proteins and RNAs identified in exosomes [205]. Noncoding RNAs in exosomes show their potential as detection tools to provide more distinct and complementary information about the tumor phenotype.

The molecular content of exosomes are protected by a lipid membrane, so they are more stable and resistant to RNase enzymatic activity compared to other types of miRNAs. Noncoding RNAs meet the basic conditions for utility as biomarkers that may be measured repeatedly and noninvasively, and microRNAs (miRNAs) are the most extensively studied noncoding RNAs. Here, we will focus on the diagnosis of exosomal microRNAs.

15.4.5.1 Exosome miRNAs as Biomarkers of Cancer Diagnosis

Numerous studies indicate that expression of miRNAs in exosomes is different in the normal condition and pathological conditions such as tumor, which shows a good promise to serve as biomarkers for diagnosis. In addition, exosome-associated miRNAs in biofluids have been suggested as potentially minimally invasive biomarkers for multiple human diseases.

Samsonov, R. et al. identified three miRNAs (miR-574-3p, miR-141-5p, miR-21-5p) with significant upregulation in urinary exosomes associated with prostate cancer [206]. Cazzoli, R. et al. selected 14 microRNAs from a wide-range analysis of 746 microRNAs, analyzed the level of these miRNAs in circulating exosomes, and identified four miRNAs (miR-378a, miR-379, miR-139-5p, and miR-200b-5p) as screening markers that can distinguish lung adenocarcinoma patients from healthy former smokers. They also selected six miRNAs (miR-151a-5p, miR-30a-3p, miR-200b-5p, miR-629, miR-100, and miR-154-3p) for segregating lung adenocarcinoma patients and lung granuloma patients [207]. Singh, R. et al. found that the expression level of miR-10b in metastatic MDA-MB-231 breast cancer cells is higher than non-metastatic MCF-7 [208], suggesting that miR-10b may be the potential biomarker to distinguish two groups: metastatic breast cancer and non-metastatic breast cancer [208]. They also demonstrated that exosome-based miRNA analyses in blood samples are better to distinguish malignant and non-malignant lesions. It was reported that the exosome-based miRNA-30a and miRNA-192 are significantly increased in the plasma of alcoholic hepatitis patients, showing promising values for diagnosis of alcoholic hepatitis [209]. Taken together, these indicate that exosome-based miRNAs are superior to traditional diagnostic biomarkers.

15.4.5.2 Methods of Detecting Exosome-Based Noncoding RNA

Although exosomes contain a large amount of miRNAs, quantitative analysis of their abundance and stoichiometry are lacking. Recently, a number of efficient methods have been established to detect miRNAs, such as quantitative real-time PCR (RT-PCR), northern blotting, microarray-based analysis, digital PCR, and next-generation sequencing (NGS). Among them, the two most common ones to validate certain miRNAs are RT-PCR and northern blotting. Analysis of microarray-based miRNA expression can be quick and highly efficient to screen for interesting miRNAs, but it is relatively less accurate and reproducible compared to qRT-PCR [210]. Array-based technologies are able to analyze a large number of miRNAs, but the accuracy needs to be improved. Digital RT-PCR can be used for clinical diagnostic purpose for absolute quantification but only suitable for small-scale samples [211]. Therefore, improvement of these methods will promote multiplexing analysis of miRNA, which is needed in the future. NGS has gained particular attention to its potential applicability on biofluids. Matullo et al. showed that NGS is suitable for analyzing miRNA profiles in urine for biomarker discovery in urothelial carcinoma. Using NGS technology, they identified 66–184 miRNAs in exosomes and detected miR-3648 and miR-4516 specifically in cell-free urine [212]. NGS also represents a unique tool to investigate different layers of transcriptome complexity at an incredible level of resolution [213]. With recent reduction in the cost of NGS, it becomes more feasible to analyze a larger number of samples. However, additional studies are also needed to improve nucleic acid extraction and the procedure to prepare small RNA libraries.

15.4.5.3 Challenges with Exosomes for Diagnosis

Exosome-based analysis reduces the complexity of techniques with biofluids, so we can detect low-abundant biomolecules more specifically and more sensitively. Although there is growing interest in the potential use of exosomal protein and miRNAs as tumor biomarkers, isolation of exosomes is extremely challenging. Today, many methods have been developed to isolate exosomes including differential centrifugation coupled with ultracentrifugation, immunoaffinity capture, density gradient separation, affinity chromatography separation, and some commercial kits [214]. As we all know, ultracentrifugation is a classic and widely used method for exosome isolation. Commercial kits such as the ExoQuick are easy to handle and quick to process, which needs only a common centrifugation, but they are still too expensive to use in clinical settings. Thus, all the methods for exosome isolation have advantages and disadvantages. More importantly, there is no quantitative analysis for miRNA abundance and stoichiometry. The development and clinical application of exosomal miRNA-based assays are challenged by the lack of knowledge about exosome structure and biological functions, as well as the absence of standard preparation protocols.

15.4.5.4 The Future of Exosome-Mediated Diagnosis

Recently, more and more studies showed that the biological roles of exosomes are crucial in cancer progression. Given the fact that exosome-based miRNAs are identified as one of the major components stably present in exosomes, they can act as stable sources of biomarkers in various tumors that reflect the different stages. It would be of great interest for future studies to develop other potential applications of exosome-based miRNAs. Therefore, analyses of exosome-based miRNA biomarkers have a huge potential for clinical use. Exosome-based miRNAs have opened up a new field for cancer diagnosis because they simply require body fluid samples for detection, which is noninvasive and convenient. However, as the field is still in the early stage of development, robust methods are needed to harness the true potential of exosome-based miRNAs in the clinical scenario.

15.5 ncRNAs in Non-blood Body Fluids

In addition to hematologic ncRNAs, the presence of ncRNAs in non-blood body fluids may also represent a gold mine of noninvasive biomarkers in cancer diagnosis.

15.5.1 ncRNAs in Saliva

Saliva is a kind of complex body fluid produced by salivary glands (parotid, submandibular, and sublingual). Several recent studies reported that miRNAs in tissue, plasma, and saliva share similar expression profiles. Saliva analysis is thought to be an ideal assay, as it provides a noninvasive method for early detection of diseases.

In 2009, Park et al. [215] measured 314 miRNAs using RT-PCR in 12 healthy controls. Selected miRNAs were validated in 50 oral squamous cell carcinoma patients. Two miRNAs, miR-125a and miR-200a, show significantly lower levels in the saliva of oral squamous cell carcinoma patients than those in control subjects. The authors claimed that saliva miRNAs can be used for oral cancer detection.

In 2013, Xie et al. [216] found distinctive miRNAs for esophageal cancer in both whole saliva and saliva supernatant. miR-10b, miR-144, and miR-451 in whole saliva and miR-10b, miR-144, miR-21, and miR-451 in saliva supernatant are significantly upregulated in esophageal cancer patients with sensitivities of 89.7 %, 92.3 %, 84.6 %, 79.5 %, 43.6 %, 89.7 %, and 51.3 %, respectively, and specificities of 57.9 %, 47.4 %, 57.9 %, 57.9 %, 89.5 %, 47.4 %, and 84.2 %, respectively. These miRNAs show discriminatory power for detection of esophageal cancer.

Changes in microRNA expression have been reported in pancreatic cancer. Xie et al. [217] used an Agilent microarray and validated candidate biomarkers in 40 patients with pancreatic cancer, 20 with benign pancreatic tumors (BPT), and 40 healthy controls. miR-3679-5p showed significant downregulation in the pancreatic cancer group within the three categories ($P=0.008$, 0.007, and 0.002, respectively), whereas miR-940 showed significant upregulation in pancreatic cancer ($P=0.006$, 0.004, and 0.0001, respectively). In logistic regression models, a combination of two salivary miRNAs could diagnose pancreatic cancer within the three categories with the sensitivities of 72.5 %, 62.5 %, and 70.0 % and the specificities of 70.0 %, 80.0 %, and 70.0 %, respectively. These studies all suggest that microRNAs in saliva could be a biomarker for pancreatic cancer.

15.5.2 Urinary ncRNAs as Biomarkers in Urologic Cancers

Noncoding RNAs possess the diagnostic potential with virtue of the highly tissue- and cancer-specific expression in urologic malignancies. Urologic cancers are generally divided into three classes: the prostate cancer (PCa), renal cell carcinoma

(RCC), and bladder cancer (BCa). Detection of highly stable urinary miRNAs can be used as noninvasive procedure for early diagnosis. Since the FDA approved the long noncoding PCA3 [218] RNA-based urine test for diagnosis of PCa patients, more researchers have started to focus on exploring more novel noncoding RNAs which can serve as biomarkers associated with urologic cancers.

15.5.2.1 PCa

PCa antigen 3 (PCA3), a prostate-specific noncoding RNA, is an approved diagnostic urinary biomarker for PCa and shows a superior diagnostic potential to biopsy. Kok et al. [218] identified DD3PCA3 as one of the most specific prostate-cancer-associated genes in 2002. Expression of the DD3PCA3gene is a very sensitive and specific marker for detection of prostate tumor cells with a high background of normal (prostate) cells. After this groundbreaking discovery, Gils et al. [219] tested PCA3 scores between men with negative and positive biopsy findings. The sensitivity for the detection of PCa by the urine PCA3 assay was 65 %. Meanwhile, the specificity was 66 % (compared with 47 % for the serum PSA test) and the negative predictive value was 80 %. PCA3 ncRNA urine test can improve the specificity in prostate cancer diagnosis and prevent many unnecessary prostate biopsies. Therefore, PCA3 ncRNAs are so far the most widely used measurement for prostatic clinical application in body fluids such as blood, ejaculate, and urine.

Recently, high miRNA levels have been reported in the urine of PCa patients. Yamada et al. [220] reported that the expression levels of miR-96 and miR-183 in the urine samples are significantly higher in 100 urologic cancers (UC) than in healthy controls. Each microRNA has good sensitivity and specificity to distinguish UC patients from non-UC patients (miR-96, 71.0 %, and 89.2 %; miR-183, 74.0 %, and 77.3 %). In particular, upon combination of miR-96 and urinary cytology data, sensitivity of diagnosis could rise from 43.6 to 78.2 %. In another study, Lewis et al. [221] indicated a correlation of miR-888 and disease progression. They found that miR-888 expression in EPS urine (expressed prostatic secretions in urine) correlates with high-grade prostate cancer patients. miR-888 can be an ideal biomarker to diagnose high-grade prostate cancer.

15.5.2.2 BCa

Although urethrocystoscopy is considered as the "gold" standard, a new noninvasive diagnostic biomarker is highly needed in patients with BCa. Hanke et al. [222] identified that the RNA ratio of microRNA-126/microRNA-152 enables detection of BCa in the urine with a specificity of 82 % and a sensitivity of 72 %. Another study demonstrated that the expression levels of miR-96 and miR-183 in the urine samples were significantly higher in 100 UC patients than in healthy controls.

15.5.2.3 RCC

The absence of specific biomarkers makes it difficult to early diagnose RCC. Several studies described the important role of miRNAs in the histological classification of RCC and in the prediction of prognosis. Changes in miRNA expression have been observed in many types of cancer. miRNAs have been shown as promising biomarkers for alimentary tract-specific screening of RCC with high sensitivity and specificity.

15.5.3 ncRNAs in Feces

15.5.3.1 The Novel Biomarkers in Colorectal Cancer

Colorectal cancer (CRC) is a malignant disease ranking as the fourth most common cause of cancer-related deaths in the world. Thus, early detection is required urgently. However, current screening methods including colonoscopies, fecal occult blood tests (FOBT), and fecal immunochemical tests (FIT) lack sensitivity and convenience.

In 2009, Ahmed et al. [223] analyzed miRNA expression in stool and tissue samples of 15 CRC patients. Seven upregulated miRNAs in CRC were observed, which include miR-21, miR-106a, miR-96, miR-203, miR-20a, miR-326, and miR-92. miR-21, miR-20a, miR-106a, and miR-203 were also described in two important papers by Schetter et al. and Yantiss et al. [224]. In another study enrolling 206 CRC patients and 134 healthy volunteers, Koga et al. conducted miRNA expression analysis of exfoliated colonocytes isolated from feces for CRC screening and the sensitivity and specificity of miRNA expression assay were 74 % and 79 %, respectively [225]. After that, Kalimutho et al. [226] reported that miR-144 was overexpressed in paired CRC tissues using RT-qPCR analysis. The level of miRNAs detected in feces strongly reflects the changes occurring in corresponding tissues. miR-144 represents a novel fecal-based diagnostic marker for CRC screening.

15.5.3.2 The Noninvasive Biomarkers of Pancreatic Cancer

Pancreatic cancer (PCA) is a major leading cause of gastrointestinal cancer-related deaths in Europe and the USA. Therefore, searching for potential biomarkers is important for early detection and prevention of malignancies.

In 2012, Link et al. [227] analyzed fecal miRNA expression in patients and healthy controls with chronic pancreatitis and PCA. They revealed that the expression levels of miR-216a, −196a, −143, and −155 were higher in controls and lower in PCA. The fecal miRNA-based approach to screen for pancreatic tumors could be easily translated into clinical practice, compared to the currently available invasive PCA diagnostic tests.

miRNA detection is a promising method for cancer screening. In addition, it is a noninvasive method for cancer screening and the materials are easy to obtain. In recent years, microRNA diagnostic markers have been shown as a viable new screening method for gastrointestinal cancer.

References

1. Anker P, Mulcahy H, Stroun M. Circulating nucleic acids in plasma and serum as a noninvasive investigation for cancer: time for large-scale clinical studies? Int J Cancer. 2003;103:149–52. doi:10.1002/ijc.10791.
2. Perez DS, et al. Long, abundantly expressed non-coding transcripts are altered in cancer. Hum Mol Genet. 2008;17:642–55. doi:10.1093/hmg/ddm336.
3. Ponting CP, Belgard TG. Transcribed dark matter: meaning or myth? Hum Mol Genet. 2010;19:R162–8. doi:10.1093/hmg/ddq362.
4. Place RF, Noonan EJ. Non-coding RNAs turn up the heat: an emerging layer of novel regulators in the mammalian heat shock response. Cell Stress Chaperones. 2014;19:159–72. doi:10.1007/s12192-013-0456-5.
5. Gomes AQ, Nolasco S, Soares H. Non-coding RNAs: multi-tasking molecules in the cell. Int J Mol Sci. 2013;14:16010–39. doi:10.3390/ijms140816010.
6. Salzman J, Gawad C, Wang PL, Lacayo N, Brown PO. Circular RNAs are the predominant transcript isoform from hundreds of human genes in diverse cell types. PLoS ONE. 2012;7, e30733. doi:10.1371/journal.pone.0030733.
7. Weick EM, Miska EA. piRNAs: from biogenesis to function. Development. 2014;141:3458–71. doi:10.1242/dev.094037.
8. Rinn JL, Chang HY. Genome regulation by long non-coding RNAs. Annu Rev Biochem. 2012;81:145–66. doi:10.1146/annurev-biochem-051410-092902.
9. Hodgkinson A, Chen Y, Eyre-Walker A. The large-scale distribution of somatic mutations in cancer genomes. Hum Mutat. 2012;33:136–43. doi:10.1002/humu.21616.
10. Cheetham SW, Gruhl F, Mattick JS, Dinger ME. Long non-coding RNAs and the genetics of cancer. Br J Cancer. 2013;108:2419–25. doi:10.1038/bjc.2013.233.
11. Pin Wang YX, Han Y, Lin L, Wu C, Xu S, Jiang Z, Xu J, Liu Q, Cao X. The STAT3-binding long non-coding RNA lnc-DC controls human dendritic cell differentiation. Science. 2014;18:310–3. doi:10.1126/science.1251456.
12. Qi P, Du X. The long non-coding RNAs, a new cancer diagnostic and therapeutic gold mine. Mod Pathol. 2013;26:155–65. doi:10.1038/modpathol.2012.160.
13. Mercer TR, Dinger ME, Mattick JS. Long non-coding RNAs: insights into functions. Nat Rev Genet. 2009;10:155–9. doi:10.1038/nrg2521.
14. Taft RJ, Pang KC, Mercer TR, Dinger M, Mattick JS. Non-coding RNAs: regulators of disease. J Pathol. 2010;220:126–39. doi:10.1002/path.2638.
15. Ma L, Bajic VB, Zhang Z. On the classification of long non-coding RNAs. RNA Biol. 2013;10:925–33. doi:10.4161/rna.24604.
16. St Laurent G, Wahlestedt C, Kapranov P. The landscape of long non-coding RNA classification. Trends Genet. 2015;31:239–51. doi:10.1016/j.tig.2015.03.007.
17. Gardner PP, Bateman A, Poole AM. SnoPatrol: how many snoRNA genes are there. J Biol. 2010;9:4. doi:10.1186/jbiol1211.
18. Terns MP, Terns RM. Small nucleolar RNAs: versatile trans-acting molecules of ancient evolutionary origin. Gene Expr. 2002;10:17–39.
19. Mourtada-Maarabouni M, Pickard MR, Hedge VL, Farzaneh F, Williams GT. GAS5, a non-protein-coding RNA, controls apoptosis and is down-regulated in breast cancer. Oncogene. 2009;28:195–208. doi:10.1038/onc.2008.373.

20. Dong XY, et al. SnoRNA U50 is a candidate tumor-suppressor gene at 6q14.3 with a mutation associated with clinically significant prostate cancer. Hum Mol Genet. 2008;17:1031–42. doi:10.1093/hmg/ddm375.
21. Sasaki T, Shiohama A, Minoshima S, Shimizu N. Identification of eight members of the Argonaute family in the human genome. Genomics. 2003;82:323–30.
22. Aravin A, et al. A novel class of small RNAs bind to MILI protein in mouse testes. Nature. 2006;442:203–7. doi:10.1038/nature04916.
23. Kwon C, et al. Detection of PIWI and piRNAs in the mitochondria of mammalian cancer cells. Biochem Biophys Res Commun. 2014;446:218–23. doi:10.1016/j.bbrc.2014.02.112.
24. Moyano M, Stefani G. piRNA involvement in genome stability and human cancer. J Hematol Oncol. 2015;8:38. doi:10.1186/s13045-015-0133-5.
25. Rosalind C, Lee RLF, Ambrost V. The C. elegans heterochronic gene lin-4 encodes small RNAs with antisense complementarity to & II-14. Cell. 1993;75:843–54. doi:10.1016/0092-8674(93)90529-Y.
26. Calin GA, et al. Frequent deletions and down-regulation of micro- RNA genes miR15 and miR16 at 13q14 in chronic lymphocytic leukemia. Proc Natl Acad Sci U S A. 2002;99:15524–9. doi:10.1073/pnas.242606799.
27. Michael MZ, O' Connor SM, van Holst Pellekaan NG, Young GP, James RJ. Reduced accumulation of specific microRNAs in colorectal neoplasia. Mol Cancer Res. 2003;1:882–91.
28. Meng F, et al. Involvement of human micro-RNA in growth and response to chemotherapy in human cholangiocarcinoma cell lines. Gastroenterology. 2006;130:2113–29. doi:10.1053/j.gastro.2006.02.057.
29. Johnson SM, et al. RAS is regulated by the let-7 microRNA family. Cell. 2005;120:635–47. doi:10.1016/j.cell.2005.01.014.
30. Blenkiron C, Miska EA. miRNAs in cancer: approaches, aetiology, diagnostics and therapy. Hum Mol Genet. 2007;16(Spec No 1):R106–13. doi:10.1093/hmg/ddm056.
31. Marques AC, Ponting CP. Intergenic lncRNAs and the evolution of gene expression. Curr Opin Genet Dev. 2014;27:48–53. doi:10.1016/j.gde.2014.03.009.
32. White NM, Canbanski CR, Silva-Fisher JM, Dang HX, Ramaswamy G, Maher CA. Transcriptome sequencing reveals altered long intergenic non-coding RNAs in lung cancer. Genome Biol. 2014;15(8):429.
33. Hangauer MJ, Vaughn IW, McManus MT. Pervasive transcription of the human genome produces thousands of previously unidentified long intergenic non-coding RNAs. PLoS Genet. 2013;9, e1003569. doi:10.1371/journal.pgen.1003569.
34. Hung T, et al. Extensive and coordinated transcription of non-coding RNAs within cell-cycle promoters. Nat Genet. 2011;43:621–9. doi:10.1038/ng.848.
35. Iacobucci I, et al. A polymorphism in the chromosome 9p21 ANRIL locus is associated to Philadelphia positive acute lymphoblastic leukemia. Leuk Res. 2011;35:1052–9. doi:10.1016/j.leukres.2011.02.020.
36. Chu C, Qu K, Zhong FL, Artandi SE, Chang HY. Genomic maps of long non-coding RNA occupancy reveal principles of RNA-chromatin interactions. Mol Cell. 2011;44:667–78. doi:10.1016/j.molcel.2011.08.027.
37. Beckedorff FC, et al. The intronic long non-coding RNA ANRASSF1 recruits PRC2 to the RASSF1A promoter, reducing the expression of RASSF1A and increasing cell proliferation. PLoS Genet. 2013;9, e1003705. doi:10.1371/journal.pgen.1003705.
38. Bertozzi D, et al. Characterization of novel antisense HIF-1α transcripts in human cancers. Cell Cycle. 2014;10:3189–97. doi:10.4161/cc.10.18.17183.
39. Leygue E. Steroid receptor RNA activator (SRA1): unusual bifaceted gene products with suspected relevance to breast cancer. Nucl Recept Signal. 2007;5, e006. doi:10.1621/nrs.05006.
40. Georges St Laurent DS, Tackett MR, Yang Z, Eremina T, Wahlestedt C, Urcuqui-Inchima S, Seilheimer B, McCaffrey TA, Kapranov P. Intronic RNAs constitute the major fraction of the non-coding RNA in mammalian cells. BMC Genomics. 2012;13:504.

41. Memczak S, et al. Circular RNAs are a large class of animal RNAs with regulatory potency. Nature. 2013;495:333–8. doi:10.1038/nature11928.
42. Barnhill LM, et al. High expression of CAI2, a 9p21-embedded long non-coding RNA, contributes to advanced-stage neuroblastoma. Cancer Res. 2014;74:3753–63. doi:10.1158/0008-5472.CAN-13-3447.
43. Li Y, et al. Circular RNA is enriched and stable in exosomes: a promising biomarker for cancer diagnosis. Cell Res. 2015;25:981–4. doi:10.1038/cr.2015.82.
44. Li L, et al. The diagnostic efficacy and biological effects of microRNA-29b for colon cancer. Technol Cancer Res Treat. 2015. doi:10.1177/1533034615604797.
45. Valeri N, et al. MicroRNA-135b promotes cancer progression by acting as a downstream effector of oncogenic pathways in colon cancer. Cancer Cell. 2014;25:469–83. doi:10.1016/j.ccr.2014.03.006.
46. Kunsbaeva GB, Gilyazova IR, Pavlov VN, Khusnutdinova EK. The role of miRNAs in the development of prostate cancer. Russ J Genet. 2015;51:627–41. doi:10.1134/s102279541507008x.
47. Volinia S, et al. A microRNA expression signature of human solid tumors defines cancer gene targets. Proc Natl Acad Sci U S A. 2006;103:2257–61. doi:10.1073/pnas.0510565103.
48. Ambs S, et al. Genomic profiling of microRNA and mRNA reveals deregulated microRNA expression in prostate cancer. Cancer Res. 2008;68:6162–70. doi:10.1158/0008-5472.CAN-08-0144.
49. Martens-Uzunova ES, et al. Diagnostic and prognostic signatures from the small non-coding RNA transcriptome in prostate cancer. Oncogene. 2012;31:978–91. doi:http://www.nature.com/onc/journal/v31/n8/suppinfo/onc2011304s1.html.
50. Larne O, et al. miQ—A novel microRNA based diagnostic and prognostic tool for prostate cancer. Int J Cancer. 2013;132:2867–75. doi:10.1002/ijc.27973.
51. Moltzahn F, et al. Microfluidic based multiplex qRT-PCR identifies diagnostic and prognostic microRNA signatures in sera of prostate cancer patients. Cancer Res. 2011;71:550–60. doi:10.1158/0008-5472.CAN-10-1229.
52. Bryant RJ, et al. Changes in circulating microRNA levels associated with prostate cancer. Br J Cancer. 2012;106:768–74. doi:10.1038/bjc.2011.595.
53. Srivastava A, et al. MicroRNA profiling in prostate cancer – the diagnostic potential of urinary miR-205 and miR-214. PLoS ONE. 2013;8, e76994. doi:10.1371/journal.pone.0076994.
54. Haj-Ahmad TA, Abdalla MAK, Haj-Ahmad Y. Potential urinary miRNA biomarker candidates for the accurate detection of prostate cancer among benign prostatic hyperplasia patients. J Cancer. 2014;5:182–91. doi:10.7150/jca.6799.
55. Schaefer A, et al. Diagnostic and prognostic implications of microRNA profiling in prostate carcinoma. Int J Cancer. 2010;126:1166–76. doi:10.1002/ijc.24827.
56. Samantarrai D, Dash S, Chhetri B, Mallick B. Genomic and epigenomic cross-talks in the regulatory landscape of miRNAs in breast cancer. Mol Cancer Res. 2013;11:315–28. doi:10.1158/1541-7786.mcr-12-0649.
57. Rasheed SAK, et al. MicroRNA-31 controls G protein alpha-13 (GNA13) expression and cell invasion in breast cancer cells. Mol Cancer. 2015;14:67. doi:10.1186/s12943-015-0337-x.
58. Zheng H-B, Zheng X-G, Liu B-P. miRNA-101 inhibits ovarian cancer cells proliferation and invasion by down-regulating expression of SOCS-2. Int J Clin Exp Med. 2015;8:20263–70.
59. Wilczynski M, Danielska J, Dzieniecka M, Malinowski A. The role of miRNA in endometrial cancer in the context of miRNA 205. Ginekol Pol. 2015;86:856–61.
60. Chen S, Sun K-X, Liu B-L, Zong Z-H, Zhao Y. MicroRNA-505 functions as a tumor suppressor in endometrial cancer by targeting TGF-α. Mol Cancer. 2016;15:11. doi:10.1186/s12943-016-0496-4.
61. Lukiw WJ, Cui JG, Li YY, Culicchia F. Up-regulation of micro-RNA-221 (miRNA-221; chr Xp11.3) and caspase-3 accompanies down-regulation of the survivin-1 homolog BIRC1 (NAIP) in glioblastoma multiforme (GBM). J Neuro-Oncol. 2009;91:27–32. doi:10.1007/s11060-008-9688-0.

62. Li W, et al. miR-124 Acts as a tumor suppressor in glioblastoma via the inhibition of signal transducer and activator of transcription 3. Mol Neurobiol. 2016. doi:10.1007/s12035-016-9852-z.
63. Ji WG et al. miRNA-155 modulates the malignant biological characteristics of NK/T-cell lymphoma cells by targeting FOXO3a gene. J Huazhong Uni Sci Technol. Medical sciences=Hua zhong ke ji da xue xue bao. Yi xue Ying De wen ban=Huazhong keji daxue xuebao. Yixue Yingdewen ban, 2014;34:882–88. doi:10.1007/s11596-014-1368-z.
64. Roisman A, et al. SOXC And MiR17-92 gene expression profiling defines two subgroups with different clinical outcome in mantle cell lymphoma. Genes Chromosomes Cancer. 2016. doi:10.1002/gcc.22355.
65. Li C, Xu N, Li Y-Q, Wang Y, Zhu Z-T. Inhibition of SW620 human colon cancer cells by upregulating miRNA-145. World J Gastroenterol. 2016;22:2771–8. doi:10.3748/wjg.v22.i9.2771.
66. Yiu AJ, Yiu CY. Biomarkers in colorectal cancer. Anticancer Res. 2016;36:1093–102.
67. Lima CR, Geraldo MV, Fuziwara CS, Kimura ET, Santos MF. MiRNA-146b-5p up-regulates migration and invasion of different Papillary Thyroid Carcinoma cells. BMC Cancer. 2016;16:108. doi:10.1186/s12885-016-2146-z.
68. Mutalib NS, et al. MicroRNAs and lymph node metastasis in papillary thyroid cancers. Asian Pac J Cancer Prev. 2016;17:25–35.
69. Volinia S, et al. A microRNA expression signature of human solid tumors defines cancer gene targets. Proc Natl Acad Sci U S A. 2006;103:2257–61. doi:10.1073/pnas.0510565103.
70. Lowery AJ, Miller N, McNeill RE, Kerin MJ. MicroRNAs as prognostic indicators and therapeutic targets: potential effect on breast cancer management. Clin Cancer Res. 2008;14:360–5. doi:10.1158/1078-0432.ccr-07-0992.
71. Lu J, et al. MicroRNA expression profiles classify human cancers. Nature. 2005;435:834–8. doi:10.1038/nature03702.
72. Cannistraci A, Di Pace AL, De Maria R, Bonci D. MicroRNA as new tools for prostate cancer risk assessment and therapeutic intervention: results from clinical data set and patients' samples. BioMed Res Int. 2014;2014:146170. doi:10.1155/2014/146170.
73. Tsuchiya N, et al. Biomarkers for the early diagnosis of hepatocellular carcinoma. World J Gastroenterol. 2015;21:10573–83. doi:10.3748/wjg.v21.i37.10573.
74. Zhou J, et al. Plasma microRNA panel to diagnose hepatitis B virus-related hepatocellular carcinoma. J Clin Oncol. 2011;29:4781–8. doi:10.1200/JCO.2011.38.2697.
75. Forner A. Hepatocellular carcinoma surveillance with miRNAs. Lancet Oncol. 2015;16:743–5. doi:10.1016/s1470-2045(15)00014-5.
76. Berindan-Neagoe I, Monroig Pdel C, Pasculli B, Calin GA. MicroRNAome genome: a treasure for cancer diagnosis and therapy. CA Cancer J Clin. 2014;64:311–36. doi:10.3322/caac.21244.
77. Wang J, et al. Elevated expression of miR-210 predicts poor survival of cancer patients: a systematic review and meta-analysis. PLoS ONE. 2014;9, e89223. doi:10.1371/journal.pone.0089223.
78. Wang B, Zhang Q. The expression and clinical significance of circulating microRNA-21 in serum of five solid tumors. J Cancer Res Clin Oncol. 2012;138:1659–66. doi:10.1007/s00432-012-1244-9.
79. Eichelser C, Flesch-Janys D, Chang-Claude J, Pantel K, Schwarzenbach H. Deregulated serum concentrations of circulating cell-free microRNAs miR-17, miR-34a, miR-155, and miR-373 in human breast cancer development and progression. Clin Chem. 2013;59:1489–96. doi:10.1373/clinchem.2013.205161.
80. Cheng H, et al. Circulating plasma miR-141 is a novel biomarker for metastatic colon cancer and predicts poor prognosis. PLoS ONE. 2011;6, e17745. doi:10.1371/journal.pone.0017745.
81. Huang Z, et al. Plasma microRNAs are promising novel biomarkers for early detection of colorectal cancer. Int J Cancer. 2010;127:118–26. doi:10.1002/ijc.25007.

82. Pu XX, et al. Circulating miR-221 directly amplified from plasma is a potential diagnostic and prognostic marker of colorectal cancer and is correlated with p53 expression. J Gastroenterol Hepatol. 2010;25:1674–80. doi:10.1111/j.1440-1746.2010.06417.x.
83. Toiyama Y, et al. Serum miR-200c is a novel prognostic and metastasis-predictive biomarker in patients with colorectal cancer. Ann Surg. 2014;259:735–43. doi:10.1097/SLA.0b013e3182a6909d.
84. Wang J, Du Y, Liu X, Cho WC, Yang Y. MicroRNAs as regulator of signaling networks in metastatic colon cancer. BioMed Res Int. 2015;2015:823620. doi:10.1155/2015/823620.
85. Zhu C, et al. A five-microRNA panel in plasma was identified as potential biomarker for early detection of gastric cancer. Br J Cancer. 2014;110:2291–9. doi:10.1038/bjc.2014.119.
86. Huang Y-K, Yu J-C. Circulating microRNAs and long non-coding RNAs in gastric cancer diagnosis: an update and review. World J Gastroenterol. 2015;21:9863–86. doi:10.3748/wjg.v21.i34.9863.
87. Zhang Y-C, Xu Z, Zhang T-F, Wang Y-L. Circulating microRNAs as diagnostic and prognostic tools for hepatocellular carcinoma. World J Gastroenterol. 2015;21:9853–62. doi:10.3748/wjg.v21.i34.9853.
88. Meng W, et al. MicroRNA-31 predicts the presence of lymph node metastases and survival in lung adenocarcinoma patients. Clin Cancer Res. 2013;19:5423–33. doi:10.1158/1078-0432.CCR-1113-0320.
89. Guz M, et al. MicroRNAs-role in lung cancer. Dis Markers. 2014;2014:218169. doi:10.1155/2014/218169.
90. Rodriguez M, et al. Different exosome cargo from plasma/bronchoalveolar lavage in non-small-cell lung cancer. Genes Chromosomes Cancer. 2014;53:713–24. doi:10.1002/gcc.22181.
91. Yang C, et al. Identification of seven serum microRNAs from a genome-wide serum microRNA expression profile as potential noninvasive biomarkers for malignant astrocytomas. Int J Cancer. 2013;132:116–27. doi:10.1002/ijc.27657.
92. Westermann AM, et al. Serum microRNAs as biomarkers in patients undergoing prostate biopsy: results from a prospective multi-center study. Anticancer Res. 2014;34:665–9.
93. Zheng H, et al. Plasma miRNAs as diagnostic and prognostic biomarkers for ovarian cancer. PLoS ONE. 2013;8, e77853. doi:10.1371/journal.pone.0077853.
94. Mercer TR, Dinger ME, Sunkin SM, Mehler MF, Mattick JS. Specific expression of long non-coding RNAs in the mouse brain. Proc Natl Acad Sci U S A. 2008;105:716–21. doi:10.1073/pnas.0706729105.
95. Maass PG, Luft FC, Bahring S. Long non-coding RNA in health and disease. J Mol Med. 2014;92:337–46. doi:10.1007/s00109-014-1131-8.
96. Yarmishyn AA, Kurochkin IV. Long non-coding RNAs: a potential novel class of cancer biomarkers. Front Genet. 2015;6:145. doi:10.3389/fgene.2015.00145.
97. Bussemakers MJ, et al. DD3: a new prostate-specific gene, highly overexpressed in prostate cancer. Cancer Res. 1999;59:5975–9.
98. Petrovics G, et al. Elevated expression of PCGEM1, a prostate-specific gene with cell growth-promoting function, is associated with high-risk prostate cancer patients. Oncogene. 2004;23:605–11. doi:10.1038/sj.onc.1207069.
99. Chung S, et al. Association of a novel long non-coding RNA in 8q24 with prostate cancer susceptibility. Cancer Sci. 2011;102:245–52. doi:10.1111/j.1349-7006.2010.01737.x.
100. Ylipaa A, et al. Transcriptome sequencing reveals PCAT5 as a novel ERG-regulated long non-coding RNA in prostate cancer. Cancer Res. 2015;75:4026–31. doi:10.1158/0008-5472.CAN-15-0217.
101. Crea F, et al. Identification of a long non-coding RNA as a novel biomarker and potential therapeutic target for metastatic prostate cancer. Oncotarget. 2014;5:764–74. doi:10.18632/oncotarget.1769.
102. Wei G, et al. Transcriptome profiling of esophageal squamous cell carcinoma reveals a long non-coding RNA acting as a tumor suppressor. Oncotarget. 2015;6:17065–80. doi:10.18632/oncotarget.4185.

103. Li X, et al. Long non-coding RNA HOTAIR, a driver of malignancy, predicts negative prognosis and exhibits oncogenic activity in oesophageal squamous cell carcinoma. Br J Cancer. 2013;109:2266–78. doi:10.1038/bjc.2013.548.
104. Li W, et al. Increased levels of the long intergenic non-protein coding RNA POU3F3 promote DNA methylation in esophageal squamous cell carcinoma cells. Gastroenterology. 2014;146(1714–1726), e1715. doi:10.1053/j.gastro.2014.03.002.
105. Zhang X, et al. Elevated expression of CCAT2 is associated with poor prognosis in esophageal squamous cell carcinoma. J Surg Oncol. 2015;111:834–9. doi:10.1002/jso.23888.
106. Nie Y, et al. Long non-coding RNA HOTAIR is an independent prognostic marker for nasopharyngeal carcinoma progression and survival. Cancer Sci. 2013;104:458–64. doi:10.1111/cas.12092.
107. Chen Z, et al. MiRNA expression profile reveals a prognostic signature for esophageal squamous cell carcinoma. Cancer Lett. 2014;350:34–42. doi:10.1016/j.canlet.2014.04.013.
108. Yang F, et al. Long non-coding RNA CCAT1, which could be activated by c-Myc, promotes the progression of gastric carcinoma. J Cancer Res Clin Oncol. 2013;139:437–45. doi:10.1007/s00432-012-1324-x.
109. Mei D, et al. Up-regulation of SUMO1 pseudogene 3 (SUMO1P3) in gastric cancer and its clinical association. Med Oncol. 2013;30:709. doi:10.1007/s12032-013-0709-2.
110. Sana J, Faltejskova P, Svoboda M, Slaby O. Novel classes of non-coding RNAs and cancer. J Transl Med. 2012;10:103. doi:10.1186/1479-5876-10-103.
111. Kogo R, et al. Long non-coding RNA HOTAIR regulates polycomb-dependent chromatin modification and is associated with poor prognosis in colorectal cancers. Cancer Res. 2011;71:6320–6. doi:10.1158/0008-5472.can-11-1021.
112. Panzitt K, et al. Characterization of HULC, a novel gene with striking up-regulation in hepatocellular carcinoma, as non-coding RNA. Gastroenterology. 2007;132:330–42. doi:10.1053/j.gastro.2006.08.026.
113. Gutschner T, et al. The non-coding RNA MALAT1 is a critical regulator of the metastasis phenotype of lung cancer cells. Cancer Res. 2013;73:1180–9. doi:10.1158/0008-5472.can-12-2850.
114. Zhang XQ, Leung GK. Long non-coding RNAs in glioma: functional roles and clinical perspectives. Neurochem Int. 2014;77:78–85. doi:10.1016/j.neuint.2014.05.008.
115. Chisholm KM, et al. Detection of long non-coding RNA in archival tissue: correlation with polycomb protein expression in primary and metastatic breast carcinoma. PLoS ONE. 2012;7, e47998. doi:10.1371/journal.pone.0047998.
116. Huang J, et al. Long non-coding RNA UCA1 promotes breast tumor growth by suppression of p27 (Kip1). Cell Death Dis. 2014;5, e1008. doi:10.1038/cddis.2013.541.
117. Qiu JJ, et al. Overexpression of long non-coding RNA HOTAIR predicts poor patient prognosis and promotes tumor metastasis in epithelial ovarian cancer. Gynecol Oncol. 2014;134:121–8. doi:10.1016/j.ygyno.2014.03.556.
118. Zhang Q, Su M, Lu G, Wang J. The complexity of bladder cancer: long non-coding RNAs are on the stage. Mol Cancer. 2013;12:101. doi:10.1186/1476-4598-12-101.
119. Wang L, et al. Genome-wide screening and identification of long non-coding RNAs and their interaction with protein coding RNAs in bladder urothelial cell carcinoma. Cancer Lett. 2014;349:77–86. doi:10.1016/j.canlet.2014.03.033.
120. Cao Q, Wang N, Qi J, Gu Z, Shen H. Long non-coding RNAGAS5 acts as a tumor suppressor in bladder transitional cell carcinoma via regulation of chemokine (CC motif) ligand 1 expression. Mol Med Rep. 2016;13:27–34. doi:10.3892/mmr.2015.4503.
121. Ploussard G, et al. Prostate cancer antigen 3 score accurately predicts tumour volume and might help in selecting prostate cancer patients for active surveillance. Eur Urol. 2011;59:422–9. doi:10.1016/j.eururo.2010.11.044.
122. Walsh AL, Tuzova AV, Bolton EM, Lynch TH, Perry AS. Long non-coding RNAs and prostate carcinogenesis: the missing 'linc'? Trends Mol Med. 2014;20:428–36. doi:10.1016/j.molmed.2014.03.005.

123. Ifere GO, Ananaba GA. Prostate cancer gene expression marker 1 (PCGEM1): a patented prostate- specific non-coding gene and regulator of prostate cancer progression. Recent Pat DNA Gene Seq. 2009;3:151–63.
124. Li W, et al. Suppressing H19 modulates tumorigenicity and stemness in U251 and U87MG glioma cells. Cell Mol Neurobiol. 2016. doi:10.1007/s10571-015-0320-5.
125. Leucci E, et al. Melanoma addiction to the long non-coding RNA SAMMSON. Nature. 2016;531:518–22. doi:10.1038/nature17161.
126. Gibb EA, et al. Long non-coding RNAs are expressed in oral mucosa and altered in oral premalignant lesions. Oral Oncol. 2011;47:1055–61. doi:10.1016/j.oraloncology.2011.07.008.
127. Gao W, Chan JY, Wong TS. Differential expression of long non-coding RNA in primary and recurrent nasopharyngeal carcinoma. Biomed Res Int. 2014;2014:404567. doi:10.1155/2014/404567.
128. Meseure D, Drak Alsibai K, Nicolas A, Bieche I, Morillon A. Long non-coding RNAs as new architects in cancer epigenetics, prognostic biomarkers, and potential therapeutic targets. Biomed Res Int. 2015;2015:320214. doi:10.1155/2015/320214.
129. Kroh EM, Parkin RK, Mitchell PS, Tewari M. Analysis of circulating microRNA biomarkers in plasma and serum using quantitative reverse transcription-PCR (qRT-PCR). Methods. 2010;50:298–301. doi:10.1016/j.ymeth.2010.01.032.
130. Jemal A, et al. Cancer statistics, 2009. CA Cancer J Clin. 2009;59:225–49. doi:10.3322/caac.20006.
131. Chen X, et al. Characterization of microRNAs in serum: a novel class of biomarkers for diagnosis of cancer and other diseases. Cell Res. 2008;18:997–1006. doi:10.1038/cr.2008.282.
132. Bianchi F, et al. A serum circulating miRNA diagnostic test to identify asymptomatic high-risk individuals with early stage lung cancer. EMBO Mol Med. 2011;3:495–503. doi:10.1002/emmm.201100154.
133. Zheng D, et al. Plasma microRNAs as novel biomarkers for early detection of lung cancer. Int J Clin Exp Pathol. 2011;4:575–86.
134. Le HB, et al. Evaluation of dynamic change of serum miR-21 and miR-24 in pre- and post-operative lung carcinoma patients. Med Oncol. 2012;29:3190–7. doi:10.1007/s12032-012-0303-z.
135. Shen J, et al. Plasma microRNAs as potential biomarkers for non-small-cell lung cancer. Lab Investig. 2011;91:579–87. doi:10.1038/labinvest.2010.194.
136. Wei J, et al. Identification of plasma microRNA-21 as a biomarker for early detection and chemosensitivity of non-small cell lung cancer. Chin J Cancer. 2011;30:407–14.
137. Abd-El-Fattah AA, Sadik NA, Shaker OG, Aboulftouh ML. Differential microRNAs expression in serum of patients with lung cancer, pulmonary tuberculosis, and pneumonia. Cell Biochem Biophys. 2013;67:875–84. doi:10.1007/s12013-013-9575-y.
138. Hennessey PT, et al. Serum microRNA biomarkers for detection of non-small cell lung cancer. PLoS ONE. 2012;7, e32307. doi:10.1371/journal.pone.0032307.
139. Chen X, et al. Identification of ten serum microRNAs from a genome-wide serum microRNA expression profile as novel noninvasive biomarkers for nonsmall cell lung cancer diagnosis. Int J Cancer. 2012;130:1620–8. doi:10.1002/ijc.26177.
140. Boeri M, et al. MicroRNA signatures in tissues and plasma predict development and prognosis of computed tomography detected lung cancer. Proc Natl Acad Sci U S A. 2011;108:3713–8. doi:10.1073/pnas.1100048108.
141. Polyak K. Heterogeneity in breast cancer. J Clin Invest. 2011;121:3786–8. doi:10.1172/JCI60534.
142. Zhu W, Qin W, Atasoy U, Sauter ER. Circulating microRNAs in breast cancer and healthy subjects. BMC Res Notes. 2009;2:89. doi:10.1186/1756-0500-2-89.
143. Heneghan HM, et al. Circulating microRNAs as novel minimally invasive biomarkers for breast cancer. Ann Surg. 2010;251:499–505.

144. Wu Q, et al. Next-generation sequencing of microRNAs for breast cancer detection. J Biomed Biotechnol. 2011;2011:597145. doi:10.1155/2011/597145.
145. Asaga S, et al. Direct serum assay for microRNA-21 concentrations in early and advanced breast cancer. Clin Chem. 2011;57:84–91. doi:10.1373/clinchem.2010.151845.
146. Hu Z, et al. Serum microRNA profiling and breast cancer risk: the use of miR-484/191 as endogenous controls. Carcinogenesis. 2012;33:828–34. doi:10.1093/carcin/bgs030.
147. Mar-Aguilar F, et al. Serum circulating microRNA profiling for identification of potential breast cancer biomarkers. Dis Markers. 2013;34:163–9. doi:10.3233/DMA-120957.
148. McDermott AM, et al. Identification and validation of oncologic miRNA biomarkers for luminal A-like breast cancer. PLoS ONE. 2014;9, e87032. doi:10.1371/journal.pone.0087032.
149. Schwarzenbach H, Milde-Langosch K, Steinbach B, Muller V, Pantel K. Diagnostic potential of PTEN-targeting miR-214 in the blood of breast cancer patients. Breast Cancer Res Treat. 2012;134:933–41. doi:10.1007/s10549-012-1988-6.
150. Chen W, Cai F, Zhang B, Barekati Z, Zhong XY. The level of circulating miRNA-10b and miRNA-373 in detecting lymph node metastasis of breast cancer: potential biomarkers. Tumour Biol. 2013;34:455–62. doi:10.1007/s13277-012-0570-5.
151. Zhao R, et al. Plasma miR-221 as a predictive biomarker for chemoresistance in breast cancer patients who previously received neoadjuvant chemotherapy. Onkologie. 2011;34:675–80.
152. Rao X, et al. MicroRNA-221/222 confers breast cancer fulvestrant resistance by regulating multiple signaling pathways. Oncogene. 2011;30:1082–97. doi:10.1038/onc.2010.487.
153. Wang H, et al. MiR-125b as a marker predicting chemoresistance in breast cancer. PLoS ONE. 2012;7, e34210. doi:10.1371/journal.pone.0034210.g001.
154. Jung EJ, et al. Plasma microRNA 210 levels correlate with sensitivity to trastuzumab and tumor presence in breast cancer patients. Cancer. 2012;118:2603–14. doi:10.1002/cncr.26565.
155. Enache LS, et al. Circulating RNA molecules as biomarkers in liver disease. Int J Mol Sci. 2014;15:17644–66. doi:10.3390/ijms151017644.
156. Li LM, et al. Serum microRNA profiles serve as novel biomarkers for HBV infection and diagnosis of HBV-positive hepatocarcinoma. Cancer Res. 2010;70:9798–807. doi:10.1158/0008-5472.CAN-10-1001.
157. Xu J, et al. Circulating microRNAs, miR-21, miR-122, and miR-223, in patients with hepatocellular carcinoma or chronic hepatitis. Mol Carcinog. 2011;50:136–42. doi:10.1002/mc.20712.
158. Li J, Wang Y, Yu W, Chen J, Luo J. Expression of serum miR-221 in human hepatocellular carcinoma and its prognostic significance. Biochem Biophys Res Commun. 2011;406:70–3. doi:10.1016/j.bbrc.2011.01.111.
159. Qu KZ, Zhang K, Li H, Afdhal NH, Albitar M. Circulating microRNAs as biomarkers for hepatocellular carcinoma. J Clin Gastroenterol. 2011;45:355–60.
160. Liu AM, et al. Circulating miR-15b and miR-130b in serum as potential markers for detecting hepatocellular carcinoma: a retrospective cohort study. BMJ Open. 2012;2, e000825. doi:10.1136/bmjopen-2012-000825.
161. Borel F, Konstantinova P, Jansen PL. Diagnostic and therapeutic potential of miRNA signatures in patients with hepatocellular carcinoma. J Hepatol. 2012;56:1371–83. doi:10.1016/j.jhep.2011.11.026.
162. Cheng G. Circulating miRNAs: roles in cancer diagnosis, prognosis and therapy. Adv Drug Deliv Rev. 2015;81:75–93. doi:10.1016/j.addr.2014.09.001.
163. Tsujiura M, et al. Circulating microRNAs in plasma of patients with gastric cancers. Br J Cancer. 2010;102:1174–9. doi:10.1038/sj.bjc.6605608.
164. Wang M, et al. Circulating miR-17-5p and miR-20a: molecular markers for gastric cancer. Mol Med Rep. 2012;5:1514–20. doi:10.3892/mmr.2012.828.
165. Zheng Y, et al. MicroRNA-21 is a new marker of circulating tumor cells in gastric cancer patients. Cancer Biomarkers. 2011;10:71–7. doi:10.3233/CBM-2011-0231.
166. Kim SY, et al. Validation of circulating miRNA biomarkers for predicting lymph node metastasis in gastric cancer. J Mol Diagn. 2013;15:661–9. doi:10.1016/j.jmoldx.2013.04.004.

167. Komatsu S, et al. Prognostic impact of circulating miR-21 in the plasma of patients with gastric carcinoma. Anticancer Res. 2013;33:271–6.
168. Li BS, et al. Plasma microRNAs, miR-223, miR-21 and miR-218, as novel potential biomarkers for gastric cancer detection. PLoS ONE. 2012;7, e41629. doi:10.1371/journal.pone.0041629.
169. Huang YK, Yu JC. Circulating microRNAs and long non-coding RNAs in gastric cancer diagnosis: an update and review. World J Gastroenterol. 2015;21:9863–86. doi:10.3748/wjg.v21.i34.9863.
170. Ng EK, et al. Differential expression of microRNAs in plasma of patients with colorectal cancer: a potential marker for colorectal cancer screening. Gut. 2009;58:1375–81. doi:10.1136/gut.2008.167817.
171. Wang LG, Gu J. Serum microRNA-29a is a promising novel marker for early detection of colorectal liver metastasis. Cancer Epidemiol. 2012;36:e61–7. doi:10.1016/j.canep.2011.05.002.
172. Wang J, et al. Identification of a circulating microRNA signature for colorectal cancer detection. PLoS ONE. 2014;9, e87451. doi:10.1371/journal.pone.0087451.
173. Kanaan Z, et al. Plasma miR-21: a potential diagnostic marker of colorectal cancer. Ann Surg. 2012;256:544–51. doi:10.1097/SLA.0b013e318265bd6f.
174. Toiyama Y, et al. Serum miR-200c is a novel prognostic and metastasis-predictive biomarker in patients with colorectal cancer. Ann Surg. 2014;259:735–43. doi:10.1097/SLA.0b013e3182a6909d.
175. Wang Q, et al. Plasma miR-601 and miR-760 are novel biomarkers for the early detection of colorectal cancer. PLoS ONE. 2012;7, e44398. doi:10.1371/journal.pone.0044398.
176. Zhang J, et al. Circulating microRNA expressions in colorectal cancer as predictors of response to chemotherapy. Anticancer Drugs. 2014;25:346–52.
177. Lawrie CH. MicroRNAs in hematological malignancies. Blood Rev. 2013;27:143–54. doi:10.1016/j.blre.2013.04.002.
178. Lawrie CH, et al. Detection of elevated levels of tumour-associated microRNAs in serum of patients with diffuse large B-cell lymphoma. Br J Haematol. 2008;141:672–5. doi:10.1111/j.1365-2141.2008.07077.x.
179. Chen W, et al. Clinical significance and detection of microRNA-21 in serum of patients with diffuse large B-cell lymphoma in Chinese population. Eur J Haematol. 2014;92:407–12. doi:10.1111/ejh.12263.
180. Fayyad-Kazan H, et al. Circulating miR-150 and miR-342 in plasma are novel potential biomarkers for acute myeloid leukemia. J Transl Med. 2013;11:31.
181. Moussay E, et al. MicroRNA as biomarkers and regulators in B-cell chronic lymphocytic leukemia. Proc Natl Acad Sci U S A. 2011;108:6573–8. doi:10.1073/pnas.1019557108.
182. Han YC, et al. microRNA-29a induces aberrant self-renewal capacity in hematopoietic progenitors, biased myeloid development, and acute myeloid leukemia. J Exp Med. 2010;207:475–89. doi:10.1084/jem.20090831.
183. Felli N, et al. MicroRNAs 221 and 222 inhibit normal erythropoiesis and erythroleukemic cell growth via kit receptor down-modulation. Proc Natl Acad Sci U S A. 2005;102:18081–6. doi:10.1073/pnas.0506216102.
184. Pekarsky Y, et al. Tcl1 expression in chronic lymphocytic leukemia is regulated by miR-29 and miR-181. Cancer Res. 2006;66:11590–3. doi:10.1158/0008-5472.CAN-06-3613.
185. Chen RW, et al. Truncation in CCND1 mRNA alters miR-16-1 regulation in mantle cell lymphoma. Blood. 2008;112:822–9. doi:10.1182/blood-2008-03-142182.
186. Xiao C, et al. Lymphoproliferative disease and autoimmunity in mice with increased miR-17-92 expression in lymphocytes. Nat Immunol. 2008;9:405–14. doi:10.1038/ni1575.
187. Yamanaka Y, et al. Aberrant overexpression of microRNAs activate AKT signaling via down-regulation of tumor suppressors in natural killer-cell lymphoma/leukemia. Blood. 2009;114:3265–75. doi:10.1182/blood-2009-06-222794.
188. Wang J, Song YX, Wang ZN. Non-coding RNAs in gastric cancer. Gene. 2015;560:1–8. doi:10.1016/j.gene.2015.02.004.

189. Cui L, et al. Detection of circulating tumor cells in peripheral blood from patients with gastric cancer using piRNAs as markers. Clin Biochem. 2011;44:1050–7. doi:10.1016/j.clinbiochem.2011.06.004.
190. Law PT, et al. Deep sequencing of small RNA transcriptome reveals novel non-coding RNAs in hepatocellular carcinoma. J Hepatol. 2013;58:1165–73. doi:10.1016/j.jhep.2013.01.032.
191. Liao J, et al. Small nucleolar RNA signatures as biomarkers for non-small-cell lung cancer. Mol Cancer. 2010;9:198. doi:10.1186/1476-4598-9-198.
192. Mei YP, et al. Small nucleolar RNA 42 acts as an oncogene in lung tumorigenesis. Oncogene. 2012;31:2794–804. doi:10.1038/onc.2011.449.
193. Gee HE, et al. The small-nucleolar RNAs commonly used for microRNA normalisation correlate with tumour pathology and prognosis. Br J Cancer. 2011;104:1168–77. doi:10.1038/sj.bjc.6606076.
194. Appaiah HN, et al. Persistent up-regulation of U6:SNORD44 small RNA ratio in the serum of breast cancer patients. Breast Cancer Res. 2011;13:R86. doi:10.1186/bcr2943.
195. Hansen TB, et al. Natural RNA circles function as efficient microRNA sponges. Nature. 2013;495:384–8. doi:10.1038/nature11993.
196. Lukiw WJ. Circular RNA (circRNA) in Alzheimer's disease (AD). Front Genet. 2013;4:307. doi:10.3389/fgene.2013.00307.
197. Koh W, et al. Noninvasive in vivo monitoring of tissue-specific global gene expression in humans. Proc Natl Acad Sci. 2014;111:7361–6. doi:10.1073/pnas.1405528111.
198. Memczak S, Papavasileiou P, Peters O, Rajewsky N. Identification and characterization of circular RNAs as a new class of putative biomarkers in human blood. PLoS ONE. 2015;10, e0141214. doi:10.1371/journal.pone.0141214.
199. Rybak-Wolf A, et al. Circular RNAs in the mammalian brain are highly abundant, conserved, and dynamically expressed. Mol Cell. 2015;58:870–85. doi:10.1016/j.molcel.2015.03.027.
200. Li P, et al. Using circular RNA as a novel type of biomarker in the screening of gastric cancer. Clin Chim Acta. 2015;444:132–6. doi:10.1016/j.cca.2015.02.018.
201. Qu S, et al. Microarray expression profile of circular RNAs in human pancreatic ductal adenocarcinoma. Genomics Data. 2015;5:385–7. doi:10.1016/j.gdata.2015.07.017.
202. Conde-Vancells J, et al. Characterization and comprehensive proteome profiling of exosomes secreted by hepatocytes. J Proteome Res. 2008;7:5157–66. doi:10.1021/pr8004887.
203. Street JM, et al. Identification and proteomic profiling of exosomes in human cerebrospinal fluid. J Transl Med. 2012;10:5. doi:10.1186/1479-5876-10-5.
204. Gourzones C, et al. Extra-cellular release and blood diffusion of BART viral micro-RNAs produced by EBV-infected nasopharyngeal carcinoma cells. Virol J. 2010;7:271. doi:10.1186/1743-422X-7-271.
205. Mathivanan S, Simpson RJ. ExoCarta: a compendium of exosomal proteins and RNA. Proteomics. 2009;9:4997–5000. doi:10.1002/pmic.200900351.
206. Samsonov R, et al. Lectin-induced agglutination method of urinary exosomes isolation followed by mi-RNA analysis: application for prostate cancer diagnostic. Prostate. 2015. doi:10.1002/pros.23101.
207. Cazzoli R, et al. microRNAs derived from circulating exosomes as noninvasive biomarkers for screening and diagnosing lung cancer. J Thorac Oncol. 2013;8:1156–62. doi:10.1097/JTO.0b013e318299ac32.
208. Singh R, Pochampally R, Watabe K, Lu ZH, Mo YY. Exosome-mediated transfer of miR-10b promotes cell invasion in breast cancer. Mol Cancer. 2014;13:256. doi:10.1186/1476-4598-13-256.
209. Momen-Heravi F, et al. Increased number of circulating exosomes and their microRNA cargos are potential novel biomarkers in alcoholic hepatitis. J Transl Med. 2015;13:261. doi:10.1186/s12967-015-0623-9.
210. Chen C, et al. Real-time quantification of microRNAs by stem-loop RT-PCR. Nucleic Acids Res. 2005;33, e179. doi:10.1093/nar/gni178.
211. Conte D, et al. Novel method to detect microRNAs using chip-based QuantStudio 3D digital PCR. BMC Genomics. 2015;16:849. doi:10.1186/s12864-015-2097-9.

212. Matullo G, Naccarati A, Pardini B. microRNA expression profiling in bladder cancer: the challenge of Next Generation Sequencing in tissues and biofluids. Int J Cancer. 2015. doi:10.1002/ijc.29895.
213. Mestdagh P et al. Evaluation of quantitative miRNA expression platforms in the microRNA quality control (miRQC) study (vol 11, pg 809, 2014). Nat Methods, 2014;11:971–971
214. Kalra H, et al. Comparative proteomics evaluation of plasma exosome isolation techniques and assessment of the stability of exosomes in normal human blood plasma. Proteomics. 2013;13:3354–64. doi:10.1002/pmic.201300282.
215. Park NJ, et al. Salivary microRNA: discovery, characterization, and clinical utility for oral cancer detection. Clin Cancer Res. 2009;15:5473–7. doi:10.1158/1078-0432.CCR-09-0736.
216. Xie Z, et al. Salivary microRNAs as promising biomarkers for detection of esophageal cancer. PLoS ONE. 2013;8, e57502. doi:10.1371/journal.pone.0057502.
217. Xie ZJ, et al. Salivary microRNAs show potential as a noninvasive biomarker for detecting resectable pancreatic cancer. Cancer Prev Res. 2015;8:165–73. doi:10.1158/1940-6207.CAPR-14-0192.
218. Kok JB. DD3PCA3, a very sensitive and specific marker to detect prostate tumors. Cancer Res. 2002;62(9):2695–8.
219. van Gils MP, et al. The time-resolved fluorescence-based PCA3 test on urinary sediments after digital rectal examination; a Dutch multicenter validation of the diagnostic performance. Clin Cancer Res. 2007;13:939–43. doi:10.1158/1078-0432.CCR-06-2679.
220. Yamada Y, et al. MiR-96 and miR-183 detection in urine serve as potential tumor markers of urothelial carcinoma: correlation with stage and grade, and comparison with urinary cytology. Cancer Sci. 2011;102:522–9. doi:10.1111/j.1349-7006.2010.01816.x.
221. Lewis H, et al. miR-888 is an expressed prostatic secretions-derived microRNA that promotes prostate cell growth and migration. Cell Cycle. 2014;13:227–39. doi:10.4161/cc.26984.
222. Hanke M, et al. A robust methodology to study urine microRNA as tumor marker: microRNA-126 and microRNA-182 are related to urinary bladder cancer. Urol Oncol. 2010;28:655–61. doi:10.1016/j.urolonc.2009.01.027.
223. Ahmed FE, et al. Diagnostic microRNA markers for screening sporadic human colon cancer and active ulcerative colitis in stool and tissue. Cancer Genomics Proteomics. 2009;6:281–95.
224. Yantiss RK, et al. Clinical, pathologic, and molecular features of early-onset colorectal carcinoma. Am J Surg Pathol. 2009;33:572–82. doi:10.1097/PAS.0b013e31818afd6b.
225. Koga Y, et al. MicroRNA expression profiling of exfoliated colonocytes isolated from feces for colorectal cancer screening. Cancer Prev Res (Phila). 2010;3:1435–42. doi:10.1158/1940-6207.CAPR-10-0036.
226. Kalimutho M, et al. Differential expression of miR-144* as a novel fecal-based diagnostic marker for colorectal cancer. J Gastroenterol. 2011;46:1391–402. doi:10.1007/s00535-011-0456-0.
227. Link A, Becker V, Goel A, Wex T, Malfertheiner P. Feasibility of fecal microRNAs as novel biomarkers for pancreatic cancer. PLoS ONE. 2012;7, e42933. doi:10.1371/journal.pone.0042933.

Chapter 16
Therapeutic Potentials of Noncoding RNAs: Targeted Delivery of ncRNAs in Cancer Cells

Yang Liu and Jun Wang

Abstract Knowledge of multiple actions of short noncoding RNAs (ncRNAs) has truly allowed for viewing DNA, RNA, and protein in novel ways. The ncRNAs are an attractive new class of therapeutics, especially against undruggable targets for the treatment of cancer and other diseases. Despite the potential of ncRNAs in cancer therapy, many challenges remain, including rapid degradation and clearance, poor cellular uptake, off-target effects, and immunogenicity. Rational design, chemical modifications, and delivery carriers offer significant opportunities to overcome these challenges. In this chapter, the development of ncRNAs as cancer therapeutics from early stages to clinical trials and strategies for ncRNA-targeted delivery to cancer cells will be introduced.

Keywords RNA interference • Small interfering RNA (siRNA) • MicroRNA (miRNA) • Cancer therapeutics • Delivery system

16.1 Introduction

Cancer is a genetic disease resulting from the dysregulation of the gene networks that maintain normal cellular identity, growth, and differentiation. A key development in unraveling the complex genetics of cancer may be the shift in focus from exclusively investigating the protein-coding components of the genome to considering the role of variation in regulatory elements [1]. Cancer in particular has been a major focus of noncoding RNA, especially microRNA research over the past decade, and many studies have demonstrated the importance of microRNAs in

Y. Liu • J. Wang (✉)
Hefei National Laboratory for Physical Sciences at Microscale, and School of Life Sciences, University of Science and Technology of China, Hefei, Anhui 230027, People's Republic of China
e-mail: jwang699@ustc.edu.cn

E. Song (ed.), *The Long and Short Non-coding RNAs in Cancer Biology*, Advances in Experimental Medicine and Biology 927,
DOI 10.1007/978-981-10-1498-7_16

cancer biology through controlling the expression of their target mRNAs to facilitate tumor growth, invasion, angiogenesis, and immune evasion. Additionally, tumor microRNA profiles can define relevant subtypes, patient survival, and treatment response [2].

MicroRNA dysregulation in cancer was first reported in 2002 when a cluster of two microRNAs (miR-15 and miR-16) was identified at 13q14.3, a region frequently deleted in chronic lymphocytic leukemia (CLL) [3]. This miRNA deletion was correlated with higher expression of the antiapoptotic target B-cell lymphoma 2 (BCL2). MicroRNAs have since been documented in roles in all of the cancer hallmarks including sustaining proliferative signaling, evading growth suppressors, resisting cell death, enabling replicative immortality, inducing angiogenesis, and activating invasion and metastasis, reprogramming of energy metabolism, and evading immune destruction [4].

MicroRNA is dysregulated through the following mechanisms: genetic alterations, epigenetic mechanisms, miRNA suppression by oncogenic transcription factors (such as Myc and KRAS), and miRNA downregulation by loss of tumor-suppressor transcription factors. MicroRNA-155 overexpression is associated with many cancer types including hematopoietic cancers, breast, lung, and colon cancer [5]. Overexpression of miR-155 is implicated in facilitating tumor cell growth and invasion and has attracted considerable interest as a putative therapeutic target [6]. MicroRNA-21 was the first miRNA to be coined an oncomiR due to the rather universal overexpression of this miRNA in cancer [7]. Studies in miR-21 knockout mice have demonstrated reduced lung tumor burden following activation of a mutant KrasG12D allele, and, in accordance, a miR-21 transgene resulted in increased tumor outgrowth [8]. MicroRNA-34a is a tumor-suppressor microRNA downstream of p53. Its replacement in cancer cells antagonizes key hallmarks including self-renewal, migratory potential, and chemoresistance [9].

Along with miRNA, another type of small regulatory ncRNA known as exogenous small interfering RNA is also involved in gene regulation and genome defense and shares components of the cellular pathways of RNA interference (RNAi). Small interfering RNAs (siRNAs) are 20–28-nt-long RNA molecules that can specifically cleave mRNA through a cytoplasmic pathway known as RNA interference (RNAi). Due to its special advantages such as unique specificity, unlimited range of targets, and high efficiency, siRNA has emerged as a powerful tool for cancer therapeutic gene silencing since its initial discovery in 1998 [10]. Due to the special mechanism of siRNA, it has four advantages as a potential cancer therapeutic strategy compared with traditional chemotherapy. The first is its high degree of safety. siRNA acts on the posttranslational stage of gene expression. Thus, it does not interact with DNA and avoids the mutation and teratogenicity risks of gene therapy. The second advantage of using siRNA is its high efficacy. In a single cancer cell, siRNA can cause dramatic suppression of gene expression with just several copies. Compared to other small molecule drugs or antibody-based drugs, the greatest advantages of siRNA are the unrestricted choice of targets and specificity determined by the principle of complementary base pairing. This strategy also benefits from rapid devel-

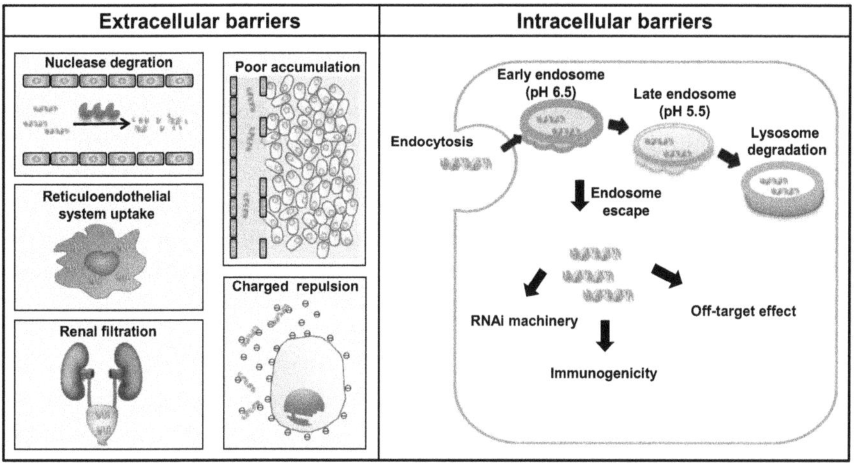

Fig. 16.1 Barriers encountered by ncRNAs following systemic administration

opments in molecular biology and whole-genome sequencing. In addition, comprehensive nucleotide sequence databases have been established, including human genomic databases, cDNA databases, and disease gene databases, which have laid a solid foundation for siRNA drug development. The basic strategy of a siRNA drug is to treat cancer by silencing the specific cancer-promoting gene with a rationally designed siRNA.

The miRNA-based therapeutics could include anti-miRNA antisense oligodeoxyribonucleotide (also known as antagomirs), and miRNA replacement therapy with synthetic miRNA or miRNA mimics [11]. Along with synthetic siRNAs, ncRNA-based therapeutics are usually short nucleotides (~20 nt), and there are multiple challenges for ncRNA-based therapeutics in vivo (see Fig. 16.1), such as off-target effects, delivery barriers, and immunogenicity [12].

MicroRNAs bind and block translation of their target mRNAs having partial complementary sites typically located in the 3′-UTR, which may cause some off-target effects associated with miRNA-based therapeutics [13]. Studies have shown that siRNAs may also silence an unknown number of unintended genes. There are two mechanisms suggested to explain this off-target effect. First, siRNAs can tolerate several mismatches at the mRNA target and retain their ability to silence targets with imperfect complementarity [14]. The second mechanism involves the promiscuous entry of siRNAs into endogenous miRNA machinery [15]. MicroRNAs recognize targets with perfect complementarity to their “seed regions” composed of 2–8 nucleotides. Complementarity of the remaining nucleotides has less importance for recognition. Because siRNAs are very nearly identical to the related class of miRNAs, they can recognize mRNAs with their seed region and lead to degradation of an unpredictable number of mRNAs in a miRNA-like manner.

RNAi is a mechanism involved in the innate immune response to protect cells from invasion by nucleic acids of pathogens such as viruses and bacteria. Several studies have demonstrated that ncRNAs themselves can activate innate immunity by inducing interferon expression, even at low concentrations [16]. Protein kinase R (PKR) and Toll-like receptor (TLR) 3 signaling pathways may be involved in sequence-independent immune activation by siRNAs. Some sequence motifs, such as 5′-UGUGU-3′ [17] or 5′-GUCCUUCAA-3′ [18], secondary structures, and uridine content of the sequence have been identified as important factors for immune activation by these pathways. However, the exact rules of sequence-dependent immune activation are not yet known. Hence, potential therapeutic siRNAs must be tested for an immune response prior to clinical applications.

The systemic delivery and in vivo application of ncRNAs are further hampered by many additional anatomical and physiological defensive barriers presented by the human body, which must be overcome for ncRNAs to reach their sites of action. The ncRNAs are easily filtered from the glomerulus and rapidly excreted from the kidney [19]. Together with rapid excretion kinetics, the susceptibility to degradation by nucleases is a major problem leading to the short half-life (15 min to 1 h) of ncRNAs in plasma, potentially limiting the use of noncoding RNA drugs administrated by intravenous injection [20]. In addition to circulating nucleases and renal clearance, another major barrier to effective in vivo delivery of noncoding RNA drugs is the clearance by the reticuloendothelial system (RES) [21]. The RES is composed of phagocytic cells, including circulating monocytes and tissue macrophages, whose physiological function is to clear the body of foreign pathogens, remove cellular debris that is generated during tissue remodeling, and clear cells that have undergone apoptosis. Phagocytic cells of the RES, particularly the abundant Kupffer cells in the liver and splenic macrophages, also detect and phagocytose noncoding RNAs, as well as nanoparticle carriers that may be used to enhance their delivery. The unfavorable physicochemical properties such as negative charge, large molecule weight, and size complicate passive diffusion of noncoding RNAs through the cell membrane, which makes their cellular uptake to be one of the major hurdles [22]. Once the noncoding RNA drugs are transferred into the tumor cells, the intracellular release which is always associated with endosomal escape is the crucial challenge of efficient gene delivery [23]. The intracellular trafficking of ncRNA delivered by different reagents generally begins in early endosomes. These early endosomes subsequently fuse with sorting endosomes, which in turn transfer their contents into late endosomes. The endosomal compartments of cells are significantly acidic (pH 5.0 ~ 6.2), while the cytosol or intracellular space is neutral. Endosomes are then relocated to lysosomes, which are further acidified (pH ~4.5) and contain various nucleases that promote the degradation of ncRNA. The intracellular stability is another barrier for ncRNA therapeutics.

16.2 Preclinical and Clinical Development of ncRNA-Based Therapeutics

16.2.1 Preclinical Development of ncRNA-Based Therapeutics

To overcome the abovementioned challenges, many efforts have made in preclinical studies of ncRNA-based therapeutics. The rational design enables improvement of efficacy, specificity, and off-target profiles of siRNAs. The backbone length, secondary structures, and nucleotide sequences of siRNAs have effects on these properties, and several rules have been formulated for the rational design of siRNAs: 2-nt overhangs at each 3′-end (typically UU or TT) are important for recognition of siRNAs by the RNAi machinery [24]; the GC content of the sequence determines the thermodynamic stability of siRNAs and should ideally be between 30 and 70 % [25]; the target sequences are generally chosen 75–100 bases downstream of the start codon to avoid nucleotide sequences occupied by regulatory or translational proteins and exon–exon junctions [26]; and inclusion or exclusion of specific nucleotides at particular positions (e.g., A/U at positions 10 and 19, a G/C at position 1) is also considered important for the specificity and efficacy of designed siRNAs [27].

Chemical modifications at the sequence or structural level can help alleviate major obstacles for therapeutic use of siRNAs [28]. A variety of chemical modifications of siRNA have been developed to improve the nuclease stability of siRNAs: modification of the 2′-position of the ribose (such as 2′-*O*-methyl, 2′-OMe) can decrease susceptibility of internucleotide phosphate linkages to nuclease cleavage and increase the stability of the duplex [29]; 2′-fluoro (2′-F) modifications are known to increase nuclease resistance without causing a significant compromise in efficiency [30]; modification with locked nucleic acids (LNAs) is another strategy to increase stability and nuclease resistance [31]; and another alternative strategy to increase stability while retaining potency is the substitution of DNA bases into siRNAs [32]. The replacement of the guide-strand seed region by deoxynucleotides, placing a single 2′-OMe residue at position +2 of the guide strand, selective placement of LNA residues, and modification of the 5′-phosphate group were commonly used chemical modifications to reduce off-target effects [33]. siRNA-induced immune activation can be limited by the replacement of uridines with their 2′-F-, 2′-deoxy-, or 2′-OMe-modified counterparts. The 2′-OMe-modified siRNAs inhibit production of TNF-alpha induced by their unmodified immunostimulatory counterparts even at very low concentrations. To minimize stability issues and reduce off-target effects, these chemical modifications are likely to be transferrable to miRNAs due to their similar structures.

Rational design strategies and chemical modifications have substantially improved some of the problems involved with ncRNA-based therapeutics. However, poor cellular uptake remains an important issue that requires the use of carriers to facilitate ncRNA uptake into the cells. Viral vectors have the advantages in terms of gene transfer efficiency as a result of optimized receptor-mediated internalization, efficient cytosolic release, directed and fast intracellular transport toward target compartments, and immediate disassembly [34]. However, the need for long circulation in the blood and the accumulation in the target site in addition to safety concerns including carcinogenesis, immunogenicity, and broad tissue tropism limit the application of viral vectors [35] and have motivated the exploration of nonviral vectors such as nanocarriers. Nanocarriers are small particles (ranging from 1 to 300 nm) that can carry and deliver drugs, oligonucleotides, peptides, or desired cargos to target tissues. Various nanocarriers have been used for ncRNA delivery in biomedical applications. Based on surface charge, size and hydrophobicity, they have unique tissue biodistribution, toxicity, and tumor cell uptake profiles [36]. The nanomaterials used in the fabrication process, such as natural or synthetic lipids (e.g., liposomes, micelles) and polymers (e.g., chitosan, polylactic-co-glycolic acid, polylactic acid, polyethylenimine), determine the attributes of the resulting carrier [37]. Recently, it has been reported that direct conjugation of small drug molecules, aptamers, lipids, peptides, proteins, or polymers to ncRNA can improve the in vivo pharmacokinetic behavior of ncRNAs. Such ncRNA bioconjugates, either with or without forming nanocomplexes with cationic carriers, can significantly enhance biological half-life with a concomitant increase of delivery efficiency to the target tissue while maintaining sufficient gene-silencing activity [38].

16.2.2 Clinical Development of ncRNA-Based Therapeutics

After validation using in vivo models, siRNA-based therapies were introduced into clinical trials. Since the discovery of RNAi, there have been more than 50 clinical trials involving 26 different siRNAs. Although many of the earlier studies have not reached the clinical stage due to safety concerns and poor efficacy, ncRNA-based therapeutics are still being pursued (see Table 16.1).

TKM-PLK1 (solid lipid-based stable nucleic acid lipid particles, SNALP-carried siRNA), targeting polo-kinase-1, was tested in solid tumors with liver involvement by Tekmira Pharmaceuticals. The drug was well tolerated in phase I of the trial. Currently, two distinct phase II trials are recruiting participants to determine the safety and efficacy in hepatocellular carcinoma or neuroendocrine tumors and adrenocortical carcinoma. Alnylam Pharmaceuticals develops ALN-VSP02, with two distinct siRNAs targeting kinesin spindle protein (KSP) and VEGF, in a partnership with Tekmira for the use of SNALP as carrier. In phase I, ALN-VSP02 was well tolerated, and an anti-VEGF effect was observed in patients with advanced solid tumors with liver involvement. An extension study was then initiated in patients who responded to therapy in phase I, in order to collect long-term safety data. The siRNA siG12D, which targeted mutant KRAS (KRASG12D), was designed by Silenseed Ltd. for pancreatic ductal adenocarcinoma [39]. siG12D was encapsulated in a biodegradable polymer local drug eluter (LODER) for controlled and prolonged delivery. A phase II study to assess the efficacy of siG12D LODER in combination with gemcitabine or FOLFIRINOX chemotherapy was announced in

Table 16.1 ncRNA-based drug in clinical trials

Drug	Target	Delivery system	Disease	Phase	Status	Company	ClinicalTrials.gov identifier
Bevasiranib	VEGF	Naked siRNA	Macular degeneration	I	Completed	Opko Health, Inc.	NCT00306904
				II	Completed		NCT00306904
				III	Terminated		NCT00499590
AGN211745	VEGFR1	Naked siRNA	Age-related macular degeneration, choroidal neovascularization	I/II	Completed	Allergan	NCT00363714
SYL040012	β2-AR	Naked siRNA	Glaucoma, ocular hypertension	I	Completed	Sylentis, S.A.	NCT00990743
				I/II	Completed		NCT01227291
ALN-RSV01	RSV-N gene	Naked siRNA	RSV infections	IIb	Completed	Alnylam Pharmaceuticals	NCT01065935
DCR-MYC	Myc	LNP	Hepatocellular carcinoma	Ib/II	Recruiting	Dicerna Pharmaceuticals	NCT02314052
siRNA-EphA2-DOPC	EphA2	LNP	Advanced cancers	I	Recruiting	MD Anderson Cancer Center	NCT01591356
Atu027	PKN3	Cationic lipoplex	Pancreatic ductal carcinoma	Ib/IIa	Active	Silence Therapeutics GmbH	NCT01808638
PRO-040201	ApoB	SNALP	Hypercholesterolemia	I	Terminated	Tekmira Pharmaceuticals	NCT00927459
TKM-080301	PLK1	SNALP	Solid tumors with liver involvement	I	Completed	National Cancer Institute	NCT01437007
				I/II	Completed		NCT01262235
			Neuroendocrine tumors			Tekmira Pharmaceuticals	
			Adrenocortical carcinoma				
			Hepatocellular carcinoma	IIa	Recruiting		NCT02191878

(continued)

Table 16.1 (continued)

Drug	Target	Delivery system	Disease	Phase	Status	Company	ClinicalTrials.gov identifier
CALAA-01	RRM2	Cyclodextrin	Solid tumor	I	Terminated	Calando Pharmaceuticals	NCT00689065
siG12D LODER	KRAS G12D	LODER polymer	Pancreatic ductal adenocarcinoma, pancreatic cancer	I	Completed	Silenseed Ltd.	NCT01188785
				II	Not yet recruiting		NCT01676259
APN401	E3 ubiquitin ligase Cbl-b	Mononuclear cells	Melanoma, pancreatic cancer, renal cell cancer	I	Recruiting	Wake Forest University	NCT02166255
ALN-TTR02	TTR	LNP	TTR-mediated amyloidosis	III	Recruiting	Alnylam Pharmaceuticals	NCT01960348
ALN-TTRSC	TTR	siRNA–GalNAc	TTR-mediated FAC	III	Recruiting	Alnylam Pharmaceuticals	NCT02319005
ALN-VSP02	KSP and VEGF	LNP	Solid tumors	I	Completed	Alnylam Pharmaceuticals	NCT01158079
MRX34	miR-34a	LNP	Liver cancer	I	Recruiting	Mirna Therapeutics, Inc.	NCT01829971
Miravirsen	miR-122	LNA	HCV	II	Completed	Santaris Pharma A/S	NCT01200420
			HCV with no response to PEGylated interferon alpha	II	Active		NCT01727934

Abbreviation
VEGF vascular endothelial growth factor, *VEGFR 1* vascular endothelial growth factor 1, *β2-AR* β2 adrenergic receptor, *RSV-N* respiratory syncytial virus nucleocapsid gene, *EphA2* ephrin type-A receptor 2, *ApoB* apolipoprotein B, *LNP* lipid nanoparticle, *PLK1* polo-kinase-1, *LODER* local drug eluter, *SNALP* stable nucleic acid lipid particle, *RRM2* M2 subunit of ribonucleotide reductase, *KRASG12D* K-ras G12D mutant, *FAC* familial amyloidotic cardiomyopathy, *KSP* kinesin spindle protein, *TTR* transthyretin, *HCV* hepatitis C virus

patients with unresectable, locally advanced pancreatic cancer. Silence Therapeutics designed AtuPLEX, which was a cationic lipoplex with negatively charged nucleic acids. Atu027, a siRNA targeting protein kinase N3 (PKN3) carried in AtuPLEX, was shown to cause stabilization or regression of disease with no dose-dependent toxicities in patients with advanced solid tumors. A phase Ib/IIa trial is currently being conducted to evaluate the safety and activity of Atu027 in combination with standard gemcitabine treatment in patients with advanced or metastatic pancreatic adenocarcinoma [40]. Dicerna Pharmaceuticals announced two distinct trials in 2014 for DCR-MYC, an LNP carrying siRNA against MYC for hepatocellular carcinoma and solid tumors, multiple myeloma, or non-Hodgkin lymphoma. Lastly, a phase I clinical trial is underway with siRNA-EphA2-DOPC in patients with ovarian cancer (OC) at the MD Anderson Cancer Center.

As to miRNA-based therapeutics, there are two drugs in clinical trials that may shed light on the clinical application. Miravirsen (or SPC3649) is an LNA-modified oligonucleotide designed to inhibit miR-122 developed by Danish firm Santaris Pharma [41]. Miravirsen has gone through two phase I clinical trials, successfully demonstrating that the drug is safe even in humans (NCT00688012, NCT00979927) and one phase IIa clinical trial (NCT01200420). This phase IIa trial enrolled 38 patients with treatment-naïve chronic HCV infection to monitor safety, tolerability, pharmacokinetics, and efficacy on HCV viral titer. Multiple dosage of miravirsen administered subcutaneously to patients gave promising outcomes with a mean reduction of HCV RNA levels by two to three logarithmic levels. Further, almost half of the patients treated by the highest dose displayed undetectable levels of HCV RNA within 4 weeks. As for miRNA replacement therapy, MRX34, a miR-34a mimic compound, will probably be the first miRNA replacement compound to reach clinical stages [42]. miR-34a represents one of the most documented tumor suppression-associated miRNAs, being a transcriptional product of the transcription factor and genome guardian p53. Mirna Therapeutics has developed custom nanoparticle liposomes to increase stability, enhance delivery, and prevent immune response effects, and the upcoming clinical trial in phase I is recruiting patients with non-respectable primary liver cancer or metastatic cancer such as melanoma with liver involvement.

ncRNAs serve as therapeutic drugs for cancer treatment, while effective strategies for short ncRNA delivery into cancer cells in vivo are being extensively explored. The recent strategies to deliver ncRNAs as therapeutic molecules for cancer treatment will be introduced in the following section (see Table 16.2).

16.3 Conjugate Delivery Systems

One strategy for improving the function of ncRNAs in vivo is to make dramatic changes to conjugate the ncRNAs to small molecules or peptides which are designed to increase binding to proteins or cellular uptake. The concept behind ncRNA conjugates is simple. One part of the conjugate is siRNA or miRNA, which provides

Table 16.2 Current strategies for ncRNA in vivo delivery

Delivery system	Target gene	Indications	Route	Ref.
Cholesterol conjugate	ApoB	Hypercholesterolemia	Intravenous injection	[43, 44]
Cholesterol conjugate	Let-7a	Hepatocellular carcinoma	Intravenous injection	[45]
R-tocopherol conjugate	ApoB	Hypercholesterolemia	Intravenous injection	[46]
PTD-DRBD conjugate	EGFR and Akt20	Glioblastoma	Intracerebral injection	[48]
RVG-9R peptide conjugate	FvE	Viral encephalitis	Intravenous injection	[97]
TRA conjugate	Luciferase	Brain	Intravenous injection	[51]
scFvCD7-9R complex	CCR5	HIV	Intravenous injection	[52]
F105-P complex	c-Myc/MDM2/VEGF	Melanoma	Intravenous injection	[53]
F5-P complex	PLK1	Breast cancer	Intravenous injection	[54]
A10 aptamer–siRNA chimera	PLK1	Prostate cancer	Intratumoral injection	[59]
A10 aptamer–siRNA chimera	PLK1	Prostate cancer	Intraperitoneal injection	[60]
GL21.T aptamer–let-7 g chimera	Let-7 g	Lung adenocarcinoma	Intravenous injection	[64]
Neutral liposome	EphA2	Ovarian cancer	Intraperitoneal injection	[69]
Neutral liposome	PAR-1	Melanoma	Intravenous injection	[70]
Cationic liposome	TNFα	Sepsis	Intraperitoneal injection	[71]
Cationic liposome	MCL-1	Non-small cell lung cancer	Intravenous injection	[72]
Cationic liposome	c-Myc/MDM2/VEGF	Melanoma	Intravenous injection	[74]
SNALP	HBsAg	Hepatitis B virus	Intravenous injection	[81]
SNALP	ApoB	Hypercholesterolemia	Intravenous injection	[82]

(continued)

Table 16.2 (continued)

Delivery system	Target gene	Indications	Route	Ref.
SNALP	PLK1/KSP	Hepatocellular carcinoma	Intravenous injection	[83]
CDP	EWS-FLI1	Metastatic Ewing's sarcoma	Intravenous injection	[89]
CC9-PC	miR-34a	Pancreatic cancer model	Intravenous injection	[92]
LMW-PEI	HER-2	Ovarian carcinoma	Intravenous injection	[93]
PEI	miR145	Lung adenocarcinoma	Intravenous injection	[96]
PEI	miR-145	Glioblastomas	Intracranial injection	[97]
RVG–SSPEI	miR-124a	Brain	Intravenous injection	[99]
PLGA	ERK2	Infectious disease	Vaginal instillation	[103]
PEG-PLA	PLK1	Breast cancer	Intravenous injection	[104]
PEG-PLA	GATA2	Non-small cell lung cancer	Intravenous injection	[105]
PAMAM	TAT/REV	HIV	Intravenous injection	[110]
PEG-PAMAM	GFP	GFP-transgenic mouse model	Intramuscular injection	[113]
Amphiphilic PAMAM	Hsp27	Prostate cancer	Intravenous injection	[114]
PPI	EGFR	Glioblastoma	Convection-enhanced delivery	[116]

Abbreviation
ApoB apolipoprotein B, *PTD-DRBD* peptide transduction domains and double-stranded RNA-binding domain, *EGFR* epidermal growth factor receptor, *RVG-9R* rabies virus glycoprotein-conjugated oligo-9-arginine, *TRA* transferrin receptor antibody, *scFvCD7-9R* CD7-specific single-chain antibody-conjugated oligo-9-arginine, *CCR5* C-C chemokine receptor type 5, *F105-P* Fab antibody (F105) fragment directed against HIV-1 envelope fused to protamine, *MDM2* mouse double minute 2 homolog, *VEGF* vascular endothelial growth factor, *F5-P* Fab antibody (F5) fragment directed against Her2 fused to protamine, *A10 aptamer* aptamer against the extracellular domain of the prostate-specific membrane antigens, *GL21.T aptamer* anti-Axl receptor inhibitory aptamer, *PAR-1* protease-activated receptor-1, *TNFα* tumor necrosis factor alpha, *MCL1* myeloid cell leukemia 1 protein, *HBsAg* hepatitis B surface antigen, *SNALP* stable nucleic acid lipid particle, *CDP* cyclodextrin polymer, *CC9-PC* CC9 peptide-conjugated β-cyclodextrin-polyethylenimine, *LMW-PEI* low molecular weight polyethylenimine, *PLGA* poly(lactide-co-glycolide), *PEG-PLA* polyethylene glycol–polylactic acid, *PAMMA* polyamidoamine, *GATA2* GATA-binding protein 2, *GFP* green fluorescent protein, *PPI* poly(propylenimine)

specificity for the target mRNA sequence. The other part of the conjugate is a molecule optimized for improving biodistribution, cellular uptake, or other in vivo properties. A major strength of the approach is that the two portions of the conjugate can be developed as separate modules and then coupled to create hybrid molecules that combine the strengths of the two parts. A weakness is that the synthesis of novel conjugates is complicated by the need to couple a molecule to the ncRNAs, making a large and complex ncRNA even larger and more complex. The ncRNA bioconjugates could be lipophile–ncRNA conjugates, peptide–ncRNA conjugates, antibody–ncRNA conjugates, and aptamer–ncRNA conjugates.

16.3.1 Lipophile Conjugates

Cholesterol was covalently conjugated to siRNA for systemic delivery [43]. It was conjugated to the 3′-terminus of the sense strand of siRNA via a pyrrolidone linkage. The cholesterol-modified siRNAs could silence an endogenous gene encoding apolipoprotein B (ApoB) after intravenous injection in mice. The administration of chemically modified siRNAs resulted in silencing of the ApoB messenger RNA in the liver and jejunum, decreased plasma levels of ApoB protein, and reduced total cholesterol levels [43]. In addition to the chol–siRNA conjugate, a series of lipophilic siRNA conjugates, including siRNA conjugates with bile acids and lipids, were synthesized by Wolfrum et al. [44]. The degree of hydrophobicity, which directly related to the length of the alkyl chain, seemed to be a major determinant for the affinity of siRNA–fatty acid conjugates to lipoproteins. The siRNA conjugates with higher affinity to lipoproteins (i.e., the ones with longer fatty acid chains) showed enhanced gene-silencing capabilities, suggesting that lipoproteins may facilitate the cellular uptake of the conjugates. When systemically administered, chol–siRNA bound to HDL demonstrated fivefold higher cleavage of the target RNA transcript (ApoB) in mice, compared to unbound chol–siRNA at the same concentration. Liu et al. [45] have recently demonstrated antitumor efficacy of cholesterol-conjugated let-7a mimics (Chol-let-7a) in vitro and in vivo and verified for the first time that Chol-let-7a effectively carries let-7a to orthotopic tumors in the liver and successfully inhibits tumor growth in a preclinical model when delivered systemically. The results show that Chol-let-7a downregulates all three human Ras at transcriptional and translational levels and primarily functions in the cytoplasm, thus, suggesting that the use of cholesterol-conjugated miRNAs is a promising tool for HCC systemic therapy. Another lipophile–siRNA conjugate, R-tocopherol (vitamin E)-siRNA, was introduced for systemic siRNA delivery to the liver [46]. The α-tocopherol was covalently bound to the antisense strand of 27/29-mer siRNA at the 5′-end (Toc-siRNA). The 27/29-mer Toc-siRNA was designed to be cleaved by Dicer, producing a mature form of 21/21-mer siRNA after releasing α-tocopherol. Using this new vector, intravenous injection of 2 mg/kg of Toc-siRNA, targeting apolipoprotein B (ApoB), achieved efficient reduction of endogenous ApoB messenger RNA (mRNA) in the liver.

16.3.2 Peptide Conjugates

Cell-penetrating peptides (CPPs) are another conjugate materials used for siRNA transfection efficacy improvement. A well-known CPP is the TAT trans-activator protein from human immunodeficiency virus type 1 (HIV-1). TAT has been conjugated to the 3′-terminus of the antisense strand of a siRNA using a heterobifunctional cross-linker (HBFC), such as sulfosuccinimidyl-4-(p-maleimidophenyl) butyrate [47]. The extent of cellular uptake showed a direct relationship with the amount of conjugate used for the transfection and the time elapsed after transfection. An alternative peptide–siRNA complexation approach utilizes a recombinant fusion of the HIV Tat protein PTD with a double-stranded RNA-binding domain (DRBD) that binds to siRNA and neutralizes its negative charges. The PTD-DRBD peptide vector has shown excellent cellular delivery of siRNA into various primary and transformed cells. PTD-DRBD has been used to package two siRNAs simultaneously (against EGFR and Akt20) to induce tumor-specific apoptosis in a glioblastoma model after intracerebral injection and to also substantially increase mouse survival [48]. Stearylated peptide vectors have also been used successfully to deliver siRNA. A TP10-derived lipopeptide (PF6) was designed to aid endosomal release through the attachment of four pH titratable trifluoromethylquinoline moieties to a lysine side chain of TP10. It was shown to form nanoparticles with siRNA and knockdown HPRT1 mRNA production in a range of cell types as well as in the kidney, lung, and liver of mice upon tail vein infusion at 1 mg/kg [49]. Most of the anti-miR oligonucleotide types utilize 2′-OMe, LNA, or 2′-fluoro analogues usually as mixmers of more than one analogue type or with DNA. In vivo applications all utilize PS linkages. Some naked oligonucleotide analogues may have the ability not only to enter cells through endocytosis but also to efficiently block miRNA activity without the need for any enhancement of transfection by peptides [50].

16.3.3 Antibody Conjugates

The targeted delivery of different therapeutic ncRNA formulations to desired tissues/cells may be a prerequisite for the clinical use of the drugs. Antibody-mediated targeted drug delivery systems have attracted much attention due to their superior stability and high specificity. Xia et al. have delivered siRNAs to the brain in vivo with the combined use of a receptor-specific monoclonal antibody delivery system and avidin–biotin technology. The siRNA was mono-biotinylated on the terminus of the sense strand, in parallel with the production of a conjugate of the targeting MAb and streptavidin. Following the formation of intracranial tumors, the rats were treated with a single intravenous injection of 270 mg/kg of biotinylated siRNA attached to a transferrin receptor antibody via a biotin–streptavidin linker. The intravenous administration of the siRNA caused a 69–81 % decrease in luciferase gene expression in the intracranial brain cancer in vivo [51]. Kumar et al. have used a

CD7-specific single-chain antibody conjugated to oligo-9-arginine peptide (scFvCD7-9R) for T cell-specific siRNA delivery in NOD/SCIDIL2rγ$^{-/-}$ mice reconstituted with human lymphocytes (Hu–PBL) or CD34+ hematopoietic stem cells (Hu–HSC). In HIV-infected Hu–PBL mice, treatment with anti-CCR5 (viral co-receptor) and antiviral siRNAs complexed to scFvCD7-9R controlled viral replication and prevented the disease-associated CD4 T cell loss. This treatment also suppressed endogenous virus and restored CD4 T cell counts in mice reconstituted with HIV+ peripheral blood mononuclear cells. Moreover, scFvCD7-9R could deliver antiviral siRNAs to naive T cells in Hu–HSC mice and effectively suppress viremia in infected mice [52].

The delivery of siRNA by targeting the single-chain variable fragment (scFv) on the cell surface is rapid. scFv can accurately identify intracellular and extracellular antigens and achieve the precise positioning. Song et al. have designed a protamine–antibody fusion protein to deliver siRNA to HIV-infected or envelope-transfected cells. The fusion protein (F105-P) was designed with the protamine-coding sequence linked to the C-terminus of the heavy-chain Fab fragment of an HIV-1 envelope antibody. siRNAs bound to F105-P induced silencing only in cells expressing HIV-1 envelope. Additionally, siRNAs targeted against the HIV-1 capsid gene *gag* inhibited HIV replication in hard-to-transfect, HIV-infected primary T cells. Intratumoral or intravenous injection of F105-P-complexed siRNAs into mice targeted HIV envelope-expressing B16 melanoma cells, instead of normal tissue or envelope-negative B16 cells. Injection of F105-P with siRNAs targeting c-myc, MDM2, and VEGF inhibited envelope-expressing subcutaneous B16 tumors. Furthermore, an ErbB2 single-chain antibody fused with protamine delivered siRNAs specifically into ErbB2-expressing cancer cells [53]. Yao et al. have used a protamine peptide fused to a scFv that binds ERBB2 (F5-P) to specifically deliver a siRNA targeting PLK1 into ERBB2 (also known as HER2)-expressing breast cancer cells. F5-P-mediated delivery of PLK1 siRNAs effectively reduced PLK1 expression and proliferation and increased apoptosis of ERBB2+ breast cancer cell lines and primary breast cancer cells in vitro. F5-P was also capable of delivering PLK1 siRNAs to ERBB2^{+} cell lines or primary breast tumor cells grown as xenografts in nude mice. In these models, tail vein injection of PLK1 siRNAs in a complex with F5-P twice a week for 4 weeks significantly slowed tumor growth (followed for 7 weeks). Meanwhile, ERBB2^{-} tumors were insensitive to this treatment [54].

16.3.4 Aptamer Conjugates

Aptamers are short, structured, single-stranded RNA or DNA ligands that bind to target molecules with high specificity and affinity. Since their discovery in the 1980s, aptamers have been generated that target the extracellular domain of transmembrane receptors overexpressed in tumors, thus becoming (along with monoclonal antibodies) ideal tools for the specific recognition of cancer cell surfaces. Aptamers are generated from high-complexity pools through a combinatorial process named systematic evolution of ligands by exponential enrichment (SELEX) to tightly bind to their proper targets [55]. The use of aptamers offers the possibility to overcome insertional mutagenesis and immunogenicity of viral vectors [56] and possible limited effectiveness and toxicity of nanoparticles, enabling the specific accumulation of ncRNAs in target tumor cells in a safe and effective manner.

Aptamers against the extracellular domain of the prostate-specific membrane antigens (PSMAs), A9 and A10 aptamers [57], have been extensively characterized for siRNA delivery by developing different approaches based either on noncovalent or covalent conjugation. Chu et al. [58] have developed a multivalent RNA aptamer–siRNA chimeric structure in which two biotinylated anti-PSMA aptamers (A9) are linked to two biotinylated anti-lamin A/C siRNAs using streptavidin as a connector. To enhance siRNA release after internalization, a reducible disulfide linker was designed between the sense strand of the siRNA and the biotin group. By using such a streptavidin connector, this RNA aptamer–streptavidin–siRNA conjugate was efficiently internalized by the PSMA-positive LNCaP cells and mediated a rapid inhibition of gene expression. McNamara et al. [59] have described the first-generation aptamer–siRNA chimera. An aptamer that specifically bound to PSMA was covalently linked to the passenger strands of siRNAs, followed by annealing of the guide strands of the siRNAs to the passenger strands to create a functional siRNA duplex. The chimeras mediated targeted silencing in prostate cancer cells expressing PSMA and efficiently promoted cell death. When the chimera was injected intratumorally, the tumor volume in a xenograft mouse model of prostate cancer was decreased. Subsequently, Dassie et al. [60] have optimized the aptamer–siRNA chimera for systemic administration, leading to second-generation chimeras. They reduced the aptamer portion of the chimera, designed a 2-nt (UU) overhang at the 3′-end of the siRNA duplex and swapped the passenger and guide strands of the siRNA. They also appended a PEG (MW = 20 kDa) onto the siRNA passenger. As a result of these modifications, the optimized second-generation, aptamer–siRNA chimeras displayed a clear regression of PSMA-expressing tumors in nude mice following intraperitoneal administration. Several other aptamers against cell surface proteins overexpressed on cancer cells have been used for siRNA delivery. For example, aptamers specific for Her-2-positive breast cancer cells were covalently conjugated to BCL-2 siRNA, generating a chimera able to sensitize cells to chemotherapy [61].

Given the progress in the design of aptamer-based strategies for siRNA delivery, the use of aptamers as delivery moieties for microRNAs has recently been explored. A second-generation aptamer against PSMA (A10-3.2) was conjugated to a polyamidoamine (PAMAM)-based microRNA (miR-15a and miR-16-1) using PEG as a spacer. The construct demonstrated selective delivery of the miRNA moiety into LNCaP (PSMA-positive) prostate cancer cells, inducing cell death in vitro [62]. Hao et al. used the same aptamer as recognition ligand in an atelocollagen (ATE)-based microRNA (miRNA, miR-15a and miR-16-1) vector to target prostate cancer bone metastasis. The anticancer efficacy of miRNA/ATE-APT was superior to that of other treatments in a human PCa bone metastasis mice model [63]. Esposito et al. have recently combined the anti-Axl receptor inhibitory aptamer named GL21.T with the tumor-suppressor let-7g miRNA. The conjugate combined the miRNA activity with the aptamer function (Axl signaling inhibition), resulting in an effective inhibition of cell migration and survival in vitro and of tumor growth in vivo [64]. The selective delivery of anti-miRs to target cancer cells is still in its infancy; nevertheless, the development of aptamer-mediated approaches represents a concrete possibility to achieve this goal.

16.4 Lipid-Based Delivery Systems

Liposomes are one of the most commonly used transfection reagents in vitro. Usually, liposomes are formed in an aqueous environment, in which a lipid bilayer forms a sphere with an aqueous core. For example, one set of polar head groups can create the outer surface of the nanocomplex, while another set of polar head groups faces the interior hydrophilic core, which houses the nucleic acid payload [12]. Liposomes can be created using single or multiple types of synthetic or natural lipids, which allows for additional flexibility when optimizing the physical and chemical properties of the nanoparticle [65]. Almost 50 years after the discovery of liposomes, the US FDA has approved 13 liposome-based products for human use, and a large number of liposomal products are in different phases of clinical trials [66].

Liposomes offer several advantages as a ncRNA delivery system due to their ability to (1) prevent degradation of the payload, (2) accumulate preferentially in tumor tissues (passive targeting/delivery) and deliver high concentrations of the payload, (3) specifically target to tumor cells and the microenvironment with high-affinity ligands (active targeting), and (4) provide safe and effective systemic delivery platforms in animals and humans depending on the lipid content [67].

However, safe and efficacious delivery in vivo is rarely achieved due to toxicity, nonspecific uptake, and unwanted immune response [68]. Much of the nonspecific response and toxicity is directly linked to the positive charge on the surface of the particles necessary for the binding of oligonucleotides. In recent years, a significant effort has been dedicated to modifying the composition and chemical structure of liposomes for pharmaceutical drug delivery. For robust and successful ncRNA delivery with lipid-based systems, optimization of lipid composition, drug-to-lipid ratio, particle size, charge, surface-targeting moieties, payload encapsulation efficiency, and the manufacturing process are required.

16.4.1 Cationic Liposomes

Cationic liposomes have been traditionally the most commonly used nonviral delivery systems for oligonucleotides, including plasmid DNA, antisense oligos, and ncRNAs. Cationic lipids, such as 1,2-dioleoyl-3-trimethylammonium-propane

(DOTAP) and *N*-[1-(2,3-dioleoyloxy)propyl]-*N,N,N*-trimethyl-ammonium methyl sulfate (DOTMA), can form complexes or lipoplexes with negatively charged siRNA or miRNA to form nanoparticles by electrostatic interaction, providing high in vitro transfection efficiency [69, 70]. Sorensen et al. have used cationic DOTAP liposomes to deliver siTNF-α, and the lethal reaction to LPS injection in a mouse model of sepsis is suppressed [71]. To maintain an overall positive surface charge for adsorption through the cell membrane and to reduce the possible clearance caused by positive charge, the N/P (nitrogen-to-phosphate) ratio usually ranges from 2 to 3. Pre-miR-133b contains DOTMA–cholesterol. In the previous study, TPGS lipoplexes were prepared by adding pre-miR-133b to the empty liposomes. The in vitro transfection efficiency and in vivo biodistribution of lipoplex formulations were compared with siPORT NeoFX transfection agent. In vitro, the lipoplexes transfected pre-miR-133b more efficiently than siPORT NeoFX, a commercially available lipid-based agent, in A549 non-small cell lung cancer cells. The mature miR-133b level in lungs following *i.v.* administration of pre-miR-133b-containing lipoplexes was approximately 52-fold higher than that in untreated mice [72]. Polycationic liposome–hyaluronic acid (LPH) nanoparticles have also been described by several investigators [73]. A tumor-targeting GC4 single-chain antibody fragment-modified LPH (scFv-LPH) nanoparticles systemically co-delivered siRNA and miR-34a into experimental lung metastasis of murine B16F10 melanoma. The scFv-LPH nanoparticles encapsulating combined siRNAs against c-Myc, MDM2, and VEGF and miR-34a decreased the metastasis tumor growth to approximately 20 % of the untreated control. When treated with scFv-LPH nanoparticles containing only combined siRNAs or miR-34a, the reduction was approximately 30 and 50 %, respectively, of the untreated control, suggesting that the effects were mediated through different mechanisms. The advantage of such a system lies in the potential to deliver siRNA and/or miRNA together to simultaneously target several different oncogenic pathways [74].

Cationic liposomes, while efficiently taking up and condensing ncRNAs, have had limited success for in vivo gene downregulation, perhaps because of their stable intracellular nature and resultant failure to release siRNA or miRNA contents [75]. In addition, toxicity of cationic lipids is the major issue following systemic administration preventing them from being a major candidate for ncRNA delivery. The use of cationic liposomes in in vivo mouse models elicits dose-dependent toxicity and pulmonary inflammation, hepatotoxicity, and a systemic interferon type I response, which is attributed in part to the activation of TRL4 [68]. Cationic lipids also activate the complement system and cause their rapid clearance by macrophages of the RES. It has been demonstrated that cationic lipids are highly toxic to macrophages and other immune cells (ED50 < 50 nm/L) [76]. Different lengths of hydrocarbon chains can also influence the cytotoxicity of cationic lipids [77]. Toxicity of cationic lipids is linked to induction of reactive oxygen species (ROS) and increased intracellular calcium levels. In addition, DOTAP-based particles accumulate near the vasculature and are preferentially taken up by the liver and spleen, limiting their effectiveness in systemic or antitumor therapy [69]. Overall, although cationic lipid-based delivery systems offer some advantages as an ncRNA delivery system, potential toxicities need to be addressed before their translation in clinical trials. Careful selection of lipids and formulation strategies may help reduce the potential toxicities.

16.4.2 Neutral Liposomes

Because the surface charge of all biological membranes is negative, electronegative or neutral liposomes are more biocompatible than cationic liposomes and have superior pharmacokinetics in general. DOPC (1,2-dioleoylsn-glycero-3-phosphatidylcholine) is a kind of neutral lipid which has been used to improve the siRNA entrapment efficiency. In 2005, Landen et al. developed the oncoprotein EphA2 targeting DOPC-encapsulated siRNA liposomes, which was highly effective in reducing EphA2 expression 48 h after the administration of a single dose in an orthotopic model of ovarian carcinoma [69]. Currently, the EphA2 targeting DOPC-encapsulated siRNA liposome (siRNA-EphA2-DOPC) is in a phase I clinical trial initiated by the MD Anderson Cancer Center. As for miRNA, miR-34a and let-7 were delivered with a type of neutral liposome to treat non-small cell lung cancer (NSCLC). The treatment with miR-34a or let-7 significantly decreased the lung tumor burden to approximately 40 % of the mice treated with miRNA controls, and the expression level of miR-34a and let-7 in lungs was also significantly higher than groups treated with miRNA mimic controls [78]. These findings demonstrate the potential of developing ncRNA therapy formulations with neutral liposomes as novel therapies for lung cancer patients.

Coating liposomes with lipid-anchored PEG can reduce particle size, prevent aggregation during storage, increase circulatory half-life, and reduce uptake by the reticuloendothelial system (RES), such as red blood cells and macrophages [79]. However, using PEG is not always advantageous because the steric effect and charge effect of PEG block the interaction between the liposome and the endosomal membrane and prevent the liposome from escaping the endosome. Many studies have been performed to improve the efficacy of PEGylated nanoparticles, including rationally designed PEG length and density or incorporation of pH-sensitive bonds linking PEG to the liposome. How to achieve the best outcome with modulation of PEG length and density remains controversial. However, pH-sensitive modified PEG with ionic interactions, such as the HEMA–histidine–methacrylic acid-modified PEG liposome, has been shown to be effective. At neutral pH, the PEG copolymer has a net negative charge, whereas the liposomal core consisting of DOPE and cholesterol has a net positive charge. In the endosome, imidazole and methacrylic acid residues become protonated, and the net charge of the PEG becomes positive, which results in PEG release and positively charged liposomal membrane exposure, after which the liposome can fuse with the endosome and escape successfully [80].

16.4.3 Stable Nucleic Acid Lipid Particles (SNALPs)

To date, 12 clinically tested siRNA-based therapeutics have been administered by the *i.v.* route. All but one of these siRNAs has been carried by synthetic carriers, mostly SNALPs, which are a type of lipid nanoparticle that encapsulates siRNAs and delivers them to their target cells. SNALPs are microscopic particles approximately 120 nm in diameter. They have been used to deliver siRNAs therapeutically to mammals in vivo. In SNALPs, the siRNA is surrounded by a lipid bilayer containing a mixture of cationic and fusogenic lipids, coated with diffusible polyethylene glycol [81]. With enhanced permeability and retention due to prolonged circulation time in the blood, SNALPs are highly bioavailable, which leads to the accumulation of SNALPs at the sites of vascular leakage, especially at cancer growth sites. After accumulation, SNALPs are easily endocytosed by cancer cells and deliver the siRNAs into cells successfully. SNALPs have been used for the treatment of many diseases, including hepatitis B viral infection, dyslipidemia, and Ebola (Zaire) [12, 82]. Judge et al. described the preclinical development of chemically modified siRNAs targeting the essential cell-cycle proteins polo-like kinase 1 (PLK1) and kinesin spindle protein (KSP) in mice. The siRNAs formulated in stable nucleic acid lipid particles (SNALPs) displayed potent antitumor efficacy in both hepatic and subcutaneous tumor models. This was correlated with target gene silencing following a single intravenous administration that was sufficient to cause extensive mitotic disruption and tumor cell apoptosis. Their siRNA formulations induced no measurable immune response, minimizing the potential for nonspecific effects. Additionally, RNAi-specific mRNA cleavage products were found in tumor cells, and their presence correlated with the duration of target mRNA silencing. Histological biomarkers confirmed that RNAi-mediated gene silencing effectively inhibited the target's biological activity [83].

Another lipid-like delivery system is lipoid nanoparticles, which are comprised of cholesterol and PEG-modified lipids specific for siRNA delivery [84]. To improve SNALP-mediated delivery, a new class of lipid-like delivery molecules was described, termed lipidoids, as delivery agents for RNAi therapeutics [85]. Chemical methods were developed to allow the rapid synthesis of a large library of over 1200 structurally diverse lipidoids. From this library, they identified lipidoids that facilitated high levels of specific silencing of endogenous gene transcripts when formulated with either double-stranded small interfering RNA (siRNA) or single-stranded antisense 2-O-methyl (2′-OMe) oligoribonucleotides targeting microRNA (miRNA). The safety and efficacy of lipidoids were evaluated in three animal models: mice, rats, and nonhuman primates. One of the most potential lipidoid drugs was the lipidoid-based siRNA formulation $98N_{12}$-5, which led to a 75–90 % reduction in ApoB or FVII factor expression in hepatocytes in nonhuman primates and mice. In addition, mice injected intraperitoneally with thioglycollate as a sterile inflammation stimulus followed by injection of $98N_{12}$-5-formulated siCD45 showed a 65 % reduction of CD45 protein expression in the peritoneal macrophage population. In the end, the potential of $98N_{12}$-5 to facilitate the delivery of anti-miRs was tested. The results demonstrated that $98N_{12}$-5-formulated anti-miR122 dosed at 5 mg/kg on three consecutive days in mice resulted in greater miR-122 repression than the cholesterol-conjugated version of the same oligoribonucleotide (antagomir122) dosed at 80 mg/kg on three consecutive days.

16.5 Polymer-Based Delivery Systems

Polymer-mediated delivery systems, usually called polymeric nanoparticles, are solid, biodegradable, colloidal systems which have been widely studied as drug vesicles. According to the material used, polymeric nanoparticles are classified into two major categories, natural polymers and synthetic polymers. Natural polymers for siRNA delivery include cyclodextrin and chitosan [86]. Of the synthetic polymers, polyethylenimine (PEI), poly(lactide-co-glycolide) (PLGA), and dendrimers have been intensively investigated [87].

16.5.1 Cyclodextrin

Cyclodextrins are natural polymers generated during the bacterial digestion of cellulose, and they possess defined geometric (~70 nm) and cationic structural characteristics that offer advantages for cationic siRNA and miRNA payloads to form inclusion complexes. Additionally, each cyclodextrin molecule may contain covalently bound polyethylene glycol (PEG), which acts to stabilize the nanoparticle and avoid nonspecific interaction with blood and extracellular elements under physiological conditions [88]. It was first introduced for the delivery of plasmid DNA in 1999 and later optimized for siRNA delivery. Less than a decade later, cyclodextrin polymer (CDP)-based nanoparticles were moved into clinical trials for siRNA delivery. The cyclodextrin-containing polycation system was developed for the targeted delivery of siRNA [89]. This system consists of a cyclodextrin-containing polymer, PEG for stability, and human transferrin as the targeting ligand for binding to transferrin receptors, which are often overexpressed on cancer cells. This targeted nanoparticle system, called CALLA-01, which targets the M2 subunit of ribonucleotide reductase (R2) to inhibit tumor growth was developed for the first siRNA phase I trial by Calando Pharmaceuticals (Pasadena, CA, USA) [90].

The β-cyclodextrin-PEI (PEI-CD) carrier was developed for delivery of the tumor-suppressor miR-34a mimic to pancreatic cancer cells. The PEI-CD nanoparticles were conjugated with CC9, a specific tumor-homing and tumor-penetrating bifunctional peptide via its CRGDK motif, which binds to neuropilin-1 (NRP-1) [91]. This delivery system could greatly upregulate the miR-34a level in the PANC-1 cell line and substantially inhibit the target gene expressions such as E2F3, Bcl-2, c-myc, and cyclin D1, inducing cell cycle arrest and apoptosis and suppressing migration. More importantly, the in vivo evaluation of the antitumor activity indicated that the delivery of miR-34a significantly inhibited tumor growth and induced cancer cell apoptosis [92].

16.5.2 Polyethylenimine (PEI)

PEI, a commonly used cationic polymeric drug carrier with high transfection efficiency, has been widely investigated for siRNA and miRNA delivery. PEI's high charge density enables the formation of small and compact structures with nucleic acid delivery, facilitating endosomal escape via the proton sponge effect. The PEI–siRNA/miRNA complexes protect siRNA/miRNA from nuclease degradation, resulting in prolonged half-life. In addition, complete encapsulation of siRNA/miRNA prevents off-target effects such as immune activation by a Toll-like receptor-dependent mechanism. Polyethylenimine (PEI) has been used successfully for nucleic acid delivery under both in vitro and in vivo conditions. Urban–Klein et al. have showed that the noncovalent complexation of synthetic siRNAs with low molecular weight PEI efficiently stabilizes siRNAs and delivers siRNAs into cells where they display full bioactivity at completely nontoxic concentrations. More importantly, in a subcutaneous mouse tumor model, the systemic (intraperitoneal, *i.p.*) administration of complexed, but not of naked, siRNAs leads to the delivery of intact siRNAs into the tumors [93]. High molecular weight PEIs provide high transfection efficiency but also have high toxicity, while low molecular weight PEIs are more biocompatible and are much less efficient. A type of micelle-like nanoparticle (MNP) has been reported that is based on the combination of a covalent conjugate between a phospholipid and low molecular weight PEI (1.8 kDa) with PEG-stabilized liposomes as the outer layers [94]. MNPs have been shown to have the capacity for siRNA delivery and gene silencing with improved biocompatibility properties. The MNP delivery system was further utilized in silencing P-gp to overcome doxorubicin resistance in MCF-7 human breast cancer cells. The presence of P-gp on the surface of resistant cells decreased after treating cells with MNP-loaded siRNAs targeting MDR-1, which effectively inhibited the drug efflux activity [94]. PEI has also been used to construct ligand-targeted, sterically stabilized nanoparticles for systemic siRNA delivery. The PEGylated nanoparticles were conjugated with an Arg–Gly–Asp peptide ligand attached at the distal end of PEG to target integrin-expressing tumor neovasculature. The resulting nanoparticles, upon intravenous administration to tumor-bearing mice, successfully delivered siRNAs in a tumor-selective manner, inhibited vascular endothelial growth factor (VEGF) receptor-2 expression, inhibited tumor angiogenesis, and slowed tumor growth [95].

Polyethylenimine has also been utilized for the delivery of miRNAs. Using a polyurethane–short-branch polyethylenimine (PU–PEI) as a carrier, miR-145 was delivered to treat cancer stem cell (CSC)-derived lung adenocarcinoma (LAC). The LAC–CSC xenograft tumors did not respond to the combination of ionizing radiation (IR) and cisplatin during the 30-day experimental course. However, PU–PEI-bound miR-145 delivery moderately reduced tumor growth. Most importantly, the miR-145 delivery combined with IR and cisplatin led to significant tumor growth inhibition [96]. When administered to orthotopic CSC-derived glioblastoma tumors, intracranially delivered PU–PEI-miR-145 significantly suppressed tumorigenesis. When used in combination with radiotherapy and temozolomide, synergistic effects and improved survival rates were achieved [97]. The significant inhibitory effect of PU–PEI-miR-145 on lung adenocarcinoma and glioblastoma CSC-induced tumors demonstrated the potential of miRNA therapy in overcoming tumor chemoradioresistance, preventing cancer relapse and achieving cancer eradication. Beyond traditional delivery approaches, PEI-based systems have been modified for transport across the blood–brain barrier (BBB). The BBB is the most significant physiologic obstruction of systemic drug or gene delivery to the brain parenchyma and central nervous system (CNS) [98]. Using a short peptide derived from rabies virus glycoprotein (RVG), the PEI–RVG bound specifically to nicotinic acetylcholine receptors on neuronal cells. RVG was coupled to PEI *via* disulfide bonds (RVG–SSPEI) to deliver miR-124a, a neuron-specific miRNA that could potentially promote neurogenesis [99, 100]. To overcome the size limitation of PEI vector transport across the BBB, mannitol was used to permeabilize the BBB. After administration, a much higher accumulation of miR-124a in the brain was observed in the RVG-mediated SSPEI delivery group compared to that in the miR-124a/SSPEI group as determined by tracking the Cy5.5-labeled miR-124a. However, the functional activities of miR-124a in promoting neurogenesis were not tested. The modification of PEI using RVG decreased the toxicity associated with PEI and achieved remarkable targeted delivery to neuronal cells. The RVG–SSPEI could be a useful system to deliver miRNA therapeutics for the treatment of brain diseases. Although this system did show greater accumulation in the brain, the use of permeabilizing agents limits the utility. The combination of delivery strategies that improve the activity of the miRNA has great potential. However, the complexity of the systems at times can counterbalance the improvements [92].

16.5.3 Poly(lactide-co-glycolide) (PLGA)

Poly(D,L-lactide) (PLA) and poly(lactide-co-glycolide) (PLGA) have also demonstrated the potential for sustained nucleic acid delivery. The advantages of PLGA- or PLA-based siRNA delivery include high stability, facile cellular uptake by endocytosis, ability to target specific tissues or organs by adsorption or ligand binding, biodegradability, low toxicity, sustained release characteristics, and multiple surface modifications [101, 102]. In 2009, Saltzman and coworkers reported that PLGA nanoparticles could be densely loaded with siRNA in the presence of spermidine and, when applied topically to the vaginal mucosa, led to efficient and sustained gene silencing [103]. Yang et al. reported a cationic lipid-assisted polymeric nanoparticle system with a stealthy property for efficient siRNA encapsulation and delivery, which was fabricated with poly(ethylene glycol)-b-poly(D,L-lactide), siRNA, and a cationic lipid, using a double emulsion-solvent evaporation technique.

By incorporation of the cationic lipid, the encapsulation efficiency of siRNA into the nanoparticles was greater than 90 %. Systemic delivery of specific siRNA by nanoparticles significantly inhibited luciferase expression in an orthotopic murine liver cancer model and suppressed tumor growth in an MDA-MB-435s murine xenograft model, suggesting its therapeutic promise in disease treatment [104]. Using the same cationic lipid-assisted polymeric nanoparticle system, GATA2 siRNA was delivered to non-small cell lung cancer (NSCLC) harboring oncogenic KRAS mutations and successfully inhibited tumor growth in a mouse model [105].

Using a miR-155 Cre-lox*P* tetracycline-controlled knockin mouse model, pre-B-cell tumors were dependent on high miR-155 expression where withdrawal of miR-155 using doxycycline caused rapid tumor regression. Systemic delivery of anti-miR-155 peptide nucleic acids (PNAs) using PLGA polymeric nanoparticles exhibited enhanced delivery efficiency and achieved therapeutic effects. The surface of the nanoparticles was modified with penetratin, a cell-penetrating peptide [6]. The pre-B-cell tumors had an approximately 50 % decrease in growth relative to control-treated tumors after systemic delivery of 1.5 mg/kg anti-miR-155 PNAs loaded in ANTP–NP for 5 days, which was approximately 25-fold less than the naked anti-miR dosage needed. There was a need, in this case, to protect the PLGA particle using steric stabilization (i.e., PEGylation) and also add a cell penetration enhancer. PLGA particles were typically nonspecifically cleared, and the PEGylation diminished the ability of the particles to enter cells. This type of particle is readily adaptable, but still does not have significantly more than 5 % accumulation in the diseased organ due to passive accumulation. Important to the future development of miRNAs is the assertion by the authors that RNA degradation occurs in or around the endosomal and lysosomal compartments [106, 107]. If this proves true, much of the design criteria for miRNAs will be altered. The fact that Ago2 resides in the membrane of the endolysosomal compartment can explain the better performance of NP170-PFCE formulation [108]. The future development of miRNA (and possibly siRNA) will rely on the rational design of systems that take advantage of the complex biology of the disease and molecule being delivered.

16.5.4 Dendrimers

Dendrimers are synthetic, highly branched monodisperse and usually highly symmetric, spherical macromolecules with three-dimensional nanometric structures. The unique structural features such as tunable structure and molecular size, large number of accessible terminal functional groups, and ability to encapsulate cargos add to their potential as drug carriers [109]. Polycationic dendrimers such as poly(amidoamine) (PAMAM) and poly(propylenimine) (PPI) dendrimers have been studied for siRNA delivery in recent years. PAMAM dendrimers have become the most used dendrimer-based carriers for gene delivery because of the ease of synthesis and commercial availability. Rossi JJ's group has reported on the generation 5 (G5) dendrimer for functional delivery of siRNAs that inhibit HIV infection and replication by targeting HIV genes *tat* and *rev* and host dependency factors CD4 and transportin-3 (TNPO3). The G5 dendrimer–siRNA complexes demonstrated effective inhibition of HIV-1 replication in T lymphocytes in vitro and in a humanized mouse model [110]. However, PAMAMs were demonstrated to be cytotoxic, predominately related to apoptosis mediated by mitochondrial dysfunction [111]. Cytotoxicity could be reduced by various modifications without compromising gene silencing. Surface-modified and cationic PAMAM dendrimers showed very low cytotoxicity, even at high concentrations and efficiently penetrated cancer cells in vitro and in vivo [112–114]. PPI dendrimers were also used to formulate siRNA nanoparticles, and these nanoparticles showed efficient gene silencing [115]. Dendrimer-conjugated magnetofluorescent nanoworms (dendriworms) were developed to achieve siRNA delivery in a transgenic murine model of glioblastoma [116]. Dendriworms were well tolerated after 7 days of convection-enhanced delivery to the mouse brain, and in an EGFR-driven transgenic model of glioblastoma, anti-EGFR dendriworms led to specific and significant suppression of EGFR expression. For targeted delivery, dendrimers could be easily conjugated with one or multiple targeting ligands. For example, the 9-mer luteinizing hormone-releasing hormone (LHRH) peptide was conjugated to PAMAM dendrimers, whose internal amino group was quaternalized for siRNA loading [117]. The cellular uptake was observed to be dependent on the targeting peptide. Similarly, a 53-mer epidermal growth factor (EGF) peptide was grafted with generation 4 (G4) PAMAM dendrimers for siRNA delivery [118]. Few applications of miRNA delivery with dendrimers have been reported. Co-delivery of anti-miR-21 and 5-fluorouracil (5-FU) to U251 glioblastoma cells using poly(amidoamine) (PAMAM) dendrimer increased apoptosis of U251 cells markedly. Migration of tumor cells was decreased compared with cells that were only treated with 5-FU [119]. Although not in vivo, this suggests that dendrimers may be amenable for in vivo miRNA delivery. Dendrimers are capable of binding miRNAs and aiding in the entry into cells. However, the entry is nonspecific in nature. This alternate mechanism of cell entry should not be overlooked due to the potential to protect miRNAs by avoiding the endosomal and lysosomal compartments [120].

16.6 Summary

In this chapter, the therapeutic potentials and delivery strategies of noncoding RNAs have been introduced. The main challenges facing ncRNA-based cancer therapeutics, including off-target effects, immunogenicity, and the most difficult delivery barriers, have been described in detail. Rational design, chemical modification, and nanoparticle delivery carriers offer significant opportunities to overcome these challenges. Multiple approaches to the delivery of ncRNAs as therapeutic molecules for cancer treatment and ncRNA-based clinical trials are systemically summarized. Improvements in delivery strategies hold great potential to make the translational process of ncRNA-based drugs faster and more effective for cancer clinical applications.

References

1. Cheetham SW, Gruhl F, Mattick JS, Dinger ME. Long non-coding RNAs and the genetics of cancer. Br J Cancer. 2013;108(12):2419–25.
2. Hayes J, Peruzzi PP, Lawler S. MicroRNAs in cancer: biomarkers, functions and therapy. Trends Mol Med. 2014;20(8):460–9.
3. Calin GA, Dumitru CD, Shimizu M, et al. Frequent deletions and down-regulation of micro-RNA genes miR15 and miR16 at 13q14 in chronic lymphocytic leukemia. Proc Natl Acad Sci U S A. 2002;99(24):15524–9.
4. Hanahan D, Weinberg RA. Hallmarks of cancer: the next generation. Cell. 2011;144(5):646–74.
5. Faraoni I, Antonetti FR, Cardone J, Bonmassar E. miR-155 gene: a typical multifunctional microRNA. Biochim Biophys Acta Mol Basis Dis. 2009;1792(6):497–505.
6. Babar IA, Cheng CJ, Booth CJ, Liang XP, Weidhaas JB, Saltzman WM, Slack FJ. Nanoparticle-based therapy in an in vivo microRNA-155 (miR-155)-dependent mouse model of lymphoma. Proc Natl Acad Sci U S A. 2012;109(26):E1695–704.
7. Volinia S, Calin GA, Liu CG, Ambs S, Cimmino A, Petrocca F, Visone R, Iorio M, Roldo C, Ferracin M, Prueitt RL, Yanaihara N, Lanza G, Scarpa A, Vecchione A, Negrini M, Harris CC, Croce CM. A microRNA expression signature of human solid tumors defines cancer gene targets. Proc Natl Acad Sci U S A. 2006;103(7):2257–61.
8. Hatley ME, Patrick DM, Garcia MR, et al. Modulation of K-Ras-dependent lung tumorigenesis by microRNA-21. Cancer Cell. 2010;18(3):282–93.
9. Bader AG. MiR-34 – a microRNA replacement therapy is headed to the clinic. Front Genet. 2012;3:120.
10. Fire A, Xu SQ, Montgomery MK, et al. Potent and specific genetic interference by double-stranded RNA in Caenorhabditis elegans. Nature. 1998;391(6669):806–11.
11. Zhang Y, Wang ZJ, Gemeinhart RA. Progress in microRNA delivery. J Control Release. 2013;172(3):962–74.
12. Whitehead KA, Langer R, Anderson DG. Knocking down barriers: advances in siRNA delivery. Nat Rev Drug Discov. 2009;8(2):129–38.
13. Hammond SM, Caudy AA, Hannon GJ. Post-transcriptional gene silencing by double-stranded RNA. Nat Rev Genet. 2001;2(2):110–19.
14. Snove O, Holen T. Many commonly used siRNAs risk off-target activity. Biochem Bioph Res Commun. 2004;319(1):256–63.
15. Doench JG, Petersen CP, Sharp PA. siRNAs can function as miRNAs. Genes Dev. 2003;17(4):438–42.
16. Judge A, Maclachlan I. Overcoming the innate immune response to small interfering RNA. Hum Gene Ther. 2008;19(2):111–24.
17. Judge AD, Sood V, Shaw JR, Fang D, McClintock K, MacLachlan I. Sequence-dependent stimulation of the mammalian innate immune response by synthetic siRNA. Nat Biotechnol. 2005;23(4):457–62.
18. Hornung V, Guenthner-Biller M, Bourquin C, Ablasser A, Schlee M, Uematsu S, Noronha A, Manoharan M, Akira S, de Fougerolles A, Endres S, Hartmann G. Sequence-specific potent induction of IFN-alpha by short interfering RNA in plasmacytoid dendritic cells through TLR7. Nat Med. 2005;11(3):263–70.
19. Juliano R, Alam MR, Dixit V, Kang H. Mechanisms and strategies for effective delivery of antisense and siRNA oligonucleotides. Nucleic Acids Res. 2008;36(12):4158–71.
20. Bartlett DW, Davis ME. Effect of siRNA nuclease stability on the in vitro and in vivo kinetics of siRNA-mediated gene silencing. Biotechnol Bioeng. 2007;97(4):909–21.
21. Akerman ME, Chan WCW, Laakkonen P, et al. Nanocrystal targeting in vivo. Proc Natl Acad Sci U S A. 2002;99(20):12617–21.
22. Zuhorn IS, Kalicharan R, Hoekstra D. Lipoplex-mediated transfection of mammalian cells occurs through the cholesterol-dependent clathrin-mediated pathway of endocytosis. J Biol Chem. 2002;277(20):18021–8.

23. Pezzoli D, Candiani G. Non-viral gene delivery strategies for gene therapy: a “menage a trois” among nucleic acids, materials, and the biological environment Stimuli-responsive gene delivery vectors. J Nanopart Res. 2013;15(3):1523.
24. Walton SP, Wu M, Gredell JA, Chan C. Designing highly active siRNAs for therapeutic applications. FEBS J. 2010;277(23):4806–13.
25. Yuan BB, Latek R, Hossbach M, et al. siRNA selection server: an automated siRNA oligonucleotide prediction server. Nucleic Acids Res. 2004;32:W130–4.
26. Hajeri PB, Singh SK. siRNAs: their potential as therapeutic agents – part I. Designing of siRNAs. Drug Discov Today. 2009;14(17–18):851–8.
27. Jagla B, Aulner N, Kelly PD, et al. Sequence characteristics of functional siRNAs. RNA. 2005;11(6):864–72.
28. Behlke MA. Chemical modification of siRNAs for in vivo use. Oligonucleotides. 2008;18(4):305–19.
29. Choung S, Kim YJ, Kim S, et al. Chemical modification of siRNAs to improve serum stability without loss of efficacy. Biochem Biophys Res Commun. 2006;342(3):919–27.
30. Richardson FC, Tennant BC, Meyer DJ, et al. An evaluation of the toxicities of 2 '-fluorouridine and 2 '-fluorocytidine-HCl in F344 rats and woodchucks (Marmota monax). Toxicol Pathol. 1999;27(6):607–17.
31. Fluiter K, ten Asbroek ALMA, de Wissel MB, et al. In vivo tumor growth inhibition and biodistribution studies of locked nucleic acid (LNA) antisense oligonucleotides. Nucleic Acids Res. 2003;31(3):953–62.
32. Hogrefe RI, Lebedev AV, Zon G, et al. Chemically modified short interfering hybrids (siHYBRIDS): nanoimmunoliposome delivery in vitro and in vivo for RNAi of HER-2. Nucleos Nucleotides Nucleic Acids. 2006;25(8):889–907.
33. Ozcan G, Ozpolat B, Coleman RL, et al. Preclinical and clinical development of siRNA-based therapeutics. Adv Drug Deliv Rev. 2015;87:108–19.
34. Kay MA, Glorioso JC, Naldini L. Viral vectors for gene therapy: the art of turning infectious agents into vehicles of therapeutics. Nat Med. 2001;7(1):33–40.
35. Thomas CE, Ehrhardt A, Kay MA. Progress and problems with the use of viral vectors for gene therapy. Nat Rev Genet. 2003;4(5):346–58.
36. Mintzer MA, Simanek EE. Nonviral vectors for gene delivery. Chem Rev. 2009;109(2):259–302.
37. Pack DW, Hoffman AS, Pun S, Stayton PS. Design and development of polymers for gene delivery. Nat Rev Drug Discov. 2005;4(7):581–93.
38. Jeong JH, Mok H, Oh YK, Park TG. siRNA conjugate delivery systems. Bioconjug Chem. 2009;20(1):5–14.
39. Khvalevsky EZ, Gabai R, Rachmut IH, et al. Mutant KRAS is a druggable target for pancreatic cancer. Proc Natl Acad Sci U S A. 2013;110(51):20723–8.
40. Strumberg D, Schultheis B, Traugott U, et al. Phase I clinical development of Atu027, a siRNA formulation targeting PKN3 in patients with advanced solid tumors. Int J Clin Pharm Ther. 2012;50(1):76–8.
41. Lanford RE, Hildebrandt-Eriksen ES, Petri A, et al. Therapeutic silencing of microRNA-122 in primates with chronic hepatitis C virus infection. Science. 2010;327(5962):198–201.
42. Liu C, Kelnar K, Liu BG, et al. The microRNA miR-34a inhibits prostate cancer stem cells and metastasis by directly repressing CD44. Nat Med. 2011;17(2):211–15.
43. Soutschek J, Akinc A, Bramlage B, et al. Therapeutic silencing of an endogenous gene by systemic administration of modified siRNAs. Nature. 2004;432(7014):173–8.
44. Wolfrum C, Shi S, Jayaprakash KN, et al. Mechanisms and optimization of in vivo delivery of lipophilic siRNAs. Nat Biotechnol. 2007;25(10):1149–57.
45. Liu YM, Xia Y, Dai W, et al. Cholesterol-conjugated let-7a mimics: antitumor efficacy on hepatocellular carcinoma in vitro and in a preclinical orthotopic xenograft model of systemic therapy. BMC Cancer. 2014;14(1):889.
46. Nishina K, Unno T, Uno Y, et al. Efficient in vivo delivery of siRNA to the liver by conjugation of alpha-tocopherol. Mol Ther. 2008;16(4):734–40.

47. Chiu YL, Ali A, Chu CY, et al. Visualizing a correlation between siRNA localization, cellular uptake, and RNAi in living cells. Chem Biol. 2004;11(8):1165–75.
48. Eguchi A, Meade BR, Chang YC, et al. Efficient siRNA delivery into primary cells by a peptide transduction domain-dsRNA binding domain fusion protein. Nat Biotechnol. 2009;27(6):567–71.
49. Andaloussi SEL, Lehto T, Mager I, et al. Design of a peptide-based vector, PepFect6, for efficient delivery of siRNA in cell culture and systemically in vivo. Nucleic Acids Res. 2011;39(9):3972–87.
50. Jarver P, Coursindel T, Andaloussi SEL, et al. Peptide-mediated cell and in vivo delivery of antisense oligonucleotides and siRNA. Mol Ther-Nucleic Acids. 2012;1(6), e27.
51. Xia CF, Zhang Y, Zhang Y, et al. Intravenous siRNA of brain cancer with receptor targeting and avidin-biotin technology. Pharm Res. 2007;24(12):2309–16.
52. Kumar P, Ban HS, Kim SS, et al. T cell-specific siRNA delivery suppresses HIV-1 infection in humanized mice. Cell. 2008;134(4):577–86.
53. Song EW, Zhu PC, Lee SK, et al. Antibody mediated in vivo delivery of small interfering RNAs via cell-surface receptors. Nat Biotechnol. 2005;23(6):709–17.
54. Yao YD, Sun TM, Huang SY, et al. Targeted delivery of PLK1-siRNA by scFv suppresses Her2(+) breast cancer growth and metastasis. Sci Transl Med. 2012;4(130), 130ra48.
55. Tuerk C, Gold L. Systematic evolution of ligands by exponential enrichment: RNA ligands to bacteriophage-T4 DNA-polymerase. Science. 1990;249(4968):505–10.
56. Kay MA. State-of-the-art gene-based therapies: the road ahead. Nat Rev Genet. 2011;12(5):316–28.
57. Lupold SE, Hicke BJ, Lin Y, Coffey DS. Identification and characterization of nuclease-stabilized RNA molecules that bind human prostate cancer cells via the prostate-specific membrane antigen. Cancer Res. 2002;62(14):4029–33.
58. Chu TC, Twu KY, Ellington AD, Levy M. Aptamer mediated siRNA delivery. Nucleic Acids Res. 2006;34(10), e73.
59. McNamara JO, Andrechek ER, Wang Y, et al. Cell type-specific delivery of siRNAs with aptamer-siRNA chimeras. Nat Biotechnol. 2006;24(8):1005–15.
60. Dassie JP, Liu XY, Thomas GS, et al. Systemic administration of optimized aptamer-siRNA chimeras promotes regression of PSMA-expressing tumors. Nat Biotechnol. 2009;27(9):839–49.
61. Thiel KW, Hernandez LI, Dassie JP, et al. Delivery of chemo-sensitizing siRNAs to HER2(+)-breast cancer cells using RNA aptamers. Nucleic Acids Res. 2012;40(13):6319–37.
62. Wu X, Ding BY, Gao J, et al. Second-generation aptamer-conjugated PSMA-targeted delivery system for prostate cancer therapy. Int J Nanomed. 2011;6:1747–56.
63. Hao Z, Fan W, Hao J, et al. Efficient delivery of micro RNA to bone-metastatic prostate tumors by using aptamer-conjugated atelocollagen in vitro and in vivo. Drug Deliv. 2015;17(1):1–8.
64. Esposito CL, Cerchia L, Catuogno S, et al. Multifunctional aptamer-miRNA conjugates for targeted cancer therapy. Mol Ther. 2014;22(6):1151–63.
65. Torchilin VP. Recent advances with liposomes as pharmaceutical carriers. Nat Rev Drug Discov. 2005;4(2):145–60.
66. Chang HI, Yeh MK. Clinical development of liposome-based drugs: formulation, characterization, and therapeutic efficacy. Int J Nanomed. 2012;7:49–60.
67. Ozpolat B, Sood AK, Lopez-Berestein G. Liposomal siRNA nanocarriers for cancer therapy. Adv Drug Deliv Rev. 2014;66:110–16.
68. Lv HT, Zhang SB, Wang B, et al. Toxicity of cationic lipids and cationic polymers in gene delivery. J Control Release. 2006;114(1):100–9.
69. Landen CN, Chavez-Reyes A, Bucana C, et al. Therapeutic EphA2 gene targeting in vivo using neutral liposomal small interfering RNA delivery. Cancer Res. 2005;65(15):6910–18.
70. Villares GJ, Zigler M, Wang H, et al. Targeting melanoma growth and metastasis with systemic delivery of liposome-incorporated protease-activated receptor-1 small interfering RNA. Cancer Res. 2008;68(21):9078–86.

71. Sorensen DR, Leirdal M, Sioud M. Gene silencing by systemic delivery of synthetic siRNAs in adult mice. J Mol Biol. 2003;327(4):761–6.
72. Wu Y, Crawford M, Yu B, Mao YC, Nana-Sinkam SP, Lee LJ. MicroRNA delivery by cationic lipoplexes for lung cancer therapy. Mol Pharm. 2011;8(4):1381–9.
73. Medina OP, Zhu Y, Kairemo K. Targeted liposomal drug delivery in cancer. Curr Pharm Des. 2004;10(24):2981–9.
74. Chen YC, Zhu XD, Zhang XJ, et al. Nanoparticles modified with tumor-targeting scFv deliver siRNA and miRNA for cancer therapy. Mol Ther. 2010;18(9):1650–6.
75. Dokka S, Toledo D, Shi XG, et al. Oxygen radical-mediated pulmonary toxicity induced by some cationic liposomes. Pharm Res. 2000;17(5):521–5.
76. Filion MC, Phillips NC. Toxicity and immunomodulatory activity of liposomal vectors formulated with cationic lipids toward immune effector cells. Biochim Biophys Acta Biomembr. 1997;1329(2):345–56.
77. Kim HR, Kim IK, Bae KH, et al. Cationic solid lipid nanoparticles reconstituted from low density lipoprotein components for delivery of siRNA. Mol Pharm. 2008;5(4):622–31.
78. Trang P, Wiggins JF, Daige CL, et al. Systemic delivery of tumor suppressor microRNA mimics using a neutral lipid emulsion inhibits lung tumors in mice. Mol Ther. 2011;19(6):1116–22.
79. Bao YJ, Jin Y, Chivukula P, et al. Effect of PEGylation on biodistribution and gene silencing of siRNA/lipid nanoparticle complexes. Pharm Res. 2013;30(2):342–51.
80. Lin SY, Zhao WY, Tsai HC, et al. Sterically polymer-based liposomal complexes with dual-shell structure for enhancing the siRNA delivery. Biomacromolecules. 2012;13(3):664–75.
81. Morrissey DV, Lockridge JA, Shaw L, et al. Potent and persistent in vivo anti-HBV activity of chemically modified siRNAs. Nat Biotechnol. 2005;23(8):1002–7.
82. Zimmermann TS, Lee ACH, Akinc A, et al. RNAi-mediated gene silencing in non-human primates. Nature. 2006;441(7089):111–14.
83. Judge AD, Robbins M, Tavakoli I, et al. Confirming the RNAi-mediated mechanism of action of siRNA-based cancer therapeutics in mice. J Clin Invest. 2009;119(3):661–73.
84. Shen H, Sun T, Ferrari M. Nanovector delivery of siRNA for cancer therapy. Cancer Gene Ther. 2012;19(6):367–73.
85. Akinc A, Zumbuehl A, Goldberg M, et al. A combinatorial library of lipid-like materials for delivery of RNAi therapeutics. Nat Biotechnol. 2008;26(5):561–9.
86. Wang Y, Li ZG, Han Y, et al. Nanoparticle-based delivery system for application of siRNA in vivo. Curr Drug Metab. 2010;11(2):182–96.
87. Yuan XD, Naguib S, Wu ZQ. Recent advances of siRNA delivery by nanoparticles. Expert Opin Drug Deliv. 2011;8(4):521–36.
88. Davis ME, Brewster ME. Cyclodextrin-based pharmaceutics: past, present and future. Nat Rev Drug Discov. 2004;3(12):1023–35.
89. Hu-Lieskovan S, Heidel JD, Bartlett DW, et al. Sequence-specific knockdown of EWS-FLI1 by targeted, nonviral delivery of small interfering RNA inhibits tumor growth in a murine model of metastatic Ewing's sarcoma. Cancer Res. 2005;65(19):8984–92.
90. Davis ME, Zuckerman JE, Choi CHJ, et al. Evidence of RNAi in humans from systemically administered siRNA via targeted nanoparticles. Nature. 2010;464(7291):1067–70.
91. Sugahara KN, Teesalu T, Karmali PP, et al. Coadministration of a tumor-penetrating peptide enhances the efficacy of cancer drugs. Science. 2010;328(5981):1031–5.
92. Hu QL, Jiang QY, Jin X, et al. Cationic microRNA-delivering nanovectors with bifunctional peptides for efficient treatment of PANC-1 xenograft model. Biomaterials. 2013;34(9):2265–76.
93. Urban-Klein B, Werth S, Abuharbeid S, et al. RNAi-mediated gene-targeting through systemic application of polyethylenimine (PEI)-complexed siRNA in vivo. Gene Ther. 2005;12(5):461–6.
94. Navarro G, Sawant RR, Biswas S, et al. P-glycoprotein silencing with siRNA delivered by DOPE-modified PEI overcomes doxorubicin resistance in breast cancer cells. Nanomedicine. 2012;7(1):65–78.

95. Schiffelers RM, Ansari A, Xu J, et al. Cancer siRNA therapy by tumor selective delivery with ligand-targeted sterically stabilized nanoparticle. Nucleic Acids Res. 2004;32(19), e149.
96. Chiou GY, Cherng JY, Hsu HS, et al. Cationic polyurethanes-short branch PEI-mediated delivery of Mir145 inhibited epithelial-mesenchymal transdifferentiation and cancer stem-like properties and in lung adenocarcinoma. J Control Release. 2012;159(2):240–50.
97. Yang YP, Chien Y, Chiou GY, et al. Inhibition of cancer stem cell-like properties and reduced chemoradioresistance of glioblastoma using microRNA145 with cationic polyurethane-short branch PEI. Biomaterials. 2012;33(5):1462–76.
98. Kumar P, Wu HQ, McBride JL, et al. Transvascular delivery of small interfering RNA to the central nervous system. Nature. 2007;448(7149):39–43.
99. Hwang DW, Son S, Jang J, et al. A brain-targeted rabies virus glycoprotein-disulfide linked PEI nanocarrier for delivery of neurogenic microRNA. Biomaterials. 2011;32(21):4968–75.
100. Maiorano NA, Mallamaci A. Promotion of embryonic cortico-cerebral neuronogenesis by miR-124. Neural Dev. 2009;4(1):40.
101. Singha K, Namgung R, Kim WJ. Polymers in small-interfering RNA delivery. Nucleic Acid Ther. 2011;21(3):133–47.
102. Jalil RU. Biodegradable poly(lactic acid) and poly (lactide-co-glycolide) polymers in sustained drug delivery. Drug Dev Ind Pharm. 1990;16(16):2353–67.
103. Woodrow KA, Cu Y, Booth CJ, et al. Intravaginal gene silencing using biodegradable polymer nanoparticles densely loaded with small-interfering RNA. Nat Mater. 2009;8(6):526–33.
104. Yang XZ, Dou S, Sun TM, et al. Systemic delivery of siRNA with cationic lipid assisted PEG-PLA nanoparticles for cancer therapy. J Control Release. 2011;156(2):203–11.
105. Shen S, Mao CQ, Yang XZ, et al. Cationic lipid-assisted polymeric nanoparticle mediated GATA2 siRNA delivery for synthetic lethal therapy of KRAS mutant non-small-cell lung carcinoma. Mol Pharm. 2014;11(8):2612–22.
106. Gibbings D, Voinnet O. Control of RNA silencing and localization by endolysosomes. Trends Cell Biol. 2010;20(8):491–501.
107. Gibbings DJ, Ciaudo C, Erhardt M, Voinnet O. Multivesicular bodies associate with components of miRNA effector complexes and modulate miRNA activity. Nat Cell Biol. 2009;11(10):1272–1272.
108. Gomes RSM, das Neves RP, Cochlin L, et al. Efficient pro-survival/angiogenic miRNA delivery by an MRI-detectable nanomaterial. ACS Nano. 2013;7(4):3362–72.
109. Kesharwani P, Tekade RK, Gajbhiye V, et al. Cancer targeting potential of some ligand-anchored poly(propylene imine) dendrimers: a comparison. Nanomed Nanotechnol. 2011;7(3):295–304.
110. Zhou JH, Neff CP, Liu XX, et al. Systemic administration of combinatorial dsiRNAs via nanoparticles efficiently suppresses HIV-1 infection in humanized mice. Mol Ther. 2011;19(12):2228–38.
111. Lee JH, Cha KE, Kim MS, et al. Nanosized polyamidoamine (PAMAM) dendrimer-induced apoptosis mediated by mitochondrial dysfunction. Toxicol Lett. 2009;190(2):202–7.
112. Patil ML, Zhang M, Taratula O, et al. Internally cationic polyamidoamine PAMAM-OH dendrimers for siRNA delivery: effect of the degree of quaternization and cancer targeting. Biomacromolecules. 2009;10(2):258–66.
113. Tang Y, Li YB, Wang B, et al. Efficient in vitro siRNA delivery and intramuscular gene silencing using PEG-modified PAMAM dendrimers. Mol Pharm. 2012;9(6):1812–21.
114. Yu TZ, Liu XX, Bolcato-Bellemin AL, et al. An amphiphilic dendrimer for effective delivery of small interfering RNA and gene silencing in vitro and in vivo. Angew Chem Int Ed. 2012;51(34):8478–84.
115. Taratula O, Garbuzenko OB, Kirkpatrick P, et al. Surface-engineered targeted PPI dendrimer for efficient intracellular and intratumoral siRNA delivery. J Control Release. 2009;140(3):284–93.
116. Agrawal A, Min DH, Singh N, et al. Functional delivery of siRNA in mice using dendriworms. ACS Nano. 2009;3(9):2495–504.

117. Kim SH, Jeong JH, Lee SH, et al. LHRH receptor-mediated delivery of siRNA using polyelectrolyte complex micelles self-assembled from siRNA-PEG-LHRH conjugate and PEI. Bioconjug Chem. 2008;19(11):2156–62.
118. Yuan QA, Lee E, Yeudall WA, Yang H. Dendrimer-triglycine-EGF nanoparticles for tumor imaging and targeted nucleic acid and drug delivery. Oral Oncol. 2010;46(9):698–704.
119. Ren Y, Kang CS, Yuan XB, et al. Co-delivery of as-miR-21 and 5-FU by poly(amidoamine) dendrimer attenuates human glioma cell growth in vitro. J Biomat Sci Polym Ed. 2010;21(3):303–14.
120. Cheng CJ, Saltzman WM. Polymer nanoparticle-mediated delivery of microRNA inhibition and alternative splicing. Mol Pharm. 2012;9(5):1481–8.

GPSR Compliance
The European Union's (EU) General Product Safety Regulation (GPSR) is a set of rules that requires consumer products to be safe and our obligations to ensure this.

If you have any concerns about our products, you can contact us on

ProductSafety@springernature.com

In case Publisher is established outside the EU, the EU authorized representative is:

Springer Nature Customer Service Center GmbH
Europaplatz 3
69115 Heidelberg, Germany

www.ingramcontent.com/pod-product-compliance
Ingram Content Group UK Ltd.
Pitfield, Milton Keynes, MK11 3LW, UK
UKHW020227290926
15470UKWH00001B/409
9789811014963